THE OXIDE HANDBOOK

Second Edition

IFI DATA BASE LIBRARY

Recent volumes in the series:

RAPIDLY QUENCHED METALS
A Bibliography, 1973—1979
C. Suryanarayana

MOSFET TECHNOLOGIES
A Comprehensive Bibliography
A. H. Agajanian

PARENT – CHILD SEPARATION
Psychosocial Effects on Development
An Abstracted Bibliography
Faren R. Akins, Dianna L. Akins, and Gillian S. Mace

THE BEREAVED CHILD
Analysis, Education, and Treatment
An Abstracted Bibliography
Gillian S. Mace, Faren R. Akins, and Dianna L. Akins

THE HOSPITALIZED CHILD
Psychosocial Issues
An Abstracted Bibliography
Dianna L. Akins, Gillian S. Mace, and Faren R. Akins

PREPARATION OF NUCLEAR TARGETS
A Comprehensive Bibliography
Jozef Jaklovsky

THE OXIDE HANDBOOK
Second Edition
Edited by G. V. Samsonov

THE OXIDE HANDBOOK
Second Edition

Edited by G. V. Samsonov

Institute of Problems in Materials Science
Academy of Sciences of the Ukrainian SSR
Kiev, USSR

Translated from Russian by
Robert K. Johnston

SPRINGER SCIENCE+BUSINESS MEDIA, LLC

Library of Congress Cataloging in Publication Data

Main entry under title:

The Oxide handbook.

A rev. and corr. translation of Fiziko-khimicheskie svoistva okislov.
Includes index.
1. Oxides—Tables, etc. I. Samsonov, Grigorii
Valentinovich.
QD181.01F5813 1981 546'.7212 80-23223
ISBN 978-1-4757-1615-3 ISBN 978-1-4757-1613-9 (eBook) AACR 1
DOI 10.1007/978-1-4757-1613-9

This Russian edition, published by Metallurgiya Press in Moscow in 1978, has been revised and corrected for this edition. This translation is published under an agreement with the Copyright Agency of the USSR (VAAP).

Г. В. САМСОНОВ

ФИЗИКО-ХИМИЧЕСКИЕ СВОЙСТВА ОКИСЛОВ

FIZIKO-KHIMICHESKIE SVOISTVA OKISLOV

Originally published by Plenum Publishing Corporation, New York in 1982

PREFACE TO THE SECOND EDITION

In the comparatively short amount of time since the publication of the first edition of this handbook in 1969, the authors continued to work on the collection and treatment of data on oxide properties that have appeared in the literature. The numerous responses that we have received during this time have given indisputable evidence of the value of such a reference book to scientists, production engineers, and instructors, and these comments have also provided grounds for publishing a new edition of this book.

In preparing the second edition, the authors reviewed critically the material in the first edition of the handbook, updated obsolete information, and added new material to certain chapters. Of major assistance in defining the scope of the revision were the comments of Prof. S. G. Tresvyatskii, Prof. A. I. Avgustinik, Prof. É. K. Keler, Prof. K. K. Strelov, Dr. A. N. Borisenko, Dr. D. S. Rutman, and many others, to whom the authors wish to express their deep gratitude.

The structure of the second edition of this handbook has not undergone any major changes; however, certain corrections and improvements have been made in the content of certain chapters. In the first chapter, for example, we have eliminated certain information that was not directly related to oxide properties. In other chapters, we have deleted information that is set forth quite fully in other publications. Some of the sections have been expanded and are now presented as separate chapters. For each property of an oxide, we have usually listed the single numerical value considered to be most reliable, i.e., the value recommended for use. In other cases, when it was not possible to select any single most reliable value, several values are given.

The material of this handbook is arranged in the form of tables that are grouped in the following order: general data, stoichiometry, and crystal-chemical properties; thermodynamic and thermal properties; molecular properties; mechanical properties; electrical properties; magnetic properties; optical properties; nuclear properties and radiation effects; chemical and catalytic properties; refractory properties; phase diagrams of binary element/oxygen systems.

The oxides in the tables are listed in increasing order of atomic number of the element. If an element forms several oxides, they are listed in order of increasing number of oxygen atoms per atom of element. In the material taken from Reference 9, Nos. 1–7 for nonstoichiometric compounds of a metal with a nonmetal, the metal is generally given with an integral subscript, the nonmetal with a fractional (decimal) subscript. The integral subscript is used for an element forming a complete sublattice. If the incompleteness is the same in both cases, no fractional subscripts are given ($Ti_{0.85}O_{0.85} \rightarrow TiO$). The constants are given for compositions representing the limits of the regions of homogeneity, and also for certain intermediate compositions.

The following abbreviations are used in the tables when necessary to denote the state of the oxide:

am	amorphous
ap	amorphous powder
c or cr	crystalline
cm	compact mass
crp	crystalline powder
g	gaseous
gl	glassy
l	liquid
ml	molten
oi	oily
p	powder
rs	resinous
s	solid
sf	scaly flakes

If possible, the characteristics of the specimens used in determining the corresponding values are described more specifically. Values that are considered doubtful are either placed in parentheses, or a question mark is shown along with the value. The only calculated values (given in parentheses) are those from Reference 75.

For the convenience of the reader, this handbook includes an index of elements and their oxides, this index being arranged not in order of increasing atomic number, but alphabetically by name of the element.

The values listed in this handbook are expressed in units of the SI system. Two temperature scales are used, Celsius and Kelvin. In a number of cases, the temperature value 298.15°K (25°C) has been shortened to 298°K. In converting numerical values from the units of one system to the units of other systems, the values have been rounded off as required.

In Chapter I we present general data on the properties of oxides. In the table "Geometric Constants of Atoms and Ions," in the case of metals the atomic radii for a coordination number of 12 have been used, for nonmetals the covalent radii according to Pauling, and for ions the radii for a coordination number of 6. In the table "Composition of Oxides," data are listed on the formula compositions of oxide phases, their molecular masses, and the oxygen contents, in atomic and mass percentages. In the table "Crystal Structure," basic data are given on the structure of oxides as determined for single-crystal or polycrystalline specimens; it should be noted that this information is incomplete for many oxides, since data are missing on either the structural type or the space group and lattice constants.

In the tables included in this and subsequent chapters, we have used the following abbreviations to denote the crystal system:

Cub.	Cubic
Hex.	Hexagonal
Monocl.	Monoclinic
Rhom.	Orthorhombic
Tetr.	Tetragonal
Tricl.	Triclinic
Trig.	Trigonal

In Chapter II, information has been assembled on the thermodynamic and thermal properties of oxides. In preparing the second edition, the material of this chapter has been revised and supplemented. In the table "Melting and Boiling Points," the values that are listed were determined for the most part at normal pressure. In those cases in which the values refer to a pressure other than normal, this is so stated. The values for the heats of fusion in this table refer to the melting point at normal pressure; for the heats of vaporization, the temperatures and pressures are indicated. The values for the change in entropy upon fusion refer to the melting point. The temperature and pressure are indicated when these values refer to vaporization of a liquid. Data on the heats of sublimation and the change in entropy upon sublimation are combined in a single section.

Significant changes have been made in the presentation of material in the table "Parameters of Polymorphic Transitions." Major use has been made of data published in the handbook "Thermal Constants of Substances," edited by Academician V. P. Glushko [9]. Together with the parameters of polymorphic transitions, values are listed for the parameters of second-order phase transitions, transitions at the Curie and Néel points, and other phase transitions. In order to differentiate between the actual vapor pressure of the given oxide and the external pressure, the values of the external pressure according to Reference 9, Nos. 1–7 are listed in parentheses. When a given crystalline oxide has several modifications, roman numerals are assigned to these modifications in accordance with Reference 9, Nos. 1–7, beginning with the highest-temperature modification. In accordance with Reference 9, Nos. 6–7, the description of the phase transitions of oxides in condensed states has been given in greater detail, as follows:

$c \to l$	melting that is congruent in nature;
$c \to l + c_1$	incongruent melting with the formation of a liquid phase (l) and another crystalline phase (c_1);
$c_1 + c_2 \to c$ $c \to c_1 + c_2$ $c \to l_1 + l_2$	phase transitions in which the given crystalline phase (c) is formed from two other crystalline phases, or decomposes into two crystalline or liquid phases;
c III → c II c II → c I etc.	polymorphic transitions (first-order phase transitions);
c I → c I′ c I′ → c I″ etc.	second-order phase transitions.

The values listed for the heat capacity, if the state of the oxide is not indicated, refer to the solid state. Since the thermal conductivity is highly dependent on the state of the specimen, we have wherever possible emphasized the specific features of the specimens investigated.

The activation energy for diffusion of oxygen is given in kJ/mole and electron volts. In converting the activation energy from kcal/mole to electron volts (referred to one atom), we used Avogadro's number ($6.02486 \cdot 10^{23}$ mole^{-1}) and the conversion factors 1 eV = $1.60207 \cdot 10^{-19}$ J and 1 cal = 4.1868 J, giving an overall conversion factor

$$4.1868 \cdot 10^3/(1.60207 \cdot 10^{-19} \cdot 6.02486 \cdot 10^{23}) = 0.04338 \text{ (eV} \cdot \text{mole)/kcal.}$$

In the table "Parameters of Diffusion of Elements into Oxides," we encounter expressions of the form $4.19\,(+2.45; -1.55) \cdot 10^{-4}$; this expression means that the most reliable value of the sought quantity lies in the interval between $(4.19-1.55) \cdot 10^{-4}$ and $(4.19+2.45) \cdot 10^{-4}$.

In the table "Vapor Pressure" we have listed both calculated and experimental values corresponding to the molecular regime of vaporization. There are generally rather large deviations between these values. The information given on dissociation pressure is mainly in the form of functional equations $p_{O_2} = f(T)$. The equations and numerical values for vapor pressure and dissociation that were taken from Reference 195 were obtained by calculation. In order to convert from a pressure expressed in Pa* to a pressure expressed in mm Hg, the number 2.1244896 (or the appropriate rounded-off value) is deducted from the temperature-independent term in the equation; in order to convert to pressure in kgf/cm^2, the number 5.0056 (or a rounded value) is deducted. For example:

$$\lg p = 13.74 - 24{,}944/T \text{ (p, Pa);}$$
$$\lg p = 13.74 - 2.12 - 24{,}044/T = 11.62 - 24{,}044/T \text{ (p, mm Hg)}$$
$$\lg p = 13.74 - 5.01 - 24{,}044/T = 8.73 - 24{,}044/T \text{ (p, kgf/cm}^2\text{).}$$

Quantities related to the molecular structure of oxides are grouped in a separate Chapter III, "Molecular Properties."

* 1 Pa (Pascal) corresponds to 1 N/m^2; 1 kPa (kilopascal) = 10^3 Pa; and 1 MPa (megapascal) = 10^6 Pa.

The data in Chapter IV, "Mechanical Properties," refer primarily to the solid state. The values listed for mechanical properties show a considerable scatter, owing to differences in the state of the specimens tested, the character and purity of the specimens, the treatment of the specimens, and a number of other factors that are difficult to take into account. The temperature dependences of the properties of certain oxides were taken from plots on a logarithmic scale; hence, errors are entirely possible. This situation is indicated by notes. Wherever possible, characteristics of the specimens are given: porosity, density, conditions of sintering and firing, etc. The hardness of oxides on the mineralogical scale is given for the most part for natural minerals.

In Chapter V, "Electrical and Magnetic Properties," information is given on the following: temperature dependences of the electrical resistivity and coefficient of thermoelectromotive force; values for the work function, Hall constant, and carrier mobility; data on the magnetic susceptibility and effective magnetic moments; values for the relative permittivity (dielectric constant), width of the forbidden band, and activation energy. For a number of oxides the data are taken from graphs, and this fact is indicated. The values listed for the work function are taken mainly from the handbooks of V. S. Fomenko, "Emission Properties of Elements and Chemical Compounds" (1965) and "Emission Properties of Materials" (1970), where more detailed information may be found on the literature sources, methods of measurement, etc. The values for the specific magnetic susceptibility can be readily converted to molecular values by multiplying by the molecular mass of the oxide.

In Chapter VI, "Optical Properties," the information on the refractive index and emission characteristics has been refined and supplemented. Here in the second edition, for the first time, graphical relationships are presented for the spectra of certain optical constants, as well as a description of characteristic features in the optical properties of oxides. The designations used for the polymorphic modifications in this chapter are kept the same as in the original sources of data. The refractive indices of the oxides were determined (unless otherwise indicated) for the wavelength of the Fraunhofer D-line of sodium at room temperature. We have retained the generally accepted designations of the principal indices of refraction n_g, n_m, and n_p (for biaxial crystals). The values of the refractive index for extraordinary n_e and ordinary n_0 rays in the case of a monoaxial crystal are given in the n_g and n_p columns, respectively. For isotropic crystals and any other crystals with a known average index of refraction, the data are given in the n_m column. In the table "Emissivity," the coverage has been expanded (in comparison with the first edition) on the integral normal emissivity ε_{tn} and monochromatic normal emissivity $\varepsilon_{\lambda n}$.

In a new table, "Specific Features of Optical Properties," data are presented on various optical properties of oxides that were not included in preceding sections.

In Chapter VII, "Nuclear Properties and Radiation Effects," information is given on the absorption and scattering cross sections, the nuclear properties of oxides used as moderators, the threshold energies of reactions that lead to the formation of new elements in oxides, and certain characteristics of isotopes formed in oxides by irradiation. In other tables in this chapter, data are presented on the effects of radiation on the volume of oxides, the density, lattice constants, thermal conductivity, and mechanical, electrical, and optical properties. Data are also presented on stored energy and internal friction, and also on radiation defects and radiation resistance.

Chapter VIII, "Chemical and Catalytic Properties," consists of two tables. In the first table, "Chemical Properties," data are listed on the character of interaction between oxides and mineral and organic acids, mixtures of acids, caustic solutions, salt solutions and melts, and gaseous media. In the second table, "Catalytic Properties," the oxides of elements are listed in order of increasing atomic number of the element, the same as in the preceding table. The oxides of those elements for which there is no information on catalytic properties in the literature that has been reviewed are omitted from this table. In column 1, information is given on the qualitative composition of the catalysts, i.e., which compounds are used in mixture with the particular oxide. If the oxide acts not as a catalyst but as a promoter, this is indicated by the letter "p" in parentheses; if the oxide acts as a support, this is indicated by the letter "s." In column 2 we have listed the temperature and pressure (if the pressure is greater than 98.1 kPa or 1 kgf/cm^2) at which the reaction is carried out; in column 3 we have listed the type of reaction catalyzed.

In Chapter IX, "Refractory Properties," the first table contains data on the interaction of a number of oxides with simple substances and binary compounds in the solid phase. In the second table, information is presented on the wetting of oxides by liquid metals. The resistance of oxides to the action of molten metals, alloys, and slags is described in the third table.

Phase diagrams for binary element/oxygen systems are given in Chapter X. In those cases in which several different diagrams are available for the same system, covering the same region of temperature and concentration, we have chosen the more detailed diagrams, constructed from the greater number of experimental points. As a rule, these are diagrams that have been published more recently, representing either a refinement or extension of earlier diagrams. In some cases, for the same system (such as Ce–O, Ti–O, V–O, and Zr–O), two or three diagrams are presented, supplementing each other. In view of the great complexity of the U–O system, two variants of this system are presented (Figs. 84 and 85).

During the time that has elapsed since preparation of the first edition of this handbook, information has been reported on new phase diagrams for element–oxygen systems (Al–O, Am–O, Cm–O, Eu–O, K–O, Rb–O, Sc–O, Ta–O, Th–O), as well as on refined or replotted diagrams of systems for which earlier diagrams were available (for example, Cu–O, Cr–O, Nb–O, Ti–O, U–O, V–O, W–O, Zr–O). As a result of this new information, the content of Chapter X has been changed quite substantially. New diagrams have been added, some diagrams appearing in the first edition have been revised, others have been replaced by more detailed and accurate diagrams, and some have been omitted as lacking in general interest. The diagrams of the Pr–O and Tb–O systems are considered to be only tentative by the particular investigators; those of the Sn–O and W–O systems are considered to be schematic, and the Cm–O diagram is a hypothetical diagram, certain lines of which (peritectoid phase transitions i and δ) have been drawn by analogy with the corresponding lines in Me–O systems for lanthanides. In a number of Me–O systems, suboxides have been discovered recently, these being oxides with compositions that do not match up with the valence ratios of atoms, but with an ordered structure. In those cases in which the existence of suboxides has been specifically confirmed, they have been included in the phase diagram.

Such diagrams, refined in the region of solid solutions, are given for the systems Ti–O (Fig. 83), V–O (Fig. 88), and Zr–O (Fig. 94). For the convenience of the reader, the diagrams are arranged in alphabetical order of the chemical symbol for the first component in each diagram.

Those taking part in preparation of the new edition included staff members of the Refractory Materials Branch of the Institute of Problems in Material Science, the Institute of Nuclear Research of the Academy of Sciences of the Ukrainian SSR, the Physics Department of Zhdanov Metallurgical Institute, and the Kiev Automotive Highway Institute.

The authors involved in the present work plan to continue efforts toward a further improvement in the content of the handbook, reflecting new and more reliable information on various properties of simple oxides; they also plan to work toward structural improvement of the handbook so that the great diversity of information on complex properties of oxides can be presented in a condensed and concrete form. As before, the authors will be grateful for all critical comments and recommendations from the readers.

TABLE OF CONTENTS

CHAPTER I

GENERAL DATA STOICHIOMETRY, AND CRYSTAL-CHEMICAL PROPERTIES

1. GEOMETRIC CONSTANTS OF ATOMS AND IONS [1]

Element	Atomic radius according to Belov and Bokii, nm		Ionic radius, nm, according to				Ionic charge
	covalent	metallic	Pauling	Goldschmidt	Rice	Belov and Bokii	
1	2	3	4	5	6	7	8
H	0,028	0,046	0,208	0,154	0,205	0,136	1—
He	0,122	—	—	—	—	—	—
Li	0,133	0,155	0,060	0,078	0,059	0,068	1+
Be	0,100	0,113	0,032	0,034	0,043	0,034	2+
B	0,083	0,091	0,020	—	0,034	0,020	3+
C	0,077	—	0,015	0,020	0,029	0,020	4+
			0,260	0,260	0,414	0,260	4—
N	0,070	—	0,011	0,015	0,025	0,015	5+
			0,171	0,247	0,247	0,148	3—
O	0,066	—	0,009	0,009	0,022	—	6+
			0,140	0,132	0,176	0,136	2—
F	0,071	—	0,007	—	0,019	—	7+
			0,136	0,133	—	0,133	1—
Ne	0,160	—	—	—	—	—	—
Na	0,154	0,189	0,095	0,098	0,095	0,098	1+
Mg	0,138	0,160	0,065	0,078	0,082	0,074	2+
Al	0,126	0,143	0,050	0,057	0,072	0,057	3+
Si	0,117	—	0,041	0,039	0,065	0,039	4+
			0,271	0,198	0,384	—	4—
P	0,104	—	0,034	0,035	0,059	0,035	5+
			0,212	—	0,279	0,186	3—
S	0,104	—	0,029	0,034	0,053	0,029	6+
			0,184	0,174	0,219	0,182	2—
Cl	0,099	—	0,026	—	0,049	0,026	7+
			0,181	0,181	—	0,181	1—
Ar	0,192	—	—	—	—	—	—
K	—	0,236	0,133	0,133	0,133	0,133	1+
Ca	—	0,197	0,098	0,106	0,118	0,104	2+
Sc	—	0,164	0,081	0,083	0,106	0,083	3+
Ti	—	0,146	0,068	0,064	0,096	0,064	4+
			0,069	0,069	—	0,069	3+
			—	0,080	—	0,078	2+
V	—	0,134	0,059	0,040	—	0,040	5+
			—	0,061	0,088	0,061	4+
			0,066	0,065	—	0,067	3+
			—	0,072	—	0,072	2+
Cr	—	0,127	0,052	0,035	0,081	0,035	6+
			0,064	—	—	0,064	3+
			—	—	—	0,083	2+

1 – Continued

1	2	3	4	5	6	7	8
Mn	—	0,130	0,046	—	0,075	0,046	7+
			0,050	0,052	—	0,052	4+
			0,062	0,070	—	0,070	3+
			0,080	0,091	—	0,091	2+
Fe	—	0,126	0,060	0,067	—	0,067	3+
			0,075	0,083	—	0,080	2+
Co	—	0,125	—	—	—	0,064	3+
			0,072	0,082	—	0,078	2+
Ni	—	0,124	0,069	0,078	—	0,074	2+
Cu		0,128	—	—	—	0,080	2+
			0,096	—	0,096	0,098	1+
Zn	0,131	0,139	0,074	0,083	0,088	0,083	2+
Ga	0,127	0,139	0,062	0,062	0,081	0,062	3+
Ge	0,122	—	0,053	0,044	0,076	0,044	4+
			—	—	—	0,065	2+
			0,272	—	—	—	4—
As	0,121	—	0,047	—	0,071	0,047	5+
			—	0,069	—	0,069	3+
			0,222	—	—	0,191	3+
Se	0,117	—	0,042	0,035	0,066	0,035	6+
			—	—	—	0,069	4+
			0,198	0,191	0,232	0,193	2—
Br	0,114	—	0,039	—	0,062	0,039	7+
			0,195	0,196	0,195	0,196	1+
Kr	0,198	—	—	—	—	—	—
Rb	—	0,248	0,148	0,149	0,148	0,149	1+
Sr	—	0,215	0,113	0,127	0,132	0,120	2+
Y	—	0,181	0,093	0,106	0,120	0,097	3+
Zr	—	0,160	0,080	0,087	0,109	0,082	4+
Nb	—	0,145	0,070	0,069	0,100	0,066	5+
			0,077	—	—	0,067	4+
Mo	—	0,139	0,062	—	0,093	0,065	6+
			0,066	0,068	—	0,068	4+
Tc	—	0,136	—	—	—	—	—
Ru	—	0,134	0,063	0,065	—	0,062	4+
Rh	—	0,134	—	—	—	0,065	4+
			—	0,068	—	0,075	3+
Pd	—	0,137	—	—	—	0,064	4+
Ag	—	0,144	0,126	0,113	0,126	0,113	1+
Cd	0,148	0,156	0,097	0,103	0,114	0,099	2+
In	0,144	0,166	0,081	0,092	0,104	0,092	3+
			—	—	—	0,130	1+
Sn	0,140	0,158	0,071	0,074	0,096	0,067	4+
			—	—	—	0,102	2+
			0,294	0,215	—	—	4—
Sb	0,140	0,161	0,062	—	0,089	0,062	5+
			—	0,090	—	0,090	3+
			0,245	—	—	0,208	3—
Te	0,137	—	0,056	—	0,082	0,056	6+
			0,081	0,089	—	0,089	4+

1 – Continued

1	2	3	4	5	6	7	8
			0,221	0,211	0,250	0,211	2–
I	0,133	—	0,050	—	0,077	0,050	7+
			—	0,094	—	—	5+
			0,216	0,220	—	0,220	1–
Xe	0,218	—	—	—	—	—	—
Cs	—	0,268	0,169	0,165	0,169	0,165	1+
Ba	—	0,221	0,135	0,143	0,153	0,138	2+
La	—	0,187	—	—	—	0,090	4+
			0,115	0,122	0,139	0,104	3+
Ce	—	0,183	0,101	0,102	0,127	0,088	4+
			—	0,118	—	0,102	3+
Pr	—	0,182	0,092	0,100	—	—	4+
			—	0,116	—	0,100	3+
Nd	—	0,182	—	0,115	—	0,099	3+
Pm	—	—	—	—	—	0,098	3+
Sm	—	0,181	—	0,113	—	0,097	3+
Eu	—	0,202	—	0,113	—	0,101	3+
Gd	—	0,179	—	0,111	—	0,094	3+
Tb	—	0,177	—	0,089	—	—	4+
			—	0,109	—	0,089	3+
Dy	—	0,177	—	0,107	—	0,088	3+
Ho	—	0,176	—	0,105	—	0,086	3+
Er	—	0,175	—	0,104	—	0,085	3+
Tm	—	0,174	—	0,104	—	0,085	3+
Yb	—	0,193	—	0,100	—	0,081	3+
Lu	—	0,174	—	0,099	—	0,080	3+
Hf	—	0,159	—	—	—	0,082	4+
Ta	—	0,146	—	—	—	0,066	5+
W	—	0,140	—	—	—	0,065	6+
			0,066	0,068	—	0,068	4+
Re	—	0,137	—	—	—	0,052	6+
Os	—	0,135	0,065	0,067	—	0,065	4+
Ir	—	0,135	0,064	0,066	—	0,065	4+
Pt	—	0,138	—	—	—	0,064	4+
Au	—	0,144	0,137	—	0,137	0,137	1+
Hg	—	0,160	0,110	0,112	0,125	0,112	2+
Tl	—	0,171	0,095	0,105	0,115	0,105	3+
			0,144	0,149	—	0,136	1+
Pb	—	0,175	0,084	0,084	0,106	0,076	4+
			0,121	0,132	—	0,126	2+
			—	0,215	—	—	4–
Bi	—	0,182	0,074	—	0,098	0,074	5+
			0,116	—	—	0,120	3+
			—	—	—	0,213	3–
Fr	—	0,280	—	—	—	—	—
Ra	—	0,235	—	0,152	—	0,144	2+
Ac	—	0,203	—	—	—	0,111	3+
		0,180	0,102	0,110	—	0,095	4+
Th	—		—	—	—	0,108	3+
		0,162	—	—	—	0,091	4+

1 — Continued

1	2	3	4	5	6	7	8
Pa	—		—	—	—	0,106	3+
		0,153	0,097	0,105	—	0,089	4+
U	—		—	—	—	0,104	3+

2. COMPOSITION OF OXIDES

Oxide	Molecular mass	Oxygen content, %		Oxide	Molecular mass	Oxygen content, %	
		atomic	mass			atomic	mass
1	2	3	4	1	2	3	4
H_2O	18,0153	33,33	88,88	VO_2	82,94	66,67	38,58
H_2O_2	34,0147	50	94,11	V_2O_5	181,8810	71,40	43,98
Li_2O	29,8774	33,33	53,56	CrO	67,9954	50	23,53
Li_2O_2	45,8768	50	69,76	Cr_2O_3	151,9902	60	31,58
BeO	25,0116	50	63,97	CrO_3	99,9942	75	48,00
BO	26,8104	50	59,67	MnO	70,9375	50	22,55
B_2O_3	69,6202	60	68,94	Mn_3O_4	228,8119	57,12	27,97
CO	28,0105	50	57,12	Mn_2O_3	157,8744	60	30,40
CO_2	44,0099	66,67	72,71	MnO_2	86,9369	66,67	36,80
N_2O	44,0128	33,33	36,36	FeO*	71,8464	51,26—54,57	23,56—25,60
NO	30,0061	50	53,32	Fe_3O_4	231,5386	57,12	27,64
N_2O_3	76,0116	60	63,15	Fe_2O_3	159,6925	60	30,05
N_2O_4	92,0110	66,67	69,55	CoO	74,9326	50	21,35
N_2O_5	108,0104	71,40	74,06	Co_3O_4	240,7972	57,12	26,57
F_2O	53,9962	33,33	29,63	Co_2O_3	165,8646	60	28,94
Na_2O	61,9790	33,33	25,80	NiO	74,7094	50	21,41
Na_2O_2	77,9784	50	41,04	Cu_2O	143,0794	33,33	11,18
NaO_2	54,9886	66,67	58,20	CuO	79,5394	50	20,11
MgO	40,3114	50	39,69	ZnO	81,3694	50	19,66
Al_2O_3	101,9612	60	47,07	Ca_2O	153,4394	33,33	10,29
SiO	30,0061	50	53,33	CaO	85,7194	50	18,66
SiO_2	46,0055	66,67	69,56	Ce_2O_3	187,4382	60	25,60
P_4O_6	219,8016	60	56,34	CeO	88,5894	50	18,06
P_2O_5	141,9446	71,40	56,36	CeO_2	104,5888	66,67	30,59
P_4O_{10}	283,8892	71,40	56,36	As_2O_3	197,8414	60	24,26
SO_2	64,0628	66,67	49,95	As_2O_5	154,22	71,40	51,64
SO_3	80,0622	75	59,96	SeO_2	110,9588	66,67	28,83
Cl_2O	86,9054	33,33	18,41	Rb_2O	186,9494	33,33	8,55
ClO	51,4524	50	31,09	Rb_2O_3	218,9600	60	21,92
ClO_2	67,4518	66,67	47,44	RbO_2	117,4738	66,67	27,24
Cl_2O_7	182,9018	77,78	61,23	SrO	103,6194	50	15,44
K_2O	94,2034	33,33	16,98	SrO_2	119,6188	66,67	26,75
K_2O_4	142,2016	66,67	45,00	Y_2O_3	225,8082	60	21,25
CaO	56,0794	50	26,75	ZrO_2	123,2188	66,67	25,96
Sc_2O_3	137,9102	60	34,80	NbO	108,90	50	14,81
Ti_2O	111,80	33,33	14,31	Nb_2O_5	265,8090	71,40	30,09
TiO	63,90	50	25,03	MoO_2	127,9388	66,67	25,01
Ti_2O_3	143,7982	60	33,37	MoO_3	143,9382	75	33,34

1	2	3	4	1	2	3	4
Ti_3O_5	223,6970	62,5	35,76	TcO_2	129	66,67	24,81
TiO_2	79,8988	66,67	40,05	RuO_2	133,0688	66,67	24,04
VO	66,9414	50	23,90	RhO	118,9044	50	13,45
V_2O_3	149,8822	60	32,02	Rh_2O_3	253,8082	60	18,91
V_3O_5	232,8230	62,5	34,36	PdO	122,7	50	13,11
				Ag_2O	231,7394	33,33	6,90
Ag_2O_2	247,7388	50	12,91	Tu_2O_3	385,8662	60	12,43
CdO	128,3994	50	12,46	Yb_2O_3	394,0782	60	12,18
In_2O_3	277,6382	60	17,28	Lu_2O_3	397,9382	60	12,06
SnO	134,6894	50	11,88	HfO_2	210,4888	66,67	15,20
Sn_3O_4	420,0676	57,12	15,24	TaO	196,9474	50	8,12
SnO_2	150,6888	66,67	21,23	Ta_2O_3	409,8942	60	11,71
SbO	137,7494	50	11,62	TaO_2	212,9468	66,67	15,02
Sb_2O_3	291,4982	60	16,46	Ta_2O_5	441,8930	42,84	18,10
Sb_2O_4	307,4976	66,67	20,81	WO_2	215,8488	66,67	14,82
Sb_2O_5	323,4970	71,40	24,73	W_3O_8	679,5452	77,72	18,83
TeO	143,5994	50	11,14	WO_3	231,8482	75	20,70
TeO_2	150,5988	66,67	20,05	OsO_4	254,1976	80	25,17
I_2O_5	337,8058	71,40	23,68	IrO_2	224,1988	66,67	14,27
Cs_2O	281,8094	33,33	5,67	Au_2O_3	441,9322	60	10,86
Cs_2O_2	297,8188	20	10,74	Hg_2O	417,1794	33,33	3,83
CsO_2	164,9038	66,67	19,40	HgO	216,5894	50	7,38
BaO	153,3394	50	10,43	TlO	220,3694	50	7,26
BaO_2	169,3388	66,67	18,89	Pb_2O	430,3794	33,33	3,72
La_2O_3	325,8182	60	14,73	PbO	223,1894	50	7,16
CeO_2	172,1188	66,67	18,59	Pb_3O_4	685,5676	57,12	7,00
Pr_2O_3	329,8122	60	14,56	PbO_2	239,1888	66,67	13,38
Nd_2O_3	336,4782	60	14,26	BiO	224,9794	50	7,11
Pm_2O_3	341,9982	60	14,03	Bi_2O_3	465,9582	60	10,30
Sm_2O_3	348,6982	60	13,76	Fr_2O	461,9994	33,33	3,46
Eu_2O_3	351,9182	60	13,63	RaO_2	258,0488	66,67	12,11
Cd_2O_3	362,4982	60	13,24	ThO_2	264,0368	66,67	12,11
Tb_2O_3	365,8462	60	13,12	Pa_2O_5	541,9970	71,40	14,76
Dy_2O_3	372,9982	60	12,86	UO_2	270,0288	66,67	11,85
Ho_2O_3	377,8582	60	12,70	U_2O_5	556,1570	71,40	14,38
Er_2O_3	382,5582	60	12,54	U_3O_8	842,0852	77,72	14,01
				UO_3	286,1282	75	16,77

• Wustite; variable in composition; exists at 560–1424°C.

3. REGIONS OF HOMOGENEITY [1–4]

Oxide	Oxygen content, %		Oxide	Oxygen content, %	
	atomic	mass		atomic	mass
TiO	47,08—54,54	22,91—28,61	Nb_2O_5	70,50—71,42	29,15—30,09
Ti_2O_3	59,4—60,8	32,8—34,12	MoO_3	72,6—74,4	30,64—32,64
TiO_2	65,5—66,7	38,8—40,09	Ce_2O_3	60,0—60,31	14,62—14,78
VO	46,23—55,34	21,26—28,51	WO_2	52,38—70,58	8,73—18,08
V_2O_3	58,33—60,78	30,53—32,73	W_2O_5	68,75—73,68	16,06—19,58
VO_2	64,4—66,67	36,23—38,56	γ-WO_3	72,77—73,1	18,81—19,13
V_2O_5	71,1—71,4	43,58—43,95	β-WO_3	74,0—74,4	19,85—20,19
CrO_2	62,8—65,5	34,1—36,87	α-WO_3	74,85—75,00	20,58—20,71
MnO_2	65,8—66,67	35,91—36,81	$Pb_{12}O_{17}$	59,4—60,1	10,15—10,41
FeO	50,0—54,4	22,24—25,44	$Pb_{12}O_{19}$	60—62,6	10,36—11,44
NbO	48,45—50,98	14,01—15,18	PbO_2	65,1—66,67	12,59—13,38
NbO_2	65,98—67,63	25,04—26,46	UO_2	66,4—67,4	11,72—12,20

4. CRYSTAL STRUCTURE [1, 2, 7–12, 14, 16, 22]

Oxide	System	Space group	Type of structure	Lattice constants, nm			
				a	b	c	c/a
1	2	3	4	5	6	7	8
H_2O*1	Hex.	$C_{6v}^4 - P6_3mc$	—	0,782	—	0,736	0,941
H_2O*2	Rhom.	$D_2^5 - C222_1$	—	0,780	0,450	0,556	0,714
H_2O*3	"	—	—	1,020	0,587	0,717	0,702
BeO	Hex.	$C_{6v}^4 - P6_3mc$	ZnS	0,269	—	0,437	1,62
N_2O*4	Cub.	$T_h^6 - Pa3$	—	—	—	—	—
NO*5	Monocl.	$C_{2h}^5 - P2_1/c$	—	—	—	—	—
N_2O_4*6	Cub.	$T_h^5 - Im3$	—	—	—	—	—
Na_2O	"	—	—	—	—	—	—
Na_2O_2	Hex.	$D_{3h}^3 - P\bar{6}2m$	—	0,622	—	0,447	0,718
NaO_2	Rhom.	$D_{2h}^{12} - Pnnm$	—	0,426	0,554	0,344	0,819
MgO	Cub.	$O_h^5 - Fm3m$	NaCl	0,4208	—	—	—
α-Al_2O_3	Trig.	$D_{3d}^6 - R\bar{3}c$	α-Al_2O_3	0,4758	—	1,2991	2,72
β-Al_2O_3	Hex.	$D_{6h}^4 - P6_3/mmc$	β-Al_2O_3	0,564	—	2,265	4,02
γ-Al_2O_3	Cub.	$O_h^7 - Fd3m$	$MgAl_2O_4$	0,790	—	—	—
δ-Al_2O_3	Hex.	—	—	0,57	0,290	1,18	2,07

4 — Continued

Oxide	System	Space group	Type of structure	Lattice constants, nm			
				a	b	c	c/a
1	2	3	4	5	6	7	8
ε-Al_2O_3	Cub.	—	—	0,795	—	—	—
SiO_2^{*7}	Hex.	—	—	0,3464	—	0,4382	1,27
SiO_2^{*8}	Trig.	$D_3^4 — P3_121$	α-SiO_2	0,4913	—	0,5405	1,10
SiO_2^{*9}	Rhom.	—	—	0,988	1,71	1,63	1,65
SiO_2^{*10}	Tetr.	$D_4^8 — P4_32_12$	—	0,4971	—	0,6918	1,39
SiO_2^{*11}	Cub.	$T^4 — P2_13$	—	0,718	—	—	—
P_4O_{10}	Rhom.	$D_{2h}^{16} — Pnma$	—	—	—	—	—
P_4O_{10}	Trig.	$C_{3v}^6 — R3c$	—	—	—	—	—
P_2O_5	Rhom.	$C_{2v}^{19} — Fdd2$	—	—	—	—	—
α-KO_2	Tetr.	$D_{4h}^{17} — I4/mmm$	—	0,4034	—	0,6699	1,67
β-KO_2	Cub.	—	—	0,609	—	—	—
KO_3	Tetr.	—	—	0,430	—	0,354	0,82
CaO	Cub.	$O_h^5 — Fm3m$	NaCl	0,4799	—	—	—
CaO_2	Tetr.	$D_{4h}^{17} — I4/mmm$	—	0,354	—	0,591	1,67
Sc_2O_3	Cub.	$T^5 — I2_13$	—	0,9845	—	—	—
Ti_2O	Hex.	—	—	0,2959	—	0,4845	1,637
TiO	Cub.	$O_h^5 — Fm3m$	NaCl	0,4172	—	—	—
Ti_2O_3	Trig.	$D_{3d}^6 — R\bar{3}c$	α-Al_2O_3	0,515	—	1,361	2,64

Ti_3O_5*[12]	Monocl.	—	—	0,9757	0,3802	0,9452	0,968
Ti_5O_9*[13]	Tricl.	—	—	0,5369	0,7120	0,8865	1,245
TiO_2*[14]	Tetr.	$D_{4h}^{19}-I4_1/amd$	—	0,3733	—	0,937	2,51
TiO_2*[15]	Rhom.	$D_{2h}^{15}-Pbca$	—	0,5436	0,9166	0,5135	0,944
TiO_2*[16]	Tetr.	$D_{4h}^{14}-P4_2/mnm$	TiO_2	0,4584	—	0,2953	0,644
V_4O	"	—	—	0,294	—	3,49	11,87
VO	Cub.	O_h^5-Fm3m	NaCl	0,4093	—	—	—
V_2O_3	Trig.	$D_{3d}^6-R\bar{3}c$	α-Al_2O_3	0,4933	—	1,394	2,82
V_3O_5	Monocl.	C_{2h}^6-C2/c	—	0,9991	0,5033	0,9845	0,985
VO_2*[18]	"	$C_{2h}^5-P2_1/c$	—	0,5346	0,4518	0,5739	1,07
V_2O_5	Rhom.	$D_{2h}^{13}-Pmmn$	V_2O_5	11,51	0,3559	0,4371	0,379
Cr_3O	Cub.	O_h^3-Pm3n	β-W	4,544	—	—	—
Cr_2O_3	Trig.	$D_{3d}^6-R\bar{3}c$	α-Al_2O_3	0,4950	—	1,3665	2,76
CrO_2	Tetr.	$D_{4h}^{16}-P4_2/ncm$	SnO_2	0,4421	—	0,2916	0,658
CrO_3	Rhom.	$C_{2v}^{16}-Ama2$	CrO_3	0,4789	0,8557	0,5743	1,19
MnO	Cub.	O_h^5-Fm3m	NaCl	0,4425	—	—	—
α-Mn_3O_4	Tetr.	$D_{4h}^{19}-I4_1/amd$	—	0,575	—	0,942	1,64
γ-Mn_3O_4	Cub.	—	—	0,87	—	—	—
α-Mn_2O_3	Tetr.	—	—	0,885	—	0,995	1,12

4 – Continued

1	2	3	4	5	6	7	8
γ-MnO_2	Rhom.	D_{2h}^{16} — $Pnma$	—	0,4533	0,927	0,2866	0,63
ε-MnO_2	Hex.	—	NiAs	0,279	—	0,441	1,58
FeO	Cub.	O_h^5 — $Fm3m$	NaCl	0,4311	—	—	—
Fe_3O_4	"	O_h^7 — $Fd3m$	$MgAl_2O_4$	0,838	—	—	—
α-Fe_2O_3	Trig.	D_{3d}^6 — $R\bar{3}c$	α-Al_2O_3	0,543	—	—	—
γ-Fe_2O_3	Cub.	O_h^7 — $Fd3m$	$MgAl_2O_4$	0,835	—	—	—
δ-Fe_2O_3	Hex.	—	—	0,509	—	0,441	0,866
CoO	Cub.	O_h^5 — $Fm3m$	NaCl	0,42581	—	—	—
Co_3O_4	"	O_h^7 — $Fd3m$	$MgAl_2O_3$	0,8084	—	—	—
Co_2O_3	Trig.	D_{3d}^6 — $R\bar{3}c$	α-Al_2O_3	0,464	—	0,575	1,24
β-NiO	"	—	—	0,29518	—	0,7243	2,45
α-NiO	Cub.	O_h^5 — $Fm3m$	NaCl	0,41946	—	—	—
Cu_2O	"	O_h^4 — $Pn3m$	Cu_2Mg	0,42696	—	—	—
CuO*[19]	Monocl.	C_{2h}^6 — $C2/c$	—	0,4684	0,3425	0,5129	1,09
ZnO	Hex.	C_{6v}^4 — $P6_3mc$	—	0,325	—	0,5205	1,60
α-Ga_2O_3	Trig.	D_{3d}^6 — $R\bar{3}c$	α-Al_2O_3	0,4979	—	1,3429	2,70
β-Ga_2O_3	Monocl.	C_{2h}^3 — $C2/m$	—	0,580	0,306	1,223	2,1

δ-Ga_2O_3	Cub.	$T^5 — I2_13$	—	1,000	—	—	—
CeO_2	Tetr.	$D_{4h}^{16} — P4_2/ncm$	SnO_2	0,43963	—	0,28626	0,65
CeO_2	Trig.	$D_3^4 — P3_1 21$	α-SiO_2	0,4987	—	0,5652	1,13
As_4O_6	Cub.	$O_h^7 — Fd3m$	Sb_4O_6	—	—	—	—
As_4O_6	Monocl.	$C_{2h}^5 — P\,2_1/c$	As_4O_6	—	—	—	—
SeO_2	Tetr.	—	—	0,835	—	—	—
Rb_2O	Cub.	—	—	0,6742	—	—	—
Rb_2O_2	Rhom.	—	—	0,4201	0,7075	0,5983	1,42
Rb_2O_3	Cub.	—	—	0,930	—	—	—
RbO_2	Tetr.	$D_{4h}^{17} — I4/mmm$	—	0,600	—	0,703	1,17
SrO	Cub.	$O_h^5 — Fm3m$	—	0,5160	—	—	—
SrO_2	Tetr.	$D_{4h}^{17} — I4/mmm$	—	0,3508	—	0,6616	1,89
Y_2O_3	Cub.	$T_h^7 — Ia3$	Mn_2O_3	1,0605	—	—	—
ZrO_2[*20]	Monocl.	$C_{2h}^5 — P2_1/c$	—	0,517	0,526	0,530	1,02
ZrO_2	Tetr.	—	—	0,507	—	0,516	1,017
ZrO_2	Cub.	—	—	0,507	—	—	—
NbO	"	$O_h^5 — Fm3m$	—	0,42013	—	—	—
Nb_2O_3	Trig.	$D_3^2 — P321$	La_2O_3	0,384	—	0,601	1,56
Nb_2O_3	Cub.	$T_h^7 — Ia3$	Mn_2O_3	1,107	—	—	—
NbO_2	Tetr.	—	—	1,371	—	0,5985	0,43
α-Nb_2O_5	Monocl.	—	—	2,134	0,3816	1,947	0,91
δ-Nb_2O_5	Hex.	—	—	0,3607	—	0,3925	1,09

4 – Continued

1	2	3	4	5	6	7	8
γ-Nb_2O_5	Rhom.	—	—	0,619	0,365	0,394	0,63
MoO_2^{*}[21]	Monocl.	$C_{2h}^5 — P2_1/c$	MoO_2	0,561	0,484	0,553	0,98
MoO_3	Rhom.	$D_{2h}^{16} — Pnma$	MoO_3	0,39	1,38	0,37	0,94
TcO_2^{*}[22]	Monocl.	$C_{2h}^5 — P2_1/c$	MoO_2	0,553	0,479	0,553	1,0
RuO_2	Tetr.	$D_{4h}^{14} — P4_2/mnm$	TiO_2	0,4519	—	0,3116	0,68
PdO	"	$D_{4h}^{9} — P4_2/mmc$	—	0,3036	—	0,534	1,76
Ag_2O	Cub.	$O_h^4 — Pn3m$	Cu_2O	0,4727	—	—	—
CdO	"	$O_h^5 — Fm3m$	NaCl	0,46943	—	—	—
In_2O_3	"	$T^5 — I2_13$	—	1,0118	—	—	—
SnO	Tetr.	$D_{4h}^7 — P4_2/ncm$	PbO	0,3802	—	0,4836	1,27
SnO_2	"	$D_{4h}^{16} — P4_2/ncm$	SnO_2	0,4737	—	0,3185	0,673
Sb_4O_6	Rhom.	$D_{2h}^{10} — Pccn$	Sb_4O_6	0,4914	1,2468	0,5421	1,10
Sb_4O_6	Cub.	$O_h^7 — Fd3m$	Sb_4O_6	1,1152	—	—	—
Sb_2O_4	Rhom.	$C_{2v}^9 — Pna2_1$	Sb_2O_4	0,5436	0,4810	1,176	2,16
Sb_2O_5	Cub.	$O_h^7 — Fd3m$	Sb_2O_5	1,0305	—	—	—

TeO_2	Rhom.	D_{2h}^{15} — *Pbca*	—	0,5607	1,2034	0,5463	0,97
TeO_2	Tetr.	D_4^4 — $P4_12_12$	—	0,4810	—	0,7613	1,58
Cs_3O	Hex.	D_{6h}^3 — $P6_3/mcm$	—	0,878	—	0,752	0,86
Cs_2O	Trig.	D_{3d}^5 — $R\bar{3}m$	—	0,4256	—	1,899	4,46
Cs_2O_2	Rhom.	—	—	0,4322	0,7517	0,6430	1,48
Cs_2O_3	Cub.	—	—	0,986	—	—	—
CsO_2	Tetr.	D_{4h}^{17} — $I4/mmm$	—	0,628	—	0,724	1,15
BaO	Cub.	O_h^5 — $Fm3m$	NaCl	0,5542	—	—	—
BaO_2	Tetr.	D_{4h}^{17} — $I4/mmm$	—	0,3816	—	0,6851	1,79
La_2O_3	Trig.	D_3^2 — $P321$	La_2O_3	0,39373	—	0,61299	1,557
La_2O_3	Cub.	T_h^7 — $Ia3$	Mn_2O_3	1,142	—	—	—
Ce_2O_3	Trig.	D_3^2 — $P321$	La_2O_3	0,388	—	0,606	1,56
Ce_2O_3	Cub.	T_h^7 — $Ia3$	Mn_2O_3	1,117	—	—	—
CeO_2	"	O_h^5 — $Fm3m$	CaF_2	0,5395	—	—	—
Pr_2O_3	Trig.	D_3^2 — $P321$	La_2O_3	0,386	—	0,6024	1,55
Pr_2O_3	Cub.	T_h^7 — $Ia3$	Mn_2O_3	1,114	—	—	—

4 – Continued

1	2	3	4	5	6	7	8
Nd_2O_3	Trig.	$D_3^2 — P321$	La_2O_3	0,384	—	0,601	1,56
Nd_2O_3	Cub.	$T_h^7 — Ia3$	Mn_2O_3	1,1078	—	—	—
SmO	"	$T_d^2 — F\bar{4}3m$	—	0,49883	—	—	—
$Sm_2O_3^{*23}$	Monocl.	—	—	1,4177	0,3633	0,8847	0,62
EuO	Cub.	$O_h^5 — Fm3m$	NaCl	0,51439	—	—	—
Eu_2O_3	"	—	—	1,084	—	—	—
Cd_2O_3	"	$T_h^7 — Ia3$	Mn_2O_3	1,079	—	—	—
Dy_2O_3	"	$T_h^7 — Ia3$	Mn_2O_3	1,0667	—	—	—
Ho_2O_3	"	—	—	1,058	—	—	—
Er_2O_3	"	$T_h^7 — Ia3$	Mn_2O_3	1,0550	—	—	—
Tu_2O_3	"	—	—	1,046	—	—	—
Yb_2O_3	"	$T_h^7 — Ia3$	Mn_2O_3	1,0435	—	—	
HfO_2^{*24}	Monocl.	$C_{2h}^5 — P2_1/c$	—	0,511	0,514	0,528	1,03
HfO_2^{*25}	"	—	—	0,521	0,515	0,543	1,04
HfO_2^{*26}	Tetr.	—	—	0,514	—	0,5288	1,02

Ta_4O	Rhom.	—	—	0,7194	0,3266	0,3204	0,44
Ta_2O	"	—	—	0,529	0,692	0,305	0,57
TaO	Cub.	$O_h^5 - Fm3m$	NaCl	0,4422	—	—	—
σ-Ta_2O_5	Tetr.	$D_{4h}^{14} - P4_2/mnm$	—	0,9934	—	0,7613	0,82
TaO_2	"	$D_{4h}^{14} - P4_2/mnm$	TiO_2	0,4709	—	0,3065	0,651
Ta_2O_5	Rhom.	—	—	0,620	0,367	0,390	0,63
W_3O	Cub.	—	—	0,5036	—	—	—
WO_2*[27]	Monocl.	$C_{2h}^5 - P2_1/c$	MoO_2	0,556	0,4884	0,5546	0,99
WO_3*[29]	"	$C_{2h}^5 - P2_1/c$	—	0,7285	0,7517	0,3835	0,53
WO_3*[29]	Tetr.	$D_{4h}^7 - P4/nmm$	ReO_3	0,5250	—	0,3915	0,746
ReO_2*[30]	Monocl.	$C_{2h}^5 - P2_1/c$	MoO_2	0,5562	0,4838	0,5561	1,0
ReO_2	Tetr.	$D_{4h}^7 - P4/nmm$	ReO	—	—	—	—
OsO_2	"	$D_{4h}^{16} - P4_2/ncm$	SnO_2	—	—	—	—
OsO_4	"	$D_{4h}^{14} - P4_2/mnm$	TiO_2	0,4519	—	0,321	0,71
IrO_2	"	$D_{4h}^{14} - P4_2/mnm$	TiO_2	0,450	—	0,315	0,70
PtO	"	—	PtS	0,3046	—	0,5348	1,75

4 – Continued

1	2	3	4	5	6	7	8
HgO	Rhom.	D_{2h}^{16} — $Pnma$	—	0,6608	0,5518	0,3519	0,53
HgO	Trig.	D_3^3 — $P3_112$	—	0,3577	—	0,8681	2,43
Tl_2O_3	Cub.	T^5 — $I2_13$	—	1,0543	—	—	—
PbO	Rhom.	C_{2v}^8 — $Pba2$	—	0,5489	0,4755	0,5891	1,07
PbO	Tetr.	D_{4h}^7 — $P4/nmm$	PbO	0,39759	—	0,5025	1,26
Pb_3O_4	"	D_{4h}^{14} — $P4_2/mnm$	—	0,8815	—	0,6563	0,74
Pb_2O_3	Cub.	—	—	0,550	—	—	—
α-PbO_2	Rhom.	D_{2h}^{14} — $Pbcn$	—	0,4938	0,5939	0,5486	1,11
β-PbO_2	Tetr.	D_{4h}^{14} — $P4_2/mnm$	—	0,493	—	0,337	0,68
Bi_2O_3	Monocl.	C_{2h}^5 — $P2_1/c$	Bi_2O_3	0,583	0,814	0,748	1,28
Bi_2O_3	Tetr.	$D\frac{7}{2d}$ — $C\bar{4}b2$	Bi_2O_3	—	—	—	—
Bi_2O_3	Cub.	$O\frac{4}{h}$ — $Pn3m$	Bi_2O_3	1,0245	—	—	—
Bi_2O_3	"	T^3 — $I23$	Bi_2O_3	—	—	—	—
Bi_2O_{3-x}	Tetr.	$D\frac{17}{4h}$ — $I4/mmm$	—	—	—	—	—

ThO_2	Cub.	$O\frac{5}{h} - Fm3m$	CaF_2	0,559	—	—	—
PaO	"	—	—	0,4961	—	—	—
UO	"	$O\frac{5}{h} - Fm3m$	NaCl	0,492	—	—	—
α-UO_2	"	$O\frac{5}{h} - Fm3m$	CaF_2	0,54682	—	—	—
β-UO_2	"	—	—	0,5440	—	—	—
γ-UO_2	Tetr.	—	—	0,5388	—	0,5561	1,03
U_3O_7	"	—	—	0,545	—	0,540	0,99
U_2O_5	Rhom.	—	—	0,829	3,171	0,673	0,81
U_3O_8	"	—	U_3O_8	0,67198	0,3983	0,41462	0,62
α-UO_3	Hex.	—	—	0,3963	—	0,4160	1,05
UO_3	Cub.	—	—	0,4146	—	—	—

4 – Continued

1	2	3	4	5	6	7	8
UO_3	Rhom.	—	—	1,301	1,072	0,751	0,58
NPO	Cub.	—	—	0,500	—	—	—
NPO_2	"	—	—	0,5425	—	—	—
PuO	"	—	—	0,494	—	—	—
PuO_2	"	—	—	0,5386	—	—	—

*1 Ice. *2 At −55°C. *3 At −155°C. *4 At < 91°C. *5 At < 163°C. *6 At < 11.2°C. *7 β-Quartz. *8 α-Quartz. *9 α-Tridymite. *10 α-Cristobalite. *11 β-Cristobalite. *12 β = 93.11°. *13 α = 97.55°, β = 112.34°, γ = 108.50°. *14 Anatase. *15 Brookite. *16 Rutile. *17 β = 138.83°. *18 β = 122.18°. *19 β = 99.46°. *20 β = 80.16°. *21 β = 119.6°. *22 β = 120°. *23 β = 99.96°. *24 β = 99.73° at < 1640°C. *25 β = 98.80° at 1640-1920°C. *26 At 1920°C. *27 β = 118.93°. *28 β = 90.90° at < 700°C. *29 At > 700°C. *30 β = 120.87°

5. DENSITY [1, 2, 6, 10, 14, 17–21]

Oxide	Density, kg/m^3	Oxide	Density, kg/m^3	Oxide	Density, kg/m^3
1	2	1	2	1	2
H_2O	1000	V_2O_5	3357	Sb_2O_4	3800—4000
H_2O_2	1463	Cr_2O_3	5210	Sb_2O_5	3780
BeO	3030	CrO_3	2700	I_2O_5	4800
B_2O_3	1844	MnO	5430—5460	BaO	5720
CO	1250*4	Mn_3O_4	4700	BaO	4960
CO_2	1970*4	Mn_2O_3	4940	La_2O_3	6510
N_2O	1977*4	MnO_2	5026	Ce_2O_3	6900—7000
NO	1340*4	FeO	5870	CeO_2	7130
N_2O_3	1477*4	Fe_3O_4	5000—5400	Sm_2O_3	7400
N_2O_4	1491*4	Fe_2O_3	5240	Eu_2O_3	7400
N_2O_5	1642*4	CoO	5700—6700	Gd_2O_3	7407
F_2O	1900	Co_3O_4	6070	Dy_2O_3	7407
Na_2O_2	2805	Co_2O_3	5180	Er_2O_3	2650
MgO	3650	NiO	7450	Yb_2O_3	9170
α-Al_2O_3	3970	Gu_2O	5800—6110	HfO_2*10	10010
δ-Al_2O_3	2400	CuO	6400—6450	HfO_2*6	9680
SiO_2*1	2651	ZnO	5660	Ta_2O	15500
SiO_2*2	2260	Ga_2O_3*5	6440	TaO_2	9950
SiO_2*3	2320	Ga_2O_3*6	5880	Ta_2O_5	8730
P_4O_6	2135	GeO_2	4700—6300	W_2O	14800
P_2O_5	2390	As_2O_3	3870	WO_2	11400
SO_2	2390*4	As_2O_5	4090	WO_3	6470
SO_3	2750*4	SrO	4700	ReO_3	6974
Cl_2O_2	3890*4	Y_2O_3	4840	ReO_4	8400
ClO_2	3090*4	ZrO_2*6	5560	OsO_2	7910
K_2O	2320	ZrO_2 (Tr)	6270	OsO_4	4910
CaO	3370	NbO	7260	HgO	11140
Sc_2O_3	3864	NbO_2	5980	PbO	9530
TiO	4930—5530	Nb_2O_5	4950	Pb_3O_4	9100
TiO_2*7	3830	MoO_2	4110	PbO_2	9375
TiO_2*8	4170	MoO_3	4690	Bi_2O_3	8900
TiO_2*9	4240	CdO	6950	ThO_2	10000
VO	5200—5758	SnO	6450	UO_2	10960
V_2O_3	4870	SnO_2	6950	U_3O_8	8100—8300
V_2O_5	4550	Sb_2O_3	5200	UO_3	5920—7540
VO_2	4339				

*1 Quartz. *2 Tridymite. *3 Cristobalite. *4 At temperature 20°C and pressure 101.325 kPa.
*5 Hexagonal. *6 Monoclinic. *7 Anatase. *8 Brookite. *9 Rutile. *10 Tetragonal.

THERMODYNAMIC AND THERMAL PROPERTIES

1. STANDARD HEAT OF FORMATION (STANDARD ENTHALPY CHANGE) AND STANDARD ENTROPY [1; 9, Nos. 1–7; 19; 28; 34; 35; 36; 41; 75; 52; 66, Vol. 1; 79, Vols. 1 and 2; 81, Vol. 5; 92; 101; 144; 156; 183; 189; 190; 206]

Oxide	State	Standard heat of formation $-\Delta H^0_{298,15} \cdot 10^{-3}$, kJ/kmole	Standard entropy $S^0_{298,15}$, kJ/(kmole · °K)
1	2	3	4
H_2O	s	292,046	(39,356)
	l	286,0208±0,0401	70,13±0,21
	g	241,980±0,042	188,850±0,050
HO	g	−38,983±1,256	183,759±0,033
H_2O_2	*l*	187,90±0,08	109,57±0,42
	g	136,24±0,17	233,037±0,50
H_2O_3	—	46,1±20,9	—
HO_2	g	−20,545±8,374	227,59±1,67
H_2O_4	*l*	24,7±16,7	—
	g	−25,5±16,7	266,28±4,19*1
D_2O	*l*	294,797±0,100	75,95±0,21
	g	249,370±0,105	198,375±0,050
DO	g	−36,019	189,696±0,033
D_2O_2	*l*	196,499±0,419	—
	g	144,122±0,377	239,82±0,63
DO_2	g	−16,785±8,374	231,57±1,67
T_2O	*l*	298,653±0,46	—
	g	252,598±0,209	204,266±0,063
TO	g	−35,651±1,26	193,263±0,037
TO_2	g	−15,165±8,374	234,21±1,88
HDO	*l*	290,145±0,209	79,34
	g	245,476±0,209	199,539±0,05
HDO_2	g	140,170±0,461	242,42±0,84
HTO	*l*	293,83±1,26	—
	g	246,921±0,209	202,721±0,063
DTO	g	250,936±0,209	207,121±0,063
HDO_2	g		
Li_2O	s	596,912	37,93
	g	133,304±12,56	238,80
LiO	g	−79,679±20,9	206,65
Li_2O_2	s	634,30±16,75	69,08
	g	115,137±25,12	—

1	2	3	4
LiO_2	s	(238,6)	—
Be_2O	g	33,49±41,87	—
BeO	s	599,13±4,19	14,11
	g	—126,44	197,64
Be_2O_2	g	418,68±79,55	—
Be_3O_3	g	1038,33±37,68	—
Be_4O_4	g	1553,30±50,24	—
Be_5O_5	g	2068,28±75,36	—
Be_6O_6	g	2612,56±92,11	—
B_2O	g	—	231,11±2,51
BO	g	—22,23±15,91	203,478±0,042
B_2O_2	g	456,36±8,37	242,83±0,84
B_2O_3	s	1273,75±1,21	54,01±0,29
	am	1254,83±1,72	80,81±1,26
	gl	1245,99	77,66
	g	842,028±8,37	279,26±2,93
BO_2	g	293,08±20,9	229,646±0,419
BO_3	—	1461,19	—
C_2O	g	—289,73±12,56	233,20±0,84
C_3O_2	g	33,5±20,9	257,794
CO	g	110,598±0,129	197,676±0,029
CO_2	*l*	399,052±0,837	—
	g	393,777±0,046	213,824±0,042
CO_3	g	—	259,2±1,26
N_2O	g	—81,70±0,42	220,02±0,13
NO	g	—90,31±0,29	210,72±0,08
N_2O_2	g	—168,73	—
N_2O_3	*l*	—49,40	—
	g	—83,32±1,26	307,31±1,26
NO_2	g	—33,49±0,84	240,32±0,16
N_2O_4	*l*	19,05	209,34
	g	—9,63±1,67	303,96±0,84
N_2O_5	s	42,71±2,51	178,36
	g	—11,72±2,09	355,88±2,09
NO_3	g	—71,18±20,9	252,88±2,09
O	g	—249,349±0,126	161,054±0,021
O_2	g	0,000	205,174±0,033
O_3	g	—142,35±2,09	238,98±0,21
O_4	g	0,544±0,837	—
F_2O	g	—25,12±8,4	247,15±0,08
FO	g	—136,07	218,011±0,84
F_2O_2	g	—19,80±1,26	—
F_2O_3	g	—26,13±3,14	—
FO_2	g	77,456	—
F_2O_5	g	(224,41)	—
Na_2O	s	416,16	72,85
	g	154,91	—
Na_2O_2	s	511,2±2,1	94,622
NaO_2	s	260,0±2,9	115,974
	g	163,29	—

1 — Continued

1	2	3	4
MgO	s	602,23±0,42	26,921
	g	29,73	213,296
MgO_2	s	623,0	85,829
Al_2O	g	128,12	259,58±1,67
AlO	g	—89,60	218,425±0,042
Al_2O_2	g	391,47±20,9	—
Al_2O_3	s	1676,81±1,26*2	50,95±0,08
	s	1662,2*3	—
	am*6	1603,5	—
	g	833,17	—
SiO	s	437,94	27,214
	g	103,41±6,28	211,60±0,08
Si_2O_2	g	404,44±33,5	—
SiO_2 *4	s	911,55±1,42	41,868±0,419
*5	s	908,87±1,55	42,663±0,419
	s	906,02	43,54±0,84
	gl	902,17±1,51	46,89
	am	897,44±1,51	—
*7	s	906,48±1,55	40,40±0,21
*8	s	862,10±1,88	27,80±0,21
PO	g	326,6±33,5	227,76±1,67
P_2O_3	g	27,855	222,82±0,13
P_4O_6	*l*	1130,4±37,7	142,351
	s	1641,23	—
	g	1594,58	347,09±0,84
PO_2	s	272,14	48,15
	g	387,70	253,72±3,35
P_4O_8	g	2407,510	385,2±8,4
P_2O_5	s	1531,53	140,26
P_4O_{10}	s	2986,03±4,19	240,32±0,08
	am	3044,64	—
	gl	3084,0	—
	g	2896,43±5,86	394,82±6,28
S_2O	g	109,32±5,0	267,33±0,63
SO	g	—0,41±1,26	221,98±0,17
SO_2	*l*	322,38	—
SO_3 *9	g	297,10±0,21	248,24±0,08
*10	s	462,73	—
	s	449,872	—
	l	439,28±1,26	—
	g	396,11±0,38	256,86±0,21
S_3O_9	g	1314,66	—
S_2O_7	s	813,50	—
Cl_2O	g	—76,62±1,26	266,402±0,17
ClO	g	—101,40±0,42	226,669±0,84
ClO_2	g	—104,67±6,28	257,195±0,08
ClO_3	g	—154,9±8,4	—
Cl_2O_7	*l*	321,13±1,26	—
	g	287,13±1,34	—

1	2	3	4
K_2O	s	361,739	96,3
	g	58,62	—
K_2O_2	s	494,04±41,87	113,0
	g	159,10	—
K_2O_3	s	561,03	140,26
KO_2	s	283,03±3,35	116,81
K_2O_4	s	561,03	—
KO_3	s	260,00±3,77	—
CaO	s	635,514±0,879	39,78±0,84
	g	59,402±16,75	219,761
CaO_2	s	654,82±4,19	—
Sc_2O	g	102,58±46,05	—
ScO	g	66,15*11	—
Sc_2O_3	s	1918,8	103,83
$TiO_{0,1}$	s	57,36	—
$TiO_{0,2}$	s	114,72	—
$TiO_{0,3}$	s	172,08	—
$TiO_{0,4}$	s	229,44	—
$TiO_{0,8}$	s	434,17	—
$TiO_{0,9}$	s	480,23	—
TiO	s	526,28	34,79±0,21*12
	g	−57,31±9,21	227,76±7,12
$TiO_{1,01}$	s	530,89	35,96±0,42*12
$TiO_{1,1}$	s	572,34	—
$TiO_{1,20}$	s	618,39	—
$TiO_{1,22}$	s	627,60	—
Ti_2O_3	s	1518,97	77,37±0,21
Ti_3O_5	s	2457,65	127,15±0,42
Ti_4O_7	s	3389,63	194,94±0,63
Ti_5O_9	s	4319,52	249,11±0,84
Ti_6O_{11}	s	5248,15	301,45±0,84
Ti_7O_{13}	s	6178,04	356,72
Ti_8O_{15}	s	7105,00	405,70
Ti_9O_{17}	s	8035,73	459,71
$Ti_{10}O_{19}$	s	8967,71	504,09±1,26
TiO_2*13	s	944,50±0,96	50,37±0,21
*14	s	939,27±1,26	49,95±0,42
*15	s	921,1	—
	am	900,2	—
	g	297,397±21,35	259,16±4,19
$VO_{0,86}$	s	—	32,28±0,42
VO	s	432,08±4,61	33,62±0,42
	g	−151,386±41,87	231,11±2,09

1 – Continued

1	2	3	4
$VO_{1,24}$	s	523,77±5,86	39,31±0,42
$VO_{1,30}$	s	—	41,95±0,42
V_3O_4	s	1473,75±58,62	—
V_2O_3	s	1219,87±2,64	98,39±1,26
V_3O_5	s	1942,68±3,77	154,91±4,19
V_4O_7	s	2663,64±5,44	211,43±0,84
V_5O_9	s	3380,42±7,12	258,74±6,28
V_6O_{11}	s	4096,78±8,40	322,38±8,37
V_7O_{13}	s	4811,47±9,63	351,69±16,75
V_8O_{15}	s	5526,16±10,89	401,93±16,75
VO_2	s	715,94±8,37	—
	g	226,09±29,31	—
V_2O_4	s	1427,70±5,86	96,72±1,26
V_4O_8	g	2457,7±25,1	—
V_6O_{13}	s	4440,52±12,6	289,73±16,75
V_3O_7	s	2327,9	—
V_2O_5	s	1553,30±9,63	131,05±1,26
V_4O_{10}	s	2826,1±20,9	—
CrO	s	389,37	—
	g	272,14	—
Cr_2O_3	s	1141,32±1,67	81,22±1,26
*16	s	1510,18	—
*17	s	1848,47	—
*18	s	2176,30	—
$CrO_{1,98}$	s	582,80	—
CrO_2	s	588,66	48,15
	g	61,127	—
Cr_5O_{12}	s	2888,9	—
Cr_8O_{21}	s	4743,64	—
CrO_3	s	590,76±3,35	73,27
	g	295,081±83,7	270,89±12,6
$(CrO_3)_3$	g	1628,7±41,9	—
$(CrO_3)_4$	g	2210,6±41,9	—
$(CrO_3)_5$	g	2771,7±41,9	—
MnO	s	385,35±0,46	—
	s	—	61,55±2,09*19
	s	—	226,50±6,28*20
	g	—123,866±12,56	—
Mn_3O_4	s	1388,55±1,05	154,91±8,37
Mn_2O_3	s	958,36±0,84	110,53±2,09
MnO_2	s	521,84±0,84	53,17±0,42
Mn_2O_7	*l*	726,83±2,5	—
$Fe_{0,947}O$	s	265,44±1,67	57,61±0,42
FeO	s	265,02±4,19	60,79±0,42
	g	—255,767	272,14±2,93
$FeO_{1,05}$	s	278,84±4,19	—
$FeO_{1,056}$	s	280,306±1,758	60,83±0,42

1	2	3	4
$FeO_{1,10}$	s	292,66±4,19	—
$FeO_{1,15}$	s	306,06±4,19	—
$FeO_{1,20}$	s	319,87±4,19	—
Fe_3O_4	s	1117,88±2,09	146,29±0,84
	s	—	145,70±0,84*21
	s	—	144,19±0,84*22
	s	—	142,48±0,84*23
Fe_2O_3	s	822,71±4,19	87,50±2,09
	am	688,73	—
CoO	s	239,07±1,26	52,75±0,42
	am	211,43	—
	g	—309,652±21,353	242,0±8,4
Co_3O_4	s	887,6±8,4	103,0±0,8
NiO	s	239,90±0,54	38,02±0,42
	g	—313,901±20,93	241,16±8,4
Ni_2O_3	s	485,7	—
Cu_2O	s	173,292±1,340	92,99±0,21
CuO*24	s	157,131±0,879	—
*25	s	162,113±1,256	—
	s	—	42,66±0,21
	g	—320,215±41,87	234,88±2,1
Zn_2O	g	⩾21,269	—
ZnO	s	350,862±0,25	43,67±0,42
	g	—	224,83±2,09
Ga_2O	s	355,9	94,2
	g	85,83±6,28	283,87±2,09
GaO	s	—141,292±17,166	—
	g	—	230,99±0,21
Ga_2O_3	s	1089,82±3,35	85,03±0,42
	g	516,23±25,1	—
GeO	s	255,4±20,9	52,34
	g	30,702	223,95±0,21
Ge_2O_2	g	247,0±20,9	—
Ge_3O_3	g	460,5±33,5	—
GeO_2*26	s	580,54±1,42	39,73±0,13
*27	s	555,09±1,00	55,31±0,42
	am	539,39±1,17	—
AsO	g	—57,326±18,00	230,32±0,33
As_2O_3	s	654,82	107,18
As_4O_6*28	s	1332,45±2,51	245,35
*29	s	1335,59±2,51	233,62±8,37
	g	1230,9±8,4	—
AsO_2	s	367,601	—
As_2O_4	s	800,10±1,26	—
As_2O_5	s	927,04±1,67	105,5±1,67
SeO	s	—	46,06
	g	—54,625±8,40	234,46±0,21
SeO_2	s	225,67±2,09	62,8
	g	126,86±4,61	265,02±0,42

1 — Continued

1	2	3	4
Se_2O_5	s	415,33	—
SeO_3	s	172,91	—
	g	92,110	268,0
BrO	g	—125,713±2,51	237,446±0,92
BrO_3	s	—52,34	—
Rb_2O	s	330,338	113,04
Rb_2O_2	s	424,96±41,87	115,14
Rb_2O_3	s	508,696	136,07
RbO_2	s	284,70±2,51	130,18±0,63*30
Rb_2O_4	s	528,374	—
Sr_2O	g	236,97	—
SrO	s	590,76±8,37	54,43
	g	59,620±8,79	230,111
SrO_2	s	641,42±16,7	62,0
Y_2O	g	—19,26±33,49	—
YO	g	48,15*11	—
Y_2O_2	g	514,98	—
Y_2O_3	s	1758,5±37,7	123,5
$ZrO_{0,079}$	s	41,87±2,51	—
$ZrO_{0,201}$	s	119,32±2,51	—
$ZrO_{0,333}$	s	194,69±2,51	—
$ZrO_{0,46}$	s	268,37±25,5	57,36±15,07
ZrO	s	724,3±41,9	—
	g	91,394±41,9	227,76±0,42
Zr_2O_3	s	1808,7	—
ZrO_2	s	1101,30±0,63*31	50,41±0,42
	s	1090,24±2,09*32	—
	g	295,931±20,9	267,96±5,02
NbO	s	406,12±4,19	50,24
	g	—199,924±41,9	239,07±1,26
NbO_2	s	795,91±0,84	54,55±0,42
Nb_2O_5	s	1899,13±1,67	137,33±0,84
MoO_2	s	589,50±0,84	46,31±0,21
	g	13,40±20,9	—
Mo_4O_{11}	s	2809,3	—
Mo_8O_{23}	s	5823,8	—
Mo_9O_{26}	s	6585,8	—
MoO_3	s	745,67±0,42	77,79±0,42
	g	363,0±20,9	279,26±12,6
$(MoO_3)_2$	g	1201,6±29,3	—
$(MoO_3)_3$	g	1925,9±62,8	—
$(MoO_3)_4$	g	2595,8±41,9	—
$(MoO_3)_5$	g	3286,6±41,9	—
TcO_2	s	434,3	54,43±4,19
TcO_3	s	540,1±25,1	77,46±4,19
Tc_2O_7	s	1115,36±11,72	191,76±8,37
	g	983,06±11,72	—
RuO	g	—412,643±41,87	237,81±6,28

1	2	3	4
RuO_2	s	305,22±6,28	58,62±6,28
RuO_3	g	78,29±10,47	276,3±12,6
RuO_4	s	239,48±5,44	141,1
	l	228,60±5,02	177,52
	g	184,22±5,02	290,86±0,42
Rh_2O	s	95,04	106,8
RhO	s	94,20	50,24
	g	—409,921±62,8	229,86±8,37
Rh_2O_3	s	355,9	106,34±8,37
RhO_2	g	—188,4	—
PdO	s	115,56±3,35	38,94±4,19
	g	—337,310	248,70±8,37
Ag_2O	s	31,150	121,42±0,29
AgO	s	—	57,82
	g	—292,716	245,77±0,42
Ag_2O_2	s	24,70±2,09	85,4
AgO_2	s	—	63,85
CdO	s	259,16±1,26	54,85±1,26
	am	238,6	—
	g	—81,123	233,20±2,1
In_2O	s	167,47	117,23
	g	55,387±30,56	298,52±2,51
InO	s	272,14	60,7
In_2O_3	s	926,41±1,67	108,02±3,35
	s	—	134,92±0,92
	g	904,35±29,31	—
SnO	s	286,168±0,67	56,52±2,09
	g	—20,867	232,162±0,042
$(SnO)_2$	g	238,6±25,1	—
$(SnO)_3$	g	510,8±37,7	—
$(SnO)_4$	g	787,1±41,9	—
SnO_2	s	581,17±3,35	52,34±1,26
SbO	g	—103,569±83,7	238,40±0,25
Sb_2O_3	s	699,61	123,1
	s	705,48[*33]	—
Sb_4O_6[*20]	s	—	265,44±8,37
[*33]	s	1418,07±5,86	282,19±8,37
SbO_2	s	447,99	63,64
Sb_2O_4	s	908,12±4,6	127,28±8,37
Sb_6O_{13}	s	2806,83	—
Sb_2O_5	s	1008,18	125,19±8,37
TeO	s	234,46	54,43
	g	—174,016±6,7	240,74±0,42
TeO_2	s	321,96±5,0	58,62±8,37
	g	51,92±9,63	273,40±1,67
IO	g	—168,041±25,1	245,447±1,05
I_2O_5	g	183,38	—
Cs_2O	s	317,78	96,3
Cs_2O_2	s	401,9±62,8	167,5
Cs_2O_3	s	465,6	196,78

1 – Continued

1	2	3	4
CsO_2	s	289,73±2,09	—
Ba_2O	s	615,460	98,39
	g	298,94	—
BaO	s	558,94	70,338
	g	136,49	235,516
Ba_2O_2	g	638,91	—
BaO_2	s	635,93±1,05	77,46
La_2O	g	—51,92±33,49	—
LaO	g	116,81*[11]	—
La_2O_2	g	510,79	—
La_2O_3	s	1799,00±5,44	127,40
CeO	g	93,37*[11]	—
Ce_2O_3	s	1821,26±25,12	152,40
CeO_2	s	1089,32±1,382	74,11
PrO	g	139,42*[11]	—
Pr_2O_3	s	1824,61±0,67	158,68
$PrO_{1,703}$	—	935,75±3,35	—
$PrO_{1,717}$	—	938,56	—
$PrO_{1,804}$	—	948,31±3,35	—
Pr_6O_{11}	s	5717,494±20,097	488,6
PrO_2	s	962,96	71,18
NdO	g	139,42*[11]	—
A-Nd_2O_3	s	1784,0±1,26	158,68
SmO	g	144,44*[11]	—
C-Sm_2O_3*[34]	s	1827,5±7,1	—
B-Sm_2O_3*[35]	s	1833,4±7,9	—
Sm_2O_3	s	—	152,82
EuO	s	607,92±17,17	86,67±12,56
	g	136,91	—
C—Eu_2O_3*[34]	s	1734,6±6,7	—
B—Eu_2O_3*[35]	s	1726,64±5,11	—
GdO	g	74,53*[11]	—
B—Gd_2O_3	s	1825,44±1,80	—
Gd_2O_3	s	—	150,72
TbO	g	83,74*[11]	—
Tb_2O_3	s	1828,79±8,37	158,26
$TbO_{1,71}$	s	934,91±4,19	—
$TbO_{1,80}$	s	947,891±4,19	—
DyO	g	84,15*[11]	—
Dy_2O_3	s	1866,64±3,89	149,89
HoO	g	99,65*[11]	—
Ho_2O_3	s	1882,0±5,0	158,26
ErO	g	71,59*[11]	—
Er_2O_3	s	1899,0±1,9	153,24
TuO	g	87,09*[11]	—
Tu_2O_3	s	1890,0±6,0	153,24
YbO	g	⩾—30,14*[11]	—

1	2	3	4
Yb_2O_3	s	1816,0±44,1	133,14
LuO	g	23,03*11	—
Lu_2O_3	s	1896,0±13,8	110,033
$HfO_{0,104}$	s	59,45±5,86	—
$HfO_{0,152}$	s	87,92±5,86	—
$HfO_{0,199}$	s	115,14±5,86	—
$HfO_{0,255}$	s	153,24±5,86	—
HfO	g	—120,316±25,54	237,72±0,84
HfO_2 *36	s	1118,29±2,09	59,37±0,63
*32	s	1127,09±2,09	—
	g	224,965±20,9	275,91±5,02
TaO	g	236,387±41,87	241,16±0,08
TaO_2	s	—	385,2
	g	184,2±20,9	279,68±1,67
Ta_2O_5	s	2048,6±1,67	143,19±1,67
W_3O	s	316,52	83,7±8,4
WO_2	s	589,92±0,84	50,58±0,42
	g	—76,20±29,31	—
W_2O_5	s	1402,6	—
W_3O_8	g	1710,31±41,9	—
$WO_{2,72}$ *37	s	776,23±8,37	68,66±1,26
$WO_{2,90}$ *38	s	818,94±8,37	73,27±0,84
$WO_{2,96}$ *39	s	838,59±8,37	74,94±0,84
WO_3	s	843,22±0,84	76,0±0,5
	g	299,77±29,3	—
$(WO_3)_2$	g	1168,1±41,9	415,75
$(WO_3)_3$	g	2026,4±41,9	504,93
$(WO_3)_4$	g	2805,2±41,9	605,41
ReO	g	—	242,42±4,19
Re_2O_3	s	519,58±7,54*32	—
ReO_2	s	451,76±3,77	56,52±8,37

1 — Continued

1	2	3	4
	s	1001,48±5,02*40	—
ReO_3	s	593,27±4,19	80,81±8,37
Re_2O_7	s	1272,79±8,39	207,37±0,84
	g	1116,2±9,2	—
ReO_4	s	645,6	144,45
OsO	g	—	254,14±8,37
OsO_2	s	259,58±41,87	71,2±8,4
OsO_3	g	280,1±12,6	—
OsO_4*41	s	383,93	145,3
*42	s	391,047	124,95
	s	393,98±8,37	136,9±8,4
	g	336,41±8,79	293,83±0,33
IrO	g	—563,598±20,9	257,49±8,37
Ir_2O_3	s	284,70	110,95
IrO_2	s	254,98±1,67	58,6±12,6
	g	—216,04±12,56	263,8
IrO_3	g	—13,40±6,28	288,9
PtO	s	71,18	56,52
	g	—442,641	256,65±8,37
Pt_3O_4	s	267,96±100,48	167,5±16,8
$PtO_{1,39}$	s	56,94±4,61	—
$Pt_3O_{4,17}$	s	170,82±13,82	—
PtO_2	s	133,98±41,87	69,08
	g	—170,82±6,28	259,6
AuO	g	—	254,98±1,67
Au_2O_3	s	12,98±2,51	125,6
Hg_2O	s	92,11	125,6
HgO*42	s	90,52±0,13	71,335
*43	s	90,937±0,100	70,34±0,21
*27	s	90,31±0,29	71,531
	g	—	241,58±2,09
Tl_2O	s	167,47±7,54	99,65
	g	—8,79±6,28	314,00±2,93
Tl_2O_3	s	390,63±4,19	140,26

1	2	3	4
Tl_2O_4	s	448,0	—
PbO*42	s	217,76±0,71	68,75±0,21
*43	s	219,43±0,63	66,15±1,26
	g	–68,89±5,44	240,07±0,13
Pb_2O_3	g	122,67	—
Pb_3O_3	g	315,68	—
Pb_4O_4	g	561,03	—
Pb_3O_4	s	723,90±1,88	211,43±6,7
PbO_2	s	276,75±1,26	76,62±2,1
BiO	s	209,3	62,8
	g	–118,97±83,7	246,23±0,21
Bi_2O_3	s	578,2±4,2	151,6±4,2
PoO_2	s	251,2	71,18±8,4
RaO	s	544,28	71,18
Ac_2O_3	s	1858,94	152,82
ThO	s	607,09	67,0
ThO_2	s	1227,57	65,272±0,209
	g	≥510,37	—
PaO_2	s	1029,95	74,53
Pa_2O_5	s	2093,40	157,0
UO	s	540,1	67,0
U_3O_5	s	3573,43	276,33
UO_2	s	1086,48±8,37	78,00
U_2O_5	s	2386,48	—
U_3O_8	s	3579,71	281,8
UO_3	s	1226,73	98,68
	g	569,40	—
NpO_2	s	1029,95	80,35
Np_2O_5	s	2135,27	180,03
PuO	s	481,48	83,74
Pu_2O_3	s	1666,35±41,87	159,10
PuO_2	s	1058,7±1,60	82,48
Am_2O_3	s	1758,46	154,91
AmO_2	s	1004,41	83,74

*1 $S^0_{253.15}$. *2 Corundum, hexagonal. *3 Tetragonal. *4 α-Quartz. *5 α-Cristobalite. *6 Finely dispersed. *7 Coesite. *8 Stishovite. *9 Difficultly fusible, fibrous. *10 Readily fusible, fibrous. *11 $\Delta H^0_{f,0}$. *12 Value cited is $S^0_{298.15} - S^0_0$. *13 Rutile. *14 Anatase. *15 Hydrated, precipitated. *16 $Cr_2O_3 \cdot H_2O$. *17 $Cr_2O_3 \cdot 2H_2O$. *18 $Cr_2O_3 \cdot 3H_2O$. *19 Hexagonal. *20 Cubic. *21 1% cation vacancies. *22 2% cation vacancies. *23 4% cation vacancies. *24 Microcrystalline. *25 Macrocrystalline. *26 Tetragonal. *27 Hexagonal. *28 Claudetite. *29 Arsenolite. *30 00.8% purity. *31 Baddeleyite. *32 Hydrated. *33 Orthorhombic. *34 Low-temperature modification. *35 High-temperature modification. *36 Monoclinic. *37 (1/18) $W_{18}O_{49}$ *38 (1/10) $W_{10}O_{29}$. *39 (1/50) $W_{50}O_{148}$ *40 $ReO_2 \cdot 2H_2O$. *41 White. *42 Yellow. *43 Red.

2. TEMPERATURE DEPENDENCE OF HEAT OF FORMATION ($-\Delta H_T^0 \cdot 10^{-3}$, kJ/kmole [19, 75]

Oxide	State	Temperature, °K									
		298,15	400	600	800	1000	1200	1400	1600	1800	2000
1	2	3	4	5	6	7	8	9	10	11	12
H_2O	*l*, g	286,0	243,0	244,9	246,6	248,1	249,1	250,0	251,4	251,4	252,3
H_2O_2	*l*, g	187,4	187,4	139,8	140,7	141,1	141,1	140,9	—	—	(135,7)
Li_2O	s	596,9	573,2	580,3	579,7	578,0	575,1	571,7	—	—	—
BeO	s	599,2	599,4	599,2	597,9	596,9	595,0	(593,3)	—	—	(598,3)
B_2O_3	s, *l*	1278,6	1278,6	1277,0	1249,1	1243,1	1237,6	1234,7	1233,0	—	—
CO	g	110,5	110,1	110,3	111,0	112,0	113,0	114,5	115,8	117,2	119,1
CO_2	g	393,8	393,8	394,2	394,4	395,2	394,8	394,8	396,5	396,9	397,1
N_2O	g	—81,6	—81,2	—81,4	—82,1	—82,9	—83,7	—85,0	—86,2	—87,1	—87,9
NO	g	—90,4	—90,4	—90,4	—90,4	—90,4	—90,6	—90,6	—90,9	—90,9	—90,9
N_2O_3	g	—73,3	(—73,3)	—	—	—	—	—	—	—	—
NO_2	g	—33,9	—33,5	—32,7	—32,4	—33,1	—33,7	—33,5	—34,3	—34,8	—34,3
N_2O_4	g	—9,6	—9,2	—10,5	—13,0	—16,3	—	—	—	—	—
Na_2O	s, *l*	416,2	421,6	421,4	420,4	417,8	(578,2)	(569,8)	(561,4)	(553,1)	(544,7)
Na_2O_2	s, *l*	511,2	(516,2)	(514,1)	(485,7)	—	—	—	—	—	—
NaO_2	s	260,0	(259,6)	(255,4)	(251,2)	(247,0)	—	—	—	—	—
MgO	s	601,7	601,7	601,7	610,5	601.7	609,6	740,5	737,7	735,1	(732,9)
Al_2O	g	140,3	(144,4)	(148,6)	(154,9)	(180,0)	(184,2)	(188,4)	(192,6)	(196,8)	—
AlO	g	41,9	(39,8)	(37,7)	(35,6)	(23,0)	(20,9)	(18,8)	(16,7)	(14,7)	—
$Al_2O_3^{*1}$	s	1676,0	1676,5	1675,6	1673,5	1693,2	1690,0	(1686,0)	(1683,0)	(1680,0)	—
SiO_2^{*2}	s	878,8	878,8	878,0	876,3	872,8	871,1	(869,6)	(868,0)	(913,0)	—
SiO_2^{*3}	s	877,2	877,2	875,3	873,6	871,5	870,0	(868,4)	(866,7)	(911,7)	(910,7)
SiO_2^{*4}	s	876,8	876,5	875,5	873,8	872,2	870,3	(868,6)	(866,9)	(911,9)	—

SiO_2*5	s	847,9	847,9	847,4	846,4	844,5	842,4	(840,3)	(837,6)	(880,7)	(877,6)
PO	g	39,8	(41,9)	(54,4)	(56,5)	(56,5)	(58,6)	(58,6)	—	—	—
P_4O_{10}	s, g	3014,0	3019,1	3059,3	2971,4	2960,5	2948,8	2939,1	—	—	—
K_2O	s, l	420,6	(366,8)	(367,2)	(367,6)	(368,0)	(522,1)	(516,2)	—	—	—
K_2O_2	s, l	494,0	(498,2)	(496,1)	(468,9)	(460,5)	(609,2)	(598,7)	—	—	—
K_2O_3	s, l	523,3	(525,4)	(521,3)	(483,6)	(475,2)	—	—	—	—	—
KO_2	s, l	283,0	(283,4)	(279,7)	(257,5)	(250,8)	—	—	—	—	—
KO_3	s, l	(259,6)	(259,6)	—	—	—	—	—	—	—	—
CaO	s	635,5	635,2	634,5	634,3	633,9	(642,3)	(641,4)	—	—	—
CaO_2	s	(655,2)	(653,1)	—	—	—	—	—	—	—	—
Sc_2O_3	s	(1720,8)	(1720,8)	(1722,9)	(1722,9)	(1722,9)	(1722,9)	(1720,8)	(1718,7)	(1751,1)	(1748,0)
TiO	s	518,8	518,4	517,1	515,8	514,4	516,2	509,6	505,6	501,4	—
Ti_2O_3	s	1519	1518	1511	1505	1499	1500	1494	1487	1480	—
Ti_3O_5	s	2458	2454	2437	2432	2429	2433	2426	—	—	—
TiO_2*6	s	944,6	943,7	942,1	940,2	938,7	941,2	939,4	937,2	935,2	—
VO	s	410,0	409,0	407,8	405,5	403,0	400,0	397,6	395,0	—	—
V_2O_3	s	1239	1238	1233	1228	1223	!218	1212	1207	1201	—
V_2O_4	s	1432	1421	1416	1410	1404	1397	1390	1387	1384	—
V_2O_5	s, l	1560	1557	1551	1545	1473	1464	1457	—	—	—
Cr_2O_3	s	1142,0	1142,0	1136,0	1134,0	1132,0	1131,0	1130,1	1130,5	1131,5	—
CrO_2	s	596,6	596,6	596,5	—	—	—	—	—	—	—
CrO_3	s, l	(586,2)	(582,0)	(556,8)	—	—	—	—	—	—	—
MnO	s	385,4	385,2	384,9	384,2	384,4	386,2	389,0	406,3	407,8	—
Mn_3O_4	s	1387	1387	1385	1382	1381,9	1387	1390	1418	1419	—
Mn_2O_3	s	959,7	958,4	957,0	955,0	956,9	954,2	(961,0)	—	—	—
MnO_2	s	521,0	520,9	520,0	518,6	(517,3)	—	—	—	—	—
FeO	s, l	267,1	266,7	264,8	264,4	265,4	267,5	266,9	266,1	234,0	—
Fe_3O_4	s	1121	1119	1111	1098	1092	1097	1093	1091	1092	—
Fe_2O_3	s	824,0	822,3	817,3	811,4	806,4	808,9	806,0	803,5	—	—
CoO	s	240,0	239,0	236,8	235,5	235,0	235,3	236,0	236,0	249,8	—

2 — Continued

1	2	3	4	5	6	7	8	9	10	11	12
Co_3O_4	s	866,7	867,1	865,8	864,2	859,8	—	—	—	—	—
NiO	s	239,9	239,3	237,2	236,1	235,1	234,1	233,0	232,2	248,7	—
Cu_2O	s, *l*	170,8	170,2	168,7	167,1	165,4	162,4	(184,6)	(123,9)	—	—
CuO	s	157,0	156,4	155,0	152,8	150,7	147,8	(157,8)	(154,5)	—	—
ZnO	s	348,5	348,3	347,2	354,0	353,2	(467,5)	(464,0)	(461,0)	—	—
$As_2O_3^{*7}$	s, *l*, g	657,3	(656,3)	(614,8)	(578,0)	(644,8)	(645,0)	(645,6)	(646,4)	(647,5)	(648,7)
$As_2O_3^{*8}$	s, *l*, g	640,2	(638,9)	(614,6)	(578,2)	(644,8)	(645,2)	(645,6)	(646,4)	(647,3)	(649,0)
As_2O_4	s, *l*,	734,8	(734,8)	(730,6)	(722,2)	(772,5)	(749,4)	(722,2)	—	—	—
As_2O_5	s, *l*, g	914,8	(914,8)	(916,9)	(914,8)	(973,4)	(965,1)	(952.5)	(940,0)	(923,2)	(904,3)
SrO	s	590,4	(590,0)	(588,7)	(587,8)	(587,4)	(592,0)	—	—	—	—
SrO_2	s	631,4	(629,3)	(627,2)	(623,0)	(618,8)	—	—	—	—	—
Y_2O_3	s	1907	(1907)	(1907)	(1905)	(1903)	(1901)	(1899)	(1895)	(1884)	(1922)
ZrO_2	s	1095	1094	1093	1092	1090	1092	1089	(1081)	(1080)	—
NbO_2	s	(797,0)	(796,0)	(794,0)	(792,6)	(790,0)	(789,0)	(788,0)	(785,9)	(784,0)	(782,0)
Nb_2O_5	s, *l*	1905,0	1903,2	1898,0	1892,0	1887,0	1879,0	1872,4	1866,0	1755,0	(1738,0)
MoO_2	s	(548,5)	(546,4)	(546,4)	(544,3)	(542,2)	(542,2)	(540,1)	(538,0)	(538,0)	(535,9)
MoO_3	s, *l*, g	755,0	753,8	750,3	745,9	740,4	(682,2)	(675,5)	—	—	(513,3)
Ag_2O	s, *l*	30,1	(29,3)	(27,8)	(26,4)	(24,9)	—	—	—	—	—
Ag_2O_2	s, g	26,0	(25,5)	—	—	—	—	—	—	—	—
CdO	s	256,2	(255,8)	(261,7)	(261,7)	(261,3)	—	—	—	—	—
SnO	s	286,2	285,7	292,4	291,8	289,9	288,3	—	—	—	—
SnO_2	s	581,2	580,7	586,6	584,1	581,2	577,4	572,8	—	—	—
$Sb_2O_3^{*7}$	s, *l*, g	705,5	(704,6)	(701,7)	(696,7)	(672,0)	(661,5)	(651,0)	(641,0)	(990,6)	(993,1)
$Sb_2O_3^{*9}$	s, *l*, g	711,3	(710,5)	(707,6)	(702,5)	(672,0)	(661,5)	(651,0)	(641,0)	(990,6)	(993.1)
Sb_2O_4	s	(875,0)	(875,0)	(870,9)	(866,7)	(902,3)	(893,9)	—	—	—	—

Sb_2O_5	s	(958,8)	(958,8)	(952,5)	—	—	—	—	—	—	—
BaO	s	558,5	557,7	555,6	553,5	563,6	(559,0)	(559,0)	(556,9)	(554,8)	(699,2)
BaO_2	s, *l*	636,4	(636,4)	(632,2)	(630,1)	(636,4)	(634,3)	(630,1)	—	—	—
La_2O_3	s	1794,0	1794,0	1791,0	1788,0	(1784,0)	(1801,0)	—	—	—	—
Ce_2O_3	s	(1821,0)	(1821,0)	(1819,0)	(1817,0)	(1815,0)	(1834,0)	(1832,0)	(1830,0)	(1830,0)	—
CeO_2	s	1089,0	1088,0	1087,0	1087,0	(1087,0)	—	—	—	—	(1093,0)
Pr_2O_3	s	1830	(1830)	(1828)	(1826)	(1823)	(1821)	(1844)	(1842)	(1840)	(1838)
PrO_2	s	965,1	(965,1)	(963,0)	(965,1)	(965,1)	(967,2)	—	—	—	—
Nd_2O_3	s	1809,4	1808,8	1806,7	1804,6	(1804,6)	(1829,3)	—	—	—	—
Sm_2O_3	s	1817,0	(1817,0)	(1815,0)	(1813,0)	(1781,0)	(1779,0)	(1778,9)	(1777,0)	(1836,0)	(1834,0)
HfO_2	s	1114	1113	1111	1108	1105	1102	1098	1094	1090	(1082,5)
Ta_2O_5	s	2047	2045	2039	2032	2025	2016	2008	1999	—	—
WO_2	s	(573,6)	(573,2)	(572,8)	(572,3)	(571,9)	(571,5)	(570,7)	—	—	—
WO_3	s, *l*	(840,9)	(828,4)	(815,2)	(797,2)	(777,3)	(757,2)	(736,0)	(713,6)	(639,5)	(603,5)
ReO_3	s, *l*	(615,5)	(613,4)	(579,9)	(569,4)	(556,8)	—	—	—	—	—
Re_2O_7	s, *l*, g	1243,5	(1237,2)	(1157,7)	(1078,1)	(1080,2)	(1082,3)	(1082,3)	—	—	—
Re_2O_8	s, *l*, g	(1291,6)	(1283,3)	(1237,2)	—	—	—	—	—	—	—
Au_2O_3	s, g	(3349)	(1465)	—	—	—	—	—	—	—	—
Hg_2O	s, g	91,3	(90,0)	(87,9)	(201,0)	(196,8)	—	—	—	—	—
HgO	s, g	90,77	(90,2)	(88,3)	(143,8)	(139,0)	(133,6)	(127,1)	—	—	—
PbO*10	s, *l*	218,0	217,5	216,0	218,8	215,4	(199,7)	(195,7)	(191,5)	—	—
PbO*11	s	219,4	218,6	217,3	—	—	—	—	—	—	—
Pb_3O_4	s	734,8	(734,8)	(730,6)	(739,0)	(724,3)	—	—	—	—	—
PbO_2	s, *l*	276,7	(275,9)	(274,2)	(276,3)	(272,1)	—	—	—	—	—
BiO	s, *l*	208,7	(207,2)	(215,6)	(215,6)	(213,5)	(211,4)	(209,3)	(207,2)	—	—
Bi_2O_3	s, *l*	577,4	575,9	594,5	591,6	(587,4)	(550,6)	(540,1)	(529,6)	—	—
ThO_2	s	1228,0	1227,2	1226,6	1225,5	1224,7	1223,6	1223,2	(1223,0)	(1223,2)	—
UO_2	s	1085	1084	1082	1079	1081	1083	—	—	—	—
U_3O_8	s	3574	(3570)	(3562)	(3556)	(3560)	(3567)	(3559)	—	—	—
UO_3	s	1221	1210	1217	1215	—	—	—	—	—	—

*1 Alumina. *2 Quartz. *3 Cristobalite. *4 Tridymite. *5 Quartz glass. *6 Rutile. *7 Orthorhombic. *8 Monoclinic. *9 Cubic. *10 Yellow. *11 Red.

3. TEMPERATURE DEPENDENCE OF ENTHALPY (HEAT CONTENT) ($H_T - H_{298.15}$, kJ/mole) [27, 53, 56, 95, 141, 163, 164, 189, 194, 632, 633]

Oxide	State	Temperature, °K							
		400	600	800	1000	1200	1400	1600	1800
1	2	3	4	5	6	7	8	9	10
H_2O*1	g	47 500	54 540	62 050	70 070	78 570	87 570	96 460	106 050
H_2O_2*2	g	8 918	65 680	77 180	89 600	102 580	116 670	—	—
Li_2O	s	30 580	44 460	59 950	76 690	94 550	113 240	—	—
*3	s	$H_T - H_{298,15} = 117,216\,T + 13,339\cdot10^{-3}T^2 + 76,598\cdot10^6/T - 15352$ for 980—1700 K							
*3	l	$H_T - H_{298,15} = 116,404\,T - 33755$ for 1700—2015 K							
BeO*4	s	$H_T - H_{298,15} = 39,627\,T + 4,372\cdot10^{-3}T^2 - 14811$ for 1200—2820 K							
	s	3051	10 790	19 730	29 350	39 450	—	—	—
	s	$H_T - H_{298,15} = 57,062\,T + 39,955\cdot10^{-5}T^2 - 34709$ for 2000—2828 K							
	l	$H_T - H_{298,15} = 89,953\,T - 39096$ for 2828—3200 K							
B_2O_3	s, l	6866	24 535	70 380	95 920	121 450	146 900	172 550	185 320*5
*6	l	$H_T - H_{298,15} = 127,779\,T - 49873$ for 1000—2200 K							
CO*7	g	2970	8 950	15 150	21 710	28 470	35 040	41 950	49 000
CO_2*8	g	4010	12 900	22 850	33 460	44 590	56 020	66 360	78 250
N_2O*9	g	4140	13 290	23 380	34 100	45 280	56 750	68 430	80 260
NO*10, *11	g	3040	9 160	15 570	22 220	29 000	35 880	42 920	50 070
NO_2*12	g	4020	12 790	22 480	32 780	43 440	54 320	65 420	76 620

N_2O_4	g	8620	40 780	50 160	73 520	—	—	—	—
Na_2O*13	s, l	7 330	23 030	39 150	56 520	(104 870)	(125 190)	(145 280)	(165 800)
Na_2O_2	s, l	(10 890)	(29 730)	(76 620)	—	—	—	—	—
NaO_2	s	(9 210)	(25 540)	(42 290)	(59 030)	—	—	—	—
MgO*14	s	4 040	12 640	21 880	31 740	42 080	52 630	63 180	73 730
*4	s	$H_T - H_{298,15} = 44,267\,T + 3,473\cdot 10^{-3}\,T^2 + 0,628\cdot 10^6/T - 16502$ for 1000—2500 K							
Al_2O	g	(2340)	(12 060)	(20 980)	(32 970)	(43 880)	(54 930)	(66 070)	(77 330)
AlO	g	(1930)	(8 370)	(15 490)	(23 240)	(30 350)	(37 680)	(45 000)	(52 340)
Al_2O_3*15, *18	s	9211	30 230	53 760	78 340	103 600	129 200	154 900	180 800
*16	s	$H_T - H_{298,15} = 126,913\,T + 3,516\cdot 10^{-3}\,T^2 + 11,782\cdot 10^6/T - 63932$ for 1100—2322 K							
*16	l	$H_T - H_{298,15} = 144,963\,T + 37028$ for 2322—2500 K							
*17	s	$H_T - H_{298,15} = 41,730\,T + 2,6268\cdot 10^{-2}\,T^2 + 19615$ for 2000—2323 K							
*17	l	$H_T - H_{298,15} = 199,346\,T - 12,389\cdot 10^{-4}\,T^2 - 90502$ for 2323—3100 K							
SiO_2*19	s	5 024	16 920	30 650	45 720	59 660	73 906	88 350	103 1 00
*20	s	5 070	18 050	31 200	44 930	58 950	73 310	87 880	102 700
*21	s	5 320	17 500	30 600	44 340	58 370	72 730	87 300	102 100
*22	s	5 150	16 540	29 270	43 040	57 360	72 180	87 840	104 300
PO	g	(2 510)	(10 260)	(15 280)	(22 820)	(29 300)	(35 800)	—	—
P_4O_{10}	s, g	23 240	82 480	218 970	280 720	342 270	403 820	—	—
K_2O	s, l	(7 540)	(22 610)	(37 470)	(52 340)	(71 180)	(88 970)	—	—
K_2O_2	s, l	(9 840)	(30 560)	(76 200)	(103 620)	(131 260)	(157 000)	—	—
K_2O_3	s, l	(13 610)	(39 400)	(98 810)	(129 580)	—	—	—	—
KO_2	s, l	(8 160)	(24 280)	(59 030)	(78 500)	—	—	—	—
KO_3	s, l	(10 050)	—	—	—	—	—	—	—
CaO	s	4 606	14 240	24 370	34 630	45 220	56 230	—	—
CaO_2	s	(7 850)	—	—	—	—	—	—	—
Sc_2O_3*23	s	(9 630)	(29 520)	(50 240)	(71 600)	(93 370)	(115 560)	(139 000)	(195 100)

3 — Continued

1	2	3	4	5	6	7	8	9	10
TiO	s	4 522	14 280	24 704	36 010	48 110	64 600	78 300	92 490
Ti_2O_3	s	10 930	38 270	66 700	95 210	124 800	154 700	185 000	215 600
	s	$H_T - H_{298,15} = 153,643\,T + 27,283 \cdot 10^{-3}\,T^2 - 34102$ for 500—2040 K							
$Ti_3O_5^{*24}$	s	$H_T - H_{298,15} = 160,550\,T + 11,535 \cdot 10^{-3}\,T^2 + 7,890 \cdot 10^6/T - 74434$ for 500—2080 K							
	s	19 510	67 910	107 000	146 700	190 600	233 700	—	—
$Ti_4O_7^{*24}$	s	$H_T - H_{298,15} = 293,703\,T + 7,194 \cdot 10^{-3}\,T^2 + 10,670 \cdot 10^6/T - 123994$ for 500—1950 K							
TiO_2^{*26}	s	6 448	19 820	34 160	48 780	63 640	78 800	94 370	110 300
*24	s	$H_T - H_{298,15} = 68,329\,T + 4,500 \cdot 10^{-3}\,T^2 + 1,677 \cdot 10^6/T - 27190$ for 500—2000 K							
VO	s	4 860	15 240	26 290	38 100	50 500	63 500	76 900	—
V_2O_3	s	11 390	36 010	62 260	89 480	117 100	146 300	177 900	210 800
*25	s	$H_T - H_{298,15} = 109,205\,T + 10,539 \cdot 10^{-3}\,T^2 + 0,291 \cdot 10^6/T - 34470$ for 500—2200 K							
V_2O_4	s	22 060	50 240	80 520	112 300	145 200	178 400	211 900	246 400
V_2O_5	s, *l*	15 280	47 270	81 180	182 500	220 600	258 800	—	—
*25	s	$H_T - H_{298,15} = 122,993\,T + 30,924 \cdot 10^{-3}\,T^2 + 0,632 \cdot 10^6/T - 41540$ for 500—952 K							
*25	*l*	$H_T - H_{298,15} = 190,222\,T - 12647$ for 952—1100 K							
Cr_2O_3	s	11 470	35 090	59 600	84 740	110 700	136 800	163 000	189 200
CrO_2	s	(5 440)	(17 580)	—	—	—	—	—	—
CrO_3	s, *l*	(11 300)	(51 500)	—	—	—	—	—	—
MnO	s	4 730	14 530	24 700	35 300	46 480	57 900	69 500	81 000
Mn_3O_4	s	15 620	48 500	83 516	119 600	157 600	199 400	263 000	304 600
Mn_2O_3	s	21 860	46 000	72 000	99 400	128 300	—	—	—

MnO_2	s	6 070	19 640	34 300	(49 600)	—	—	—	—
FeO	s, *l*	5 070	15 490	26 290	37 350	48 860	60 800	73 150	117 800
Fe_3O_4	s	16 710	54 680	101 600	148 000	188 200	228 400	268 600	308 800
Fe_2O_3	s	11 510	37 720	67 540	100 600	129 200	157 600	186 500	—
CoO	s	5 400	16 160	27 100	38 200	49 490	61 300	73 700	86 900
Co_3O_4	s	13 690	44 630	78 800	118 280	—	—	—	—
NiO	s	4 878	16 500	27 220	37 980	48 990	60 500	72 440	84 830
Cu_2O	s, *l*	7 200	22 110	37 900	54 500	72 500	(92 200)	(169 200)	—
CuO	s	4 650	14 490	25 120	36 340	48 570	(61 300)	(78 720)	—
ZnO	s	4 480	14 020	24 030	34 400	45 200	56 300	67 700	—
Ga_2O_3 *28, *29	s	10 161	32 640	56 857	82 187	108 480	135 610	163 580	—
GeO_2 *31	s	5 765	18 635	32 644	47 478	63 011	75 111*30	—	—
*31		$H_T - H_{298,15} = 62,26\,T + 7,49 \cdot 10^{-3}\,T^2 + 12,98 \cdot 10^5/T - 23580$ for 298,15—1316,0 K							
*32	s	5 727	18 862	33 122	48 023	63 388	67 282*33	—	—
*32		$H_T - H_{298,15} = 70,13\,T + 3,77 \cdot 10^{-3}\,T^2 + 19,72 \cdot 10^5/T - 27859$ for 298,15—1251,8 K							
As_2O_3 *34, *35	s, *l*, g	(10 680)	(72 170)	(130 210)	(159 940)	(173 750)	(187 150)	(200 970)	(215 200)
As_2O_3 *36, *37	s, *l*, g	(10 890)	(55 270)	(112 620)	(142 770)	(156 170)	(169 980)	(183 800)	(197 620)
As_2O_4	s, *l*, g	(11 300)	(38 520)	(71 180)	(109 700)	(146 540)	(187 980)	—	—
As_2O_5 *38	s, *l*, g	(12 560)	(41 030)	(66 570)	(111 370)	(140 260)	(174 590)	(208 500)	(247 860)
SrO	s	4 786	14 930	25 560	36 490	47 650	—	—	—
SrO_2	s	(7 950)	(21 980)	(39 150)	(58 620)	—	—	—	—
Y_2O_3 *39, *40	s	(9 714)	(29 440)	(52 170)	(74 530)	(99 230)	(123 800)	(150 900)	(192 600)
ZrO_2	s	6 180	19 640	34 000	49 110	64 560	80 180	101 360	116 260
NbO_2 *41	s	(6 280)	(19 680)	(33 500)	(43 130)	(61 550)	(76 200)	(90 800)	(105 900)

3 – Continued

1	2	3	4	5	6	7	8	9	10
Nb_2O_5 *42	s, *l*	14 650	45 550	78 300	112 200	147 600	183 800	220 700	361 900
MoO_2 *43	s	(16 750)	(33 080)	(51 920)	(70 760)	(89 180)	(112 210)	(136 070)	(159 520)
MoO_3 *44	s, *l*, g	8 250	26 250	45 890	69 290	(141 390)	(164 580)	—	—
Ag_2O *45	s, *l*	(7 540)	(22 610)	(38 520)	(54 850)	—	—	—	—
CdO	s	(4 606)	(13 610)	(22 820)	(32 660)	—	(59 250) *46	—	—
SnO	s	4 730	14 490	24 830	35 920	47 190	—	—	—
SnO_2	s	6 320	20 010	35 130	51 120	67 870	85 160	—	—
Sb_2O_3 *34	s, *l*, g	(9 300)	(32 660)	(59 700)	(146 620)	(179 780)	(213 320)	(246 770)	—
Sb_2O_3 *47	s, *l*, g	(9 300)	(32 660)	(59 700)	(152 480)	(185 650)	(219 180)	(252 630)	—
Sb_2O_4	s	(11 300)	(38 520)	(66 990)	(97 130)	(131 470)	—	—	—
Sb_2O_5	s	(12 770)	(45 220)	—	—	—	—	—	—
BaO	s	5 020	15 500	26 380	37 500	—	—	(66 570) *46	(96 900)
BaO_2	s, *l*	(5 440)	(21 350)	(35 590)	(49 820)	—	—	(92 110) *46	—

La_2O_3	s	11 180	34 630	59 330	87 380	(113 300)	—	—	—
Ce_2O_3	s	(10 050)	(33 910)	(60 290)	(89 180)	—	—	—	—
CeO_2[*48]	s	6 780	20 390	34 420	48 900	—	—	86 800[*46]	—
Nd_2O_3	s	12 980	38 310	65 320	93 370	122 680	—	—	—
EuO	s	$H_T - H_{298,15} = 52,50\,T + 4,35 \cdot 10^{-3}\,T^2 - 16035$							
Dy_2O_3[*49]	s	12 272	37 736	64 142	91 105	118 654	146 622	175 008	—
HfO_2[*50]	s	6 490	20 520	35 800	51 710	68 040	84 580	101 500	118 900
Ta_2O_5	s	14 360	45 850	80 093	115 700	152 400	190 000	228 700	—
WO_2	s	(6 030)	(17 750)	(30 020)	(42 710)	(55 310)	(68 870)	(75 780)[*46]	—
WO_3[*51]	s, *l*	(12 560)	(25 750)	(43 750)	(63 640)	(83 740)	(104 880)	(127 280)	(201 390)
ReO_3	s, *l*	(9 210)	(57 360)	(83 320)	(111 790)	—	—	—	—
Re_2O_7	s, *l*, g	(22 190)	(133 980)	(247 860)	(280 930)	(315 270)	(351 270)	(371 790)[*46]	—
Re_2O_8	s, *l*, g	(25 540)	(107 180)	—	—	—	—	—	—

3 – Continued

1	2	3	4	5	6	7	8	9	10
Au_2O_3	s	(11 720)	—	—	—	—	—	—	—
Hg_2O	s, g	(8 370)	(24 700)	(40 820)	(56 730)	—	—	—	—
HgO	s, g	(4 810)	(15 320)	(26 170)	(38 520)	(51 500)	(65 650)	(73 230)*46	—
PbO*52	s, l	4 810	15 070	26 420	38 770	(63 640)	(77 250)	(90 640)	—
PbO*53	s	5 110	15 660	—	—	—	—	—	—
Pb_3O_4	s	(14 190)	(48 020)	(85 290)	(130 540)	—	—	—	—
PbO_2	s, l	(7 690)	(20 200)	(35 550)	(52 170)	—	—	—	—
BiO	s, l	(5 866)	(17 170)	(26 800)	(38 520)	(50 240)	(62 800)	(72 850)	—
Bi_2O_3	s, l	11 600	35 800	61 210	(87 920)	—	—	—	—
ThO_2	s	6 699	20 470	35 130	50 450	66 410	82 740	99 400	116 270
UO_2	s	7 030	22 400	38 700	55 600	72 940	90 520	—	—

U_3O_8	s	(24 540)	(76 750)	(132 900)	(190 100)	(248 000)	(307 500)	—	—	—
	s	$H_T - H_{273,15} = 331,79754\,T - 338,33828 \cdot 10^{-4}\,T^2 + 65,31316 \cdot 10^5/T - 111,98007 \cdot 10^3$ for 273—480 K								
	s	$H_T - H_{273,15} = 238,37956\,T + 217,56463 \cdot 10^{-4}\,T^2 - 397,93679 \cdot 10^4/T - 57,78772 \cdot 10^3$ for 480—1000 K								
	s	$H_T - H_{273,15} = 325,63055\,T - 250,92581 \cdot 10^{-4}\,T^2 + 64,51214 \cdot 10^5/T - 110,66780 \cdot 10^3$ for 273—1000 K								
UO_3	s	8750	27 300	46 700	—	—	—	—	—	—

*1 At 2000°K, 116,860; at 2500°K, 143,180. *2 At 2000°K, (163,890); at 2500°K, (209,340). *3 In Pt–Rh capsule. *4 In Mo capsule. *5 At 1700°K. *6 In capsule of Pt, Mo. *7 At 2000°K, 56,810. *8 At 2000°K, 91,770. *9 At 2000°K, 92,240. *10 At 2000°K, 57,960. *11 For N_2O_3 at 400°K, (7450); at 500°K, (15,070). *12 At 2000°K, 87,880. *13 At 2000°K, (186,940). *14 At 2000°K, (84,280). *15 Alumina. *16 Al_2O_3 99.98% in Mo capsule. *17 Single crystal of sapphire, 99.98% purity. *18 [19]: $H_T - H_0^0$ for synthetic sapphire at 17°K, 0.218; at 25°K, 1.063; at 50°K, 17.34; at 75°K, 98.82; at 100°K, 325.4; at 150°K, 1431; at 200°K, 3524; at 250°K, 6489; at 298°K, 1003; $H_T - H_{273.15}$ for melt of synthetic corundum (specimen 99.8–99.9% Al_2O_3 in Mo ampuls, in pure argon atmosphere at pressure of 10^5 Pa) at 2350°K, 15,180; at 2400°K, 15,590; at 2500°K, 16,410; at 2600°K, 17,240; at 2700°K, 18,050, at 2800°K, 18,870. *19 Quartz. *20 Cristobalite; at 2000°K, 117,700. *21 Tridymite. *22 Quartz glass; at 2000°K, 121,500. *23 At 2000°K, (220,640). *24 Measurements in argon at pressure of 1333–2666 Pa; error ± 1%. *25 In argon at 1333–2666 Pa. *26 Rutile. *27 In argon, ampuls Pt–Rh, W. *28 [19]: $H_T^0 - H_0^0$ for single crystal of β-Ga_2O_3 with purity of 98.67%; at 16°K, 1.386; at 25°K, 8.269; at 50°K, 129.1; at 75°K, 513.8; at 100°K, 1189.4; at 150°K, 3357; at 200°K, 6452; at 250°K, 10,250; at 298°K, 14,520. *29 β-Ga_2O_3, smoothed values. *30 At 1350°K. *31 Hexagonal. *32 Tetragonal. *33 At 1250°K. *34 Orthorhombic. *35 At 2000°K, (228,180). *36 Monoclinic. *37 At 2000°K, (211,850). *38 At 2000°K, 290,560. *39 At 2000°K, (218,000). *40 [19]: $H_T^0 - H_0^0$ at 16°K, 5.62; at 20°K, 11.71; at 40°K, 86.1; at 60°K, 321.4; at 80°K, 760.4; at 100°K, 1418; at 120°K, 2291; at 140°K, 3365; at 160°K, 4622; at 180°K, 6043; at 200°K, 7604; at 220°K, 9281; at 240°K, 11,056; at 260°K, 12,920; at 280°K, 12,870; at 298.16°K, 16,700. *41 At 2000°K, (122,300). *42 At 2000°K, (411,300). *43 At 2000°K, (185,890). *44 At 2000°K, (380,160). *45 Ag_2O_2 (s, g); at 400°K, (8580). *46 At 1500°K. *47 Cubic. *48 At 2000°K, 127,300. *49 Smoothed values. *50 At 2000°K, (136,500). *51 At 2000°K, (237,390). *52 Yellow. *53 Red.

4. STANDARD FREE ENERGY OF FORMATION ($-\Delta F^0_{298.15} \cdot 10^{-3}$, kJ/kmole). Temperature Dependence of Free Energy of Formation of Oxides ($-\Delta F_T \cdot 10^{-3}$ kJ/kmole [1, 19, 36, 75]

Oxide	State	Temperature, °K									
		298,15	400	600	800	1000	1200	1400	1600	1800	2000
1	2	3	4	5	6	7	8	9	10	11	12
H_2O	*l*, g	237,48	223,37	213,11	202,22	191,76	179,6	167,5	156,38	144,0	132,5
	g	228,75	—	—	—	—	—	—	—	—	—
H_2O_2	*l*, g	117,65	94,6	73,7	51,5	28,9	6,7	—15,1	—	—	(—82,1)
D_2O	*l*	243,70	—	—	—	—	—	—	—	—	—
	g	234,74	—	—	—	—	—	—	—	—	—
HDO	*l*	242,52	—	—	—	—	—	—	—	—	—
	g	233,74	—	—	—	—	—	—	—	—	—
Li_2O	s	560,82	523,77	496,97	469,13	441,71	414,7	388,95	—	—	—
Li_2O_2	s	571,5±±16,8	—	—	—	—	—	—	—	—	—
BeO^{*1}	s	569,9	559,8	539,3	520,6	501,4	482,6	(465,4)	—	—	(406,1)
BO	g	48,57	—	—	—	—	—	—	—	—	—
$B_2O_3^{*2}$	s, *l*	1199,1	1171,9	1117,9	1067,2	1023,7	979,7	(940,4)	(897,2)	—	—
CO	g	137,33	146,54	164,75	182,96	200,76	218,34	235,5	253,7	270,68	287,84
CO_2	g	394,61	394,82	395,44	395,65	396,07	396,28	396,70	396,9	397,33	397,75
N_2O	g	—103,41	—109,28	—125,81	—140,47	—155,12	—169,36	—183,8	—197,62	—211,22	—228,2
NO	g	—86,67	—85,41	—82,9	—80,39	—78,08	—75,57	—73,27	—70,76	—68,04	—61,71
N_2O_3	g	(—140,26)	—	—	—	—	—	—	—	—	—
NO_2	g	—51,92	—58,20	—70,13	—82,69	—95,67	—108,44	—120,16	—132,9	—144,86	—157,84
N_2O_4	g	—98,39	—128,74	—184,22	—247,23	—305,64	—	—	—	—	—
O_2	g	0,000	—	—	—	—	—	—	—	—	—
O_3	g	—163,536	—	—	—	—	—	—	—	—	—
F_2O	g	—40,61	—	—	—	—	—	—	—	—	—
Na_2O	s, *l*	376,6	362,58	333,27	303,12	275,07	(243,67)	(188,82)	(134,81)	(82,06)	(30,14)

Na_2O_2	s, *l*	447,99	(426,22)	(381,42)	(339,97)	—	—	—	—	—	—
NaO_2	s	218,13	(203,06)	(175,85)	(148,63)	(123,51)	—	—	—	—	—
MgO	s	569,9	559,0	537,6	516,3	494,3	471,5	447,4	404,9	362,5	(322,2)
Al_2O	g	177,94	(188,41)	(211,43)	(230,27)	(247,02)	(259,58)	(272,14)	(284,7)	(295,17)	—
AlO	g	—16,75	(—6,28)	(+8,37)	(25,12)	(39,77)	(52,34)	(62,8)	(75,36)	(87,92)	—
Al_2O_3*3	s	1583,0	1550,0	1488,0	1425,0	1362,0	1295,0	(1230,0)	(1164,0)	(1102,0)	—
SiO	g	137,201	—	—	—	—	—	—	—	—	—
SiO_2*4	s	824,4	805,6	769,4	734,2	698,0	663,2	(628,4)	(593,7)	(557,3)	—
SiO_2*	s	823,2	804,5	768,5	733,1	697,8	663,4	(628,9)	(594,34)	(559,2)	(518,3)
SiO_2*6	s	822,7	804,1	768,3	732,9	697,8	663,0	(628,0)	(593,5)	(557,3)	—
SiO_2*7	s	795,1	777,1	741,5	706,3	671,6	637,3	(602,9)	(568,6)	(533,0)	(494,1)
PO	g	(62,8)	(71,18)	(83,74)	(92,11)	(102,58)	(110,95)	(119,32)	—	—	—
P_4O_{10}	s, g	(2739,8)	(2645,2)	(2502,9)	(2276,4)	(2103,0)	(1931,4)	(1762,2)	—	—	—
SO	g	—53,507	—	—	—	—	—	—	—	—	—
SO_2	g	300,570	—	—	—	—	—	—	—	—	—
SO_3	g	370,615	—	—	—	—	—	—	—	—	—
Cl_2O	g	—93,78	—	—	—	—	—	—	—	—	—
ClO_2	g	—123,51	—	—	—	—	—	—	—	—	—
K_2O	s, *l*	(319,5)	(304,0)	(272,6)	(241,16)	(209,34)	(155,33)	(94,62)	—	—	—
K_2O_2	s, *l*	(427,05)	(404,03)	(357,97)	(314,0)	(276,33)	(215,62)	(152,82)	—	—	—
K_2O_3	s, *l*	429,15	(397,75)	(334,94)	(276,33)	(226,09)	—	—	—	—	—
KO_2	s, *l*	237,81	(221,9)	(192,17)	(167,47)	(145,7)	—	—	—	—	—
KO_3	s, *l*	(190,5)	(165,38)	—	—	—	—	—	—	—	—
CaO	s	604,4	593,9	573,4	552,9	532,6	(512,5)	(490,5)	—	—	—
CaO_2	s	(600,8)	(584,06)	—	—	—	—	—	—	—	—
Sc_2O_3	s	(1628,67)	(1595,2)	(1532,4)	(1469,6)	(1406,8)	(1339,8)	(1279,1)	(1214,2)	(1151,4)	(1084,4)
TiO	s	489,5	475,4	460,6	441,7	423,3	404,9	387,3	369,9	354,8	—
Ti_2O_3	s	1440,0	1403,0	1348,0	1295,0	1247,0	1191,0	1140,0	1091,0	1044,0	—
Ti_3O_5	s	2316,0	2269,0	2180,0	2096,0	2014,0	1930,0	1845,0	—	—	—
Ti_4O_7*8	s	3370,4	—	—	—	—	—	—	—	—	—
TiO_2*9	s	889,3	870,5	834,5	798,7	763,5	728,1	692,7	658,4	625,3	—
VO	s	382,7	373,1	355,5	338,3	321,6	305,6	290,2	274,7	—	—

4 — Continued

1	2	3	4	5	6	7	8	9	10	11	12
V_2O_3	s	1160,0	1123,0	1081,0	1031,0	981,8	936,6	888,9	844,9	800,7	—
V_2O_4	s	1323,0	1287,0	1220,0	1157,0	1094,0	1028,0	971,8	914,4	853,5	—
V_2O_5	s	1429,0	1384,0	1300,0	1215,0	1136,0	1069,0	1004,0	—	—	—
Cr_2O_3	s	1060,0	1035,0	980,0	928,3	877,0	826,0	775,0	723,5	674,0	—
CrO_2	s	(544,3)	(525,4)	(491,9)	—	—	—	—	—	—	—
CrO_3	s	(506,6)	(479,4)	(433,3)	—	—	—	—	—	—	—
MnO	s	363,0	355,7	341,4	326,6	312,0	297,3	282,4	266,0	249,0	—
Mn_3O_4	s	1290,0	1245,0	1176,0	1105,0	1036,0	965,0	894,0	822,0	750,0	—
Mn_2O_3	s	882,0	829,0	778,2	728,1	676,0	626,0	(574,9)	—	—	—
MnO_2	s	466,0	447,6	411,0	375,4	(339,8)	—	—	—	—	—
FeO	s, *l*	245,6	238,6	224,2	211,4	198,2	184,0	170,6	156,6	145,5	—
Fe_3O_4	s	1018,0	983,5	917,4	855,0	795,1	734,8	674,5	614,6	556,9	—
Fe_2O_3	s	742,8	715,1	662,8	612,1	563,2	514,2	465,2	416,2	—	—
CoO	s	216,5	208,7	194,3	180,2	166,4	152,6	138,8	124,6	111,0	—
Co_3O_4	s	763,25	727,7	658,2	589,3	521,9	—	—	(345,4[*10])	—	—
NiO[*11]	s	211,9	202,2	184,2	166,6	149,5	132,3	115,4	98,39	81,86	—
Cu_2O	s, *l*	148,4	140,9	126,6	112,8	99,4	86,5	(73,3)	(57,4)	—	—
CuO	s	129,2	119,75	101,7	84,4	67,2	50,9	(35,6)	(18,0)	—	—
ZnO	s	318,5	312,5	288,7	268,0	247,0	(223,6)	(200,5)	(176,7)	—	—
GeO	g	118,07	—	—	—	—	—	—	—	—	—
As_2O_3[*12]	s, *l*, g	576,5	(549,1)	(500,3)	(466,8)	(430,0)	(387,1)	(344,2)	(301,0)	(257,7)	(214,4)
As_2O_3[*13]	s, *l*, g	569,4	(545,5)	(500,3)	(466,8)	(430,0)	(387,3)	(344,2)	(301,0)	(257,9)	(214,4)
As_2O_4	s, *l*, g	623,8	(586,2)	(510,8)	(439,6)	(360,1)	(280,5)	(205,2)	—	—	—
As_2O_5	s, *l*, g	772,5	(724,3)	(628,0)	(531,7)	(429,1)	(320,3)	(213,5)	(108,9)	(6,3)	(—94,2)
Rb_2O_2	s	362,16±±48,15	—	—	—	—	—	—	—	—	—
RbO_2	s	262,10	—	—	—	—	—	—	—	—	—
SrO	s	560,4	(550,4)	(530,7)	(511,9)	(491,6)	(471,5)	—	—	—	—
SrO_2	s	579,0	(562,3)	(528,8)	(497,4)	(466,0)	—	—	—	—	—

Y_2O_3	s	1815,0	(1782,0)	(1723,0)	(1660,0)	(1600,0)	(1541,0)	(1480,0)	(1419,0)	(1361,0)	(1296,0)
ZrO_2	s	1037,0	1018,0	979,3	941,6	904,4	849,1	830,5	(794,3)	(757,6)	
NbO	s	368,44	—	—	—	318,2	—	—	276,33[*10]	—	238,65
NbO_2	s	(742,0)	(723,5)	(687,5)	(652,0)	(617,0)	(582,0)	(548,0)	(514,2)	(480,0)	(446,8)
Nb_2O_5	s, *l*	1771,0	1726,0	1638,0	1553,0	1469,0	1385,0	1302,6	1221,0	1144,0	(1072,7)
MoO_2	s	502,42	—	—	—	389,37	—	—	318,2[*10]	—	251,2
	s	(495,3)	(481,5)	(452,2)	(422,9)	(393,6)	(366,3)	(337,0)	(307,7)	(280,5)	(251,2)
MoO_3	s	678,39	651,05	602,5	553,7	506,2	(458,5)	(422,9)	—	—	(376,8)
RuO_2	s	170,82	—	—	—	50,24	—	—	—22,61[*10]	—	—83,74
Ag_2O	s, *l*	10,5	(3,8)	(—8,58)	(—20,5)	(—32,2)	—	—	—	—	—
Ag_2O_2	s, g	(—27,6)	(—45,6)	—	—	—	—	—	—	—	—
CdO	s	226,5	(216,5)	(185,9)	(164,1)	(136,9)	—	—	(36,43)[*10]	—	—
SnO	s	257,1	247,0	226,9	205,6	183,8	162,4	—	—	—	—
SnO_2	s	520,4	499,5	457,6	415,4	373,5	332,0	291,0	—	—	—
Sb_2O_3[*12]	s, *l*, g	624,3	(596,6)	(543,0)	(491,1)	(440,5)	(395,2)	(351,7)	(309,8)	(251,2)	(168,7)
Sb_2O_3[*14]	s, *l*, g	628,0	(599,5)	(544,7)	(491,5)	(440,5)	(395,2)	(351,7)	(309,8)	(251,2)	(168,7)
Sb_2O_4	s	(764,1)	(726,4)	(653,1)	(579,9)	(504,5)	(427,1)	—	—	—	—
Sb_2O_5	s	(818,5)	(768,3)	(676,2)	—	—	—	—	—	—	—
TeO_2	s, *l*, g	269,6	—	—	—	132,3	—	—	39,36[*10]	—	—30,56
Cs_2O	s, *l*, g	276,33	—	—	—	146,96	—	—	15,07[*10]	—	—65,7
Cs_2O_2	s	332,9± ±71,2	—	—	—	—	—	—	—	—	—
CsO_2	s	267,1	—	—	—	—	—	—	—	—	—
BaO	s	528,8	519,2	499,9	481,5	465,2	(444,0)	(425,0)	(404,0)	(387,0)	(360,0)
BaO_2	s, *l*	584,1	(567,3)	(531,7)	(500,3)	(466,8)	(433,3)	(399,8)	—	—	—
La_2O_3	s	1703,2	—	—	—	1548,7	—	—	1406,8[*10]	—	—
	s	1686,0	1646,0	1577,0	1507,0	(1434,0)	(1369,0)	—	—	—	—
Ce_2O_3	s	(1723,0)	(1689,0)	(1622,0)	(1558,0)	(1493,0)	(1426,0)	(1359,0)	(1290,0)	(1223,0)	—
CeO_2	s	1030,0	1010,0	970,8	931,4	(892,7)	—	—	(793,4)[*10]	—	(674,1)
Pr_2O_3	s	1766,0	—	—	—	1507,25	—	—	1340,2[*10]	—	1184,45
	s	(1761,1)	(1733,0)	(1679,0)	(1627,0)	(1574,0)	(1522,0)	(1465,0)	(1411,0)	(1356,0)	(1300,0)
PrO_2	s	910,6	(890,8)	(854,1)	(818,5)	(782,9)	(745,3)	—	—	—	—

4 — Continued

1	2	3	4	5	6	7	8	9	10	11	12
Nd_2O_3	s	(1727,0)	(1695,7)	(1640,0)	(1585,2)	(1529,1)	(1474,9)	—	—	—	—
Sm_2O_3	s	(1719,0)	(1692,0)	(1637,0)	(1585,0)	(1532,0)	(1480,0)	(1428,0)	(1377,5)	(1321,0)	(1266,0)
HfO_2	s	1054,0	1034,0	994,6	957,4	919,9	884,3	847,9	813,7	779,0	(743,6)
Ta_2O_5	s	1911,0	1865,9	1777,0	1692,0	1607,0	1523,0	1443,0	1364,0	—	—
WO_2	s	(521,7)	(504,1)	(469,3)	(435,0)	(401,1)	(366,8)	(332,9)	—	—	—
WO_3	s, *l*	763,9	(737,9)	(687,5)	(638,3)	(589,9)	(542,4)	(495,5)	(449,7)	(406,3)	(368,6)
ReO_3	s, *l*	(535,9)	(508,7)	(468,9)	(433,3)	(399,8)	—	—	—	—	—
Re_2O_7	s, *l*, g	(1055,1)	(992,3)	(877,1)	(803,9)	(736,9)	(667,8)	(598,7)	—	—	—
Re_2O_8	s, *l*, g	(1076,0)	(1002,7)	(875,0)	—	—	—	—	—	—	—
Au_2O_3	s, g	(—77,7)	(—105,1)	—	—	—	—	—	—	—	—
Hg_2O	s, g	(54,4)	(41,9)	(16,7)	(—37,7)	(—96,3)	—	—	—	—	—
HgO	s, g	58,6	(47,5)	(26,6)	(—9,42)	(—47,1)	(—83,95)	(—119,7)	—	—	—
Tl_2O	s	136,07	—	—	—	—	—	—	—	—	—
PbO*15	s, *l*	188,2	179,0	167,1	137,7	118,3	(102,2)	(86,0)	(70,55)	—	—
PbO*16	s	188,95	178,8	159,5	—	—	—	—	—	—	—
Pb_3O_4	s	617,6	(577,8)	(500,3)	(420,8)	(341,2)	—	—	—	—	—
PbO_2	s, *l*	(219,0)	(187,1)	(161,6)	(123,1)	(85,0)	—	—	—	—	—
BiO	s, *l*	(182,1)	(171,7)	(152,8)	(131,9)	(110,95)	(92,1)	(71,2)	(52,3)	—	—
Bi_2O_3	s, *l*	497,0	469,55	414,9	355,25	(296,6)	(242,8)	(192,6)	(144,4)	—	—
ThO	s	(577,8)	—	(561,0)*17	—	(515,0)	—	—	(473,1)*10	—	(427,05)
ThO_2	s	1170,0	1150,0	1112,0	1074,0	1036,0	998,2	960,7	(923,4)	(886,4)	—
UO_2	s	1032,0	1014,0	980,2	946,7	913,2	879,3	—	(823,6)*10	—	—
U_3O_8	s	3366,0	(3296,0)	(3160,0)	(3028,0)	(2895,0)	(2761,0)	(2627,0)	—	—	—
UO_3	s	1144,0	1117,0	1067,0	1017,0	—	—	—	—	—	—
PuO_2	s	978,04	—	—	—	837,36	—	—	795,49*10	—	711,76

*1For BeO (g), standard free energy SFE = −23.86 [1, p. 89]. *2For B_2O_3 (gl), SFE = 1173.98 [1, p. 89]. *3Alumina. *4Quartz. *5Cristobalite. *6Tridymite. *7Quartz glass. *8SFE for Ti_5O_9, Ti_6O_{11}, Ti_7O_{13}, Ti_9O_{17}, and $Ti_{10}O_{19}$ is respectively 4294.8, 5212.6, 6142.5, 7989.3, and 8915.4 [174, p. 118]. *9Rutile. *10At 1500°K. *11For NiO (g), SFE = −216.88 [1, p. 91]. *12Orthorhombic. *13Monoclinic. *14Cubic. *15Yellow. *16Red. *17At 500°K.

Oxide	Method of determination*1	Free energy function, kJ/(kmole·°K), at indicated temperature, °K					
		298,15	1000	1500	2000	2500	3000
1	2	3	4	5	6	7	8
A. Function $\left(-\dfrac{F^{\circ}-H^{0}_{298}}{T}\right)$ for gaseous oxides [1]							
SiO_2	c	228,2	253,7	270,5	283,9	294,8	304,4
	c	—	—	278,4 (at 1800 K)	280,9 (at 1900 K)	—	—
	ex	—	—	273,4 (at 1800 K)	275,1 (at 1900 K)	—	—
TiO_2	c	237,0	263,8	280,9	294,8	306,1	315,7
	c	—	—	—	291,4 (at 1800 K)	—	—
	ex	—	—	—	284,3 (at 1880 K)	—	—
VO_2	c	251,2	279,7	297,7	312,3	324,1	333,7
CrO_2	c	256,7	287,2	306,1	320,7	332,9	342,9
MnO_2	es	260,4	291,4	310,7	326,2	338,3	349,2
ZrO_2	c	245,3	272,1	289,7	303,5	315,3	324,9
	c	—	—	—	—	312,8 (at 2400 K)	—
	ex	—	—	—	—	314,0 (at 2400 K)	—
NbO_2	c	256,7	284,3	302,7	317,8	329,5	340,0
MoO_2	c	261,3	290,1	309,4	324,9	337,5	347,9
	c	—	—	312,3 (at 1600 K)	—	—	—
	ex	—	—	322,0 (at 1600 K)	—	—	—
TcO_2	es	262,9	291,8	311,1	326,6	338,7	349,6
RuO_2	es	261,3	290,1	309,4	324,9	337,0	347,9
RhO_2	ex	262,9	291,8	311,1	326,6	338,7	349,6
PdO_2	es	261,7	290,6	309,8	334,3	337,5	348,3
HfO_2	c	253,3	280,5	298,1	311,9	323,6	333,3
TaO_2	c	265,4	293,1	311,1	334,3	337,5	347,5
WO_2	c	269,6	297,7	315,7	329,9	342,5	352,5
ReO_2	es	269,2	297,3	315,3	329,5	342,1	352,1

1	2	3	4	5	6	7	8
OsO_2	es	264,2	291,8	309,8	324,1	336,2	346,2
IrO_2	es	260,4	288,1	306,1	320,3	322,4	342,5
PtO_2	ex	258,7	286,4	304,4	318,6	330,8	340,8
ThO_2	ex	265,4	294,8	313,6	328,7	341,6	352,5
PaO_2	es	268,8	298,5	317,8	333,3	346,2	357,1
UO_2	c	273,0	303,5	323,2	339,1	352,5	363,8
	c	—	—	—	332,9 (at 1800 K)	—	—
	ex	—	—	—	319,5 (at 1800 K)	—	—
B. Function $\left(-\frac{F^\circ - H_0^0}{T}\right)$							
BeO*2,*4	—	4,48	8,0 (at 400 K)	15,6 (at 600 K)	22,9 (at 800 K)	29,6 (at 1000 K)	35,6 (at 1200 K)
	—	41,5*3 (at 1400 K)	44,2*3 (at 1500 K)	—	—	—	—
MgO*5	—	9,622	15,58 (at 400 K)	26,51 (at 600 K)	35,98 (at 800 K)	44,16 (at 1000 K)	51,34 (at 1200 K)
Al_2O_3*6	—	0,004 (at 17 K)	0,013 (at 25 K)	0,110 (at 50 K)	0,404 (at 75 K)	1,025 (at 100 K)	3,440 (at 150 K)
	—	7,267 (at 200 K)	12,10 (at 250 K)	17,28 (at 298 K)	—	—	—
Ga_2O_3*7	—	0,0293 (at 16 K)	0,1089 (at 25 K)	0,875 (at 50 K)	2,692 (at 75 K)	5,343 (at 100 K)	12,15 (at 150 K)
	—	19,97 (at 200 K)	28,13 (at 250 K)	36,00 (at 298 K)	—	—	—
Y_2O_3*8	—	0,118 (at 16 K)	1,65 (at 50 K)	7,17 (at 100 K)	24,47 (at 200 K)	43,2 (at 298 K)	78,51 (at 500 K)
	—	143,00 (at 1000 K)	221,4 (at 2000 K)	249,2 (at 2500 K)	—	—	—

*1 Method of determination: c = calculation, es = estimate, ex = experiment. *2 Solid. *3 Obtained by extrapolation. *4 [1[illegible], p. 44]. *5 [19, p. 122]. *6 [19, p. 21]. *7 [19, p. 69]. Single crystal of β-Ga_2O_3, 98.67% purity. *8 [19, p. 86].

6. MELTING AND BOILING POINTS [1; 9, Nos. 1–5; 19; 39; 42; 43; 52; 62; 67; 75; 81, Vols. 1–5; 103; 113; 125; 171]

Oxide	Melting point, °C	Boiling point, °C
1	2	3
H_2O	0,00	100,0
	0,01*1	25,00*2
	—	0,00*3
H_2O_2	−0,42±0,03	152,0±0,5
	—	25,0*4
D_2O	3,813±0,005*1	101,43
D_2O_2	1,5	25,0*5
T_2O	4,49±0,02	101,6
HDO	—	100,85
	—	25,0*6
HTO	—	100,8
Li_2O	1727	2600
Li_2O_2	197*7	—
BeO	2520—2570	3900—4260
B_2O_3	450—470*8	2124
	~294*9	1250*10
C_3O_2	−108 to −107	7
CO	−205,02	−191,50
CO_2	−56,6*11	−78,515*12
N_2O	−91,0±0,5	−88,5±0,5
NO	−163,65	−151,65
N_2O_3	−101	+3,5*7
	—	−40*13
NO_2 & N_2O_4	−11,2	21*14
N_2O_5	33±1*15	47*7
	41*16	—
O_2	−218,79	−182,97
O_3	−192,5±0,4	−111,9±0,3
F_2O	−223,85	−145,05±0,5
F_2O_2	−163,4	—
F_2O_3	−189	—
Na_2O	920	1350
Na_2O_2	>596*7	(675)*17
NaO_2	552*7	—
MgO	2800	3600
MgO_2	88*7	—
Al_2O_3	2046,5*18	2980*18
	2053±4	—
SiO	2277*19	—
SiO_2	1720*20	2950*20
	1610*21	—
	1680*22	—
P_2O_3, P_4O_6	23,9±0,1	175,4
PO_2	77*7	—
P_2O_4	100*15	180
P_2O_5, P_4O_{10}	420±3*23	—

6 — Continued

1	2	3
$(P_2O_5)_n$	562±2[*24]	588
	580±2[*25]	588
SO_2	—75,46±0,01	—10,01±0,03
SO_3	—	44,5
	16,79±0,02[*26]	62,2[*15]
	32,5[*27]	44,7±0,3
	62,2[*28]	—
S_2O_7	0	—
SO_4	3[*29]	—
Cl_2O	—116±1[*30]	2,2[*30]
ClO_2	—59,8[*30]	11,1[*30]
	—59	9,7±0,3
Cl_2O_6	3,50±0,05[*30]	203[*30]
Cl_2O_7	—90±2[*30]	78,8[*30]
K_2O	707[*7]	—
K_2O_2	490	1527
K_2O_3	430	700
KO_2	380	543[*7]
KO_3	60±2[*7]	—
CaO	~2587	(3500)
	~2910	—
CaO_2	375—425[*7]	—
Sc_2O_3	2405	—
Ti_3O	1940±30	—
TiO[*31]	1750—2020	—
Ti_2O_3[*31]	1900—2130	3027
	1830±10	—
Ti_3O_5[*31]	2177	3327
[*32]	1774±20	—
TiO_2	1870±15[*33]	2927[*7]
TiO_3	670	—
VO	1830±20	3127
V_2O_3	1970±20	3027[*7]
V_3O_5	1827[*7]	—
VO_2	1545	3027[*7]
V_2O_4[*31]	1542±20	—
V_6O_{13}	662±10	—
$V_{12}O_{26}$[*34]	708±5	—
V_2O_5	670—685	2052[*7]
Cr_2O_3	2334±25	~3000
	2257—2427[*15]	—
CrO_2	427[*7]	—
CrO_3	180—202[*35]	727[*7]
MnO	1785	3127[*36]
	1842±10	—
Mn_3O_4[*31]	1560±10	2627

1	2	3
Mn_2O_3	1347*[7]	—
MnO_2	847*[7]	—
Mn_2O_7	6±1*[37]	—
$Fe_{0,947}O$	1374±5*[38]	—
FeO	—	2512
$FeO_{1,05}$	1371±5*[38]	—
$FeO_{1,056}$	1374±5*[38]	—
$FeO_{1,10}$	1388±5*[38]	—
$FeO_{1,15}$	1406±5*[38]	—
$FeO_{1,20}$	1424±5*[38]	—
Fe_3O_4	1583±2*[33]	2623*[7]
	1594±2*[39]	—
	1597±2*[40]	—
Fe_2O_3	1350*[41]	—
	1562*[7]	—
CoO	1810±10*[42]	2627
	1830±5*[43]	—
	1817±5*[44]	—
	1769±5*[45]	—
Co_3O_4	967*[7]	—
Co_2O_3	895*[7]	—
NiO	1957±20	—*[7]
	1552*[46]	—
Cu_2O	1242±5	1800*[7]
CuO	1336	—*[7]
Cu_2O_3	100*[7]	—
ZnO	1975	—
	1800*[47]	—
Ga_2O	652	727
Ga_2O_3	1725±15	—
GeO	710*[15]	—
GeO_2	1116±4*[48]	2352*[49]
	1086±5*[50]	—
As_2O_3	278±3*[51]	457
	314±2*[52]	465
As_4O_6	313[15]	461±2
AsO_2	927*[7]	—
As_2O_5	> 827*[7]	—
SeO	1102	1802
SeO_2	340—390*[15]	—
Se_2O_5	224±1*[29]	—
SeO_3	121±2*[7]	—
Br_2O	—17*[7]	—
Rb_2O	627	—*[7]
Rb_2O_2	570	—*[7]

6 – Continued

1	2	3
Rb_2O_3	489	—*7
RbO_2	412	> 567*7
SrO	~ 2430	3000
SrO_2	410—450*7	—
Y_2O_3	2410—2435	4300
Zr_3O	1975±30	—
ZrO_2	2900	4300
	2700±25	—
Nb_4O	350*7	—
NbO	1940±15	—*7
	2377	—
Nb_2O_3	1772	—
NbO_2	2080±20	3527
Nb_2O_5	1490±20	2927
MoO_2	1927	1977*7
Mo_4O_{11}	824±10*53	—
Mo_8O_{23}	> 700*7	—
Mo_9O_{26}	800±10*54	—
MoO_3	801±2	1257
TcO_2	2127	3727
TcO_3	< 927*7	—
Tc_2O_7	120±1	311±2
RuO_2	1127*7	—
RuO_4	25—25,5*55	—*7
	27*56	—
Rh_2O	1127*7	—
RhO	1121*7	—
Rh_2O_3	1115*7	—
PdO	877*7	—
Ag_2O	187*7	—
	815±20	—
CdO	826*57	—
	1084*58	—
	> 1500	—
In_2O	327	527
InO	1052	1727
In_2O_3	~2000	3327
	1910±10	—
SnO	1042*7	1527
Sn_3O_4	1100*59	—
SnO_2	1625*60	~2000
	2000	2500
	2300*59	—
Sb_2O_3	655*61	1425*62
Sb_4O_6	656	1456
Sb_2O_4	930*63	—
Sb_6O_{13}	700*64	1423
Sb_2O_5	380*63	—
TeO_2	733±1	1257
TeO_3	> 400*7	—

6 — Continued

1	2	3
I_2O_4	130	—
I_4O_9	75	—
I_2O_5	200[*7]	—
Cs_7O	4	—
Cs_4O	10[*64]	—
Cs_3O	165	—
Cs_2O	490	—[*7]
Cs_2O_2	594	960[*7, *64]
Cs_2O_3	502	870[*7, *64]
CsO_2	450	>597[*7]
Ba_2O	607	767
BaO	1920[*62]	2200
	—	2727
BaO_2	450	790—973[*7]
La_2O_3	2217±30	4200
Ce_2O_3	1690	3227
	1920	—
CeO_2	2600	—
Pr_2O_3	2127±30	3727
Pr_6O_{11}	2042±30[*65]	—
PrO_2	427[*7]	—
Nd_2O_3	2212±30	3000
Pm_2O_3	2320	—
Sm_2O_3	~2320	3527
EuO	> 1700	—
Eu_2O_3	2050±30	—
Gd_2O_3	2322±30	—
Tb_2O_3	2292±30	—
Tb_4O_7	2337±30[*66]	—
Dy_2O_3	~2385	—
Ho_2O_3	~2395	—
Er_2O_3	~2400	3000
Tu_2O_3	2392±30[*67]	—
Yb_2O_3	~2420	—
	2260	—
Lu_2O_3	2467±30[*68]	—
HfO_2	2780±30	—
Ta_2O_5[*69]	1785±30[*70]	2227
[*71]	1872±10	—
WO_2[*72]	1227—1327[*15]	1727
WO_2	1570	1852[*7]
WO_3	1473—2130	1850
ReO_2	1202	2977
	1363±20[*62, *73]	—
ReO_3	160±3	620±10

6 – Continued

1	2	3
Re_2O_7	301,5±0,5	359±1
ReO_4	147	187
Re_2O_8	150±3	—
OsO	40	—
OsO_2	650[*7]	—
OsO_4	39,50[*55]	130±2[*55, *74]
	41,0±1[*74]	—
Ir_2O_3	1177	1977
IrO_2	1100[*7]	—
PtO	507[*7]	4100
PtO_2	450	477[*7]
Au_2O	~200[*7]	—
HgO	400[*55, *7]	—
	500[*75, *7]	—
Tl_2O	579±5	—
Tl_2O_3	~717[*76]	—[*7]
	834	—
TlO_2	(490)[*77]	—
PbO	886±2[*55]	1472
	—	1535±20[*73]
Pb_3O_4	500[*7]	—
	830	—
PbO_2	290[*7]	—
BiO	902	1647
Bi_2O_3	817	1890
	825±5	1890
Bi_2O_4	305[*7]	—
PoO_2	552	—[*7]
RaO	>2227	—
Ac_2O_3	1977	—
ThO	1877	2977
ThO_2	3050	4400[*78]
PaO_2	2287	
Pa_2O_5	1777	3077
UO	2477	—
UO_2	2840±20	—
U_2O_4	2405	—
U_3O_8	2500[*7]	—
UO_3	652[*7]	—
UO_4	150[*7]	—
NpO_2	2560±50	—
Np_2O_5	527—627[*7]	—
PuO	1017	2052
Pu_2O_3	2240	2977
	2085	—

1	2	3
Pu_4O_7	2340	—
PuO_2	2390±20	3227
Am_2O_3	2205±15	3127
Cm_2O_3	2275±50	—
Bk_2O_3	1920±25	—
Cf_2O_3	1750±25	—

*1 Triple point. *2 At 3166.4 Pa, calculated data for state of vapor corresponding to standard conditions. 3 At 610.615 Pa, same. 4 At 279.98 Pa, calculated data. 5 At 2762.4 Pa, calculated data. 6 At 2933.08 Pa, calculated data. 7 Decomposes. 8 In air. 9 Amorphous glassy mass changes over to a viscous liquid. *10 266.64 Pa. *11 Triple point, pressure 535 kPa. *12 Sublimation temperature. *13 Equilibrium mixture of NO and N_2O_4. *14 Mixture of NO_2 and N_2O_4. *15 Sublimes. *16 At 142.9 kPa. *17 Decomposes completely. *18 Corundum. *19 Disproportionates at 400 –700°C. *20 Cristobalite. *21 Quartz. *22 Tridymite. *23 Hexagonal (H-form); sublimes at 360°C. *24 Orthorhombic (O-form); sublimes at 605°C. *25 Orthorhombic (O'-form); sublimes at 605°C. *26 γ, orthorhombic, S_3O_9. *27 β, monoclinic, fibrous. *28 α, monoclinic, fibrous, sublimes. *29 Melts with decomposition. *30 Explodes. *31 β. *32 Liquid with crystals of Ti_3O_2. *33 At pressure pO_2 = 101.325 kPa. *34 Liquid with VO_2. *35 Depending on degree of decomposition, which begins while still in the solid state. *36 Sublimes with dissociation. *37 Solidification point. *38 Cubic → l + s. *39 At pO_2 = 21.28 kPa. *40 At pO_2 = 5.83 kPa. *41 Hematite. *42 Melting in air. *43 Oxygen-free atmosphere p_{tot} = 101.325 kPa, composition of liquid phase $CoO_{1.000}$. *44 p_{tot} = 101.325 kPa, pO_2 = 101.325 kPa, composition of liquid phase $CoO_{1.014}$. *45 pO_2 = 101.325 kPa, composition of liquid phase $CoO_{1.014}$. *46 In vacuo. *47 Sublimes at this temperature. *48 Stable melting point, high-temperature form. *49 Appreciable vaporization. *50 Metastable melting point, low-temperature form. *51 Arsenolite, cubic. *52 Claudetite 1, monoclinic. *53 Liquid + (MoO_2 + MoO_3). *54 Liquid + (MoO_3 + Mo_4O_{11}). *55 Yellow. *56 Brown. *57 At 700°C, sublimes without melting. *58 Sublimes without melting. *59 Calculated data. *60 Cassiterite; sublimes. *61 Valentinite. *62 Sublimes. *63 Loses oxygen. *64 Calculated temperature of dissociation. *65 99.5%, in argon. *66 99.70%, in argon. *67 99.56%, in argon. *68 99.88%, in argon. *69 α. *70 Orthorhombic. *71 Triclinic. *72 δ. *73 p_{vap} = 101.325 kPa. *74 White. *75 Red. *76 At pO_2 = 101.325 kPa. *77 Dissociates. *78 Thorianite.

7. HEATS OF FUSION AND VAPORIZATION [1; 9, Nos. 1–7; 19; 42; 43; 49; 52; 66; 77; 79, Vol. 1; 81, Vol. 5; 96]

Oxide	Heat of fusion, kJ/mole	Heat of vaporization vaporization, kJ/mole	Temperature, °C	Pressure, Pa
1	2	3	4	5
H_2O	6,0131± ±0,004*1, *2	—	—	—
	6,0135± ±0,0042*3	45,079±0,013*3	0,01	610,8
	—	44,020±0,013	25	3170
	—	44,041±0,013	25	101 325

7 – Continued

1	2	3	4	5
	—	40,683	t_b = 100 -	101 325
H_2O_2	12,506± ±0,042	52,858±0,134	t_m = −0,42	50,65
	—	51,66±0,13	25	285,7
H_2O_4	—	50,24	25	—
D_2O	6,314± ±0,021	46,494±0,084	t_m = 3,813	667,6
	—	45,43±0,04[*4]	25	101 325
D_2O_2	—	52,38±0,21	25	—
T_2O	—	46,05±0,4	25	2644
HDO	—	44,67±0,25	25	2928
HTO	—	46,89±1,26	25	2897
Li_2O	58,62	234,46	t_b = 2600	—
BeO	85,41±6,28	—	—	—
	—	608,76	t_m = 2550	—
	—	489,86±43,96	t_b = 3850	—
B_2O_3	24,58±0,08	403,44±8,37	t_m = 450	$3,34 \cdot 10^{-15}$
	—	356,3±12,6	t_b = 2124	101 325
	—	322,38	t_b = 1250	266,64
CO	0,8386± ±0,0029	6,531±0,004	t_m = − 205,02	15 367
	—	6,0441±0,0042	t_b = −191,50	101 300
CO_2	8,37±0,13	16,24±0,42	t_m = −56,58	517 846
	—	5,28±0,84	25	6 422 400
N_2O	6,544	16,62	t_m = −91,0	87 827
	—	16,571±0,021	t_b = −88,5	101 325
NO	2,303± ±0,008	14,24	t_m = −163,65	22 083
	—	13,783±0,013	t_b = −151,65	101 325
N_2O_3	—	39,356	t_b	—
N_2O_4	14,662± ±0,013	—	—	—
	—	37,14[*5]	t_m = −11,2	18 740
	—	38,56[*6]	t_b = 21	101 325
	—	28,68[*7]	25	—
N_2O_5	—	56,940	t_b	—
O_2	0,4459± ±0,0008	7,653±0,021	t_m = −218,79	156
	—	6,833±0,017	t_b = −182,97	101 325
O_3	2,09	15,193	t_b = −111,9	101 325
F_2O	—	11,10	t_b = −145,05	101 325
Na_2O	29,73	—	—	—
NaO_2	25,96	117,23	t_b	—
MgO	77,46±4,19	—	—	—
	—	544,28	t_b	—
Al_2O_3	113,04±8,37	—	—	—

1	2	3	4	5
	—	485,67	t_b	—
SiO	50,24	—	—	—
SiO_2	8,541*8	—	—	—
	9,002*9	—	—	—
	7,704*10	—	—	—
	—	573,59*10	25	—
P_2O_3	—	18,84	t_b	—
P_4O_6	14,07	46,64	$t_m = 23,9$	233
	—	46,64	25	—
	—	43,46	$t_b = 175,4$	101 325
PO_2	11,30	—	—	—
P_4O_{10}*2	20,93	37,26	$t_m = 420$	491 305
	—	73,69	t_b	—
	63,64*11	—	—	—
	—	78,29*12	$t_m = 580$	73 949
	74,11*13	—	—	—
		78,29*12	$t_m = 562$	58 754
	67,41*14	—		—
	—	84,99*15	$t_m = 571$	75 975
	—	79,13*15	$t_b = 584$	101 325
SO_2	7,406±0,008	28,22±0,04	$t_m = -75,46$	15 702
	—	24,953±0,017	$t_b = -10,01$	101 325
SO_3*16	30,14	38,1±2,1	$t_m = 62,2$	234 003
SO_3*17	12,14	42,37±1,7	$t_m = 31,5$	50 144
	—	40,82±1,7	$t_b = 44,7$	101 325
SO_3*18	5,61	44,21±1,7	$t_m = 16,79$	20 463
	—	43,17±1,3	25	34 341
Cl_2O	—	26,335	$t_b = 2,2$	—
	—	25,96	$t_b = 2$	101 325
ClO_2	—	29,64	$t_b = 11,1$	—
	—	26,33±0,4	$t_b = 9,7$	101 325
Cl_2O_6	11,72±6,28	39,78±4,19	12	91,17
Cl_2O_7	—	34,00±0,4	25	12 156
	—	35,67	$t_b = 78,8$	—
	—	32,36±0,63	$t_b = 80$	101 325
K_2O	28,47	154,91	t_b	—
K_2O_2	29,31	188,41	$t_b = 1527$	—
K_2O_3	25,54	104,67	$t_b = 700$	—
KO_2	20,52	—	—	—
CaO	75,36	—	—	—
	—	625,33	t_b	—
Sc_2O_3	96,30	—	—	—
TiO	54,43± ±4,2*19	—	—	—
Ti_2O_3	100,48	—	—	—
Ti_3O_5	209,34	355,88	$t_b = 3327$	—

7 – Continued

1	2	3	4	5
TiO_2*[20]	66,99	598,71	25	—
VO	62,80	560,61	25	—
	—	293,08	$t_b = 3127$	—
V_2O_3	100,48	—	—	—
V_3O_5	175,85	—	—	—
VO_2	56,94	—	—	—
V_2O_4*[21]	113,92±1,26	—	—	—
V_2O_5	65,147± ±0,419	—	—	—
	—	263,77	t_b	—
Cr_2O_3	104,67	—	—	—
	117,23	—	—	—
CrO_3	25,54	104,67	t_b	—
MnO	43,96±6,28	—	—	—
Mn_3O_4	127,70±12,6	—	—	—
	—	314,0	t_b	—
$Fe_{0,947}O$	32,20±2,1*[22]	—	—	—
$Fe_{0,95}O$*[23]	31,4	230,3	$t_b = 2512$	—
	—	~432,42	t_b	—
$FeO_{1,05}$	34,33±2,1*[22]	—	—	—
$FeO_{1,056}$	34,04±2,1*[22]	—	—	—
$FeO_{1,10}$	32,70±2,1*[22]	—	—	—
$FeO_{1,15}$	31,07±2,1*[22]	—	—	—
$FeO_{1,20}$	29,43±2,1*[22]	—	—	—
Fe_3O_4	138,16	—	—	—
	—	298	$t_b = 2623$	—
CoO	50,24	255,4	$t_b = 2627$	—
NiO	50,66	—	—	—
Cu_2O	56,10	—	—	—
	64,27±0,8	—	—	—
CuO	37,26	—	—	—
Ga_2O	35,59	83,74	$t_b = 727$	—
Ga_2O_3	92,11	314,01	$t_b = 2627$	—
GeO	209,34	—	—	—
GeO_2	43,96	255,40	$t_b = 2352$	—
	17,17±2,1*[2]	—	—	—
As_2O_3	18,42	29,94	$t_b = 465$	—
As_4O_6*[24]	36,84	96,30	$t_b = 461$	—
As_4O_6*[24]	45,22±9,21	56,10±4,2	$t_m = 314$	10 029
As_4O_6*[25]	49,82	109,28	t_b	—
As_4O_6*[26]	48,57±9,21	56,10±4,2	$t_m = 278$	4 812

1	2	3	4	5
	—	56,10±4,2	t_b = 461 –	101 325
AsO_2	37,68	—	—	—
SeO	31,82	188,41	t_b = 1802	—
SeO_2	25,54	84,57	t_b	—
SeO_3^{*27}	7,12±0,8	30,56	t_m = 121	729,4
Rb_2O	23,87	—	—	—
Rb_2O_2	30,56	—	—	—
Rb_2O_3	31,82	—	—	—
RbO_2	17,17	—	—	—
SrO	69,92±8,37	—	—	—
	—	534,24	t_b = 3000	—
Y_2O_3	104,67	—	—	—
	—	2047,4±4,2	t_b = 4300	—
ZrO_2	87,09	752,79	25	—
	—	643,09	t_b = 4300	—
NbO	66,99	—	—	—
NbO_2	66,99	355,88	t_b = 3527	—
Nb_2O_5	117,23	334,94	t_b = 2927	—
	103,33±0,42	—	—	—
MoO_2	66,99	—	—	—
MoO_3	52,50	138,16	1155	—
	48,99	—	—	—
TcO_2	75,36	439,61	t_b = 3727	—
Tc_2O_7	47,48±2,1	78,84±1,26	t_m = 120	79,72
	—	58,82±1,26	t_b = 311	101 325
RuO_4	10,89±1,3	44,38±0,84	t_m = 25,4	1 418
Ag_2O	15,49±2,1	—	—	—
In_2O	18,84	66,99	t_b = 527	—
InO	16,75	251,21	t_b = 1727	—
In_2O_3	83,74	355,88	t_b = 3327	—
SnO	26,80	251,21	t_b = 1527	—
	—	161,1	t_b	—
SnO_2	47,69	314,01	t_b = 2000	—
Sb_2O_3	61,71	37,35	t_b = 1425	—
Sb_4O_6	110,11	74,53	t_m = 656	1114
Sb_6O_{13}	—	74,55	t_b = 1423	—
TeO_2	29,52±0,84	225,88±8,4	t_b = 733	14,18
	—	205.2 ± 12.6	t_b = 1257	101 325
Cs_2O	19,18	—	—	—
Cs_2O_2	23,03	—	—	—
Cs_2O_3	32,45	—	—	—
Ba_2O	21,77	83,74	t_b = 767	—
BaO	57,77±8,38	259,58	t_b = 2727	—
BaO_2	23,87	—	—	—
La_2O_3	75,36	—	—	—
	—	1783,6±2,9	t_b = 4200	—
Ce_2O_3	83,74	339,94	t_b = 3227	—
CeO_2	79,55	—	—	—
	44,88	—	—	—

7 – Continued

1	2	3	4	5
	—	565,2±25,1	t_b	—
Pr_2O_3	92,11	376,81	$t_b = 3727$	—
Nd_2O_3	92,11	—	—	—
Sm_2O_3	83,74	334,94	$t_b = 3527$	—
HfO_2	71,18	—	—	—
	94,20	—	—	—
Ta_2O_5	66,99	—	—	—
WO_2	48,15	—	—	—
WO_3	71,18	180,03	$t_b = 1827$	—
	73,48±0,84	—	—	—
ReO_2	50,24	334,94	$t_b = 2977$	—
ReO_3	21,77	228,18±4,19	337	0,069
Re_2O_7	66,15	74,11	$t_b = 362$	—
	63,22±2,1	75,78±1,3	$t_m = 301,5$	24 515
	—	69,92±1,3	$t_b = 359$	101 325
ReO_4	17,58	38,94	$t_b = 187$	—
Re_2O_8	12,06±1,67	19,38±0,84	$t_m = 150$	3,049
OsO_4	14,28±0,13	42,29±2,1	$t_m = 41$	3 464
	—	37,26±2,1	$t_b = 131$	101 325
Ir_2O_3	41,87	209,34	$t_b = 1977$	—
PtO_2	19,26	—	—	—
Tl_2O	30,31±0,84	—	—	—
	—	71,18	t_b	—
Tl_2O_3	51,92	—	—	—
PbO	25,54± ±0,84*[28]	—	—	—
	—	246,18±5,86	$t_m = 886$	3,24
	—	228,18±6,28	$t_b = 1535$	38 494*[29]
BiO	15,49	226,09	$t_b = 1647$	—
Bi_2O_3	28,47	—	—	—
	62,8± ±8,4*[24]	—	—	—
	16,33± ±0,84*[26]	273,82	1127	—
PoO_2	23,03	—	—	—
Ac_2O_3	83,74	—	—	—
ThO	54,43	272,14	$t_b = 2977$	—
ThO_2	75,36	752,37	25	—
	109,69	—	—	—
PaO_2	83,74	—	—	—

7 – Continued

1	2	3	4	5
Pa_2O_5	108,86	397,75	$t_b = 3077$	—
UO	58,62	573	t_b	—
UO_2	136,07	623,83	25	—
NpO_2	62,80	—	—	—
PuO	30,14	196,78	$t_b = 2052$	—
Pu_2O_3	66,99	314,01	$t_b = 2977$	—
PuO_2	62,80	376,81	$t_b = 3227$	—
Am_2O_3	71,18	355,88	$t_b = 3127$	—

*1 At 0°C and 101.325 kPa. *2 Hexagonal. *3 At triple point. *4 Data for process of vaporization of D_2O (l) to ideal gas at p = 101.325 kPa. *5 Equilibrium mixture: 16.7% NO_2 and 83.3% N_2O_4. *6 Equilibrium mixture: 28% NO_2 and 72% N_2O_4. *7 Gas N_2O_4. *8 Quartz. *9 Tridymite. *10 Cristobalite. *11 Orthorhombic I → l I. *12 l I → g. *13 Orthorhombic II → l I. *14 Orthorhombic II → l II. *15 l II → g. *16 Monoclinic α. *17 Monoclinic β. *18 Orthorhombic γ. *19 l + cr Ti_2O_3. *20 Rutile. *21 Tetragonal β. *22 Cubic → l + s. *23 Wustite. *24 Monoclinic. *25 Octahedral. *26 Cubic. *27 Equilibrium vapor under indicated conditions consists of $(SeO_3)_4$. *28 Yellow. *29 Total pressure of products from vaporization of PbO is 101.325 kPa.

8. CHANGE IN ENTROPY UPON FUSION OR VAPORIZATION [9, Nos. 1–7; 42]

Oxide	Entropy change ΔS upon fusion, kJ/(kmole·°K)	Entropy change ΔS upon vaporization of liquid, kJ/(kmole·°K)	Temperature, °C	Pressure, Pa
1	2	3	4	5
H_2O	22,014*1	165,027*1	0,01	610,8
	22,014*2, *3	—	—	—
	—	147,643	25	3 170
	—	118,876	25	101 325
	—	109,024	$t_b = 100$	101 325
H_2O_2	45,85	193,81	$t_m = -0,42$	50,65
	—	173,29	25	285,7
H_2O_4	—	16,75	25	—
D_2O	22,797	167,89	$t_m = 3,813$	667,6
	—	122,34*4	25	101 325
D_2O_2	—	175,68	25	—
T_2O	—	154,5	25	2644
HDO	—	149,85	25	2928
HTO	—	157,0	25	2897

8 – Continued

1	2	3	4	5
B_2O_3	34,00	558,02	$t_m = 450$	$3.34 \cdot 10^{-15}$
	—	148,63	$t_b = 2124$	101 325
CO	12,309	95,865	$t_m = -205,02$	15 367
	—	74,023	$t_b = -191,50$	101 325
CO_2	38,64	75,03	$t_m = -56,58$	517 846
	—	17,71	25	6 422 400
N_2O	35,84	91,27	$t_m = -91,0$	87 827
	—	89,76	$t_b = -88,5$	101 325
NO	21,02	130,00	$t_m = -163,65$	22 083
	—	113,42	$t_b = -151,65$	101 325
N_2O_4[*5]	55,98	141,77	$t_m = -11,2$	18 740
	—	131,17	$t_b = 21$	101 325
	—	96,21[*6]	25	—
O_2	8,202	140,80	$t_m = -218,79$	156
	—	75,78	$t_b = -182,97$	101 325
O_3	25,96[*7]	—	—	—
	—	94,20	$t_b = -111,9$	101 325
F_2O	—	86,62	$t_b = -145,05$	101 325
Al_2O_3	48,61	—	—	—
O_2	4,52[*8]	—	—	—
	3,85[*9]	—	—	—
P_4O_6	47,35	157,01	$t_m = -23,9$	233
	—	156,42	25	—
		96,88	$t_b = 175,4$	101 325
P_4O_{10}[*3]	30,14	53,59	$t_m = 420$	491 305
P_4O_{10}[*10]	88,76	93,78	$t_m = 562$	58 754
P_4O_{10}[*11]	79,97	100,48	$t_m = 571$	75 975
P_4O_{10}[*12]	74,5	91,69	$t_m = 580$	73 949
P_4O_{10}	—	92,53[*13]	$t_b = 584$	101 325
SO_2	37,47	142,73	$t_m = -75,46$	15 702
	—	94,83	$t_b = -10,01$	101 325
SO_3[*14]	90,02	113,46	$t_m = 62,2$	234 003
SO_3[*15]	39,77	139,09	$t_m = 31,5$	50 144
	—	128,41	$t_b = 44,7$	101 325
SO_3[*16]	19,34	152,48	$t_m = 16,79$	20 463
	—	144,78	25	34 341
Cl_2O	—	94,20	$t_b = 2$	101 300
ClO_2	—	93,11	$t_b = 9,7$	101 325
Cl_2O_6	42,29[*17]	—	—	—
	—	139,42	12	91,17
Cl_2O_7	—	113,88	25	12 156
	—	91,69	$t_b = 80$	101 300
TiO	26,50[*18]	—	—	—

8 — Continued

1	2	3	4	5
V_2O_4	62,63	—	—	—
V_2O_5	68,37	—	—	—
MnO	20,77	—	—	—
Mn_3O_4	69,67	—	—	—
$Fe_{0,947}O$	19,55*19	—	—	—
$FeO_{1,05}$	20,89*19	—	—	—
$FeO_{1,056}$	20,68*19	—	—	—
$FeO_{1,10}$	19,68*19	—	—	—
$FeO_{1,15}$	18,51*19	—	—	—
$FeO_{1,20}$	17,33*19	—	—	—
Cu_2O	42,41	—	—	—
GeO_2	12,35	—	—	—
As_4O_6*20	88,34	101,74	$t_m = 278$	4 812
As_4O_6*21	77,04	95,46	$t_m = 314$	10 029
	—	76,62	$t_b = 461$	101 325
SeO_2	38,52	—	—	—
SeO_3*22	18,0	77,46	$t_m = 121$	729,4
ZrO_2	29,31	—	—	—
Nb_2O_5	58,62	—	—	—
MoO_3	45,59	—	—	—
Tc_2O_7	120,79	200,63	$t_m = 120$	79,72
	—	100,73	$t_b = 311$	101 325
RuO_4	36,47	148,63	$t_m = 25,4$	1 418
Ag_2O	14,24	—	—	—
Sb_4O_6*23	118,49	80,39	$t_m = 656$	1114
TeO_2	29,35	224,54	$t_m = 733$	14,18
	—	133,98	$t_b = 1257$	101 325

8 – Continued

1	2	3	4	5
WO_3	42,08	—	—	—
ReO_3	—	374,05	337	0,069
Re_2O_7	110,03	131,84	$t_m = 301,5$	24 515
	—	110,62	$t_b = 359$	101 325
Re_2O_8	28,51	45,80	$t_m = 150$	3 049
OsO_4	45,47	134,69	$t_m = 41$	3 464
	—	92,24	$t_b = 131$	101 325
Tl_2O	35,588	—	—	—
PbO	22,02	212,44	$t_m = 886$	3,24
	—	126,19	$t_b = 1535$	38494*24
Bi_2O_3	7,83	195,52	1127	—

*1 At triple point. *2 At 0°C and 101.325 kPa. *3 Hexagonal. *4 For process of vaporization of D_2O (l) to ideal gas. *5 Equilibrium mixture: 16.7% NO_2 and 83.3% N_2O_4. *6 Gas N_2O_4. *7 114.50 Pa. *8 Quartz. *9 Cristobalite. *10 Orthorhombic II → l I. *11 Orthorhombic II → l II. *12 Orthorhombic I → l I. *13 l II → g. *14 Monoclinic α. *15 Monoclinic β. 16 Orthorhombic γ. *17 At 52.69 Pa. *18 Cubic → l + cr Ti_2O_3. *19 Cubic → l + s. *20 Cubic. *21 Monoclinic. *22 Equilibrium vapor under indicated conditions consists of $(SeO_3)_4$. *23 Orthorhombic. *24 Total pressure of products from vaporization of PbO is 101.325 kPa.

9. HEAT OF SUBLIMATION AND CHANGE IN ENTROPY UPON SUBLIMATION [1; 9, Nos. 1–7; 19; 29; 42; 43; 77; 180; 195]

Oxide	Heat of sublimation ΔH_{sub}, kJ/mole	Temperature, °C	Pressure, Pa	Entropy change ΔS upon sublimation, kJ/(kmole·°K)
1	2	3	4	5
H_2O	47,428±0,025	—273,15	—	—
	51,092±0,013*1	0,01	610,8	187,037*1
	51,020	0,0	—	—
H_2O_2	64,757±0,226	—273,15	—	—
D_2O	49,965±0,059	—273,15	—	—
	52,808±0,088	$t_m = 3,813$	667,6	190,67
Li_2O	460,55±12,56	—273,15	—	—

1	2	3	4	5
BeO	719,57±13,40	–273,15	—	—
	636,40±41,87	327–2727	—	—
	609,18±16,75	t_m = 2570	—	—
B_2O_3	427,05±8,37	–273,15	—	—
	431,730	25	$1{,}52\cdot10^{-59}$	1448,05
	428,02±8,37	t_m = 450	$3{,}34\cdot10^{-15}$	592,01
CO	7,3700±0,0050	t_m = –205,02	15 367	108,174
CO_2	25,246±0,021	–78,50	101 325	129,707
	24,62±0,46	t_m = –56,58	517 846	113,67
N_2O	23,07	t_m = –91,0	87 827	126,65
NO	16,54	t_m = –163,65	22 083	151,02
$N_2O_4^{*2}$	51,800	t_m = –11,2	18 740	197,74
N_2O_5	54,4	25	49 637	182,54
	54,0	t_m = 33	101 325	176,68
O_2	8,101±0,021	t_m = –218,79	156	149,05
MgO	568,99	–273,15	—	—
Al_2O_3	844,9	–273,15	—	—
SiO	322,4	–273,15	—	—
SiO_2^{*3}	565,22±20,93	–273,15	—	—
P_4O_6	60,71	t_m = 23,9	233	204,36
$P_4O_{10}^{*4}$	89,60±4,2	25	—	300,61
$P_4O_{10}^{*4}$	65,31±4,2	359	101 325	103,41
$P_4O_{10}^{*4}$	58,20±4,2	t_m = 420	491 305	83,74
$P_4O_{10}^{*5}$	152,40	t_m = 562	58 754	182,54
$P_4O_{10}^{*5}$	152,40	t_m = 571	75 975	180,45

9 — Continued

1	2	3	4	5
P_4O_{10}[*6]	141,93	t_m = 580	73 949	166,22
SO_2	35,63±0,04	t_m = —75,46	15 702	180,20
SO_3[*7]	68,24	t_m = 62,2	234 003	203,48
SO_3[*8]	54,43	t_m = 31,5	50 144	178,65
SO_3[*9]	49,82	t_m = 16,79	20 463	171,66
Cl_2O_6	51,50±2,09	—22	3,47	205,15
CaO	574,01	—273,15	—	—
TiO	563,33±2,18	1574—1695	—	—
	583,6	—273,15	—	—
TiO_2	581,55±2,09	1576—1737	—	—
	609,2	—273,15	—	—
VO_2	435,43±8,37	1027—1377	—	—
Cr_2O_3	523,98[*10]	—	—	—
	445,56[*11]	—	—	—
	589,5[*12]	—	—	—
MnO	508,3	—273,15	—	—
FeO	463,4	—	—	—
Fe_3O_4	436,2	—	—	—
CoO	510,8	—273,15	—	—
NiO	466,41	1227	—	—
	500,0	—273,15	—	—
Ga_2O_3[*13]	573,6±25,1	25	—	1923,8
	519,2±12,6	t_m = 1725	80,0	259,83
GeO_2	415,7	—273,15	—	—
As_4O_6[*14]	104,67±8,37	25	$5,37 \cdot 10^{-6}$	351,3

1	2	3	4	5
	104,67±8,37	t_m = 278	4 812	190,08
As_4O_6*13	101,32±8,37	t_m = 314	10 029	172,5
SeO_2	98,81±4,19	25	—	331,59
	91,27±4,19	337	101 325	149,47
	110,11	—	—	—
SeO_3	37,68	t_m = 121	729,4	95,46
SrO	530,64±12,14	—273,15	—	—
ZrO_2	741,06±25,12	—273,15	—	—
NbO	590,34±12,56	1400—1700	—	—
NbO_2	577,78±20,93	1400—1900	—	—
MoO_2	582,8	25	—	—
MoO_3	272,14	700	—	—
	315,68±4,19	—	—	—
Tc_2O_7	132,30±1,67	25	—	443,76
	126,32±1,26	t_m = 120	79,72	321,42
RuO_4	55,47±0,84	25	1398	185,35
	55,27±0,84	t_m = 25,4	1418	185,10
CdO	243,7	—	—	—
	225,25	—	—	—
In_2O_3	531,7±8,4	1387	0,588	320,33
SnO	307,035	25	—	1029,12
Sb_4O_6	184,64	t_m = 656	1114	198,87
TeO_2	270,05±8,37	25	—	905,60
	259,58±8,37	t_m = 733	14,18	258,03
Ba_2O	228,6	—273,15	—	—
BaO	422,87	—273,15	—	—
	372,63±2,93	900—1200	—	—

9 – Continued

1	2	3	4	5
La_2O_3	1783,6±2,9	—	—	—
Nd_2O_3	1738,8	—	—	—
ReO_2	274,65±2,1	717	0,192	277,42
Re_2O_7	156,59±4,19	25	—	525,19
	139,00±1,26	$t_m = 301,5$	24 515	241,87
Re_2O_8	31,44±1,26	$t_m = 150$	3 049	74,32
OsO_4	57,57±2,5	25	1 307	193,10
	56,56±2,5	$t_m = 41$	3 464	180,16
PbO*15	288,47±5,44	—273,15	—	—
	288,33±5,44	25	—	967,15
PbO*16	286,930±5,44	—273,15	—	—
	286,653±5,44	25	—	961,29
	271,72±5,86	$t_m = 886$	3,24	234,46
ThO	689,1	—273,15	—	—
ThO_2	711,8±4,2	25	—	—
	644,8	1877	—	—
	732,7	—273,15	—	—
UO_2	668,2	25	—	—
	574,0	1527	—	—
	581,96±20,93	1750—1900	—	—

*1 At triple point. *2 Equilibrium mixture: 16.7% NO_2 and 83.3% N_2O_4. *3 Quartz. *4 Hexagonal. *5 Orthorhombic II. *6 Orthorhombic I. *7 Monoclinic α. *8 Monoclinic β. *9 Orthorhombic γ. *10 For Cr (g). *11 For CrO (g). *12 For CrO_2 (g). *13 Monoclinic. *14 Cubic. *15 Tetragonal. *16 Orthorhombic.

10. PARAMETERS OF POLYMORPHIC TRANSITIONS AND CERTAIN OTHER PHASE TRANSITIONS [α is the low-temperature modification; β, γ, and δ are high-temperature modifications; p is the pressure, MPa- ΔH is the heat of transition, kJ/mole; ΔS is the entropy change, kJ/(kmole · °K)]

Type of transition	Temperature, °C
1	2
H_2O [9, No. 1]	
Hex. I → tetr. III*1, *2	−34.7
Hex. II → tetr. III*1, *3	
Hex. II → hex. I*1, *4	
Hex. I → tetr. III*5	−22
Hex. II → tetr. III*6, *7	−24.3
Hex. II → cr. V*6, *8	
Cr. V → tetr. III*6, *9	
Cr. V → tetr. III*10	−17
Cr. VI → cr. V*11	0.16
Cr. VII → cr. VI*12	81.6
D_2O [9, No. 1]	
Hex. I → tetr. III*13, *14	−31
Hex. II → tetr. III*13, *15	
Hex. II → hex. I*13, *16	
Hex. II → tetr. III*17, *18	−21.5
Hex. II → cr. V*17, *19	
Cr. V → tetr. III*17, *20	
Hex. I → tetr. III*21	−18.75
Cr. V → tetr. III*22	−14.5
Cr. VI → cr. IV*23	−6.2
Cr V → cr. VI*24	2.6
BeO [104]*25	
α → β*26	2107 ± 7
β → α	2097 ± 7
CO [9, No. 4]	
Cub. (α) → hex. (β)*27	−211.59
N_2O_3 [9, No. 3]	
Cr. II → cr. I	−125
O_2 [9, No. 1]	
Rhom. (α) → hex. (β)*28	−249.30
Hex. (β) → cub. (γ)*29	−229.38

10 — Continued

1	2
Al_2O_3 *30 [19, p. 6; 90, p. 434; 52, pp. 18–19]	
Cub. (γ) → α *31	> 1200
Cub. (γ) → (θ)	850-900
θ → α	1100-1150
F.c.c. (γ') → α	750-1000
Hex. (δ) → (α) *32	950
Octahed. (ξ) → α	> 1600
$\varkappa$ → α *33	–
SiO_2 [9, No. 4] *34	
Quartz IX → α-quartz	–182
α-Quartz → β-quartz *35	573
α-Quartz → β-quartz *36 }	
α-Quartz → coesite *36 }	1300 ± 100
β-Quartz → coesite *36 }	
β-Quartz → tridymite *37	867
β-Quartz → tridymite *38 }	
β-Quartz → β-cristobalite *38 }	1190 ± 20
Tridymite → β-cristobalite *38 }	
β-Quartz → β-cristobalite *39	1735 ± 50
α-Tridymite → β-tridymite *40	117
β-Tridymite → tridymite *41	163
Tridymite → tridymite *42	225
Tridymite → β-cristobalite *43	1470
α-Cristobalite → β-cristobalite *44	242
α-Cristobalite → β-cristobalite *45	200–275 *46
β-Quartz → β-cristobalite *47	~ 1050
β-Quartz → silica glass *48	1723
Silica glass → β-cristobalite *48	–
Cl_2O_7 [9, No. 1]	
Cr. II → cr. I	–100
KO_2 [1, p. 166]	
α → β	80
Ti_6O [9, No. 7]	
Hex. → cr. *49	440 ± 20
Cr. I' → cr. I *50	830 ± 20

1	2
Ti_3O [9, No. 7]	
Hex. → cr.*50, *51	540 ± 30
Ti_3O_2 [9, No. 7]	
Rhom. → $K_1 + K_2$*52	910 ± 15
TiO [9, No. 7]	
Monocl. (α) → cub. (β)*53	980 ± 10
Cub. (β) → cub.	1250 ± 10
$TiO_{1.20}$ [9, No. 7]	
Rhomb. → $K_1 + K_2$	820 ± 10
$TiO_{1.25}$ [9, No. 7]	
Tetr. → $K_1 + K_2$	720
Ti_2O_3 [9, No. 7]	
Monocl. → hex.*54, *55	160 ± 20
Ti_3O_5 [9, No. 7]	
Monocl. (α) → monocl. (β)*56	175 ± 5
TiO_2	
Rutile → cr. IV*57	400 ± 50
α-Anatase → β-anatase*58	642
β-Anatase → rutile*59	915
β-Anatase → rutile*60	800–850
Brookite → rutile*61	650
Anatase → brookite*62	800
Brookite → rutile*62	1040
α-Anatase → rutile*63	> 400
VO [9, No. 7]	
Cub. I' → cub.*64	−153 ± 10
V_2O_3 [9, No. 7]	
Monocl. → hex.*65	−105
Monocl. → hex.*66	−193 ± 10
Monocl. → hex.*67	−268.95
V_3O_5	
Monocl. I' → monocl. I*68	−140 ± 5
*69	−98

10 – Continued

1	2
V_4O_7 [9, No. 7]	
Tricl. I' → tricl. I* 70, * 71	−36 ± 4
Tricl. I' → tricl. I* 71, * 72	−35 ± 2
V_5O_9 [9, No. 7]	
Tricl. I' → tricl. I	−134 ± 5
V_6O_{11} [9, No. 7]	
Tricl. I' → tricl. I	−96 ± 5
V_8O_{15} [9, No. 7]	
Tricl. I' → tricl.* 73	70 ± 5
V_2O_4 [9, No. 7]	
Monocl. → tetr.* 74	67 ± 2
Monocl. → tetr.* 75	72
V_6O_{13} [9, No. 7]	
Monocl. I' → monocl. I* 76, * 77	−135 ± 3
Monocl. I' → monocl. I	−118
V_2O_5 [9, No. 7]	
Cr. II → rhom.* 78	23
Cr_2O_3 [9, No. 7]	
Hex. I → hex. I'* 79	32 ± 1
Hex. I → hex. I'* 80	12 ± 2
$\beta + \gamma$* 81	> 800
CrO_2 [9, No. 7]	
Tetr. I → tetr. I'* 82	116 ± 2
CrO_3 [9, No. 7]	
Rhom. I → cr. II* 83	23
MnO [9, No. 7]	
Hex. II → cr. I* 77, * 84	−155.35 ± 2.0
Hex. II → cr. I* 85	−153.85 ± 0.5
Cub. I → cr. III* 86	20
Mn_3O_4 [9, No. 7]	
Rhom. → tetr.* 77	−231.15 ± 1.0
Tetr. → cub.* 87	1160 ± 10

10 – Continued

1	2
Mn_2O_3 [9, No. 7]	
Rhom. II' → rhom. II*[88]	−193.75 ± 1.0
Rhom. II → cr. I	670 ± 100
MnO_2 [9, No. 7]	
Tetr. I' → tetr. I*[89]	−181.15 ± 0.1
$Fe_{0.947}O$ [9, No. 7]	
Cub. I' → cub. I*[77], *[90]	−87 ± 3
$FeO_{1.056}$ [9, No. 6]	
Cr. I' → cr. I*[91]	−87 ± 3
Fe_3O_4 [9, No. 6]	
Rhom. → cub.*[92]	−154.25
Cub. → cr.	578 ± 7
Rhom. → cub.*[93]	−163.45
Rhom. → cub.*[94]	−161.95
Fe_2O_3 [9, No. 6]	
Cr. II' → cr. I'*[95]	−15 ± 2
Cr. I' → cr. I*[96]	680 ± 10
CoO [9, No. 6]	
Tetr. → cub.*[77], *[97]	17 ± 3
Cub. → cr. I	985 ± 25
Co_3O_4 [9, No. 6]	
Cub. I' → cub. I*[77]	−233 ± 5
NiO [9, No. 6]	
Cub. I → cub. I'	252 ± 15
Ga_2O_3 [9, No. 5]	
Cub. (δ) → (ε)*[98]	500 ± 20
(ε) → monocl. (β)*[98]	870 ± 15
Cub. (γ) → monocl. (β)*[98]	650 ± 20
Cub. (γ) → hex. (α)*[98]	500 ± 20
Hex. (α) → monocl. (β)*[98]	625 ± 20
GeO_2 [9, No. 4]	
Tetr. → hex.*[99]	1049 ± 5
Hex. I → hex.*[95]	1000 ± 20

10 – Continued

1	2
As_2O_3 [81, Vol. 3]	
Arsenolite (cub.) → claudetite (monocl.)*[100]	–
RbO_2	
Tetr. (α) → cub. (β)*[101]	130–150
–	–258.15[102]
–	–94.85[102]
Zr_6O_3 [9, No. 7]	
Cr. → $K_1 + K_2$	940 ± 20
ZrO_2 [9, No. 7]	
Monocl. → tetr.*[103]	1175 ± 30
Monocl. → tetr.*[104]	300 ± 20
Tetr. → cub.*[105]	
Rhom. → tetr.*[105]	850 ± 100
Rhom. → cub.*[105]	
Tetr. → cub.*[106]	2350 ± 50
Rhom. → tetr.*[107]	160 ± 50
Rhom. → cub.*[108]	800 ± 50
Monocl. → tetr.*[109]	1147
Monocl. → tetr.*[110]	1000–1205
NbO_2 [9, No. 7]	
Tetr. I → tetr. I'*[50], *[111]	795 ± 10
Nb_2O_5 [9, No. 7]	
Cr. III → cr. II*[112]	780 ± 100
Cr. III → cr. II*[113]	1000 ± 100
Cr. II → monocl.*[112]	1300 ± 100
α(L) → β(M)*[114]	900
β(M) → γ(N)*[114]	1100
α → β*[115]	900
β → γ*[116]	1280
Mo_4O_{11} [9, No. 7]	
Monocl. → rhom.	655 ± 10
Mo_9O_{26} [9, No. 7]	
Tricl. → monocl.	700 ± 10
MoO_3 [9, No. 7]	
Rhom. → cr. I	467 ± 10

10 – Continued

1	2
Rh_2O_3 [9, No. 6]	
Hex. → cr. I	750
CdO [19]	
Am. → cr.*[117]	Red heat
In_2O_3 [9, No. 5]	
Cub. I → II[118]	1250
SnO_2 [9, No. 4]	
Tetr. → cr. I*[119]	425 ± 25
Sb_4O_6 [9, No. 3]	
Cub. → rhom.*[120]	572
TeO_2 [143]	
Rhom. (α) → tetr. (β)*[121]	485
CsO_2 [54]	
Tetr. → cub.*[122]	130–200
La_2O_3	
Monocl. (B) → hex. (A)*[123]	800
Hex. (A) → hex. (H)*[124]	2047
Hex. (H) → X*[124], *[125]	2107
Pr_2O_3	
Cub. (C) → hex. (A)*[126]	650
Monocl. (B_1) → hex. (A)*[126]	810
Hex. (A) → hex. (H)*[124]	1947
Hex. (H) → X*[124], *[125]	2147
Nd_2O_3 [77]	
Cub. (C) → hex. (A)	600–810
Cub. (C) → hex. (A)*[127]	750–1100*[128]
Cub. (C) → monocl. (B)	915
Monocl. (B_1) → hex..(A)	1000
Monocl. (B) → hex. (A)	1030
Hex. (A) → hex. (H)*[121]	2097
Hex. (H) → X*[124], *[125]	2197

10 – Continued

1	2
Pm_2O_3 [102][*129]	
Cub. (C) → monocl. (B)	750–800
Monocl. (B) → hex. (A)	1740
Hex. (A) → high-temperature hex. (H)	2135
High-temperature hex. (H) → high-temperature cub. (X)	2225
Sm_2O_3 [77]	
Cub. (C) → monocl. (B)	735
Monocl. (B_1) → hex. (A)	850
Cub. (C) → hex. (A)	840–1150
Cub. (C) → monocl. (B)[*130]	840–880
Cub. (C) → monocl. (B)[*131]	950
Monocl. (B) → hex. (A)[*124]	1897
Hex. (A) → hex. (H)[*124]	2097
Hex. (H) → X[*124, *125]	2247
Eu_2O_3 [77]	
Cub. (C) → monocl. (B_1)	800
Cub. (C) → monocl. (B)	1075–1350
Cub. → monocl.[*132]	1050
Cub. (C) → monocl. (B)[*130]	1100–1400
Monocl. (B) → hex. (A)[*124]	2037
Hex. (A) → hex. (H)[*124]	2137
Hex. (H) → X[*124, *125]	2267
Gd_2O_3 [77]	
Cub. (C) → monocl. (B_2	875
Cub. (C) → monocl. (B)	1025–1250
Cub. (C) → monocl. (B)[*130]	1400
Cub. (C) → monocl. (B)[*133]	1225
Monocl. (B) → hex. (H)[*124]	2167
Hex. (H) → X[*124, *125]	2357
Tb_2O_3 [77]	
Cub. (C) → monocl. (B_1)	1150
Cub. (C) → monocl. (B)	1875
Cub. (C) → monocl. (B)[*133]	1850
Monocl. (B) → hex. (H)[*134]	2157

10 — Continued

1	2
Dy_2O_3 [77]	
Cub. (C) → monocl. (B_2)	1600
Cub. (C) → monocl. (B)	2150
Monocl. (B) → hex. (H)*[134]	2197
Ho_2O_3 [137; 138; 156]	
Cub. (C) → monocl. (B)	2197
Monocl. (B) → hex. (H)	2237
Er_2O_3 [137; 138; 156]	
Cub. (C) → hex. (H)	2277
Yb_2O_3 [137; 138; 156]	
Cub. (C) → hex. (H)	2377
HfO_2 [9, No. 7]	
Rhom. → monocl.*[135]	630 ± 50
Monocl. → tetr.*[136]	
Monocl. → rhom.*[136]	900 ± 50
Rhom. → tetr.*[136]	
Rhom. → tetr.*[86]	1450 ± 50
Monocl. → tetr.*[137]	1650 ± 100
Tetr. → cub.	2500 ± 100
Ta_2O_5 [9, No. 7]	
Rhom. → tricl.	1340 ± 20
$WO_{2.96}$ [9, No. 7]*[138]	
Monocl. → cr. I	1250 ± 20
WO_3 [9, No. 7]	
Tricl. IV' → tricl.	−45 ± 5
Tricl. → monocl.	18 ± 2
Monocl. → rhom.*[139]	330 ± 20
Rhom. → tetr. I''*[140]	740 ± 10
Tetr. I'' → tetr. I'*[141]	900 ± 10
Tetr. I' → tetr. I*[142]	1220 ± 20
PbO [9, No. 4]	
Tetr. → rhom.*[143]	540
α (red) → β (yellow)*[144]	475−583
PbO_2 [52]	
α → β*[145]	50−230

10 – Continued

1	2
Bi_2O_3 [9, No. 3]	
Monocl. → cub.*[146]	730 ± 5
Tetr. (metastable) → monocl.*[147]	450
Cub. → tetr. (metastable)*[147]	650
Cub. → tetr. (metastable*[148]	650
Tetr. (metastable) → monocl.*[149]	550–450
Monocl. (α) → face-centered cub. (δ)*[150]	725–730
Face-centered cub. (δ) → tetr.*[151]	646 ± 1
Tetr. (β) → monocl. (α)*[152]	620–605
Face-centered cub. (δ) → body-centered cub. (γ)*[153]	820–800
U_2O_3 [52]	
α → β	750
U_3O_8 [52]	
Rhom. → trig.	400
Pu_2O_3 [171]	
Cub. (C) → hex. (A)	1600
Am_2O_3 [171]	
Cub. (C) → monocl. (B)	800
Monocl. (B) → hex. (A)	950
Cm_2O_3 [171]	
Cub. (C) → monocl. (B)	1000
Monocl. (B) → hex. (A)*[154]	1600
Cub. (C) → monocl. (B)*[155]	800–1300
Bk_2O_3 [171]	
Cub. (C) → monocl. (B)*[156]	1200 ± 50
Monocl. (B) → hex. (A)	> 1750
Cf_2O_3 [171]	
Cub. (C) → monocl. (B)*[157]	1400

*[1] p = (212.7). *[2] ΔH = 0.167, ΔS = 0.712. *[3] ΔH = 0.921, ΔS = 3.852. *[4] ΔH = 0.754, ΔS = 3.140. *[5] p = (207.4), ΔH = 0.381, ΔS = 1.507. *[6] p = (344.1). *[7] ΔH = 1.273, ΔS = 5.108. *[8] ΔH = 1.206, ΔS = 4.857. *[9] ΔH = 0.067, ΔS = 0.251. *[10] p = (346.0), ΔH = 0.071, ΔS = 0.293. *[11] p = (625.5), ΔH = 0.017, ΔS = 0.042. *[12] p = (2196), ΔH = 0. *[13] p = (224.5). *[14] ΔH = 0.306, ΔS = 1.256. *[15] ΔH = 0.825, ΔS = 3.391. *[16] ΔH = 0.519, ΔS = 2.135. *[17] p = (347.1). *[18] ΔH = 1.478, ΔS = 5.862. *[19] ΔH = 1.432, ΔS = 5.694. *[20] ΔH = 0.046, ΔS = 0.167. *[21] p = (220.1), ΔH = 0.322, ΔS = 1.256. *[22] p = (348.6), ΔH = 0.050, ΔS = 0.209. *[23] p = (530.4), ΔH = 0.008, ΔS = 0.042. *[24] p = (628.1), ΔH = 0.008, ΔS = 0.042. *[25] Width of temperature hysteresis 10–24°C; result independent of BeO purity in interval of 99.0–99.9% purity. *[26] ΔH = 5.8 ± 1.05. *[27] $p = 3748 \cdot 10^{-6}$, ΔH = 0.6326 ± 0.0033, ΔS = 10.279.

(cont'd.)

•[28]ΔH = 0.09387 ± 0.0004, ΔS = 3.936. •[29]ΔH = 0.7436 ± 0.0021, ΔS = 16.990. •[30]Modifications γ, γ', δ, ξ, ϰ low-temperature; upon heating they are irreversibly converted to α-Al_2O_3 (corundum). •[31]ΔH = −32.66 [90], ΔH = −22.19 [52]. •[32]ΔH = −11.30. •[33]ΔH = −15.07. •[34]Conversion of one modification or another of low-temperature variety to high-temperature variety and back again takes place rapidly, whereas conversion of one of the three main modifications to another is relatively slow [19]. •[35]ΔH = 0.628, ΔS = 0.754. •[36]p = 3390. •[37]ΔH = 0.50, ΔS = 0.461. •[38]p = (140). •[39]p = (248). •[40]ΔH = 0.29, ΔS = 0.754. •[41]ΔH = 0.17, ΔS = 0.377. •[42]ΔH = 0.21, ΔS = 0.335. •[43]ΔH = 0.21, ΔS = 0.126. •[44]ΔH = 1.30, ΔS = 2.51. •[45]For cristobalite, the $\alpha \rightleftharpoons \beta$ transition is usually observed in the interval 220–272°C upon heating and in the interval 238–198°C upon cooling [81, Vol. 2]. •[46][67]. •[47] [90]. •[48][81, Vol. 2]. •[49]ΔH = 1.63 ± 0.17, ΔS = 2.30, cr (I) → cr (I'). •[50]Transition of order–disorder type. •[51]ΔH = 1.63 ± 0.17, ΔS = 2.01. •[52]K_1 is solid solution on basis of α-Ti, K_2 on basis of TiO. •[53]ΔH = 3.43 ± 1.26, ΔS = 2.72. •[54]Transition of semiconductor–submetal type. •[55]ΔH = 0.151 ± 0.008, ΔS = 0.335. •[56]ΔH = 9.38 ± 0.84, ΔS = 20.93. •[57]p = $(1.06 \cdot 10^4)$ [9, No. 7]. •[58]Conversion proceeds rapidly [19], ΔH = 1.26 [1, p. 166]. •[59]Conversion proceeds slowly [19]. •[60]ΔH = 5.19 ± 0.80 [31], ΔH = 6.70-7.12; calculated data [32]. •[61][90]. •[62]In presence of impurities [52]. •[63]In presence of mineralizers [90]. •[64]Second-order phase transition (dielectric-metal type). •[65]First-order phase transition; transition temperature according to various sources is within range of 150–175°K, ΔH = 1.84 ± 0.21, calculated in interval 140–205°K. •[66]First-order phase transition, p = (2030). •[67]Same, p = (2530). •[68][9, No. 7]. •[69][52]. •[70]p = (4050). •[71]Phase transition of semiconductor–metal type. •[72]ΔH = 0.63 ± 0.04 calculated in interval 210–260°K. •[73]Second-order phase transition (semiconductor–metal type). •[74]ΔH = 6.28 ± 0.4, ΔS = 18.46. •[75]ΔH = 8.62 [1]. •[76]p = $(1.62 \cdot 10^4)$. •[77]Néél point (antiferromagnetic–paramagnetic transition). •[78]p = (8610). •[79]Néel point, ΔH = 0.80 ± 0.04 calculated in interval 210–350°K. •[80]p = (1520), Néel point. •[81][90]. •[82]Curie point (ferromagnetic–antiferromagnetic transition). •[83]p = $(1.16 \cdot 10^4)$. •[84]ΔH = 0.775 ± 0.084, calculated in interval 75–150°K. •[85]p = (304), Néel point. •[86]p = $(1.01 \cdot 10^4)$. •[87]ΔH = 20.9 ± 2.1, ΔS = 14.61. •[88]ΔH = 0.377 ± 0.08, calculated in interval 35–110°K. •[89]ΔH = 0.234 ± 0.013. •[90]ΔH = 0.21 ± 0.04, calculated in interval 158–204°K. •[91]Néel point, ΔH = 0.222 ± 0.04. •[92]First-order phase transition, related to ordering of Fe^{2+} and Fe^{3+} ions at octahedral lattice points; ΔH = 0.662, calculated in interval 110–125°K. •[93]1% cation vacancies; ΔH = 0.410, calculated in interval 106–113°K. •[94]2% cation vacancies; ΔH = 0.092, calculated in interval 100–120°K. •[95]Second-order phase transition. •[96]Néel point, ΔH = 3.22 ± 0.42. •[97]ΔH = 0.306, calculated in interval 240–298°K. •[98]Monotectic conversion. •[99]ΔH = 21.14 ± 2.5, ΔS = 15.99. •[100]Very slow conversion. •[102]In interval from 130 to 150°C, both phases are registered simultaneously [54]. •[102]Probably second-order phase transitions [144]. •[103]ΔH = 17.53 ± 1.26, ΔS = 4.77. •[104]p = (3040). •[105]p = (8100). •[106]ΔH = 13.0, ΔS = 4.94. •[107]p = (5100). •[108]p = (10100). •[109]ΔH = 4.75 [1, p. 167]. •[110]ΔH = 5.95 ± 0.42 [1]. •[111]ΔH = 2.9, ΔS = 2.76. •[112]p = (2030). •[113]p = (6080). •[114]According to Brauer [1, p. 167]. •[115]According to Goldschmidt [1]. •[116]According to Sheffer and Roy [1]. •[117]ΔH = 226.1. •[118]p = (6580). •[119]ΔH = 1.88, ΔS = 2.700. •[120]p = $(1216 \cdot 10^{-6})$, ΔH = 13.4, ΔS = 15.9. •[121]ΔH = −0.976 ± 0.126; irreversible transition. •[122]Tentative. •[123]At 400°C, cubic (C) and monoclinic (B) forms are detected; from 500 to 700°C, only the monoclinic (B) form; from 900 to 1300°C, only the hexagonal form (A) [99]. •[124][137, 138, 156]. •[125]X is a still unidentified phase [137, 138]; according to [102], it is a high-temperature cubic phase. •[126][138, 77]. •[127]Irreversible conversion without intermediate B-form, [154]. •[128][113]. •[129]Error in temperature measurement ±20°C; all transitions (other than C → B) reversible. •[130]Irreversible conversion [113]. •[131][137, 138]. •[132][19]. •[133][137, 138]. •[134][137, 138, 156]. •[135]p = (4050). •[136]p = (5070). •[137]Reversible conversion proceeds with hysteresis in interval 1600–1500°C. •[138]1/50 $W_{50}O_{148}$. •[139]ΔH = 1.38 ± 0.21, ΔS = 2.30. •[140]ΔH = 1.88 ± 0.21, ΔS = 1.84. •[141]ΔH = 1.17 + 0.13. •[142]ΔH - 0.50 + 0.13. •[143]ΔH = 0.4, ΔS = 0.50. •[144]ΔH = 1.68; $\beta \rightarrow \alpha$ transition slow [1]. •[145]Upon heating in air; $\alpha \rightarrow \beta$ transition also proceeds at elevated pressures and with grinding. •[146]ΔH = 41.4 ± 2.1, ΔS = 41.28. •[147]Irreversible transition. •[148]Upon cooling below 730°C [52]. •[149]Upon cooling [52]. •[150]Upon heating of specimen with 99 or 97% Bi_2O_3 by mass, ΔH = 78.29, ΔS = 75.4 [98]. •[151]Upon cooling; transition does not depend on composition of gaseous medium or temperature to which previously heated. •[152]Upon cooling; transition significantly depends on composition of gaseous medium and temperature to which previously heated. •[153]Upon cooling; conversion of Bi_2O_3 to the γ-phase is favored by an increase in oxygen partial pressure and by an increase in holding time at the indicated temperature; γ-phase is stable under normal conditions; a reduction of the oxygen partial pressure and an increase in holding time at 750°C broadens the interval of stability of the β-phase [98]. •[154][171, 179]. •[155]A−Cm_2O_3 is stable up to melting point [179]. [156]Bk_2O_3 > 99.8% (at.). •[157]Cf_2O_3 > 99 8% (at.).

Numerical values of p, ΔH, and ΔS are given in units indicated at the beginning of the table.

11. ENERGY OF ATOMIZATION OF CRYSTALLINE OXIDES AT 0°K [1; 8, Nos. 1, 3–7; 185]

Oxide	Structure	Atomization energy $D_0 \cdot 10^{-3}$, J/kmole
1	2	3
H_2O	Hex.	965761 ± 142
H_2O_2	Tetr.	1120928 ± 293
D_2O	Hex.	983195 ± 172
Li_2O	Cub.	1168100
BeO	*1	418680
	Hex.	1165600
B_2O_3	Hex.	3122164 ± 25121
Na_2O	Cub.	977200
MgO	*2	221900
	Cub.	1001100
Al_2O_3	*3	343318
	Hex.	3059525 ± 8374
SiO_2	α-Quartz	1848489 ± 7955
	α-Tridymite	1843113
	α-Cristobalite	1845868 ± 7955
	Coesite	1843381 ± 7955
		1797448 ± 7955
P_2O_3	—	2546000
P_4O_{10}	Hex.	6688061
K_2O	—	791300
CaO	Cub.	1062200
Sc_2O_3	Cub.	3151000
TiO	Cub.	1241478
$TiO_{1,01}$	Monocl.	1248705
Ti_2O_3	Monocl.	3193708
Ti_3O_5	Monocl.	5092740
Ti_4O_7	Tricl.	6988167
Ti_5O_9	Tricl.	8878938
Ti_6O_{11}	—	10768495
Ti_7O_{13}	Tricl.	12658680
Ti_8O_{15}	Tricl.	14546102
Ti_9O_{17}	Tricl.	16437334
$Ti_{10}O_{19}$	—	18329483
TiO_2	Tetr. *4	1904764
	Tetr. *5	1899543
VO	Cub.	1187829 ± 6280
$VO_{1,24}$	Cub.	1338708 ± 7118
V_2O_3	Monocl.	2979507 ± 8792
V_4O_7	Tricl.	6428237 ± 17585
V_2O_4	Monocl.	3429492 ± 10048
V_2O_5	—	3801941 ± 12560
Cr_2O_3	Hex.	2667540 ± 8374

Oxide	Structure	Atomization energy $D_0 \cdot 10^{-3}$, J/kmole
1	2	3
CrO_3	Rhom.	1733300
MnO	Hex.	915682 ± 2596
Mn_3O_4	Rhom.	3218800 ± 6280
Mn_2O_3	Rhom.	2260625 ± 4187
MnO_2	Tetr.	1294324 ± 2261
$Fe_{0,947}O$	Cub.	906208 ± 2093
FeO	Cub.	927782 ± 4605
$FeO_{1,056}$	Cub.	956939 ± 4605
Fe_3O_4	Rhom.	3344395 ± 4187
Fe_2O_3	Hex.	2386983 ± 5024
CoO	Tetr.	913388 ± 4187
Co_3O_4	Cub.	3142918 ± 15072
NiO	Cub.	911538 ± 2177
Cu_2O		1091846 ± 2847
CuO	Monocl.	738535 ± 1758
ZnO	Hex.	724852 ± 377
Ga_2O_3	Monocl. (β)	2365366 ± 9211
GeO_2	Tetr.	1444923 ± 4605
	Hex.	1421059 ± 4187
As_4O_6	Cub. *6	3956555 ± 26796
	Monocl. *7	3956513 ± 26796
As_2O_5	—	2723823 ± 13398
Rb_2O	Cub.	743600
SrO	Cub.	1003600
Y_2O_3	Cub.	3508500
ZrO_2	*8	351690
	Monocl.	2188846
NbO_2	Tetr.	2005117
Nb_2O_5	Monocl.	4563880 ± 8374
MoO_2	Monocl.	1733766 ± 3056
MoO_3	Rhom.	2136817 ± 29726
Ag_2O	Cub.	845750 ± 1675
CdO	Cub.	615878 ± 1256
In_2O_3	Cub.	2152850
SnO	Tetr.	833395 ± 2596
SnO_2	Tetr.	1371219 ± 3856
Sb_2O_3	—	1956100
Sb_2O_4	—	2422884 ± 8374
Sb_2O_4	Rhom.	3963464 ± 15072
Sb_2O_5	Cub.	2765859

11 – Continued

1	2	3	1	2	3
Cs_2O	Hex.	723500	WO_3	Tricl.	2434674 ± 4187
BaO	—	981000	ReO_3	Tetr.	2136500
La_2O_3	—	3383800	Re_2O_7	Rhom.	4541497 ± 15070
Ce_2O_3	—	3280800	OsO_2	Tetr.	1428100
CeO_2	Cub.	1943900	IrO_2	»	1352800
Pr_2O_3	—	3290000	HgO	Rhom.	397893
Nd_2O_3	—	3200400	Tb_2O	—	1035000
Sm_2O_3	—	2983500	PbO	Red	660221 ± 1047
HfO_2	Monocl.	2227625		Yellow	658680 ± 1089
Ta_2O_5	Rhom.	4845652 ± 5862	PbO_2	Tetr.	962005 ± 1507
WO_2	Monocl.	1935110 ± 4187	Pb_3O_4	»	2291385 ± 3140
$WO_{2,72}$	»	2298725	Bi_2O_3	Monocl.	1723710 ± 5862
$WO_{2,90}$	»	2385773	ThO_2	*9	464735
$WO_{2,96}$	»	2415193 ± 9211	UO_2	—	2074140
			UO_3	—	2461000

*1 $BeO (s) = Be (l) + 1/2 O_2 (g)$. *2 $MgO (s) = Mg (g) + 1/2 O_2 (g)$. *3 $1/3 Al_2O_3 (s) = 2/3 Al (l) + 1/2 O_2 (g)$. *4 Rutile. *5 Anatase. *6 Arsenolite. *7 Claudetite. *8 $1/2 ZrO_2 (s) = 1/2 Zr (s) + 1/2 O_2 (g)$. *9 $1/2 ThO_2 (s) = 1/2 Th (s) + 1/2 O_2 (g)$.

12. CRYSTAL LATTICE ENERGY [1, 36]

Oxide	Crystal lattice energy W, kJ/mole	Oxide	Crystal lattice energy W, kJ/mole	Oxide	Crystal lattice energy W, kJ/mole
Li_2O	2901,45	Cr_2O_3	15186	ZrO_2	11195,50
Li_2O_2	2593,30	MnO	3876,98	MoO_2	12363,62
LiO_2	879,2 ± 41,9	MnO_2	13067,00	Ag_2O	3056,36
BeO	4521,72–4601,29	FeO	4036,08	CdO	3876,98
B_2O_3	18254,45	CoO	4036,08	In_2O_3	14528,20
Na_2O	2579,07	NiO	4044,45	SnO	3696,94
Na_2O_2	2310,28	Cu_2O	3324,32	SnO_2	11867,02
NaO_2	799,26	CuO	3872,79	Cs_2O	2239,94
MgO	3935,59	ZnO	4107,25	Cs_2O_2	1948,96
	3893,72*1	Ga_2O_3	15608,39	CsO_2	679,10
Al_2O_3	15520,47		15185,52	BaO	3127,54
SiO_2	13134,00	GeO_2	12836,73		3081,49*1
K_2O	2327,86	Rb_2O	2294,37	La_2O_3	12585,52
K_2O_2	2115,59	RbO	3567,15	CeO_2	10621,91
KO_2	741,48	Rb_2O_2	2026,4	HgO*2	3981,65
CaO	3571,34	RbO_2	706,31	Tl_2O	2704,67
	3479,23*1	SrO	3311,76	PbO	3567,15
TiO_2	12158,47		3257,33*1	PbO_2	11748,16

*1 At 25 °C. *2 Red.

13. HEAT OF DECOMPOSITION [1] (heat of decomposition refers to formation of 1 mole of oxygen in solid phase in lowest state of oxidation)

Oxide	Heat of decomposition, kJ	Oxide	Heat of decomposition, kJ	Oxide	Heat of decomposition, kJ
Li_2O	1194,1	$Mn_2O_3(\alpha)$	212,7*13	SnO	572,8
$Li_2O_2(\alpha)$	77,5	$MnO_2(\alpha)$	160,8*14	$SnO_2(\alpha)$	580,7
BeO	1198,3	Mn_2O_7	—207,25	$Sb_2O_3(\alpha)$	465,6*21
Na_2O	843,2	FeO	529,6*15	$Sb_2O_3(\beta)$	460,5*22
Na_2O_2	186,7	Fe_3O_4	605,0	$SbO_2(\alpha)$	209,3
NaO_2	8,4	$Fe_2O_3(\alpha)$	461,4*16	Cs_2O	636,4
MgO	1203,3	$Fe_2O_3(\gamma)$	457,6	Cs_2O_2	318,2
$Al_2O_3(\alpha)$	1117,0	CoO	478,1*17	Cs_2O_3	175,8
$Al_2O_3(\gamma)$	1059,7	NiO	481,5*18	CsO_2	142,4
SiO	(833,2)	Cu_2O	334,9	BaO	1136,9*23
SiO_2	880,1*1	CuO	286,4	BaO_2	163,3
	879,2*2	ZnO	698,4	La_2O_3	1244,7
	876,3*3	Ga_2O	686,6	CeO_2	(523,4)
	875,5*4	$Ga_2O_3(\beta)$	734,8	Ta_2O_5	818,9*24
K_2O	723,5	$GeO_2(\beta)$	540,1	WO_2	589,9
K_2O_2	173,3	Rb_2O	660,7	$WO_3(\alpha)$	561,5
$KO_2(\alpha)$	117,2	Rb_2O_2	192,6	ReO_2	425,0
CaO	1268,6	Rb_2O_3	129,8	ReO_3	372,6
CaO_2	50,2	RbO_2	83,7	Re_2O_7	39,8
Sc_2O_3	(1214,2)	SrO	1180,7	OsO_2	257,5
$TiO(\alpha)$	1037,5*5	SrO_2	100,5	OsO_4	195,9*25
$TiO(\beta)$	1036,8*6	Y_2O_3	1271,1	IrO_2	221,9
$Ti_2O_3(\alpha)$	963,8	$ZrO_2(\alpha)$	1086,5	HgO	180,9*26
$Ti_2O_3(\beta)$	971,3	NbO	(816,4)	Tl_2O	355,9
$Ti_3O_5(\beta)$	782,9	NbO_2	782,9	Tl_2O_3	175,8
$TiO_{1,91-2,0}$	737,7*7	Nb_2O_5	614,6*19	$PbO(\alpha)$	438,8*26
VO	854,1*8	MoO_2	588,7	$PbO(\beta)$	441,7*25
V_2O_3	753,6	Mo_4O_{11}	311,5	Pb_3O_4	154,1*27
$VO_2(\alpha)$	401,9	MoO_3	324,9	$Pb_2O_3(\alpha)$	95,0
$VO_2(\beta)$	385,2	RuO_2	439,6	PbO_2	100,5*28
V_2O_5	251,2*9	RuO_4	0,0	$Bi_2O_3(\alpha)$	385,2
Cr_2O_3	753,6*10	RhO	181,7	ThO_2	1227,6
CrO_2	75,4	Rh_2O_3	208,5	UO_2	1084,4
CrO_3	—6,28	PdO	182,1	U_4O_9	351,7
MnO	770,4*11	Ag_2O	61,1	U_3O_8	305,6
$Mn_3O_4(\alpha)$	463,9*12	CdO	511,6*20	UO_3	233,6
$Mn_3O_4(\beta)$	422,0	In_2O_3	620,9		

*1 α-Quartz. *2 β-Quartz. *3 β-Cristobalite. *4 β-Tridymite. *5 $TiO_{0.95-1.25}$. *6 $TiO_{0.8-1.23}$. *7 Rutile. *8 $VO_{0.9-1.1}$. *9 $VO_{2.495-2.50}$. *10 $CrO_{1.5-1.54}$. *11 $MnO_{1.0-1.12}$. *12 $MnO_{1.33-1.41}$. *13 $MnO_{1.50-1.6}$. *14 $MnO_{1.96-2.0}$. *15 $FeO_{1.055-1.19}$. *16 Hematite. *17 $CoO_{1.0017}$. *18 $NiO_{1.005}$. *19 $NbO_{2.43-2.50}$. *20 $CdO_{0.999-1.000}$. *21 Senarmontite. *22 Valentinite. *23 $BaO_{0.997-1.00}$. *24 $TaO_{2.35-2.50}$. *25 Yellow. *26 Red. *27 $PbO_{1.33-1.57}$. *28 $PbO_{1.87-2.0}$.

14. ENERGY OF CHEMICAL BOND RUPTURE. Energy and Heat of Dissociation of Gaseous Oxides [1; 9, Nos. 1−7; 47; 66; 79, Vol. 1; 81, Vol. 5; 101; 156; 195; 197]

Oxide	Dissociation equation	Energy of dissociation at 0°K = $D_0 \cdot 10^{-3}$, J/kmole	Energy of dissociation at 298.15°K = $D \cdot 10^{-3}$ J/kmole
1	2	3	4
H_2O	$H_2O=OH+H$	490690±4190	495720
	$H_2O=2H+O$	918350±140	—
HO	$HO=O+H$	424370±1260	428480
H_2O_2	$H_2O_2=2OH$	197200	203900±10470
	$H_2O_2=HO_2+H$	369280	374720
	$H_2O_2=2H+2O$	1056170±356	—
HO_2	$HO_2=H+O_2$	~204320	~209340
	$HO_2=OH+O$	266280±8370	271720
	$HO_2=H+2O$	686618±8374	—
D_2O	$D_2O=OD+D$	502840±1260	508280
	$D_2O=2D+O$	933229±172	—
D_2O_2	$D_2O_2=2DO$	211430±1260	218130
	$D_2O_2=2D+2O$	1071972±376	—
	$D_2O_2=DO_2+D$	—	387280±20900
DO	$DO=D+O$	431240±1260	434260
DO_2	$DO_2=D+O_2$	—	200970±20930
	$DO_2=D+2O$	694163±8370	—
T_2O	$T_2O=OT+T$	505200±1260	512050
	$T_2O=2T+O$	939800±250	—
TO	$TO=T+O$	433330±1260	436680
TO_2	$TO_2=T+2O$	715516±8374	—
HDO	$HDO=H+D+O$	925630±167	—
HTO	$HTO=H+T+O$	928700±250	—
DTO	$DTO=D+T+O$	936470±250	—
HDO_2	$HDO_2=H+D+2O$	1064090±390	—
Li_2O	$Li_2O=LiO+Li$	282600	286380
	$Li_2O=2Li+O$	696680±12560	—
LiO	$LiO=Li+O$	326570±20934	330760
	$LiO=Li+O$	(373000±30000)	—
BeO	$BeO=Be+O$	445480±12560	449660
$(BeO)_2$	$(BeO)_2=2BeO$	665700±50240	—
$(BeO)_3$	$(BeO)_3=(BeO)_2+BeO$	745250±62800	—
$(BeO)_4$	$(BeO)_4=(BeO)_3+BeO$	640580±62800	—
$(BeO)_5$	$(BeO)_5=(BeO)_4+BeO$	640580±83740	—
$(BeO)_6$	$(BeO)_6=(BeO)_5+BeO$	669900±104670	—
BO	$BO=B+O$	793400	—
	$BO=B+O$	800520±9630	—
	$BO=B+O$	—	774560
B_2O_2	$B_2O_2=2BO$	498230±83740	502420
	$B_2O_2=BO_2+B$	728500±29300	732690
	$B_2O_2=2B+2O$	2065244±26380	—
B_2O_3	$B_2O_3=B_2O_2+O$	653140±25120	661510
	$B_2O_3=BO+BO_2$	611270±29300	619650

$Ti_3O_5^{*12}$	Monocl.	—	—	0,9757	0,3802	0,9452	0,968
$Ti_5O_9^{*13}$	Tricl.	—	—	0,5369	0,7120	0,8865	1,245
TiO_2^{*14}	Tetr.	D_{4h}^{19} — $I4_1/amd$	—	0,3733	—	0,937	2,51
TiO_2^{*15}	Rhom.	D_{2h}^{15} — $Pbca$	—	0,5436	0,9166	0,5135	0,944
TiO_2^{*16}	Tetr.	D_{4h}^{14} — $P4_2/mnm$	TiO_2	0,4584	—	0,2953	0,644
V_4O	"	—	—	0,294	—	3,49	11,87
VO	Cub.	O_h^5 — $Fm3m$	NaCl	0,4093	—	—	—
V_2O_3	Trig.	D_{3d}^6 — $R\bar{3}c$	α-Al_2O_3	0,4933	—	1,394	2,82
V_3O_5	Monocl.	C_{2h}^6 — $C2/c$	—	0,9991	0,5033	0,9845	0,985
VO_2^{*18}	"	C_{2h}^5 — $P2_1/c$	—	0,5346	0,4518	0,5739	1,07
V_2O_5	Rhom.	D_{2h}^{13} — $Pmmn$	V_2O_5	11,51	0,3559	0,4371	0,379
Cr_3O	Cub.	O_h^3 — $Pm3n$	β-W	4,544	—	—	—
Cr_2O_3	Trig.	D_{3d}^6 — $R\bar{3}c$	α-Al_2O_3	0,4950	—	1,3665	2,76
CrO_2	Tetr.	D_{4h}^{16} — $P4_2/ncm$	SnO_2	0,4421	—	0,2916	0,658
CrO_3	Rhom.	C_{2v}^{16} — $Ama2$	CrO_3	0,4789	0,8557	0,5743	1,19
MnO	Cub.	O_h^5 — $Fm3m$	NaCl	0,4425	—	—	—
α-Mn_3O_4	Tetr.	D_{4h}^{19} — $I4_1/amd$	—	0,575	—	0,942	1,64
γ-Mn_3O_4	Cub.	—	—	0,87	—	—	—
α-Mn_2O_3	Tetr.	—	—	0,885	—	0,995	1,12

14 — Continued

Oxide	Dissociation equation	Energy of dissociation at 0°K = $D_0 \cdot 10^{-3}$, J/kmole	Energy of dissociation at 298°K = $D \cdot 10^{-3}$, J/kmole
1	2	3	4
Al_2O	$Al_2O = 2Al + O$	1027215	—
	$Al_2O = AlO + Al$	544280±41870	548470
AlO	$AlO = Al + O$	484237	—
	$AlO = Al + O$	—	485670
Al_2O_2	$Al_2O_2 = 2AlO$	565200±62800	—
	$Al_2O_2 = Al_2O + O$	502400±62800	—
	$Al_2O_2 = 2Al + 2O$	1528200	—
Al_2O_3	$Al_2O_3 = Al_2O_2 + O$	682450±104670	—
	$Al_2O_3 = 2Al + 3O$	2210600	—
SiO	$SiO = Si + O$	799410±8790	—
	$SiO = Si + O$	—	808890
SiO_2	$SiO_2 = SiO + O$	464700±20900	468900
	$SiO_2 = Si + 2O$	1267700±29700	—
PO	$PO = P + O$	510790	—
	$PO = P + O$	592400±2500	595780
	$PO = P + O$	590340±8400	—
P_4O_6	$P_4O_6 = 4P + 6O$	4315670	—
PO_2	$PO_2 = PO + O$	—	607100
	$PO_2 = P + 2O$	—	(1130400±104700)
P_4O_8	$P_4O_8 = 4P + 8O$	5617128±25100	—
P_4O_{10}	$P_4O_{10} = 4P + 10O$	6594620±6280	—
S_2O	$S_2O = 2S + O$	895480±5020	—
	$S_2O = SO + S$	378490±4190	382700
	$S_2O = S_2 + O$	483580±4190	486900
SO	$SO = S + O$	517360±209	522050
SO_2	$SO_2 = SO + O$	541940±1260	546880
	$SO_2 = S + 2O$	1059290±1260	1073900±4190
SO_3	$SO_3 = SO_2 + O$	342060±628	347500
	$SO_3 = S + 3O$	1402080±1260	—
Cl_2O	$Cl_2O = ClO + Cl$	142770	146540
	$Cl_2O = 2Cl + O$	407598	—
ClO	$ClO = Cl + O$	265190±126	269340
ClO_2	$ClO_2 = ClO + O$	241160±6280	245770
	$ClO_2 = Cl + 2O$	506470±6280	516230±8370
	$ClO_2 = Cl + O_2$	15072	19260
ClO_3	$ClO_3 = ClO_2 + O$	—	199290
Cl_2O_6	$Cl_2O_6 = 2ClO_3$	—	7120±2090
KO	$KO = K + O$	(299000±25000)	—
CaO	$CaO = Ca + O$	481480±16747	485670
	$CaO = Ca + O$	489860±20900	—
	$CaO = Ca + O$	381000±6280	—
Sc_2O	$Sc_2O = 2Sc + O$	1025800±46000	—
ScO	$ScO = Sc + O$	692080±14656	—
	$ScO = Sc + O$	673656±14650	—
TiO	$TiO = Ti + O$	661500±8370	—

14 — Continued

Oxide	Dissociation equation	Energy of dissociation at $0°K = D_0 \cdot 10^{-3}$, J/kmole	Energy of dissociation at $298.15°K = D \cdot 10^{-3}$, J/kmole
1	2	3	4
TiO_2	$TiO = Ti + O$	653140±25120	657330
	$TiO = Ti + O$	602900±20930	—
	$TiO_2 = TiO + O$	—	644800±33500
	$TiO_2 = Ti + 2O$	1297900	—
	$TiO_2 = Ti + 2O$	—	1318850±20900
	$TiO_2 = Ti + 2O$	1260230±20934	—
	$TiO_2 = Ti + 2O$	1197400±41870	—
VO	$VO = V + O$	607090±41870	—
	$VO = V + O$	617560±20930	622160
VO_2	$VO_2 = VO + O$	615460±29300	—
	$VO_2 = V + 2O$	—	1247670±41870
	$VO_2 = V + 2O$	1235100±20930	—
CrO	$CrO = Cr + O$	423290±29310	427050
	$CrO = Cr + O$	452170±29310	—
CrO_2	$CrO_2 = CrO + O$	527540	—
	$CrO_2 = Cr + 2O$	—	(963000±62800)
	$CrO_2 = Cr + 2O$	950400±62800	—
CrO_3	$CrO_3 = CrO_2 + O$	476900	—
	$CrO_3 = Cr + 3O$	1427700±83700	—
MnO	$MnO = Mn + O$	401930±12560	406100
	$MnO = Mn + O$	406120±12560	—
MnO_2	$MnO_2 = Mn + 2O$	—	(921100±104670)
FeO	$FeO = Fe + O$	410300±41870	414490
	$FeO = Fe + O$	406120±12560	—
	$FeO = Fe + O$	395200±10500	—
CoO	$CoO = Co + O$	364250±20930	—
	$CoO = Co + O$	408200	—
NiO	$NiO = Ni + O$	406119	410306
	$NiO = Ni + O$	360070±20930	—
CuO	$CuO = Cu + O$	334900	339131
	$CuO = Cu + O$	263800±41870	—
ZnO	$ZnO = Zn + O$	<385190	—
	$ZnO = Zn + O$	—	⩽276330
Ga_2O	$Ga_2O = 2Ga + O$	873685±10470	—
GaO	$GaO = Ga + O$	376800±16750	—
	$GaO = Ga + O$	347500	351691
	$GaO = Ga + O$	309800±25120	—
GeO	$GeO = Ge + O$	653980±7960	—
	$GeO = Ge + O$	—	669890
GeO_2	$GeO_2 = GeO + O$	330760±50240	—
	$GeO_2 = Ge + 2O$	996460	—
AsO	$AsO = As + O$	477300±16750	—
	$AsO = As + O$	481500	—
SeO	$SeO = Se + O$	418680±83700	—
SeO_2	$SeO_2 = Se + 2O$	—	849920±20900

14 – Continued

Oxide	Dissociation equation	Energy of dissociation at 0°K = $D_0 \cdot 10^{-3}$, J kmole	Energy of dissociation at 298.15°K = $D \cdot 10^{-3}$, J/kmole
1	2	3	4
	$SeO_2 = Se + 2O$	844650 ± 12560	—
BrO	$BrO = Br + O$	231530±2512	235300
BrO_2	$BrO_2 = BrO + O$	—	≥293080
RbO	$RbO = Rb + O$	294000±25000	—
SrO	$SrO = Sr + O$	397750±6280	—
	$SrO = Sr + O$	468900±8370	472690
Y_2O	$Y_2O = 2Y + O$	1080200±33500	—
YO	$YO = Y + O$	717200±5000	—
	$YO = Y + O$	705890±11720	—
Y_2O_2	$Y_2O_2 = 2Y + 2O$	1863130±50200	—
ZrO	$ZrO = Zr + O$	753600±41870	—
	$ZrO = Zr + O$	757800±33500	762000
ZrO_2	$ZrO_2 = Zr + 2O$	1449400±33500	1452800±20900
	$ZrO_2 = Zr + 2O$	1386100±21800	—
	$ZrO_2 = ZrO + O$	690800±37680	695000
NbO	$NbO = Nb + O$	766200±41870	—
NbO_2	$NbO_2 = Nb + 2O$	1423500±20900	—
	$NbO_2 = Nb + 2O$	—	(1339800±104700)
MoO	$MoO = Mo + O$	502400±41870	—
	$MoO = Mo + O$	485670±62800	489860
MoO_2	$MoO_2 = Mo + 2O$	—	1180680±20900
	$MoO_2 = MoO + O$	611270±62800	—
MoO_3	$MoO_3 = MoO_2 + O$	623800±41870	—
	$MoO_3 = Mo + 3O$	1754947	—
TcO_2	$TcO_2 = Tc + 2O$	—	(1067600±62800)
RuO	$RuO = Ru + O$	489860±41870	—
RuO_2	$RuO_2 = Ru + 2O$	—	(262960±62800)
	$RuO_2 = Ru + 2O$	977900±41870	—
RuO_4	$RuO_4 = Ru + 4O$	1821480±7960	—
RhO	$RhO = Rh + O$	392240±62800	—
RhO_2	$RhO_2 = Rh + 2O$	—	862480±20930
PdO	$PdO = Pd + O$	280500±29300	—
PdO_2	$PdO_2 = Pd + 2O$	—	<653140
AgO	$AgO = Ag + O$	238650±41870	—
CdO	$CdO = Cd + O$	276330	—
	$CdO = Cd + O$	≤368440	372630
In_2O	$In_2O = 2In + O$	774560±29300	—
InO	$InO = In + O$	318200	—
	$InO = In + O$	314010	318200
SnO	$SnO = Sn + O$	526700±8370	—
SbO	$SbO = Sb + O$	410300±83700	—
TeO	$TeO = Te + O$	263140±420	—
TeO_2	$TeO_2 = Te + 2O$	—	908540±29300
	$TeO_2 = Te + 2O$	486983±12140	—
IO	$IO = I + O$	184220±20930	188400

Oxide	Dissociation equation	Energy of dissociation at 0°K = $D_0 \cdot 10^{-3}$, J/kmole	Energy of dissociation at 298.15°K = $D \cdot 10^{-3}$, J/kmole
1	2	3	4
XeO	XeO=Xe+O	(32347)	—
CsO	CsO=Cs+O	307000±25000	—
Ba_2O	Ba_2O=BaO+Ba	—	389370±71180
	Ba_2O=2Ba+O	963000	—
BaO	BaO=Ba+O	561000±20900	—
	BaO=Ba+O	573590±8370	577780
	BaO=Ba+O	481480±9630	—
Ba_2O_2	Ba_2O_2=2BaO	—	372625±46055
La_2O	La_2O=La+LaO	314000±29300	—
	La_2O=2La+O	1113700±33500	—
LaO	LaO=La+O	805960	—
	LaO=La+O	799680±10470	—
La_2O_2	La_2O_2=2LaO	360100±41900	—
	La_2O_2=2La+2O	1959400±46000	—
CeO	CeO=Ce+O	810150	—
	CeO=Ce+O	774560±20930	—
Ce_2O_2	Ce_2O_2=2CeO	—	1984540±62800
CeO_2	$CeO_2=CeO+\frac{1}{2}O_2$	—	1465380±62800
PrO	PrO=Pr+O	715940±29300	—
	PrO=Pr+O	762000	—
NdO	NdO=Nd+O	713000	—
	NdO=Nd+O	690820±25120	—
SmO	SmO=Sm+O	596200	—
EuO	EuO=Eu+O	560600	—
GdO	GdO=Gd+O	724300	—
TbO	TbO=Tb+O	720100	—
DyO	DyO=Dy+O	629700±2500	—
HoO	HoO=Ho+O	639300	—
ErO	ErO=Er+O	634700±3350	—
TuO	TuO=Tu+O	582800±2900	—
YbO	YbO=Yb+O	370000±3350	—
LuO	LuO=Lu+O	699200	—
HfO	HfO=Hf+O	745250±41870	—
HfO_2	HfO_2=Hf+2O	—	(1507250±83700)
	HfO_2=Hf+2O	1335950±21350	±
TaO	TaO=Ta+O	769100±16750	±
	TaO=Ta+O	812200±46060	816430
	TaO=Ta+O	795490±41870	—
TaO_2	TaO_2=TaO+O	636390±46060	—
	TaO_2=Ta+2O	—	1469570±62800
	TaO_2=Ta+2O	1461310±20930	—
WO	WO=W+O	674080±41870	—
	WO=W+O	644800±41870	648960
WO_2	WO_2=WO+O	594530±62800	—

14 – Continued

Oxide	Dissociation equation	Energy of dissociation at 298.15°K = D · 10^{-3}, J/kmole	Energy of dissociation at 298.15°K = D · 10^{-3}, J/kmole
1	2	3	4
	$WO_2=W+2O$	—	1285350±41870
WO_3	$WO_3=WO_2+O$	615460±41870	—
	$WO_3=W+3O$	1892254±29730	—
$(WO_3)_2$	$(WO_3)_2=2W+6O$	4356250±41870	—
$(WO_3)_3$	$(WO_3)_3=3W+9O$	6805350±41870	—
$(WO_3)_4$	$(WO_3)_4=4W+12O$	9178914±46060	—
ReO_2	$ReO_2=Re+2O$	—	(1172300±62800)
OsO_2	$OsO_2=Os+2O$	—	(1067600±83700)
OsO_4	$OsO_4=Os+4O$	2106800±10890	—
IrO	$IrO=Ir+O$	351690±20900	—
IrO_2	$IrO_2=Ir+2O$	—	(983900±62800)
PtO	$PtO=Pt+O$	368376±20900	—
PtO_2	$PtO_2=Pt+2O$	—	891800±20900
Tl_2O	$Tl_2O=2Tl+O$	596297±7120	—
PbO	$PbO=Pb+O$	371750±5440	—
BiO	$BiO=Bi+O$	334940±83740	—
ThO	$ThO=Th+O$	801350	—
	$ThO=Th+O$	820600	—
ThO_2	$ThO_2=Th+2O$	1574200	—
	$ThO_2=Th+2O$	—	1549100±62800
PaO_2	$PaO_2=Pa+2O$	—	(1507250±83740)
UO	$UO=U+O$	749400±71180	753624
UO_2	$UO_2=UO+O$	674080	—
	$UO_2=U+2O$	—	1482130±62800
U_2O_4	$U_2O_4=2UO_2$	—	372630 (at 2450 K)
UO_3	$UO_3=UO_2+O$	414500	—

15. STANDARD HEAT CAPACITY [1; 9, Nos. 1–7; 19; 48; 78; 194]

Oxide	State	c_p^0 298,15, J/(mole·°C)	Oxide	State	c_p^0, 298,15, J/(mole·°C)
1	2	3	1	2	3
H_2O	l	75,350±0,042	F_2O	g	43,33±0,04
	g	33,600±0,013	FO	g	31,19±0,63
D_2O	l	84,36±0,13	Na_2O	s	72,43
	g	34,361±0,021	Na_2O_2	s	89,39
T_2O	g	34,99±0,021	NaO_2	s	72,18
HO	g	29,906±0,013	MgO	s	37,82
DO	g	29,961±0,013	Al_2O	g	46,05±0,84
TO	g	29,994±0,021	AlO	g	30,903±0,008
H_2O_2	l	89,39±0,84	Al_2O_3*3	s	79,09±0,17
	l	88,55	SiO	g	29,89±0,08
	g	43,17±0,84	SiO_2*4	s	44,46±0,21
D_2O_2	g	46,64±1,05			
HO_2	g	34,92±0,04	SiO_2*5	s	44,21±0,21
DO_2	g	36,00±0,04			
TO_2	g	36,80±0,08	SiO_2*6	s	44,63±0,21
H_2O_4	g	61,13±0,42	SiO_2	gl	44,41
HDO	g	33,838±0,021			
HTO	g	33,984±0,021	SiO_2*7	s	45,43±0,13
DTO	g	34,600±0,021			
HDO_2	g	44,80±1,26	SiO_2*8	s	43,00±0,13
Li_2O	s	54,26	SiO_2	g	43,54±1,67
BeO	s	25,45	PO	g	31,78±0,04
B_2O	g	39,77±0,84	P_4O_6	g	145,7±0,8
BO	g	29,22±0,04	PO_2	g	41,45±1,26
B_2O_2	g	57,36±0,4	P_4O_8	g	167,5±4,2
B_2O_3*1	s	62,80±0,29	P_4O_{10}	s	211,85±0,42
B_2O_3*2	s	62,80±0,84		s	204,94
B_2O_3	s	63,01		g	190,92±4,2
	g	66,15±0,84	S_2O	g	44,30±0,25
BO_2	g	43,04±0,21	SO	g	30,19±0,04
C_2O	g	43,12±0,42	SO_2	g	39,90±0,04
CO	g	29,132±0,004	SO_3	g	50,70±0,08
CO_2	g	37,14±0,04	Cl_2O	g	45,43±0,04
CO_3	g	49,82±0,84	ClO	g	31,37±1,26
N_2O	g	38,64±0,04	ClO_2	g	42,00±0,04
NO	g	29,89±0,04	KO_2	g	77,58
N_2O_3	g	65,31±0,84	CaO	s	50,45
NO_2	g	37,51±0,13		s	42,83
N_2O_4	l	142,77	Sc_2O_3	s	88,34
	g	78,7±0,8	TiO	s	39,98±0,21
N_2O_5	s	143,19		g	33,08±0,42
	g	108,02	$TiO_{1,01}$	s	42,29±0,42
	g	95,04±2,9			
NO_3	g	46,9±2,9	Ti_2O_3	s	95,92±0,42
O	g	21,926±0,008	Ti_3O_5	s	151,48±0,63
O_2	g	29,370±0,008	Ti_4O_7	s	215,87±0,63
O_3	g	39,27±0,13			

15 — Continued

1	2	3
Ti_5O_9	s	275,91±0,84
Ti_6O_{11}	s	301,45±0,84
Ti_7O_{13}	s	351,69±4,19
Ti_8O_{15}	s	401,93±4,19
Ti_9O_{17}	s	452,17±4,19
$Ti_{10}O_{19}$	s	502,42±1,26
TiO_2*9	s	55,06±0,21
TiO_2*10	s	55,52±0,21
TiO_2	g	43,96±0,84
$VO_{0,86}$	s	35,67±0,42
VO	s	38,60±0,42
	g	30,56±0,84
$VO_{1,24}$	s	46,18±0,42
$VO_{1,30}$	s	50,45±0,42
V_2O_3*11	s	103,29±0,63
V_4O_7	s	229,44±0,04
V_2O_4	s	114,72±0,84
V_2O_5*12	s	127,74±0,63
Cr_2O_3	s	118,91±1,26
CrO_3	s	58,20±2,09
MnO*13	s	44,13±0,21
	g	31,65±0,21
Mn_3O_4*14	s	139,42±0,42
Mn_2O_3	s	107,60±0,42
MnO_2*15	s	54,05±0,21
$Fe_{0,947}O$*16	s	48,15±0,42
FeO	s	49,95±0,42
	g	31,40±0,13
$FeO_{1,056}$	s	50,83±0,42
Fe_3O_4*17	s	150,89±0,84
Fe_3O_4*18	s	150,31±0,84
Fe_3O_4*19	s	150,31±0,84
Fe_3O_4*20	s	150,31±0,84
Fe_2O_3*21	s	103,83±0,21
CoO	s	55,27±0,42
	g	33,49±2,09
Co_3O_4	s	122,88±0,42
NiO	s	44,34±0,42
	g	33,91±0,84
Cu_2O	s	63,68±0,21

1	2	3
	g	35,59±0,42
ZnO	s	40,28±0,21
	g	32,66±0,84
Ga_2O	g	48,15±1,67
GaO	g	32,20±0,08
Ga_2O_3*22	s	92,11±0,08
GeO	g	30,81±0,08
GeO_2*23	g	50,20±0,13
GeO_2*1	s	52,13±0,21
AsO	g	32,36±0,17
As_2O_3	s	95,71
As_4O_6*24	s	203,90±2,1
As_4O_6*22	s	221,9
As_2O_5	s	116,60±0,42
	s	117,65
SeO	g	31,15±0,08
SeO_2	g	42,71±0,42
BrO	g	32,15±1,51
SrO	s	44,56
	g	45,06
Y_2O_3	s	97,15
ZrO	g	33,91±1,67
ZrO_2*25	s	56,23±0,29
	g	45,22±0,84
NbO	s	41,29±0,42
	g	30,98±0,42
NbO_2	s	57,53±0,42
Nb_2O_5	s	132,18±0,42
MoO_2	s	56,02±0,42
MoO_3	s	75,08±0,42
	g	59,45±2,09
RuO	g	31,40±0,42
RuO_2	s	56,52±4,19
RuO_4	g	75,86±0,33
Rh_2O	s	73,28
RhO	s	48,16
	g	31,82±0,84
Rh_2O_3	s	103,83
PdO	s	31,41
	g	31,82±0,84
Ag_2O*26	s	65,90±0,21
Ag_2O*27	s	65,98±0,21
AgO	g	35,42±0,21
CdO	s	43,67±0,21
	g	33,91±0,84

1	2	3
CuO	s	42,33±0,21
In_2O_3	s	92,11±2,1
	s	93,78
SnO	s	44,39±2,51
	g	31,778±0,017
SnO_2	s	52,75±2,93
SbO	g	31,86±0,13
Sb_2O_3	s	101,34
Sb_4O_6[*24]	s	209,34±2,1
Sb_4O_6[*28]	s	224,0±2,1
Sb_2O_4	s	114,68±0,84
Sb_2O_5[*24]	s	117,69±1,26
TeO	g	31,82±0,42
TeO_2[*23]	s	64,06±2,10
	g	44,38±0,84
IO	g	32,91±1,67
BaO	s	45,31
	s	47,27
La_2O_3	s	101,34
CeO_2	s	63,23
Pr_2O_3	s	107,5
Pr_6O_{11}	s	387,80
Nd_2O_3	s	111,6
Sm_2O_3	s	117,2
Dy_2O_3	s	114,26
Er_2O_3	s	108,56
HfO	g	30,98±1,67
HfO_2	s	60,29±0,42
	g	45,22±0,84
TaO	g	30,56±0,08
TaO_2	g	44,38±0,84
Ta_2O_5	s	134,90±0,42
WO_2	s	55,77±0,42
$WO_{2,72}$[*29]	s	68,66±0,84
$WO_{2,90}$[*30]	s	72,01±0,84
In_2O	g	49,82±1,67
$WO_{2,96}$[*31]	s	73,27±0,84
WO_3	s	73,90±0,42
	g	59,45±2,09
$(WO_3)_2$	g	153,66
$(WO_3)_3$	g	223,16
$(WO_3)_4$	g	309,40
ReO	g	32,03±0,21
Re_2O_7	s	166,34±0,42
OsO	g	31,82±0,84
OsO_4	g	74,06±0,17
IrO	g	31,82±0,84
IrO_2	s	57,37
PtO	g	31,82±0,84
AuO	g	35,17±0,84
HgO[*32]	s	44,09±0,21
	g	34,33±0,84
Tl_2O	g	51,08±2,1
PbO[*32]	s	45,85±0,17
PbO[*33]	s	45,80±0,04
Pb_3O_4	s	146,96±1,26
Pb_2O_3	s	107,60±0,84
PbO_2	s	64,81±0,21
PbO_2[*32]	s	62,93
BiO	g	32,78±0,08
Bi_2O_3[*22]	s	113,88±1,26
PoO_2	s	61,55±4,19
ThO_2	s	61,96
	s	62,34
UO	s	46,89
U_3O_5	s	230,1
UO_2	s	72,43
	s	63,76
U_4O_9	s	293,54
U_3O_8	s	250,37
UO_3	s	85,00

*1 Hexagonal. *2 Amorphous. *3 Corundum. *4 α-Quartz. *5 α-Cristobalite. *6 α-Tridymite. *7 Coesite. *8 Stishovite. *9 Rutile. *10 Anatase. *11 Karelianite. *12 Shcherbinaite. *13 Manganosite. *14 Hausmannite. *15 Pyrolusite. *16 Wustite. *17 Magnetite. *18 1% cation vacancies. *19 2% cation vacancies. *20 4% cation vacancies. *21 α-Hematite. *22 Monoclinic. *23 Tetragonal. *24 Cubic. *25 Baddeleyite. *26 Macrocrystals, particle size ~ 0.02 mm. *27 Finely pulverized specimen. *28 Orthorhombic. *29 1/18 $W_{18}O_{49}$. *30 1/10 $W_{10}O_{29}$. *31 1/50 $W_{50}O_{148}$. *32 Red. *33 Yellow.

16. EQUATIONS OF MOLAR HEAT CAPACITY [1] ($c_p = a + bT - cT^{-2}$, J/(mole · °K))

Oxide	State	Coefficients in heat capacity equation: a	$b \cdot 10^3$	$c \cdot 10^{-5}$	Accuracy, %	Temperature interval, °K
1	2	3	4	5	6	7
H_2O	s	—0,197	140,26	—	—	⩽273
	l	46,89	30,02	—	—	298—373
	g	30,56	10,30	—	0,7	298—2750
D_2O	g	32,20	12,39	1,42	0,8	298—2200
T_2O	g	33,58	13,31	2,30	0,8	298—2000
HO	g	26,71	3,94	1,84	0,2	298—3000
DO	g	27,72	4,19	0,92	0,6	298—3000
TO	g	27,76	4,94	0,63	0,5	298—2200
H_2O_2	l	53,63	117,23	—	0,1	298—450
	g	49,15	14,32	9,13	0,5	298—1500
D_2O_2	g	53,59	14,57	10,05	0,4	298—1500
HDO	g	32,41	10,30	1,47	1	298—2750
HTO	g	31,65	12,06	1,17	0,7	298—2200
TO	g	32,20	13,57	1,51	0,7	298—2000
HDO_2	g	50,83	14,91	9,38	0,5	298—1500
Li_2O	s	62,59	25,48	14,16	—	298—1700
	s	47,73	22,61	—	—	298—2000
	l	87,92	—	—	—	2000—2500
	l	100,50	—	—	—	1700—5300
Li_2O_2	s	71,18	22,61	—	—	298—470
BeO	s	35,41	16,76	13,28	—	298—1200
BO	g	28,85	4,52	0,88	0,7	298—2500
B_2O_3	s	36,58	106,43	5,49	1	298—723
	am	9,55	176,26	—	2	298—723
	l	127,70	—	—	0,1	723—2000
C_3O_2	g	65,44	36,34	9,29	0,4	298—1000
CO	g	28,43	4,10	0,46	0,8	298—2500
CO_2*1	g	44,17	9,04	8,54	—	298—2500
N_2O	g	45,75	8,63	8,55	1	298—2000
NO	g	29,43	3,85	0,59	1	298—2500
NO_2	g	42,16	9,55	6,99	1	298—2000
N_2O_4	g	84,01	39,81	14,92	0,4	298—1000
O	g	20,85	0,0	—1,00	0,2	298—3000
O_2	g	29,98	4,19	1,67	0,8	298—3000
O_3	g	47,02	8,04	9,04	0,6	298—1500
F_2O	g	52,25	4,10	9,04	0,4	298—1500
Na_2O	s	65,78	22,63	—	0,7	298—1100
	l	92,11	—	—	—	1193—T_{decomp}
Na_2O_2	s	84,57	15,91	—	—	298—919
NaO_2	s	67,83	15,07	—	—	298—825
	l	96,30	—	—	—	825—1300
	g	62,80	—	—	—	1300—2500
MgO	s	45,47	5,01	8,74	—	298—2500

1	2	3	4	5	6	7
MgO*2	s	42,62	7,29	6,20	0,8	298—2100
MgO	—	56,02	0,86	—	—	2100—3075
MgO_2	s	50,66	10,05	—	—	298—361
AlO	g	34,42	1,84	3,64	0,5	298—2000
Al_2O_3	s	115,10	11,81	35,09	0,5	298—1800
Al_2O_3*3	s	109,36	18,372	30,434	—	298—2300
Al_2O_3	l	138,16	—	—	—	2320—2500
SiO	s	30,56	10,05	—	—	298—2500
	g	32,24	3,10	2,93	0,5	298—2000
SiO_2*4	s	47,01	34,36	11,31	0,1	298—848
SiO_2*5	s	60,38	8,13	—	0,1	848—1883
SiO_2*6	s	17,93	88,24	—	1,0	298—523
SiO_2*7	s	60,34	8,55	—	0,2	523—1993
SiO_2*8	s	13,70	103,91	—	0,2	298—390
SiO_2*9	s	57,15	11,06	—	0,7	390—1953
SiO_2*10	s	56,06	15,42	14,46	0,4	298—2000
SiO_2	l	83,73	—	—	—	T_m—2250
PO	g	32,95	2,51	1,72	0,6	298—2000
P_2O_3	l	144,44	—	—	—	298—448,5
	g	62,80	41,87	—	—	448,5—2500
PO_2	s	47,31	20,93	—	—	298—350
	l	83,74	—	—	—	350—T_{decomp}
P_2O_5	s	35,06	22,61	—	—	298—631
	g	154,07	—	—	—	< 1400
P_4O_{10}	s	70,13	452,17	—	0,1	298—631
	g	308,15	—	—	0,1	631—1500
SO	g	34,58	1,34	4,19	0,7	298—3000
SO_2	g	46,22	7,87	7,70	0,8	298—3000
SO_2*11	g	42,58	12,56	5,65	—	298—1800
SO_3*12	g	57,36	26,88	13,06	—	298—1200
SO_3	g	58,20	25,54	13,48	0,7	298—1500
Cl_2O	g	53,21	3,35	7,79	0,5	298—2000
ClO_2	g	48,32	7,54	7,75	0,6	298—1500
K_2O	s	66,57	26,80	—	—	298—980
	l	92,11	—	—	—	980—T_{decomp}
K_2O_2	s	87,09	22,61		—	298—763
	l	121,42	—	—	—	763—1800
	g	83,74	—	—	—	1800—2500
K_2O_3	s	79,97	97,13	—	—	298—703
	l	148,63	—	—	—	703—973
	g	83,74	20,93	—	—	973—2500

16 – Continued

1	2	3	4	5	6	7
KO_2	s	62,80	50,24	—	—	298—653
	l	100,48	—	—	—	653—T_{decomp}
CaO	s	48,40	4,53	6,54	0,3	298—2000
	s	53,76	1,72	—	—	1800—2860
	g	36,43	0,67	3,10	0,2	298—2000
ScO	g	34,42	1,84	3,64	0,5	298—2000
Sc_2O_3	s	97,08	23,63	—	—	298—2523
TiO*13	s	44,29	15,08	7,79	1,5	298—1264
TiO*14	s	49,65	12,57	—	0,1	1264—2000
TiO	g	35,09	1,09	3,64	0,7	298—5000
Ti_2O_3*15	s	30,63	224,25	—	0,8	298—473
Ti_2O_3*16	s	145,20	5,44	42,71	0,1	473—2000
	l	157,01	—	—	—	2400—2500
Ti_3O_5*17	s	148,62	12,36	—	0,1	298—450
Ti_3O_5*18	s	174,30	33,56	—	—	450—2450
	l	251,21	—	—	—	2450—2500
Ti_4O_7*19	s	238,65	40,82	—	—	1100—1323
TiO_2*20	s	71,76	4,10	14,65	—	298—1800
TiO_2*21	s	72,05	4,52	15,03	—	298—1300
	l	89,60	—	—	—	2128—2500
VO	s	47,39	13,48	5,28	0,3	298—2000
	l	60,71	—	—	—	2350—2500
	g	34,33	1,76	3,77	0,4	298—2000
V_2O_3*22	s	122,88	19,93	22,69	0,6	298—1800
	l	159,10	—	—	—	2243—2500
V_3O_5	s	150,72	125,60	—	—	298—2100
	l	232,79	—	—	—	2100—T_{decomp}
VO_2*23	s	62,63	—	—	—	298—345
VO_2*24	s	74,73	7,12	16,50	—	345—1818
	l	106,76	—	—	—	1818—2500
V_2O_4*25	s	125,23	—	—	0,1	298—345
V_2O_4*26	s	149,47	14,24	33,03	0,4	345—1818
	l	213,53	—	—	0,1	1818—2000
V_2O_5	s	194,85	16,33	55,35	—	298—943
	l	190,92	—	—	0,1	943—2000
	g	167,47	—	—	—	2325—2500
Cr_2O_3	s	108,88	16,75	—	—	273—2263
Cr_2O_3*27	s	119,45	9,21	15,66	—	350—1800
CrO_2	s	67,41	12,56	12,56	—	298—700
CrO_3	s	75,78	16,75	8,37	—	298—460
	l	113,04	—	—	—	460—1000

1	2	3	4	5	6	7
	g	83,74	—	—	—	1000—2500
MnO	s	46,52	8,12	3,69	—	298—1800
	l	56,52	—	—	—	2058—T_{decomp}
	g	35,38	1,26	3,64	0,4	298—2000
Mn_3O_4*28	s	145,03	45,30	9,21	0,5	298—1445
Mn_3O_4*29	s	210,18	—	—	0,1	1445—1800
	l	205,15	—	—	—	1863—2500
Mn_2O_3	s	103,54	35,09	13,52	0,1	298—1350
MnO_2	s	69,50	10,22	16,24	—	298—1120
FeO	s	38,81	20,10	—	—	298—1250
FeO*30	s	52,84	6,25	3,19	—	298—1600
	l	60,71	—	—	—	1651—2500
Fe_3O_4*31	s	91,61	201,80	—	0,5	298—900
Fe_3O_4*30	s	167,14	78,96	41,91	—	298—900
Fe_3O_4*32	s	200,97	—	—	0,1	900—1800
Fe_2O_3*33	s	91,61	201,80	—	—	298—950
Fe_2O_3*30	s	97,80	72,18	12,90	—	298—1000
Fe_2O_3*34	s	200,97	—	—	—	950—1050
Fe_2O_3*35	s	98,35	77,87	14,86	—	1050—1800
CoO	s	41,03	9,21	—	—	298—2078
CoO*36	s	48,32	8,54	1,67	0,6	298—2000
CoO	l	64,90	—	—	—	2078—2500
Co_3O_4	s	123,51	71,18	—	—	298—1240
Co_3O_4*36	s	129,12	71,51	23,95	0,5	298—1000
NiO*37	s	20,89	157,34	16,29	0,4	298—525
NiO*38	s	58,11	—	—	0,1	525—565
NiO*39	s	46,81	8,46	—	0,4	565—2000
	l	59,87	—	—	—	2230—T_{decomp}
	g	36,55	0,59	3,10	0,1	298—2000
Cu_2O	s	56,10	36,01	—	—	298—1503
Cu_2O*30	s	62,38	23,86	—	—	298—1200
	l	90,02	—	—	—	1503—T_{decomp}
CuO	s	60,04	25,96	—	—	298—1609
	l	92,11	—	—	—	1609—T_{decomp}
	g	36,51	0,59	3,06	0,2	298—2000
ZnO	s	49,03	5,11	9,13	0,4	298—2000
	g	35,17	1,42	3,43	0,3	298—2000
Ga_2O	s	150,72	—	—	—	298—925
	l	132,63	7,54	—	—	925—1000
	g	57,78	36,01	—	—	1000—1700
GaO	g	35,84	1,00	3,52	0,2	298—2000
Ga_2O_3	s	90,02	—	—	—	298—2013

16 — Continued

1	2	3	4	5	6	7
Ga_2O_3 *40	s	49,28	105,51	—	—	298—923
GeO	s	49,28	105,51	—	—	298—923
	g	34,25	1,93	3,56	0,5	298—2000
GeO_2 *41	s	43,54	10,89	2,09	—	298—1389
GeO_2 *30	s	46,89	30,02	—	—	298—1300
	l	34,33	1,67	0,84	—	1398—2500
AsO	g	40,78	1,26	7,20	0,4	298—2000
As_2O_3	s	35,04	203,48	—	—	298—548
	l	163,29	—	—	—	550—730
	g	90,02	—	—	—	730—2500
AsO_2	s	35,59	39,36	—	—	298—1200
	l	87,92	—	—	—	1200—T_{decomp}
As_2O_5	s	130,21	68,66	22,61	—	298—1100
SeO	s	38,10	15,91	—	—	298—1375
	l	64,90	—	—	—	1375—2075
	g	34,33	2,09	3,35	—	2075—2500
	g	34,96	1,51	3,68	0,4	298—2000
SeO_2	s	53,59	25,54	0,84	—	298—613
	g	60,71	—	—	—	613—2500
BrO	g	36,09	0,84	3,31	0,2	298—2000
Rb_2O	s	64,48	24,28	—	—	298—900
	l	92,11	—	—	—	900—T_{decomp}
Rb_2O_2	s	87,50	33,49	—	—	298—843
	l	121,42	—	—	—	843—T_{decomp}
Rb_2O_3	s	85,83	54,43	—	—	298—762
	l	142,35	—	—	—	762—T_{decomp}
RbO_2	s	57,78	26,80	—	—	298—685
	l	87,92	—	—	—	685—T_{decomp}
SrO	s	50,79	5,28	6,49	0,5	298—1800
	s	55,32	1,21	—	—	1200—2703
	—	67,00	—	—	—	2703—6000
	g	36,43	0,67	3,10	0,2	298—2000
SrO_2	s	70,34	9,21	12,56	—	298—488
YO	g	35,21	1,34	3,52	0,4	298—2000
Y_2O_3	s	108,86	34,33	9,21	—	298—2500
ZrO_2 *42, *43	s	69,67	7,54	14,07	—	298—1478
ZrO_2 *42	s	57,80	16,68	—	1	1000—1420
ZrO_2 *44	s	78,63	—	—	1	1420—2500
	s	74,54	—	—	—	1478—2950
	—	100,50	—	—	—	2950—6000
NbO	s	40,19	18,42	—	—	298—2218
NbO_2	s	71,59	6,70	11,72	—	298—2275
	l	100,48	—	—	—	2275—2500
Nb_2O_5	s	91,61	118,07	—	—	298—1783
Nb_2O_5 *45	s	151,69	23,20	20,43	—	298—1795

1	2	3	4	5	6	7
	l	185,06	—	—	—	1783—2500
MoO_2	s	67,83	12,56	12,56	—	298—2200
	l	96,30	—	—	—	2200—2250
MoO_3	s	56,94	56,52	—	—	298—1068
MoO_3*36	s	86,79	21,69	17,50	—	298—1068
	l	118,91	—	—	—	1068—1530
	g	75,78	—	—	—	1530—2500
TcO_2	s	43,54	38,52	—	—	298—2400
	l	104,67	—	—	—	2400—2500
TcO_3	s	81,22	21,77	8,37	—	298—1200
Tc_2O_7	s	163,70	77,87	10,05	—	298—392,7
	l	267,96	—	—	—	392,7—583,8
	g	104,67	117,23	—	—	583,8—2500
RuO_2	s	47,73	25,12	—	—	298—1400
RuO_4	s	83,74	—	—	—	298—300
	l	138,16	—	—	—	300—T_{decomp}
Rh_2O	s	65,27	27,09	—	—	273—1273
RhO	s	41,20	23,15	—	—	273—1273
Rh_2O_3	s	86,79	57,78	—	—	273—1273
PdO	s	13,82	59,45	—	—	298—823
Ag_2O	s	46,60	64,81	—	0,2	298—460
Ag_2O*46	s	55,52	29,48	—	—	298—460
AgO	g	37,01	0,25	2,26	0,1	298—2000
Ag_2O_2	s	68,66	51,08	—	—	298—T_{decomp}
CdO*1	s	40,40	8,71	—	—	273—1200
	s	40,95	8,46	—	—	298—2086
In_2O	s	61,55	32,66	—	—	298—600
	l	92,11	—	—	—	600—800
	g	62,80	—	—	—	800—2500
InO	s	41,87	13,40	—	—	298—1325
	l	58,62	—	—	—	1325—2000
	g	37,68	—	—	—	2000—2500
	g	36,30	0,67	3,43	0,2	298—2000
In_2O_3	s	94,62	25,12	—	—	298—2000
In_2O_3*27	s	93,78	—	—	—	273—373
	l	146,54	—	—	—	2273—2500
SnO	s	39,36	15,16	—	—	273—1273
	l	60,71	—	—	—	1315—1800
	g	37,68	—	—	—	1800—2500
	g	35,25	1,34	3,52	0,4	298—2000
SnO_2	s	73,95	10,05	21,61	0,8	298—1500
	l	94,20	—	—	—	1898—2273
SbO	g	35,46	3,52	4,14	0,9	298—2000
Sb_2O_3	s	79,98	71,61	—	—	273—928
	l	150,72	—	—	—	928—1698
	g	87,09	—	—	—	1698—2500
SbO_2	s	47,31	33,91	—	—	298—1198
Sb_2O_4	s	94,62	67,83	—	—	298—1198

16 — Continued

1	2	3	4	5	6	7
Sb_2O_5	s	93,78	98,81	—		298—T_{decomp}
TeO	g	35,34	1,34	3,48	0,3	298—2000
TeO_2	s	57,99	28,76	—	—	298—1006
	l	83,74	—	—	—	1006—T_{decomp}
Cs_2O	s	69,08	22,61	—	—	298—763
	l	92,11	—	—	—	763—T_{decomp}
Cs_2O_2	s	89,60	47,73	—	—	298—867
	l	123,51	—	—	—	867—T_{decomp}
Cs_2O_3	s	100,48	94,62	—	—	298—775
	l	146,54	—	—	—	775—T_{decomp}
Ba_2O	s	83,74	9,21	—	—	298—880
	l	92,11	—	—	—	880—1040
	g	62,80	—	—	—	1040—2500
BaO	s	53,34	4,36	8,31	—	298—2196
	l	58,20	—	—	—	2283—3000
	—	67,00	—	—	—	2283—6000
	g	36,09	0,84	2,81	0,3	298—2000
BaO_2	s	56,94	8,37	—	—	298—723
	l	87,92	—	—	—	723—1110
LaO	g	35,42	1,26	3,52	0,3	298—2000
La_2O_3	s	120,83	12,88	13,71	—	298—2573
CeO	g	35,25	1,34	3,77	0,3	298—2000
Ce_2O_3	s	96,30	37,68	—	—	298—1960
Ce_2O_3 [*47]	s	139,25	11,22	24,62	—	298—1050
	s	126,86	32,32	9,29	—	1050—1800
Ce_2O_3	*l*	154,91	—	—	—	1963—2500
CeO_2	s	62,81	10,47	—	—	298—2500
Pr_2O_3	s	121,42	16,75	16,75	—	298—2200
	l	151,72	—	—	—	—
Pr_6O_{11}	s	399,04	109,72	38,99	0,1	298—1200
PrO_2	s	73,69	14,24	11,72	—	298—700
Nd_2O_3	s	121,25	24,12	14,42	0,1	298—1500
Nd_2O_3 [*47]	s	115,85	29,81	11,89	—	298—1395
	s	155,75	—	—	—	1395—2000
Sm_2O_3 [*48]	s	134,81	5,44	—	—	273—1273
Sm_2O_3	s	108,44	29,31	—	—	298—2150
Sm_2O_3 [*49, *47]	s	128,28	21,27	16,58	—	298—1150
Sm_2O_3 [*50, *47]	s	128,74	19,43	18,00	—	298—1195
Sm_2O_3 [*50, *47]	s	154,49	—	—	—	1195—2000
Sm_2O_3	*l*	150,72	—	—	—	2593
EuO [*51]	s	52,50	8,71	—	—	298—1760
Eu_2O_3 [*52, *47]	s	136,91	14,74	16,41	—	298—1371
Eu_2O_3 [*53, *47]	s	123,93	27,13	8,71	—	298—895

1	2	3	4	5	6	7
Eu_2O_3[*53, *47]	s	130,04	17,42	—	—	895—1800
Eu_2O_3[*54]	s	147,7	—	—	—	273—873
	s	139,5	—	—	—	273—1073
Gd_2O_3	s	108,04	15,08	—	—	273—1273
Gd_2O_3[*55]	s	127,5	—	—	—	273—1273
Gd_2O_3[*56, *47]	s	120,24	11,89	16,24	—	298—1550
Gd_2O_3[*57, *47]	s	114,22	14,82	10,63	—	298—2000
Dy_2O_3[*58, *47]	s	122,88	13,23	8,46	—	298—1550
	s	148,21	—	—	—	1550—1800
Dy_2O_3[*59]	s	139,0	—	—	—	273—1073
Ho_2O_3[*60, *47]	s	121,12	10,89	10,05	—	298—2000
Er_2O_3[*61, *47]	s	124,18	8,37	18,84	—	298—2000
Er_2O_3	s	104,08	—	—	—	273—373
Er_2O_3[*62]	s	129,50	3,27	19,47	—	—
Tu_2O_3[*63, *47]	s	129,79	3,27	14,32	—	298—1680
	s	133,98	—	—	—	1680—2000
Yb_2O_3[*64, *47]	s	136,41	—2,60	19,34	—	298—1365
	s	134,81	—	—	—	1365—2000
Lu_2O_3[*65, *47]	s	123,01	7,37	24,53	—	298—1800
HfO_2	s	72,81	8,71	14,57	0,3	298—2000
Ta_2O_5	s	122,25	41,87	—	—	298—2145
Ta_2O_5[*36]	s	154,91	27,47	24,79	0,4	298—2000
Ta_2O_5	l	192,59	—	—	—	2145
WO_2	s	73,69	17,58	16,75	—	298—1843
	l	100,48	—	—	—	1843—2125
WO_3	s	72,56	32,41	—	—	298—1743
	l	125,60	—	—	—	1743—2100
	g	75,36	—	—	—	2100—2500
ReO_2	s	45,22	41,03	—	—	298—1475
	l	102,58	—	—	—	1475—2500
ReO_3	s	75,36	24,28	—	—	298—433
	l	121,42	—	—	—	433—T_{decomp}
Re_2O_7	s	175,01	61,96	12,56	—	298—569
	l	275,07	—	—	—	569—635,5
	g	159,94	—	—	—	635,5—2500
ReO_4	s	89,60	45,22	8,38	—	298—420
	l	138,16	—	—	—	420—460
	g	69,08	36,01	20,93	—	460—2500
OsO_2	s	48,15	25,12	—	—	298—923
OsO_4	s	68,66	96,72	10,05	—	298—313,3
	l	138,12	—	—	—	313,3—403

16 – Continued

1	2	3	4	5	6	7
	g	68,91	36,01	19,26	—	403—1000
	g	86,04	20,43	15,99	0,3	298—1000
Ir_2O_3	s	91,27	60,29	—	—	298—1450
	l	146,54	—	—	—	1450—2250
	g	83,74	41,87	—	—	2250—2500
IrO_2	s	38,39	63,64	—	0,5	298—1300
PtO	s	37,68	26,80	—	—	298—780
Pt_3O_4	s	128,95	72,85	—	—	298—T_{decomp}
PtO_2	s	46,47	40,19	—	—	298—723
	l	87,92	—	—	—	723—750
Au_2O_3	s	98,39	20,10	—	—	298—T_{decomp}
Hg_2O	s	30,02	10,72	0,33	—	298—T_{decomp}
HgO	s	72,81	8,71	14,57	—	298—T_{decomp}
HgO[*66, *27]	s	45,76	—	—	—	298—371
Tl_2O	s	66,15	25,12	1,26	—	298—573
	l	92,53	—	—	—	573—773
	g	57,36	—	—	—	773—2500
Tl_2O_3	s	96,30	20,93	—	—	298—990
	l	148,63	—	—	—	990—T_{decomp}
PbO[*67]	s	37,89	26,80	—	0,2	298—1159
PbO[*66]	s	44,39	16,75	—	0,1	298—900
	l	61,13	—	—	—	1160—1745
	g	33,91	1,67	—	—	1745—2500
	g	35,88	1,00	3,31	0,2	298—2000
Pb_3O_4	s	130,21	73,69	—	—	298—T_{decomp}
PbO_2	s	53,17	32,91	—	—	298—1000
BiO	s	40,61	12,56	—	—	298—1175
	l	58,62	—	—	—	1175—1920
	g	37,26	—	—	—	1920—2500
	g	36,13	0,84	3,31	0,2	298—2000
Bi_2O_3	s	103,60	33,50	—	0,3	298—800
	l	149,47	—	—	—	1090
PoO_2	s	59,87	23,45	—	—	298—825
	l	92,11	—	—	—	825—T_{decomp}
RaO	s	43,96	8,38	—	—	298—2500

1	2	3	4	5	6	7
Ac_2O_3	s	83,74	85,41	—	—	298—2250
	l	167,47	—	—	—	2250—2500
ThO	s	46,05	10,05	—	—	298—2150
	l	62,80	—	—	—	2150—2500
	s	66,32	12,06	6,70	0,4	298—2000
ThO_2	s	60,29	10,89	—	—	298—2500
PaO_2	s	118,91	47,73	—	—	298—2050
Pa_2O_5	l	200,97	—	—	—	2050—2500
UO	s	44,38	8,37	—	—	298—1500
U_3O_5	s	272,14	31,40	45,64	—	298—900
UO_2	s	80,39	6,78	16,58	0,1	298—2000
U_3O_8	s	250,37	—	—	—	275—315
U_3O_8[•12]	s	238,02	—	—	—	373—593
UO_3	s	92,49	10,63	12,43	0,1	298—1000
NpO_2	s	74,11	13,40	10,89	—	298—2600
Np_2O_5	s	135,65	52,75	—	—	298—800
PuO	s	50,24	10,05	—	—	298—1290
	l	60,71	—	—	—	1290—2325
	g	37,26	—	—	—	2325—2500
Pu_2O_3	s	88,76	76,20	—	—	298—2513
	l	167,47	—	—	—	2513
PuO_2	s	71,59	14,24	10,89	—	298—2663
	l	85,83	—	—	—	2663
Am_2O_3	s	83,74	65,31	—	—	298—2225
	l	161,19	—	—	—	2225—2500
AmO_2	s	58,62	28,47	—	—	298—T_{decomp}

[•1] [78, p. 12]. [•2] [78, p. 16]. [•3] [19, p. 19]. [•4] α-Quartz. [•5] β-Quartz. [•6] α-Cristobalite. [•7] β-Cristobalite. [•8] α-Tridymite. [•9] β-Tridymite. [•10] Glassy. [•11] [78, p. 18]. [•12] [78, p. 20]. [•13] α-TiO. [•14] β-TiO. [•15] α-Ti_2O_3. [•16] β-Ti_2O_3. [•17] α-Ti_3O_5. [•18] β-Ti_3O_5. [•19] [91]. [•20] Rutile [78, p. 20]. [•21] Anatase [78, p. 20]. [•22] Karelianite. [•23] α-VO_2. [•24] β-VO_2. [•25] α-V_2O_4. [•26] β-V_2O_4. [•27] [78, p. 14]. [•28] α-Mn_3O_4. [•29] β-Mn_3O_4. [•30] [78, p. 14]. [•31] α-Fe_3O_4. [•32] β-Fe_3O_4. [•33] α-Fe_2O_3. [•34] β-Fe_2O_3. [•35] γ-Fe_2O_3. [•36] [1, p. 448, Ref. 406]. [•37] α-NiO. [•38] β-NiO. [•39] γ-NiO. [•40] [19, p. 68]. [•41] α and β. [•42] α-ZrO_2. [•43] [78, p. 20]. [•44] β-ZrO_2. [•45] [19, p. 156]. [•46] [78, p. 10]. [•47] [156, p. 58]. [•48] [77, p. 294]. [•49] Sm_2O_3–C. [•50] Sm_2O_3–B. [•51] [189]. [•52] Eu_2O_3–C. [•53] Eu_2O_3–B. [•54] [19, p. 185]. [•55] [19, p. 187]. [•56] Gd_2O_3–C. [•57] Gd_2O_3–B. [•58] Dy_2O_3–C. [•59] [19, p. 191]. [•60] Ho_2O_3–C. [•61] Er_2O_3–C. [•62] [48]. [•63] Tu_2O_3–C. [•64] Yb_2O_3–C. [•65] Lu_2O_3–C. [•66] Red. [•67] Yellow.

17. HEAT CAPACITY c_p AT LOW TEMPERATURES [1, 19] (Symbol I in column 2 denotes values in J/(kg · °K); symbol II denotes values in J/(mole · °K)

Oxide	Code for units of measure	Heat capacity c_p at indicated temperature, °K						
		10	25	50	100	150	200	298,15
1	2	3	4	5	6	7	8	9
H_2O	I	—	171,66	435,43	883,42	1222,55	1565,86	4190,99 (l)
D_2O	I	—	175,85	431,24	891,79	1343,96	1766,83	4345,90 (l)
BeO	I	0,837	3,349	16,75	117,23	318,20	573,59	1017,39
CO	I	117,23	816,43	1549,12	1042,51 (g)	1042,51 (g)	1042,5 (g)	1042,5 (g)
CO_2	I	17,17	203,49	581,97	904,35	1092,72	738,88 (g)	845,73 (g)
N_2O	I	21,77	216,46	623,83	946,22	1159,74	762,00 (g)	879,23 (g)
NO	I	38,94	329,50	711,76	1201,61	1038,33 (g)	1013,21 (g)	996,46 (g)
N_2O_4	I	15,07	146,54	400,68	661,93	835,27	999,39	1547,02 (l)
N_2O_5	I	—	—	—	198,04	838,62	1009,02	1323,03
O_2	I	79,55	695,01	1440,26	908,54 (g)	908,54 (g)	908,54 (g)	908,54 (g)
O_3	I	—	—	—	693,48 (l)	—	—	817,35 (g)
MgO	I	<0,42	2,93	23,03	211,85	464,74	678,26	937,84
	II	(0,0)	(0,13)	0,92	8,54	18,8	27,3	37,8
Al_2O_3	I	0,42	2,09	18,42	132,30	328,66	509,12	774,98
	II	(0,04)	(0,20)	1,9	13,4	33,5	51,92	79,13
SiO_2*1	I	2,09	26,38	96,30	259,58	416,59	545,54	741,06
SiO_2*2	I	3,35	40,61	108,86	265,02	420,77	548,47	741,06
SiO_2*3	I	3,35	40,61	111,79	272,14	427,05	556,84	743,99
SiO_2*4	I	4,19	38,94	113,88	269,21	411,14	540,93	739,81
SO_2	I	18,42	178,36	481,48	748,60	903,51	1314,66 (l)	628,02 (g)
KO_2*13	II	1*5	2,2*6	2,8*7	1,9*8	2,2*9	12,3*10	1,8*11
		1,8*12	—	—	—	—	—	—
CaO	I	< 0,42	4,19	46,06	288,89	485,67	619,65	900,16
	II	(0,00)	0,34	2,51	16,2	27,2	34,8	42,83
TiO_2	I	—	242,83*14	1787,76	6472,79	10718,21	14025,78*15	18254,45*16
	II	(0,084)	(1,05)	(6,36)	19,2	31,95	43,34	56,48
V_2O_3	I	0,84	9,63	51,92	191,76	407,79	538,84	689,15
	II	(0,1256)	(1,424)	(7,79)	28,7	61,0	80,7	103,3
V_2O_4	I	0,84	10,05	58,62	222,32	401,51	511,21	706,31
	II	(0,1256)	(1,675)	(9,71)	36,84	66,49	89,81	117,1

V_2O_5	I	1,26	18,00	104,67	285,96	442,55	556,01	701,29
	II	(0,2094)	(3,26)	(19,0)	52,0	80,6	101,2	127,7
Cr_2O_3	I	0,42	3,85	30,31	159,10	338,71	495,30	794,66
	II	(0,0837)	(0,586)	(4,606)	24,2	51,1	75,4	118,8
MnO	I	25,12	31,82	147,79	464,74	473,11	535,91	607,09
	II	0,1675	2,261	10,47	32,99	33,41	38,10	43,00
Mn_3O_4	I	1,26	19,68	101,74	283,45	431,66	538,00	659,00
	II	0,2512	4,145	21,69	60,33	91,90	114,60	139,40
MnO_2	I	1,26	18,00	93,37	287,21	386,44	498,65	658,17
	II	0,1256	1,549	8,123	24,95	33,58	43,29	57,24
FeO	I	1,26	13,82	84,57	341,22	582,80	683,70	725,15
	II	(0,0837)	(1,005)	(6,07)	(24,5)	(41,87)	49,11	52,04
Fe_3O_4	I	1,26	15,49	67,41	245,35	404,86	503,67	620,90
	II	(0,251)	(3,56)	(15,58)	56,65	92,82	116,6	143,6
Fe_2O_3	I	1,26	17,58	74,94	200,97	352,53	478,97	525,44
	II	(0,21)	(3,29)	11,97	32,03	56,19	76,50	104,8
NiO	I	0,42	6,28	43,54	192,17	329,50	452,17	594,53
	II	(0,04)	(0,46)	(3,3)	14,4	24,6	33,7	44,4
Cu_2O	I	16,75	101,32	190,92	278,00	332,01	374,72	429,15
	II	(2,39)	(14,5)	(27,4)	39,8	47,40	53,59	61,51
CuO	I	0,84	14,65	79,55	210,18	329,08	443,80	531,72
	II	(0,084)	(1,17)	6,32	16,7	26,2	35,4	42,25
ZnO	I	0,84	18,42	95,46	218,13	317,78	396,91	494,04
	II	(0,084)	(1,51)	(7,79)	17,8	25,9	32,3	40,3
Ga_2O_3[*17]	II	0,347[*13]	1,327	9,676	32,71	53,34	69,50	91,90
GeO_2[*19]	II	0,0031[*20]	0,0061[*21]	0,0106[*22]	0,0305[*23]	0,0773[*24]	0,217[*25]	0,441[*26]
		0,791[*27]	1,426[*28]	2,064[*29]	2,762[*30]	3,196[*31]	—	—
As_2O_3	I	5,44	45,64	101,32	205,57	308,15	386,44	481,06
As_2O_5	I	0,42	9,21	47,73	172,08	288,89	380,16	508,28
RbO_2	II	5,359[*32]	6,104[*11]	7,432[*12]	10,011[*33]	8,039[*13]	7,666[*34]	7,783[*35]
		9,475[*36]	13,825	33,264	56,02	65,40[*37]	69,21[*38]	77,62
SrO	I	1,26	15,49	84,57	233,62	330,34	386,44	434,59
	II	(0,126)	(1,591)	(8,75)	24,20	34,21	39,98	44,80
Y_2O_3	II	1,35[*18]	3,253[*39]	11,75	39,32	58,45[*10]	81,10	102,6
Ag_2O	I	19,51	95,04	152,82	195,94	225,25	252,05	284,70
CdO	I	0,84	16,75	78,29	189,24	262,51	306,06	328,66
	II	(0,1256)	(0,214)	(9,96)	24,1	33,4	38,98	43,5
SnO	I	1,67	18,84	83,74	182,54	246,60	289,73	329,50
SnO_2	I	0,42	7,12	43,12	138,58	218,55	280,52	349,60
	II	(0,084)	(1,05)	(6,49)	20,85	32,91	42,29	52,63

17 – Continued

1	2	3	4	5	6	7	8	9
Sb_2O_3	I	3,31	31,02	79,13	172,50	241,16	292,66	347,92
Sb_2O_4	I	3,56	31,49	73,27	151,98	235,30	274,65	372,63
Sb_2O_5	I	0,84	13,31	51,08	164,12	239,07	294,33	363,41
BaO	II	0,2931	4,44	15,95	32,0	40,0	43,8	47,5
La_2O_3	II	2,20*13	3,60*36	22,94	54,18	79,89*41	91,53	108,0
Nd_2O_3	II	8,324*13	10,02*36	27,97	57,53	83,24*41	94,84	111,3
Sm_2O_3	II	1,281	8,072	27,29	57,91	80,43	95,97	114,6
Gd_2O_3	II	4,149	5,179	19,32	47,31	70,47	86,88	105,6
Ta_2O_5	I	0,84	11,47	50,24	131,47	195,52	243,25	306,06
	II	0,38	5,07	22,15	58,1	86,6	107,6	135,2
HgO	I	3,68	37,47	83,32	133,14	162,03	178,36	211,43
PbO	I	4,31	36,63	74,94	126,86	160,77	184,64	217,71
Pb_3O_4	I	—	—	—	188,83	244,09	283,03	317,36
PbO_2	I	1,26	18,21	63,64	132,30	182,96	219,81	270,89
Bi_2O_3	I	4,61	26,50	65,31	135,23	175,01	207,67	244,09
ThO_2	II	0,134	2,06	10,17	26,20	40,00	50,1	61,8
UO_2	II	—	—	—	29,13	—	52,21	64,40*42

*1 α-Quartz. *2 Cristobalite. *3 Tridymite. *4 Fused. *5 At 3°K. *6 At 5°K. *7 At 7°K. *8 At 8°K. *9 At 10°K. *10 At 12.5°K. *11 At 13°K. *12 At 14°K. *13 In interval 14 –20°K, $c_p = 4.3 \cdot 10^{-4} \cdot T^3 + 120/T^2$ [165]. *14 At 24°K. *15 At 201°K. *16 At 297.7°K. *17 β-Ga_2O_3, containing 98.67% Ga_2O_3. *18 At 16°K. *19 Fused, density 3609 kg/m³. *20 At 2.357°K. *21 At 2.964°K. *22 At 3.469°K. *23 At 4.480°K. *24 At 5.468°K. *25 At 7.444°K. *26 At 9.486°K. *27 At 11.823°K. *28 At 15.147°K. *29 At 17.991°K. *30 At 20.985°K. *31 At 22.737°K. *32 At 12°K. *33 At 15°K. *34 At 17°K. *35 At 18°K. *36 At 20°K. *37 72.77 at 178.3°K. *38 73.86 at 250°K. *39 At 30°K. *40 At 140°K. *41 At 160°K. *42 At 300°K.

18. CHARACTERISTIC TEMPERATURE [1; 40; 45; 50; 55; 56; 61; 65; 145; 165; 166; 173]

Oxide	Debye temperature, °K	Characteristic temperature, °K	
		for metal	for oxygen
BeO	1553	—	—
	830*1	—	—
O_2	—	—	62,34
MgO	630*1	—	—
Al_2O_3	590*1	—	—
KO_2	208	—	—
TiO*2	612	—	—
TiO	410	—	—
Ti_2O_3	674±12	—	—
VO	614	—	—
VO_2	750±20	—	—
CrO_2	260	—	—
GeO_2*3	870±30	—	—
Y_2O_3	451	—	—
	465	—	—
ZrO_2	340*1	—	—
MoO_2	—	309,8*4	401,6*4
MoO_3	—	213,4*4	276,7*4
Ag_2O	90	—	—
La_2O_3	433	—	—
Sm_2O_3	426	—	—
Er_2O_3	458	—	—
WO_2	—	191,1*4	343,1*4
W_2O_5	—	199,4*4	359,2*4
WO_3	—	196,6*4	354,1*4
ReO_3	—	106,45*4	176,2*4
Re_2O_7	—	122,02*4	202,0*4
PbO	~90	—	—
ThO_2	285*1	—	—

*1Calculated values, characteristic temperature. *2Cubic. *3Tetragonal. *4Calculated values.

Thermal conductivity λ, W/(m · °K)	Temperature, °K
1	2
$H_2O(l)$ [1]	
~2,3446	273,15
$H_2O(l)$*1 [1]	
0,55126	273
0,59894	293
0,63384	313
0,65942	333
0,67454	353
0,68036	363
$H_2O(v)$*1 [1]	
0,023725	373
BeO [1]*2,*3	
0,105	3
0,209	5
2,931	10
18,84	20
125,60	50
418,68	100
[19]*4	
234,2	323
[1]*4	
219,807	373
92,947	673
27,005	1073
17,250	1473
15,993	1873
15,449	2073
[77]*5	
209,340	373
19,259	1273
[77]*6	
209,340	373
88,341	673
25,749	1073
16,454	1473
14,486	1873
13,900	2073
[90]	
$\lambda = 23194,87/(T-125) + 3558,78 \cdot 10^{-36} T^{10}$	1273—2073
CO [148]*7	
0,637	22,5
0,569	25
0,479	30
0,416	35
0,317	40
0,334	45
0,303	50
0,273	55
CO (g) [1]	
0,006908	82
0,02152	273
0,02746	373
0,03835	573
0,04953	773
0,06021	973
0,06979	1173
0,07423	1273
[81, Vol. 5]*1	
0,02219	373

1	2
CO_2 (g) [1]	
0,01066	194,5
0,01373	273
0,02139	373
0,03672	573
0,05129	773
0,06452	973
0,07662	1173
0,08215	1273
N_2O (g) [1]	
0,01135	201,35
0,01472	273
0,02119	373
NO (g) [1]	
0,01742	201,75
0,02324	273
NO_2 (g) [1]	
0,03718	328
O_2 (g) [1]	
0,007205	81,75
0,017970	194,75
0,02466	273
0,03291	373
0,04802	573
0,06150	773
0,07285	973
0,08206	1173
0,08583	1273
MgO*10 [147]*2,*8	
	80
122	100
115	200
67	400
40	600
25	800
18	1000
16	1100
15	

1	2
[19]*4,*9	
36,0	373
16,5	673
8,5	1073
6,11	1473
6,87	1873
(9,46)	2073
[19]*11	
34,46	373
15,8	673
8,12	1073
5,86	1473
6,57	1873
(9,04)	2073
14,0*12	2273
[90]	
$\lambda = 7871,18/(T - 125) + 3558,78 \cdot 10^{-36} T^{10}$ or $\lambda = 8499,20/T + 1256,04 \cdot 10^{-29} T^{8}$	293—2073
Al_2O_3*17 [19]*13	
16	2
200	5
1100	10
3500	20
5500	50
400	100
40	298
[19]*4,*14	
30,3	373
13,15	673
7,20	1073
5,53	1473
6,07	1873
7,58	2073

19 — Continued

1	2
[77][*15]	
28,889	373
12,560	673
6,866	1073
5,275	1473
5,778	1873
9,0[*16]	2273
[90]	
$\lambda = 6782,62/(T-125) + 3558,78\cdot10^{-36}T^{10}$	673—2073
[169]	
23[*18]	296
17[*18]	350
25[*19]	299
17,2[*19]	343
[162][*20]	
6,0	1173
5,0±0,8	1223
10,0	1373
11,0	1473
18,0	1573
30,0±8	1673
SiO_2 [19][*21]	
0,05	2
0,12	5
0,13	10
0,15	20
0,3	50
0,6	100
1,0	Room temp.
[19][*22]	
810	8
1100	16
450	30
50	55
20	120
[1]	
13,61[*19]	273
9,00[*19]	373
7,247[*18]	273
5,581[*18]	373
[19][*23]	
0,343	50
0,686	100
1,12	200
1,38	300
[19][*2, *4, *24]	
1,55	373
1,90	673
2,85	973
5,20	1273
9,30	1473
[38][*23, *25]	
1,456	385
1,549	415
1,546	431
1,605	484,5
1,601	509
1,668	531
1,723	576
1,742	585
1,764	594
1,820	626
1,975	676
1,979	676,5
1,963	680
2,099	721
2,350	751,5
2,318	768,5
2,218	769,5
2,709	811,5
2,624	820,5
3,106	858
2,960	882,5
3,644	907
3,653	930
4,349	962,5

1	2
5,379	1026,5
5,943	1078
6,601	1090,5
[142]*23, *26	
1,88	680
2,06	790
2,22	865
2,60	1007
3,02	1137
[142]*23, *27	
1,73	680
1,80	790
1,83	865
1,95	1007
2,06	1137
[210]	
7,16±0,16*28	298
6,15±0,16*29	298
[19]*30	
0,44—0,82	Room temp.
SO_2 (g) [209]*1	
0,0084	273
0,0166	473
0,0258	673
0,0358	873
0,0463	1073
0,0576	1273
SO_2 (l) [1]	
$\lambda = 38,2971 - 0,0628T$	233—353

1	2
CaO [147]*2, *8	
109	100
45	200
20	400
14	600
11	800
10	1000
9	1100
[19]*4, *31	
15,25	373
9,17	673
7,99	1073
7,79	1273
[19]*32	
13,95	373
8,37	673
7,28	1073
7,115	1273
Sc_2O_3 [140]*2, *33	
7,9	373
6,3	473
5,0	573
4,6	673
4,6	773
Ti_6O [44]	
4,39	300—1100
Ti_5O [44]	
3,95	300—1100
Ti_3O [44]	
2,92	300—1100
Ti_2O [44]	
11,47	300—1100

19 – Continued

1	2
$TiO_{0,83}$ [45][*2]	
5,4	77
4,6	173
2,9	373
5,0	673
6,7	873
7,5	1073
TiO [44]	
3,17	300–1100
$TiO_{1,04}$ [45][*2]	
7,1	77
5,4	173
3,6	373
5,9	673
7,5	873
8,4	1073
$TiO_{1,1}$ [44]	
2,61	300–1100
$TiO_{1,13}$ [45]	
1,6	300
1,7	700
3,5	1100
[45][*2, *34]	
4,8	77
3,3	173
1,9	373
4,2	673
6,3	873
7,1	1073
[45][*35]	
0,4	300
1,1	700
3,2	1100
$TiO_{1,6}$ [44]	
2,30	300–1100
$TiO_{1,9}$ [44]	
2,36	300–1100
TiO_2[*36] [1][*37]	
6,531	373
4,995	473
3,915	673
3,617	873
3,391	1073
3,307	1273
3,307	1473
VO_2 [1]	
9,797	373
4,396	873
3,412	1273
V_2O_5]19]	
4,4	298
[19][*2]	
3,45	373
2,6	473
1,8	673
2,5	873
2,8	923
Co_2O_3 [1][38]	
0,42	321,65
NiO [1][*37, *4, *39]	
12,393	373
9,923	473
7,159	673

1	2
5,694	873
4,605	1073
4,479	1273
[19]*37, *40	
9,21	373
7,37	473
5,32	673
4,19	873
3,43	1073
3,14	1273
Ni_2O_3 [1]*38, *41	
0,938	319,35
CuO [1]*38	
1,013	318,75
ZnO [1]*38, *42	
0,595	322,85
[19]*2, *37, *4, *43	
17,05	473
11,0	673
7,0	873
5,0	1073
[151]*2, *44	
80	3,5
300	5
900	10
1000	20
500	50
300	100
120	200
[151]*2, *45	
6	1,1
30	2

1	2
200	5
600	10
800	20
400	50
200	100
100	200
70	300
[151]*2, *46	
110	3,5
180	5
600	10
800	20
400	50
180	100
80	200
60	300
SrO [147]*2, *8	
60	80
40	100
15	200
7,5	400
5,5	600
5	800
5,5	1000
Y_2O_3 [145]*47	
6,28	293
4,39	473
3,31	673
2,65	873
2,18	1073
2,01	1173
ZrO_2 [19]*4, *48	
1,95	373
2,05	673
2,20	1073
2,39	1473
2,44	1673

19 — Continued

1	2
[19]*49	
1,7	373
1,76	673
1,88	1073
2,05	1473
2,09	1673
[1]	
2,3	1773
[150]*2,50*	
3,39	293
2,76	673
2,93	973
RuO_2 [152]*2, *51	
30	5
58	10
95	20
95	20
121	30
120	40
108	50
95	60
90	70
83	80
82	90
76	100
CdO [1]*38	
0,682	319,65
SnO_2 [19]*2, *52	
30,7	323
26,8	373
22,4	423

1	2
BaO [147]*2, *8	
7,8	80
6	100
3,8	200
3,1	400
3	600
3,2	800
3,9	1000
4,8	1100
CeO_2 [51]*53	
11,78	400
8,59	500
$\lambda = (3,156\cdot10^{-4} T - 0,0413)^{-1} + 8,81\times10^{-12}\times T^3 \exp(1,85\times10^{-3}T)$	700—1500
[51]*4, *54	
12,46	400
9,11	500
5,90	700
4,37	900
3,56	1100
3,10	1300
2,96	1500
Sm_2O_3 [145]*47	
2,05	293
1,72	473
1,48	673
1,38	873
1,34	1073
1,30	1173
[19]	
2,1	1273
Eu_2O_3 [167]*55	
3,42	298
3,20	323

19 — Continued

1	2
3,15	413
3,08	523
3,01	623
2,96	753
Gd_2O_3 [19]	
2,1	1273
Er_2O_3 [145]*47	
5,19	293
4,61	373
3,48	573
2,93	773
2,57	973
2,22	1173
Lu_2O_3 [145]*47	
5,99	293
5,11	373
3.8	573
3,14	773
2,72	973
2,34	1173
HfO_2 [150]*2, *56	
4,6	293
3,3	473
2,8	673
2,5	873
2,9	1073
[19]	
1,702	323
PbO [166]*57	
$\lambda = 212\ \rho^{2/3}\ T^{-1}$	300—900
ThO_2 [19]*4, *58	
10,26	373
5,99	673
3,41	1073
2,51	1473
(2,45)	1673
[77]*59	
9,21	373
2,93	1273
[77]*60	
8,54	373
4,98	673
2,85	1073
2,09	1473
2,05	1673
[155]*61	
$\lambda^{-1} = 22{,}14 \cdot 10^{-5} \times \times T - 1{,}01 \cdot 10^{-2}$	400—1400
UO_2 [19]*4, *37, *62	
(9,8)	373
5,78	673
3,41	1273
[77]*63	
8,37	373
2,93	1273
[77]*64	
5,19	473
2,26	1273
[77]*65	
3,52	473
1,76	1273

19 — Continued

1	2
[19][*66, *37]	
(7,33)	373
4,31	673
2,76	1073
2,55	1273
[155][*67]	
7,90—8,04	333
[155][*68]	
7,92—7,35	333
[155][*69]	
4,62	80
6,03	100
8,59	200
7,39	400
6,53	500
$\lambda^{-1} = 22,85\cdot 10^{-5}$ $T + 3,45\cdot 10^{-2}$	500—1400[*61]
[149][*70]	
$\lambda = 0,0031 +$ $+ 0,239/t\ (0,4848-$ $-0,4465D)$	1073—2273
[153][*2, *71]	
3	3
10	10
1	30
5	100
10	300
[116][*2, *72]	
5,19	100
7,62	200
7,70	250
7,45	300
[158][*73]	
5,88	670
4,84	870
4,05	1070
3,33	1270

1	2
[158][*74]	
5,36	670
4,41	870
3,70	1070
3,06	1270
[158][*75]	
4,43	670
3,61	870
2,98	1070
2,50	1270
[158][*76]	
4,18	670
3,56	870
3,07	1070
2,65	1270
[158][*77]	
4,05	670
3,40	870
2,91	1070
2,46	1270
[158][*78]	
3,55	670
2,73	870
2,43	1070
2,22	1270
[158][*79]	
2,70	670
2,04	870
1,97	1070
1,90	1270
[159][*80, *81]	
$\lambda = 100/(3,77 +$ $+ 0,0258T) +$ $+ 1,1\cdot 10^{-4}\,T +$ $+ 1,01\cdot 10^{-11} \times$ $\times T^3 \exp 7,2\cdot 10^{-4}T$	350—2700
[160][*81]	
$\lambda = 11,5 - 1,14\times$ $\times 10^{-2}\,T + 4,4\times$ $\times 10^{-6}\,T^2 - 5\times$ $\times 10^{-10}\,T^3$	350—2700

1	2	1	2
U_4O_9 [116]*82		UO_3 [19]	
1,09	100	0,2805	298—423
1,34	150	0,264	433—613
1,42	200—300	0,255	583—873

*1 At 101.325 kPa. *2 From graph. *3 Specimen density 2940 kg/m^3. *4 Nonporous. *5 Total pore volume 3–7%. *6 Total pore volume 4.07–9.95%. *7 Solid polycrystalline specimens with purity 99.99%, error ±5%. *8 Single crystal, in argon. *9 Theoretical density 3580 kg/m^3. *10 Anisotropic ratio for MgO is equal to unity [19, p. 122]. *11 Porosity 2.8-8.1%, density 3290–3480 kg/m^3. *12 [1, p. 128]. *13 Single crystal of synthetic sapphire, specimen diameter 3 mm. *14 Theoretical density 3970 kg/m^3. *15 Total pore volume 4.5–7.3%. *16 [1, p. 128]. *17 Anisotropic ratio for Al_2O_3 is 0.93. *18 Perpendicular to c-axis. *19 Parallel to c-axis. *20 Al_2O_3 (Lucalox), purity 99.9%, density 3970 kg/m^3, grain size 30 μm. *21 Quartz glass. *22 From graph; single crystal of quartz, in vacuo. *23 Fused quartz. *24 Pure fused quartz. *25 Specimens of grade KV, density 2201.5 ± 0.1 kg/m^3; data cannot be used as standard at temperatures above 600°K unless a correction is introduced for the radiant component. *26 Effective thermal conductivity, error 5%. *27 True thermal conductivity, error 5%. *28 α-Quartz. *29 α-Cristobalite. *30 Quartz sand. *31 Theoretical density 3320 kg/m^3. *32 Total pore volume 8.75%, density 3030 kg/m^3. *33 Sintered, relative density 0.96-0.97 with zero open porosity. *34 Total thermal conductivity. *35 Electronic component. *36 Anisotropic ratio for TiO_2 is 0.68 [19, p. 209]. *37 Polycrystalline. *38 Pressed powder. *39 Theoretical density 6800 kg/m^3. *40 Density 5050 kg/m^3, porosity 25.7%. *41 Density 1445 kg/m^3. *42 Density 4886 kg/m^3. *43 Fused; theoretical density 5660 kg/m^3. *44 Undoped specimen, ρ = 0.052 ohm·m at 300°K, parallel to c-axis. *45 Specimen doped with lithium, ρ = 0.0042 ohm·m at 300°K, parallel to c-axis. *46 Specimen doped with lithium, $\rho = 10^4$ ohm·m at 300°K, perpendicular to c-axis. *47 Single crystal; total relative error 15–20%. *48 Theoretical density 6100 kg/m^3. *49 Density 5220–5350 kg/m^3, porosity 12.3-14.6%. *50 Impurity HfO_2 < 0.1%, other impurities < 0.16%. *51 High-purity crystals. *52 Density 6620 kg/m^3, 98% SnO_2. *53 Porous specimen, density 6700 kg/m^3. *54 Theoretical density 7100 kg/m^3. *55 Sintered, 99.85% Eu_2O_3, average density 7100–7300 kg/m^3, closed porosity no greater than 10%. *56 Impurity $ZrO_2 \approx 0.57\%$, other impurities about 0.06%. *57 Sintered, yellow, impurities below 0.25% by mass; ρ is the specimen density (from 6500 to 9600 kg/m^3; T is the temperature, °K. *58 Density 9650 kg/m^3. *59 Total pore volume 3–7%. *60 Total pore volume 16.75%. *61 Thermal resistance. *62 Density 10,900 kg/m^3. *63 Total pore volume 3–10%. *64 Total pore volume 5%. *65 Total pore volume 25.5%. *66 Density 8000 kg/m^3, porosity 26.7%. *67 Single crystal, 99.27% (at.) purity, error ±3.5%. *68 Polycrystal, 99.86% (at.) purity, O/U ratio 2.001–1.996, error ±1.5%. *69 Calculated values for UO_2 theoretical density, error ±2%. *70 t is in °C; D is the density relative to theoretical (82 ≤ D ≤ 95%). *71 Single crystal, total impurities less than 200 ppm. *72 Fused UO_2, density, 10,920 kg/m^3, relative porosity 0.004, grain size 0.5 mm. *73 With O/U = 2.00 and relative pore volume ε = 0.011. *74 With O/U = 2.00 and ε = 0.053. *75 With O/U = 2.00 and ε = 0.096. *76 With O/U = 2.02 and ε = 0.037. *77 With O/U = 2.02 and ε = 0.073. *78 With O/U = 2.05 and ε = 0.047. *79 With O/U = 2.11 and ε = 0.036. *80 For sintered polycrystalline $UO_{2.00}$. *81 Most reliable values for thermal conductivity lie between values calculated in accordance with equations listed. *82 Sintered, density 10,510 kg/m^3, porosity 0.075, grain size several micrometers.

20. LINEAR COEFFICIENT OF THERMAL EXPANSION

Linear coefficient of thermal expansion $\alpha \cdot 10^6$, °C^{-1}	Temperature, °C
1	2
BeO [19]*1	
4,8	0
5,0	50
7,2	200
9,0	400
10,0	600
10,8	800
11,4	1000
11,9	1200
12,4	1400
12,8	1600
[77]	
6,5±1,0	25—100
8,0±6,0	25—300
9,2	25—800
9,0	30—1000
9,5	20—1400
10,6	25—1700
10,9	20—1800
10,1±0.6	1000—2000
13,4	1400—2000
MgO [77]	
12,0	200
12,8	400
13,4	600
14,1	800
14,7	1000
16,2	1600
16,7	1800
17,3	2000
[1]*2	
11,7	20—100
12,1	20—300
12,8	20—500
13,4	20—700
14,0	20—900
14,2	20—1000
[19]*3	
13,8	0—1000
[1]	
14,0	20—1400
15,7	0—1500
[19]*4	
17,49	20—2000
18,60	20—2300
[19]*5	
17,18	20—2000
17,40	20—2100
[1]	
12,6±0,5	1000—2000
Al_2O_3 [77]*6	
6,2	0—300
[77]	
7,33	0—600
7,50	0—800
8,00	0—1000
8,0—8,5*7	0—1000
8,30	0—1200
8,60	0—1400
8,90	0—1600
9,00	0—1800
9,14	0—2000
7,5±0,4*8	1000—1600
[19]*9	
8,1	20—1000
9,1	20—1200
9,8	20—1400

1	2
10,3	20—1600
10,6	20—1800
SiO_2 [1]	
4,3	20—300
3,0[*10]	300—1100
0,5[*11]	20—1300
[19][*12]	
0,67	300
0,60	600
0,53	900
CaO [77][*13]	
10,2	25—300
11,6	25—600
12,8	25—900
13,6—13,7	25—1200
14,7	25—1500
14,5[*14]	300—700
13,8[*15]	0—1700
[19][*16]	
12,52	25—400
13,73	25—800
14,75	25—1200
15,71	25—1400
16,02	25—1800
16,29	25—2000
14,49	25—2100
Sc_2O_3 [77]	
8,5	20—380
[140][*17]	
7,5—7,8	20—900
7,7	20—1800
Ti_6O [1]	
4,5	20—100
Ti_3O [1]	
3,3	20—100
TiO [62]	
6,73	25
TiO_2 [1]	
8,19	0—500
[19]	
8,3[*18]	0—500
15,0[*18]	500—1000
[19]	
8,85[*19]	25—1000
8,95[*20]	25—1000
9,1—9,2[*21]	400—800
Cr_2O_3 [1]	
9,6	20—1400
MnO [1]	
11,0	Above Néel point
FeO [1]	
12,4	100—200
12,3	100—400
12,6	100—600
13,4	100—800
15,2	100—1000
12,27±0,36[*22]	22—450
Fe_3O_4 [19][*1]	
9,3	20—300
18,8	300—700
Fe_2O_3 [19][*1]	
12,5	100—1200
CoO [19][*23]	
14,0	0—400

20 – Continued

1	2
NiO [1]*1, *24	
10,0	0
13,0	200
15,0	250
[1]	
12,0	100—600
12,6	100—800
14±1	300—800
13,9*25	60—1300
[52]*2, *26	
13,2	200
13,5	400
14,1	1000
14,9	1400
16,3	1800
GeO_2 [19]	
7,59*27	25—350
9,98*28	25—480
5,36*29	25—480
7,0*29	400—800
SrO [19]	
13,9	Room temp.
13,3*1	0—1200
Y_2O_3 [139]*30	
7,2	20—1000
[19]	
8,2	0—1000
8,8*31	20—1000
9,3	0—1400
1,2*32, *1	200
2,9*32, *1	400
4,7*32, *1	600
6,3*32, *1	800
8,0*32, *1	1000
9,8*32, *1	1200
11,7*32, *1	1400
[77]	
6,9	200
7,3	400
7,7	600
8,1	800
8,3	1000
8,6	1200
8,8	1400
9,1	1600
9,3	1800
ZrO_2 [52]	
2,0	—130 to—80
8,0	—80 to —50
7,2	Below 1000
[19]*33	
7,6	300
9,2	600
10,0	900
(10,5)	1100
10,8	1200
(11,2)	1300
11,6	1500
[19]	
5,5—11,0	20—1200
10,2*34	20—1600
12,7*35	20—1600
13,0*36	20—1600
[77]*37	
10,8	25—1200
11,6	25—1500
[146]*38	
7,6	20—127
8,2	20—327
8,2	20—527
8,4	20—727
9,7	20—927
10,7	20—1127
11,3	20—1327
11,8	20—1527
12,2	20—1727
12,5	20—1927
12,7	20—2127
13,0	20—2327
[1]	
7,5	0—1000

1	2	1	2
Nb_2O_5 [19]*39, *1		CeO_2 [19]	
0,0	0—200	11,9*45	0—1200
—1,2	200—400	13,4*46	0—1200
+1,4	400—850	8,5—8,6*47	0—1000
[120]*40		Pr_2O_3 [19]	
$a=21,1848\times(1+0,56328\cdot10^{-5}T)$, $c=19,3680\times(1+0,58856\times10^{-5}T)$, $\beta=119,63\times(1+0,26456\cdot10^{-5}T)$, $V=1363,4(1+0,880152\cdot10^{-5}T)$	25—800	8,0	25—100
		[1]	
		8,3	25—1050
		$PrO_{1,70}$ [1]	
		22	25—1050
In_2O_3 [131]*41		$PrO_{1,78}$ [1]	
7,20±0,06	30—968	11	25—1050
SnO_2 [19]		Pr_6O_{11} [1]	
3,4	Room temp.	8	25—1050
3,76	22—650		
4*42	0—800		
5,28*43	—		
La_2O_3 [19]*44		Nd_2O_3 [19]*48	
5,86	100	1,53	100
7,92	150	4,26	150
8,45	200	6,91	250
9,11	250	8,90	300
9,38	300	9,97	400
9,78	350	10,60	500
10,05	400	10,99	600
10,36	450	11,44	700
10,46	500	11,35	800
10,68	550	11,23	900
10,88	600	11,37	1000
11,20	700	11,41	1050
11,61	800	[19]	
11,81	900	11,8	100—1000
12,01	1000		
12,04	1050		
[19]		Sm_2O_3 [19]*49	
12,6	Room temp.	9,9	100—1000

20 — Continued

1	2
[145]	
10,8	20—1000
Eu_2O_3 [19]	
10,5*50	30—800
10,4	0—1000
10,3	0—1200
Gd_2O_3 [19]	
10,0	30—840
10,5*51	25—1000
$TbO_{1,81}$ [1]	
3	25—1050
Dy_2O_3 [19]	
8,3	30—840
Ho_2O_3 [19]	
8,44	—
Er_2O_3 [132]*52	
5,7	100—300
[145]	
7,9—8,2	20—1000
Tu_2O_3 [132]*52	
7,1	100—300
8,2*53	—
Yb_2O_3 [132]*52	
4,9	100—300
[19]	
8,2	0—1000
9,3	25—1000
Lu_2O_3 [145]	
7,8	20—1000

1	2
HfO_2 [19, p. 71]	
6,12	—231 to +287
7,06	—231 to +1468
6,84	+287 to +847
[1]*54	
5,8	250—1300
[77]	
6,19	24—538
6,84	538—1093
7,05	24—1482
6,45	20—1700
Ta_2O_5 [19]*1	
0,8	0—400
4,0	500—800
ThO_2 [19]*55	
8,95	—230 to +287
9,9	287—847
11,07	847—1407
12,31	1407—1967
13,31	1967—2247
10,9	—230—2247
[1]	
9,1	20—800
9,67*56	100—1230
UO_2 [19]*57	
9,07	27—400
10,8	400—800
13,0	800—1260
[19]*58	
9,28	27—400
10,7	400—800
12,9	800—1260
[19]*59	
9,03	27—400
11,04	400—800
12,9	800—1260

1	2	1	2
[19]*60		Cm_2O_3 [179]	
9,76	27—400	6,6*61	500—800
10,8	400—800		
12,9	800—1260	Cm_7O_{12} [179]	
[1]		8,8	650
10,0	0—1000		
PuO_2 [1]		CmO_2 [179]*62	
9±1	20—500	8,1	Below 440
11,4	Below 1000		
15,2	125—911		

*1From graph. *2Sintered polycrystalline specimens. *3True coefficient of linear expansion. *4In air. *5In argon. 6Sintered alumina. *799% Al_2O_3 [19]. *8[1]. *9100% Al_2O_3, apparent porosity 1.7%, bulk density 3830 kg/m³, firing temperature 1730°C. *10Crystalline. *11Silicate glass. *12Fused SiO_2. *13Coarse granular specimen. *14[19]. *15[1]. *16Specimen of fused oxide. *17Sintered, relative density 0.96–0.97 with zero open porosity. *18From graph, in vacuo. *19Density 3310 kg/m³. *20Density 3960 kg/m³. *21Rutile. *22Wüstite. *23In nitrogen. *24Pressed powder. *25[19]. 26In oxygen; average value of coefficient. *27Glassy. *28Hexagonal. *29Tetragonal. *30Sintered ceramic 99.85% Y_2O_3, relative density 96.5%, zero apparent porosity. *31[145]. *32Specimen fired at 1400°C; apparent porosity 42.3%, apparent density 2860 kg/m³. *33Fused, cubic, 100% ZrO_2. *34Thermocouple sheaths, bulk density 5430 kg/m³, porosity 0.68%. *35Same, bulk density 5450 kg/m³, porosity 0.28%. *36Same, bulk density 4950 kg/m³, porosity 2.93%. *37Stabilized. *38Purity 99.8%, stabilized with 6% CaO by mass. *3999.6% Nb_2O_5, remainder Ta_2O_5. *40Thermal expansion along α-Nb_2O_5 crystallographic axes a and c given in nm (b is constant); β is in degrees, t is in °C, and B is in cubic nanometers. *41Powder sintered at 1000°C for 24 h; lattice constants a_t and a_0 are determined from equation $a_t = a_0(1 + \alpha t + \beta t)^2$, $\alpha = (7.20 \pm 0.06)10^{-6}$, $\beta = (1.15 \pm 0.08)10^{-9}$ for interval 30–968°C; a_t and a_0 are valid for t and 0°C. *42[1]. *43Average coefficient [52]. *44Impurity contents in specimen by spectroanalysis: 0.05% Sm, 0.04% Si, 0.02% Eu, 0.01% Fe, 0.0005% Mg, 0.02% Dy, and 0.01% Gd. *45Specimen 80% CeO_2 and 20% ZrO_2, zero open porosity. *46Same, open porosity 0.3%. *47[1]. *48Impurities by spectral analysis: 0.02% Sm, 0.002% Si, 0.005% Eu, 0.05% Fe, 0.0005% Mg, 0.02% Dy, and 0.01% Gd. *49After calcining at 1500°C, density 7400 kg/m³. *50Hot-pressed specimen with density 95% of theoretical; probably monoclinic oxide. *51Density 7400 kg/m³. *52Sintered 48 h at 1100°C. *53[19]. *54Monoclinic. *55Sintered at 968°C. *56Pure. *57Particle size from 0 to 5 μm; determinations made while heating. *58Same, while cooling. *59Particle size from 15 to 20 μm, while heating. *60Same, while cooling. *61C-form, in atmosphere of hydrogen with helium. *62In oxygen.

21. ANISOTROPY OF LINEAR COEFFICIENT OF THERMAL EXPANSION

Linear coefficient of thermal expansion $\alpha \cdot 10^6$, °C^{-1}		Temperature, °C
parallel to c-axis	perpendicular to c-axis	
1	2	3
BeO [19]		
5,1	5,3	18
9,8	11,0	1025
Al_2O_3 [37]*1		
6,460	5,819	127
6,983	6,303	227
7,385	6,675	327
7,713	6,975	427
7,991	7,228	527
8,232	7,447	627
8,447	7,641	727
8,641	7,814	827
8,818	7,973	927
8,983	8,119	1027
9,138	8,255	1127
9,283	8,383	1227
9,421	8,504	1327
9,552	8,618	1427
9,678	8,728	1527
9,800	8,833	1627
9,917	8,932	1727
10,031	9,032	1827

Linear coefficient of thermal expansion $\alpha \cdot 10^6$, °C^{-1}		Temperature, °C
parallel to c-axis	perpendicular to c-axis	
1	2	3
[1]		
6,0	7,1	100
8,9	9,7	500
13,8	14,2	1000
[169, p. 15]*2		
6,66	—	20—50
9,03	—	20—1000
—	5,0	50
SiO_2 [127]*3		
4,1	6,6*4	25—400
TiO_2 [133]*5		
$\alpha_{\parallel}=8,816\cdot10^{-6}+3,653\cdot10^{-9}t+6,329\cdot10^{-12}t^2$	$\alpha_{\perp}=7,249\cdot10^{-6}+2,198\cdot10^{-9}t+1,298\cdot10^{-12}t^2$	30—650
[133]*6		
$\alpha_{\parallel}=7,380\cdot10^{-6}+6,620\cdot10^{-9}t+1,771\cdot10^{-11}t^2$	$\alpha_{\perp}=3,533\cdot10^{-6}+5,610\cdot10^{-9}t+4,315\cdot10^{-12}t^2$	28—712

21 – Continued

1	2	3
	[127]*5	
10,1	8,0	25—400
10,8*7	8,3*7, *4	400—800
	VO_2 [127]	
21,4	8,5*4	100—400
	CrO_2 [127]	
—10,3	16,2	25—290
	MnO_2 [19]*8	
6,93	6,69*4	25—500
	ZnO [19]	
~5,0	~5,5*4	0—400
~4,6	~9,0*4	400—1200
	GeO_2 [23]*9	
$\alpha_{\parallel}=1,268\cdot10^{-6}+1,359\cdot10^{-8}t-1,032\cdot10^{-11}t^2$	$\alpha_{\perp}=10,558\cdot10^{-6}+8,813\cdot10^{-9}t-5,785\cdot10^{-12}t^2$	30—516
	[127]	
3,7	8,0*4	25—400
	[119, p 74]*10	
3,8	10,8*4	400—800

1	2	3
	ZrO_2 [134]*11	
14,2	8,5*4	20—800
14,2	9,7*4	800—1000
~15,8	~11,6*4	1000—1200
	Nb_2O_5 [120]*12	
5,886	5,633*4	25—800
	Rh_2O_3 [58]*13	
$\alpha_c=5,246\cdot10^{-6}+6,369\cdot10^{-9}t-7,480\cdot10^{-14}t^2$	$\alpha_a=5,350\cdot10^{-6}+1,281\cdot10^{-9}t-1,133\cdot10^{-14}t^2$	20—850
	SnO_2 [127]*14	
5,6	4,3*4	25—400
	La_2O_3 [99]*15	
17,10	10,64*4	900—1300
	HfO_2 [134]*16	
11,9	5,1*4	20—200
11,9	5,9*4	200—400
12,2	7,6*4	400—600
12,2	8,8*4	600—800
12,6	10,2*4	800—1200

*1 Composition of single crystals, % by mass: 99.992 Al_2O_3, $7\cdot10^{-3}$ Fe, $<10^{-3}$ Cu, $2\cdot10^{-4}$ Mg, $5\cdot10^{-5}$ Mn, $4\cdot10^{-3}$ Si, $3\cdot10^{-4}$ Cr, $4\cdot10^{-4}$ Ca, $7\cdot10^{-4}$ Ga, $<2\cdot10^{-4}$ V, $8\cdot10^{-5}$ Mo. *2 Corundum. *3 Stishovite. *4 Along a-axis. *5 Rutile. *6 Anatase. *7 Rutile [19]. *8 β-MnO_2. *9 Hexagonal. *10 Tetragonal. *11 Monoclinic, purity 99.7%. *12 α-Nb_2O_5, monoclinic. *13 Corundum structure. *14 Cassiterite. *15 Hexagonal A-form. *16 Monoclinic, purity 99.1% (0.8% ZrO_2).

22. PARAMETERS OF DIFFUSION OF OXYGEN INTO METALS AND NONMETALS ($D = D_0 \exp(-E/RT)$ [1; 77; 202–205; 631])

Base element	Temperature, °C	D and $D_0 \cdot 10^4$, m^2/sec (values of D_0 are given without any letter-symbol)	Activation energy $E \cdot 10^{-3}$, kJ/kmole
1	2	3	4
Si	—	0,21	240,74*1
	1573	$D = 1,8 \cdot 10^{-10}$	—
α-Ti	963—1113	$5,08 \cdot 10^{-3}$	140,26*2
	1023—1423	$4 + 13(\Delta C)^2$	$200,97 + 18,00\Delta C$*3
β-Ti	1193—1473	$3,14 \cdot 10^4$	287,63*2
	1403—1623	0,083	142,35*4
	1023—1423	0,14	138,16*5
V	343—453	0,011	121,42
	—	$(1,3 \pm 0,32) \cdot 10^{-2}$	$121,46 \pm 0,80$*6
Fe	1833—1933	$(5,59 \pm 0,80) \cdot 10^{-3}$	$81,64 \pm 3,14$*7
Cu	928—1073	4,5	179,20*8
	1073—1273	—	—*9
Ge	—	0,17	129,6*1
α-Zr	673—858	5,2	213,53
	673—858	9,4	$216,79 \pm 0,92$
	673—1773	5,4	212,69
	973	$9,13 \cdot 10^{-5}$	124,77*2
	973	69,2	235,3*2
β-Zr	1553	$5,25 \cdot 10^{-8}$	—
	1350	$D = 0,32 \cdot 10^{-7}$	—
	1450	$D = 1,18 \cdot 10^{-7}$	—
	1550	$D = 3,69 \cdot 10^{-7}$	—
Nb	—	$1,47 \cdot 10^{-2}$	115,87*10
	—	0,015	115,14
	—	0,021	112,4
	—	0,021	120,58*6
	313—423	$0,0212 \pm 0,0073$	$112,67 \pm 1,05$*6
	421—441	0,0147	115,56*6
	473—548	$2 \cdot 10^{-5}$	95,46*11
	873—1273	0,00407	104,25*12
Ag	685—1135	$2,72 \cdot 10^{-3}$	46,0*10
Hf	773—1620	0,4—1,4	233,83*2
Ta	—	0,03	121,42*2
	—	$44 \cdot 10^{-4}$	106,2
	323—423	$0,0044 \pm 0,0008$	$106,55 \pm 0,54$*6
	429—628	0,019	114,30*6
	523—723	$8 \cdot 10^{-5}$	114,72*11
	973—1673	0,015	111,79*13

1	2	3	4
W	—	1,3	100,48[*6]
	—	0,01	261,26
	1973	$D=3\cdot10^{-7}$	—[*14]

[*1]Specific methods applicable to semiconductors. [*2]Chemical method. [*3]Parameters were determined by x-ray structural analysis and measurement of microhardness when layers are removed; ΔC is the excess concentration of oxygen in relation to concentration in middle of specimen; for region of solutions up to 1% (at.) oxygen; titanium iodide in purified oxygen. [*4]At high dilution. [*5]X-ray structural analysis and measurements of microhardness when layers are removed; volume diffusion, technical titanium in air; error ±3%; practically independent of oxygen concentration; at all temperatures, $D_\alpha < D_\beta$; at temperature of polymorphic transitions, $D_\beta/D_\alpha \approx 40$. [*6]Internal friction method. [*7]Liquid iron, capillary method. [*8]Single crystal. [*9]$\log D = 0.8805 - 11985/T$, for [110] face $\log D = -9.078 - 2044/T$, for [100] face $\log D = -3.447 - 8229/T$, for [111] face $\log D = -7.427 - 3799/T$; copper purity 99.999%. [*10]Metallographic method. [*11]By measurement of specimen weight gain. [*12]By method of saturating specimen with oxygen. [*13]By measurements of microhardness. [*14]Polycrystal.

23. PARAMETERS OF DIFFUSION OF ELEMENTS INTO OXIDES ($D=D_0 \exp(-E/RT)$, m^2/sec)

Diffusing element	Temperature, °C	D, $D_0\cdot10^4$, m^2/sec *	Activation energy	
			$E\cdot10^{-3}$, kJ/kmole	E, eV[*1]
1	2	3	4	5
		BeO		
He	1080—1400	$2,5\cdot10^{-3}$	333,00	3,45[*2]
Be	—	$1,26\cdot10^{-6}$	151,14	1,57[*3]
		B_2O_3		
O	77—377	$2,49\cdot10^{-2}$	137,33±±33,49	1,42±0,35[*4]
		MgO		
	1100—1400	$4,5\cdot10^{-5}$	252,05	2,61[*5]
	1100—1400	$1,3\cdot10^{-5}$	252,05	2,61[*6]
O^{2-}	1300—1750	$2,5\cdot10^{-6}$	261,6	2,71
	—	$(11,37\pm0,26)\cdot10^{-2}$	460,5±17,6	4,78±0,18[*7]
O	1020—1450	—	233,62	2,42[*8]
	1020—1250	—	252,05	2,62[*9]
	1250—1450	—	430,4	4,46[*9]
^{18}O	1415	$D=2\cdot10^{-13}$	—	—[*10]
^{28}Mg	1400—1600	0,249	330,76	3,43[*11]

1	2	3	4	5
^{26}Mg	1450—2400	$4,19(+2,45; -1,55)\cdot 10^{-4}$	266,40±7,72	2,76±0,08 *11, *12
$^{28}Mg^{2+}$	1100—1450	0,54	308,87±9,65	3,2±0,1 *13
	1450—1750	$1,2\cdot 10^{-5}$	154,44±9,65	1,6±0,1 *14
	1450—1750	$1,6\cdot 10^{-5}$	173,74±9,65	1,8±0,1 *15
Mg^{2+}	1900—2350	$7,43(+12,17; -4,61)\cdot 10^{-2}$	333,97±±18,34	3,46±0,19
Be^{2+}	1000—1700	$1,41\cdot 10^{-5}$	154,44	1,60
	1000—1650	$1,99(+0,24; -0,22)\cdot 10^{-5}$	162,16±1,93	1,68±0,02 *16
	1800—2340	$1,99(+0,24; -0,22)\cdot 10^{-5}$	162,16±1,93	1,68±0,02 *17
Al^{3+}	1565—1900	2,20	318,2	3,3 *18
Ca^{2+}	910—1700	$2,95\cdot 10^{-5}$	205,59	2,13
	790—1850	$8,9\cdot 10^{-4}$	266,40	2,76
^{45}Ca	1850—2400	$3,43(+4,95; -2,03)\cdot 10^{-3}$	308,87±±19,30	3,2±0,2 *19
^{46}Sc	1500	$D=3\cdot 10^{-10}$ to $7,4\cdot 10^{-12}$	—	— *11
Cr^{3+}	1300—1700	$9,8\cdot 10^{-4}$	284,74	2,95
Mn^{2+}	1300—1700	$4,1\cdot 10^{-7}$	116,79	1,21
Fe^{2+}	1050—1720	$8,93\cdot 10^{-5}$	174,71	1,81
	1310—1690	$3,2\cdot 10^{-4}$	175,67	1,82
Co^{2+}	1000—1810	$5,78\cdot 10^{-5}$	198,84	2,06
Ni^{2+}	1900—2460	$1,80\cdot 10^{-5}$	202,70	2,10
^{63}Ni	1200—1300	$6\cdot 10^{-6}$	173,74	1,8 *16
$^{63}Ni^{2+}$	1900—2500	$1,4(+4,8; -1,1)\cdot 10^{-2}$	318,52±±28,96	3,3±0,3 *20
Ni	1450	$D=0,2\cdot 10^{-10}$	—	— *21
	1480	$D=0,28\cdot 10^{-10}$	—	—
	1510	$D=0,57\cdot 10^{-10}$	—	—
	1550	$D=0,74\cdot 10^{-10}$	—	—
	1480	$D=0,52\cdot 10^{-10}$	—	— *22
	1510	$D=0,73\cdot 10^{-10}$	—	—
	1550	$D=1,5\cdot 10^{-10}$	—	—
	1450	$D=1,2\cdot 10^{-10}$	—	— *23
	1510	$D=3,15\cdot 10^{-10}$	—	—
	1550	$D=6,1\cdot 10^{-10}$	—	—
	1450	$D=1,7\cdot 10^{-10}$	—	— *24
	1400	$D=1,45\cdot 10^{-10}$	—	— *25
	1450	$D=2,8\cdot 10^{-10}$	—	—
	1510	$D=5,87\cdot 10^{-10}$	—	—
	1550	$D=1,0\cdot 10^{-10}$	—	—
	1450	$D=8,0\cdot 10^{-10}$	—	— *26
	1450	$D=19\cdot 10^{-10}$	—	— *27
	1480	$D=0,25\cdot 10^{-10}$	—	— *28
	1510	$D=0,60\cdot 10^{-10}$	—	— *28
	1550	$D=0,80\cdot 10^{-10}$	—	—
	1450	$D=0,23\cdot 10^{-10}$	—	— *29
	1510	$D=0,58\cdot 10^{-10}$	—	—

1	2	3	4	5
	1550	$D=0,70\cdot10^{-10}$	—	—
Zn^{2+}	1000—1645	$1,48\cdot10^{-5}$	178,57	1,85
^{68}Ge	1850—2400	0,34(+0,70; —0,23)	386,10± ±19,30	4,0±0,2[*19]
$^{85}Sr^{2+}$	1000—1600	6,0(+10,7; $-3,8)10^{-4}$	280,88± ±12,55	2,91± ±0,13[*19, 20, *30]
Sr^{2+}	1900	$1,14\cdot10^{-2}$	337,83± ±38,61	3,5±0,4
$^{91}Y^{3+}$	1440—1760	$(2,11\pm0,18)\times10^{-2}$	299,00±6,70	3,10±0,07[*31]
Cd^{2+}	1800—2300	1,7(+6,0; $-1,4)10^{-2}$	328,18± ±28,9	3,4±0,3[*17]
Ba^{2+}	1008—1724	$7\cdot10^{-3}$	326,25	3.38[*32]
	1008—1724	$6,3\cdot10^{-5}$	178,57	1,85[*33]
$^{133}Ba^{2+}$	1900—2500	2,8(+13,2; $-2,3)10^{-2}$	337,83± ±28,9	3,5±0,3
		Al_2O_3		
O	1200—1600	$6,3\cdot10^{-8}$	241,16±14,6	2,50±0,15[*11]
	1500—1780	1900	636,4	6,59[*11]
O^{2-}	1500	$D=2\cdot10^{-16}$	—	—[*35, *3]
	1600	$D=2\cdot10^{-15}$	—	—
	1700	$D=2\cdot10^{-14}$	—	—
	1800	$D=2\cdot10^{-13}$	—	—
	1500	$D=3\cdot10^{-14}$	—	—[*34, *36, *36]
	1600	$D=2\cdot10^{-13}$	—	—
	1700	$D=1\cdot10^{-12}$	—	—
	1800	$D=8\cdot10^{-12}$	—	—
	1727	$D=8,1\cdot10^{-14}$	—	—[*11]
	1727	$D=2,7\cdot10^{-12}$	—	—[*36]
	1727	$D=7,5\cdot10^{-10}$	—	—[*37]
	1727	$D=4,6\cdot10^{-9}$	—	—
	—	$1,6\cdot10^{3}$	636,39	6,59[*11]
	—	$2,5\cdot10^{5}$	565,22	5,86[*37]
	—	$3,6\cdot10^{8}$	594,53	6,16
	—	10^{7}	690,82	7,16
Al^{3+}	1727	$D=1,4\cdot10^{-11}$	—	—[*36]
^{26}Al	1670—1905	28	477,30±63,6	4,94±0,66[*38]
Al^{3+}	1700	$D=7\cdot10^{-12}$	—	—[*39, *3]
	1800	$D=4\cdot10^{-11}$	—	—
	1900	$D=2\cdot10^{-10}$	—	—
^{22}Na	1223—1750	$(2,0\pm0,6)\cdot10^{-2}$	209,34±8,37	2,17±0,09[*40]
	1505	D=0,0158 - 0,0184	—	—[*41]
	1568	D=0,0229	—	—
	1568	D=0,0123	—	—[*42]
^{45}Ca	1397	D=30	—	—[*40]
^{59}Fe	900—1100	$1,37\cdot10^{-8}$	46,06	0,48[*43]
	900—1100	$9,18\cdot10^{-8}$	113,04	1,17[*44]
	900—1100	$3,17\cdot10^{-8}$	79,55	0,82[*45]

23 — Continued

1	2	3	4	5
		SiO_2		
D_2	385	$D=2,1\cdot10^{-7}$	—	—[*46]
	385	$D=9,7\cdot10^{-7}$	—	—[*47]
^3He	306	$D=2,88\cdot10^{-6}$	—	—[*48]
	415	$D=6,73\cdot10^{-6}$	—	—
	513	$D=12,5\cdot10^{-6}$	—	—
	602	$D=19,1\cdot10^{-6}$	—	—
^4He	306	$D=2,67\cdot10^{-6}$	—	—[*48]
	415	$D=6,14\cdot10^{-6}$	—	—
	513	$D=11,3\cdot10^{-6}$	—	—
	602	$D=17,1\cdot10^{-6}$	—	—
O	1000	$D=10^{-14}$	—	—[*11]
^{18}O	850—1250	$2\cdot10^{-9}$	121,4±0,8	1,26± ±0,009[*49]
B	—	$3,01\cdot10^{-2}$	343,62	3,56[*31]
	—	$1,61\cdot10^{-5}$	272,19	2,82
	—	$1,23\cdot10^{-4}$	327,21	3,39
	—	$7,38\cdot10^{-4}$	345,55	3,58[*50]
P	Below1150	$3,9\cdot10^{-12}$	133,98	1,39
	900—1250	$3,2\cdot10^{-9}$	105,27	1,09[*19, *31]
	—	$1,9\cdot10^{-9}$	105,27	1,09
	—	$1,86\cdot10^{-1}$	389,00	4,03[*51, *50]
Ga	—	$1,04\cdot10^{5}$	402,50	4,17[*52]
As	1100	$D=3,5\cdot10^{-15}$	—	—[*31, *53]
	1100	$D=1,5\cdot10^{-16}$	—	—[*54]
	—	$2,48\cdot10^{2}$	473,00	4,90[*31, *55]
Sb	—	$1,31\cdot10^{16}$	844,57	8,75[*56, *57]
		CaO		
Ca	900—1640	0,4	339,13	3,51[*58]
	1100	$D=2,1\cdot10^{-13}$	—	—[*11, *59]
	1200	$D=8\cdot10^{-13}$	—	—
	1300	$D=2,2\cdot10^{-12}$	—	—
	1400	$D=6,7\cdot10^{-12}$	—	—
	1500	$D=2\cdot10^{-11}$	—	—
^{45}Ca	1000—1400	$(8,75\pm1,32)\cdot10^{-8}$	144,86±3,98	1,50±0,04[*60]
	1000—1400	$(1,95\pm0,6)\cdot10^{-7}$	142,43	1,48[*61]
		Sc_2O_3		
^{18}O	1106—1298	$7,72\cdot10^{-4}$	160	1,6[*62,*11]
		TiO[*63]		
^{44}Ti	1250—1555	0,3[*64] 1,0[*65]	~272,14[*66]	~2,82[*19]

1	2	3	4	5
		TiO_2		
O	900—1000	$2,88\cdot10^{-1}$	222,74	2,31[*67]
	710—1300	$2,0\cdot10^{-3}$	251,21	2,60[*68]
	860—1025	1,1	305,64	3,17[*69]
^{18}O	950—1400	$1,7\cdot10^{-2}$	276,33	2,86[*11, *3]
	806	$D=(1,7\pm0,05)10^{-15}$	—	—[*70]
	806	$D=(0,32\pm0,01)\times\times10^{-15}$	—	—[*71]
	806	$D=(6,7\pm0,2)10^{-15}$	—	—[*72]
	806	$D=(6,2\pm0,1)10^{-15}$	—	—[*73]
	806	$D=(6,6\pm0,1)10^{-15}$	—	—[*74]
	806	$D=(6,1\pm0,2)10^{-15}$	—	—[*75]
	806	$D=(5,9\pm0,2)10^{-15}$	—	—[*76]
	806	$D=(6,3\pm0,1)10^{-15}$	—	—[*77]
	806	$D=(2,1\pm0,03)10^{-15}$	—	—[*78]
	806	$D=(7,3\pm0,2)10^{-15}$	—	—[*79]
	806	$D=(6,8\pm0,2)10^{-15}$	—	—[*80]
	806	$D=(6,3\pm0,3)10^{-15}$	—	—[*81]
	710—950	—	255,4 or 314,0	2,65 or 3,25[*82]
	950—1300	—	96,3 or 221,9	1,0 or 2,30[*82]
^{59}Fe	800—1000	$1,73\cdot10^{-5}$	142,35	1,47[*83]
	800—1000	$1,10\cdot10^{-8}$	52,96	0,55[*84]
	800—1000	$1,98\cdot10^{-2}$	230,27	2,38[*85]
	800—1000	$1,55\cdot10^{-5}$	142,35	1,47[*81]
	770—1000	$2,69\cdot10^{-4}$	142,35	1,47[*86]
	770—1000	$6,17\cdot10^{-6}$	57,78	0,60[*87]
	770—1000	0,192	231,95	2,40[*88]
	770—1000	$1,13\cdot10^{-3}$	145,28	1,50[*81]
	950—1050	$2,04\cdot10^{-2}$	139,84	1,45[*89]
		V_2O_5		
O	560—650	$2\cdot10^{4}$	255,39	2,65[*90]
		Cr_2O_3		
O	—	—	422,9	4,38
^{59}Fe	Below 1000	Very slight penetration	—	—[*91]
Cr	1000—1350	$4\cdot10^{3}$	418,68	4,34
		MnO_2		
H^+	25	$D=(8\pm5)10^{-8}$	—	—[*92]
		FeO		
O	—	—	—	—[*93]
Fe	600—910	$6\cdot10^{-5}$	104,67±12,56	1,08±0,13[*3]
Fe	1200	$D\approx10^{-7}$	—	—[*94]
S	1050	$D=4,1\cdot10^{-8}$	—	—[*95]

1	2	3	4	5
		$FeO_{1,087}$		
Fe	690—1010	0,014	126,44	1,31*3
		$FeO_{1,09}$		
Fe	—	$4,1 \cdot 10^{-3}$	116,39	1,21*96
		$FeO_{1,10}$		
Fe	1050	$D=(0,9-5) \cdot 10^{-7}$	—	—*97
		$FeO_{1,103}$		
Fe	—	0,012	124,35	1,29*97
		Fe_3O_4		
^{59}Fe	770—1200	$9,37 \cdot 10^{-2}$	204,32	2,12*86
	770—1200	$1,27 \cdot 10^{-3}$	151,56	1,57*87
	770—1200	0,250	225,67	2,34*88
	770—1200	$1,78 \cdot 10^{-2}$	188,41	1,95*81
Fe	750—1000	5,2	230,27	2,38
		Fe_2O_3		
O^{2-}	900	$D=10^{-5}$	—	—*36
	900	$D=2 \cdot 10^{-10}$	—	—*37
	—	10^{11}	590,34	6,12*36
	—	$5,5 \cdot 10^{10}$	464,73	4,81*37
		Fe_2O_3*98		
O	1150—1250	10^{11}	611,3	6,33
Fe^{3+}	900	$D=3,5 \cdot 10^{-16}$	—	—*36
	—	10^{5}	468,92	4,86*36
^{59}Fe	950—1050	$1,3 \cdot 10^{-6}$	419,52	4,35*99
Fe	930—1270	$4 \cdot 10^{4}$	468,92	4,86
		α-Fe_2O_3		
^{55}Fe	1200	$D=1,1 \cdot 10^{-12}$	—	—*100
	1200	$D=5,9 \cdot 10^{-12}$	—	—*101
	1200	$D=21,9 \cdot 10^{-12}$	—	—*102
	1300	$D=6,9 \cdot 10^{-12}$	—	—*100
	1300	$D=31,4 \cdot 10^{-12}$	—	—*101
		CoO		
Co	800—1300	$2,2 \cdot 10^{-3}$	144,45	1,50*103
	1000	$D=2,6 \cdot 10^{-9} \times (p_{O_2})^{0,35}$	—	—*104
	1350	$D=5,1 \cdot 10^{-8} \times (p_{O_2})^{0,28}$	—	—*104

1	2	3	4	5
	1150	$D=9{,}0\cdot10^{-9}\times\times(p_{O_2})^{0,3}$	—	—
^{35}S	1000—1090	$5\cdot10^{-3}$	244,51	2,53 [*11, *62]
	1050	$D=3{,}5\cdot10^{-12}$	—	— [*105]
		NiO		
H^+				
	—	$D=2\cdot10^{-10}$	—	— [*106]
^{35}S	25	$D\approx10^{-15}-10^{-16}$	—	—
	1000—1090	$3{,}87\cdot10^{-3}$	238,65	2,47 [*11, *62]
	1015	$D\approx10^{-11}$	—	— [*107]
	1050	$D=1{,}2\cdot10^{-11}$	—	—
^{63}Ni	1090	$D=3{,}8\cdot10^{-11}$	—	—
	1000—1400	$5\cdot10^{-4}$	185,06±2,51	1,92±±0,03 [*36, *62]
	1000—1400	$3{,}9\cdot10^{-4}$	185,06±0,84	1,92±±0,01 [*11, *62]
^{63}Ni				
	1182—1762	$(4{,}8\pm1{,}3)\cdot10^{-2}$	254,56±2,51	2,64±0,03 [*108]
	1182—1762	$(2{,}1\pm0{,}4)\cdot10^{-2}$	262,51±1,26	2,72±0,01 [*109]
Ni	800	$D=(6{,}0\pm1{,}5)10^{-14}$	—	— [*36, *3]
Ni	740—1400	0,017	234,46	2,43
^{51}Cr	1140—1400	$2{,}8\cdot10^{6}$	496,32	5,14
	1192—1642	$(8{,}6\pm1{,}2)10^{-3}$	282,19±4,61	2,92±±0,05 [*110]
	1000	$D=2\cdot10^{-12}$	—	— [*111]
Cr^{3+}	1300	$D=2{,}5\cdot10^{-10}$	—	— [*112]
	1330—1580	$4\cdot10^{-3}$	230,27	2,39 [*113]
	1330	$D=3{,}3\cdot10^{-10}$	—	— [*114]
	1383	$D=7{,}0\cdot10^{-10}$	—	— [*115]
	1450	$D=1{,}2\cdot10^{-9}$	—	— [*116]
^{51}Cr	1580	$D=6{,}2\cdot10^{-9}$	—	— [*117]
^{55}Fe	1100—1250	$9{,}36\cdot10^{-5}$	196,78	2,04 [*118]
^{55}Co	1000—1400	$(1{,}81\pm0{,}02)10^{-3}$	186,31±0,84	1,93±0,01 [*11, *62]
^{60}Co	1085—1649	$(9{,}12\pm0{,}58)10^{-3}$	226,770±±1,947	2,349±±0,020 [*119]
		Cu_2O		
O	—	0,0065	164,54	1,70 [*120]
Cu	800—1050	0,12	151,14	1,57
	800	$1{,}9\cdot10^{-9}$	—	— [*121]
	850	$4{,}0\cdot10^{-9}$	—	—
	900	$7{,}7\cdot10^{-9}$	—	—
	950	$1{,}4\cdot10^{-8}$	—	—
	1000	$3{,}2\cdot10^{-8}$	—	—
Cu	900—1050	$D\sim p_{O_2}^{1/3,9}$	100,48±8,37	1,04±0,1 [*3]
^{59}Fe	800	Very slight penetration	—	—

23 – Continued

1	2	3	4	5
		ZnO		
O	1100—1300	$6{,}5\cdot10^{11}$	690,32	7,15[*96, *122]
Cu	1100—1250	$2\cdot10^{7}$	460,6±9,7	4,8±0,1[*123]
Zn	950—1370	1,3	307,73	3,19[*124]
	900—1200	4,8	305,64	3,17[*125]
	—	$5{,}3\cdot10^{-4}$	53,17	0,55[*126]
		GeO_2		
He	31—253	$6{,}1\cdot10^{-6}$	20,52	0,21[*127]
Ne	37—400	$5{,}6\cdot10^{-6}$	37,26	0,39[*127]
	—	$2{,}5\cdot10^{-4}$	55,27	0,57[*128]
		SrO		
O	1300	$D\sim2{,}4\cdot10^{-5}$	—	—[*3]
^{85}Sr	1200—1400	$4{,}48(+11{,}42; -3{,}22)10^{-4}$	263,51±3,67	2,76± ±0,038[*3, *129]
	1450—1600	$2{,}52(+15{,}18; -2{,}16)10^{2}$	444,97±6,37	4,61±0,066[*129, *3]
	1500	$D\sim p_{O_2}^{1/6}$ [23]	—	—
		Y_2O_3		
^{18}O	1064—1241	$6{,}06\cdot10^{-6}$	81,97	0,8[*62, *11]
Y	1400—1670	$1{,}65\cdot10^{-2}$	289,73	3,00
		ZrO_2		
^{18}O	600—1000	$2{,}34\cdot10^{-2}$	189,66	1,96[*130, *3]
	990	$D=(1{,}9\pm0{,}3)10^{-12}$	—	—[*131]
O	—	—	234,46	2,43
		Nb_2O_5		
^{18}O	850	$D \parallel [010]\approx2{,}1\cdot10^{-8}$	—	—[*132]
^{18}O	850	$D\perp[010]\approx1{,}1\times10^{-12}$	—	—[*132]
^{18}O	900	$D \parallel [010]\approx8{,}3\times10^{-11}$	—	—[*133]
	900	$D\perp[010]\approx1{,}4\times10^{-12}$	—	—[*133]
	900	—	209,34	2,17
O	870—1000	$1{,}07\cdot10^{-2}$	179,20	1,86[*134]
	850—1200	$1{,}72\cdot10^{-2}$	206,00	2,13[*135]
		MoO_3		
O	410—500	$9{,}1\cdot10^{-8}$	84,99	0,88[*134]
	—	—	203,06	2,10
^{14}C	300	$D=6{,}5\cdot10^{-12}$	—	—[*136]
	350	$D=50\cdot10^{-12}$	—	—[*136]
	400	$D=87\cdot10^{-12}$	—	—[*136]
	450	$D=43\cdot10^{-11}$	—	—[*136]

23 — Continued

1	2	3	4	5
	300—450	$8{,}9 \cdot 10^{-4}$	88,76	0,92[*136]
^{99m}Tc	480—700	$2{,}0 \cdot 10^{-6}$	121,4±12,6	1,3±0,13
		CdO		
^{115m}Cd	1780—2270	0,017(+0,060; —0,014)	328,18± ±28,96	3,4±0,3[*137]
^{18}O	630—850	$3{,}8 \cdot 10^{6}$	386,0±17	4,0±0,2[*138]
		SnO_2		
Sn	—	10^{7}	527,54	5,46
		BaO		
Ba^{2+}	1080—1230	10^{31}	1155,56	11,97[*139]
Ba	327—1077	$10^{-9\pm1}$	41,87±2,90	0,44± ±0,03[*140]
	1077—1277	$10^{29\pm7}$	1059±212,3	11±2,2[*140]
	327—1077	$3 \cdot 10^{-10\pm1}$	28,97±4,81	0,3±0,05[*141]
	1077—1277	$10^{31\pm8}$	1158,3± ±221,9	12±2,3[*141]
^{133}Ba	900—950	$7{,}52 \cdot 10^{-8}$	101,35	1,05[*142]
	950—1100	1,20(+6,52; —1,01)	271,23± ±20,27	2,81±0,21[*142]
	1200—1400	1,35(+4,97; —1,06)	386,09± ±6,47	4,00±0,067[*142]
		CeO_2		
O	850—1330	534 (+5160; —480)	305,97× ×(+26,06; —24,13)	3,17 (+0,27; —0,25)[*62, *143]
	780—1230	$9{,}55\ (+145; -8{,}76)10^{-5}$	90,73 (+28,96; —25,10)	0,94(+0,3; —0,26)[*62, *144]
		Pr_7O_{12} (*i*)		
^{18}O	735—865	$(5{,}5\pm0{,}6)10^{-6}$	77,205± ±1,047	0,800± ±0,011[*145]
	730—800	$(9\pm3)10^{-8}$	34,374± ±3,140	0,356± ±0,033[*146]
	830—930	$(1{,}3\pm0{,}3)10^{-5}$	79,947± ±1,968	0,828± ±0,020[*146]
	—	$D=(2{,}9\pm0{,}5)\times 10^{-6} \cdot p_{O_2} \times 0{,}28\pm 0{,}01 \cdot \exp\times(-19{,}050\pm0{,}375)/RT$	—	—[*147]
	—	$D=(6\pm4)10^{-9}\times p_{O_2}^{0,25\pm0,03} \exp\times (-8{,}210\pm0{,}750)/RT$	—	—[*148]

23 — Continued

1	2	3	4	5
		$GdO_{1,485}$		
O	750—1050	$5,87 \cdot 10^{-4}$	120,5	1,2*62
		Dy_2O_3		
^{18}O	1087—1235	$1,63 \cdot 10^{-5}$	109,86	1,1*11, *62
		Ho_2O_3		
^{18}O	1050—1274	$7,18 \cdot 10^{-3}$	169,69	1,8*11, *62
		Er_2O_3		
^{18}O	1060—1292	$1,31 \cdot 10^{-4}$	126,10	1,3*11, *62
		Tu_2O_3		
^{18}O	1015—1235	$1,14 \cdot 10^{-2}$	190,75	2,0*11, *62
		Lu_2O_2		
^{18}O	1020—1297	$1,88 \cdot 10^{-4}$	124,59	1,3*11, *62
		Ta_2O_5		
O	900—1000	$4,26 \cdot 10^{-2}$	179,61	1,86*67
		PbO		
O	514	$D \approx 10^{-10}$	—	—*3
Pb	490—600	10^5	276,33	2,86
		ThO_2		
Th	1846—2045	0,35	625,93	6,48*149
		UO_2		
H	500—1000	0,037	59,87±3,77	0,62±±0,04*150
O	450—600	$2,6 \cdot 10^{-5}$	124,35	1,29*151
	160—350	$5,5 \cdot 10^{-3}$	110,12	1,14*152
	445	$0,21 \cdot 10^{-13}$	—	—*3
	465	$0,40 \cdot 10^{-13}$	—	—
	497	$1,8 \cdot 10^{-13}$	—	—
	517	$1,8 \cdot 10^{-13}$	—	—
	522	$2,7 \cdot 10^{-13}$	—	—
	535	$1,4 \cdot 10^{-13}$	—	—
	540	$1,9 \cdot 10^{-13}$	—	—
	571	$5,5 \cdot 10^{-13}$	—	—

1	2	3	4	5
	600	$6{,}8\cdot10^{-13}$	—	—
	603	$14\cdot10^{-13}$	—	—
	650	$13\cdot10^{-13}$	—	—
	686	$6{,}5\cdot10^{-13}$	—	—
	855	$11{,}5\cdot10^{-13}$	—	—
^{18}O	650—1247	0,26	247,0	2,56[*153]
	400—900	$5\cdot10^{-5}$	87,9±12,6	0,9±0,1[*154]
	400—800	10^{-4}	87,9±12,6	0,9±0,1[*155]
	500—800	$2{,}7\cdot10^{-4}$	90,0±12,6	0,93±0,1[*156]
	600—800	$5\cdot10^{-4}$	92,1	0,95[*157]
	800—910	$\sim6\cdot10^{-4}$	~92,1	~0,95[*158]
	925±5	$D=(1{,}8\pm0{,}2)\times\times10^{-12}$	—	—
	1025±5	$D=(2{,}3\pm0{,}1)\times\times10^{-12}$	—	—
	1425±5	$D=(6{,}9\pm0{,}3)\times\times10^{-12}$	—	—
N	1000—1600	$2{,}25\cdot10^{-9}$	139,84	1,45
Kr	800—1600	$4{,}5\cdot10^{-8}$	196,78±8,37	2,04±±0,09[*159]
	900—1500	$1{,}6\cdot10^{-8}$	276,33±41,87	2,86±±0,43[*160]
Kr, Xe	900—1900	—	293,08±41,87	3,04±±0,43[*161]
Kr	1100—1600	$1{,}9\cdot10^{-8}$	318,20±16,75	3,30±±0,17[*162]
	900—1400	$2{,}9\cdot10^{-3}$	334,94±16,75	3,47±±0,17[*163]
Xe	800—1100	$1{,}1\cdot10^{-14}$	100,48±4,19	1,04±±0,04[*160]
	800—1150	$2{,}5\cdot10^{-8}$	205,15±4,19	2,12±±0,04[*159]
	1000—1500	$5{,}9\cdot10^{-8}$	192,59±8,37	2,0±0,1[*164]
	800—1500	—	205,15±8,37	2,12±±0,09[*165]
	900—1450	$2{,}6\cdot10^{-9}$	217,71	2,25[*164]
	1100—1400	$2{,}0\cdot10^{-2}$	293,08±20,93	3,04±0,22
	1200—1600	$5{,}0\cdot10^{-3}$	293,08±41,87	3,04±±0,43[*160]
	800—1600	$7{,}8\cdot10^{-6}$	297,26±25,12	3,08±0,26[*160]
	800—1300	$6{,}3\cdot10^{-6}$	301,45±16,75	3,12±±0,17[*162]
	1100—1550	—	393,56±16,75	4,08±±0,17[*166]
	1100—1500	24	531,72±37,68	5,51±±0,39[*164]
	1100—1400	23	535,91±41,87	5,55±±0,43[*160]
^{133}Xe	1065—1300	2,88(+13,4;	461,38±14,48	4,78±

23 – Continued

1	2	3	4	5
Xe	1400—2200	$4\cdot10^{2}-1,6\cdot10^{3}$	579,13±9,65	6,0±0,1 [*167]
	950—1700	$5\cdot10^{-2}-5$	376,44±38,61	3,9±0,4
	600—1300	$1\cdot10^{-8}-1\cdot10^{-4}$	164,09±38,61	1,7±0,4 [*168]
U	1050—1350	$1,3\cdot10^{-6}$	230,27	2,38 [*3, *169]
^{233}U	1400	$D<10^{-17}$	—	— [*170]
U	1410	$0.6-1.3\cdot10^{-13}$	—	— [*3]
	1590	$3\cdot10^{-13}$	—	— [*3]
	1712	$6\cdot10^{-13}$	—	— [*3]
	1730	$25\cdot10^{-13}$	—	— [*3]
	1750	$43\cdot10^{-13}$	—	— [*3]
		PuO_2		
^{18}O	680—1027	$1,19\cdot10^{-3}$ —2,4)10^{3}	176,2	1,81 [*3, *171] ±0,15 [*11]

*Values for D_0 are given without any letter-symbol. [*1]In equation $D=D_0\exp(-E/kT)$ is given as calculated for 1 atom. [*2]Annealed BeO. [*3]Self-diffusion. [*4]Fused; self-diffusion; impurities: Mg 10 ppm, Si 10 ppm, traces of Na, K, and Ca. [*5]Self-diffusion; polycrystal; particle size 80 – 115 mesh. [*6]Same, particle size 170 – 200 mesh. [*7]Self-diffusion of oxygen ion. [*8]For well-sintered specimen. [*9]For freely sintered specimen. [*10]Annealed in Ar at 1.01325 kPa, polycrystal, grain size 8 μm. [*11]Single crystal. [*12]Annealed in argon, total impurities 700 ppm. [*13]Annealed in air; for single crystals of 99.99% purity of two types (M and V). [*14]Type M; impurities, ppm: 250 Ca, 30 Al, 11 Si, 200 Fe, 10 Mn, 15 Cr, 20 Na, < 5 K. [*15]Type V; impurities, ppm: 190 Ca, 20 Al, 3 Si, 200 Fe, 8 Mn, 12 Cr, 15 Na, < 5 K. [*16]Single crystal, 99.99% purity; in air. [*17]Same, in argon. [*18]With 1% (mol.) Al_2O_3 in MgO. [*19]Radioactive method (removal of layers). [*20]Single crystal, 99.99% purity. [*21]Pure MgO; chemical diffusion. [*22]0.2% (mol.) Fe_2O_3. [*23]1% (mol.) Fe_2O_3. [*24]0.2% (mol.) Cr_2O_3. [*25]0.5% (mol.) Cr_2O_3. [*26]1% (mol.) Cr_2O_3. [*27]2.5% (mol.) Cr_2O_3. [*28]1% (mol.) CoO. [*29]5% (mol.) CoO. [*30]Volume lattice diffusion. [*31]Annealed in argon. [*32]Weak penetration. [*33]Strong penetration. [*34]From graph. [*35]From graph; single crystal. [*36]Polycrystal. [*37]From experiments on sintering. [*38]Polycrystal, 99.5% Al_2O_3; annealed in air. [*39]From graph; single crystal. [*40]Commercial Al_2O_3, 95% of theoretical density, 0.15% Na_2O, < 0.03% K_2O. [*41]Single crystal; parallel to c-axis. [*42]Same, perpendicular to c-axis. [*43]Corundum; 0.133 Pa, pressed at 19.62 MPa; radioactive method (removal of layers); diffusion along grain boundaries. [*44]Same, diffusion through volume of grains. [*45]Same, averaged data. [*46]Fused, purity 99.9%, before devitrification. [*47]Same, after devitrification. [*48]Glassy. [*49]Fused in oxygen at 101.325 kPa. [*50]Annealed in nitrogen. [*51]Impurities 0.1 – 3.5% (mol.). [*52]Annealed in hydrogen with nitrogen. [*53]Impurities 5.8% (mol.). [*54]Annealed in oxygen; impurities 5% (mol.) As_2O_3. [*55]Impurities 0.8% (mol.). [*56]Annealed in oxygen with nitrogen. [*57]Impurities 3.0% (mol.). [*58]Density 91%. [*59]Estimates. [*60]Single crystal, impurity Al_2O_3 0.001% by mass; self-diffusion close to surface. [*61]Same, within crystal. [*62]Annealed in air. [*63]TiO_x, x = 0.807-1.259. [*64]With x = 0.807. [*65]With x = 1.06. [*66]Does not depend on x. [*67]pO_2 = 21.3 – 101.3 kPa. [*68]pO_2 = 1.013 Pa to 101.3 kPa. [*69]pO_2 = 50.66 kPa. [*70]Stoichiometric, perpendicular to c-axis. [*71]Same, parallel to c-axis. [*72]Perpendicular to c-axis; reduced in vacuo at 1200°C, 1 h. [*73]Same, 5 h. [*74]Same, 10 h. [*75]Same, in air, 1 h. [*76]Same, 5 h. (cont'd.)

•[77]Same, 10 h. •[78]Granular specimen, stoichiometric. •[79]Same, reduced in vacuo at 1200°C, 1 h. •[80]Same, in air. •[81]Averaged value. •[82]According to various sources. •[83]Annealed in air at atmospheric pressure; pressed at 19.62 MPa; radioactive adsorption method. •[84]Same, radioactive method (removal of layers); diffusion along grain boundaries. •[85]Same, diffusion through volume of grains. •[86]Annealed in vacuo, $p = 0.133$ Pa; pressed at 19.62 MPa; radioactive adsorption method. •[87]Same, radioactive method (removal of layers); diffusion along grain boundaries. •[88]Same, diffusion through volume of grains. •[89]Rutile, annealed in air at atmospheric pressure; pressed at 14.715 MPa; averaged data. •[90]Annealed in oxygen at $p = 16.665$ kPa in interval of oxygen pressures from 667 to 74,660 Pa, $D \sim (pO_2)^{-2/3}$. •[91]Radioactive methods: adsorption and removal of layers. •[92]γ-MnO_2. •[93]Volume diffusion of oxygen in Wustite negligibly small. •[94]Volume diffusion in Wustite. •[95]Wustite. •[96]Isotope exchange method. •[97]Chemical coefficient of self-diffusion. •[98]Hematite. •[99]Annealed in air at atmospheric pressure; pressed at 19.62 MPa; averaged data. •[100]Single crystal, $pO_2 = 101.325$ Pa. •[101]Same, $pO_2 = 10.1325$ kPa. •[102]Same, $pO_2 = 1.01325$ kPa. •[103]$pO_2 = 101.325$ kPa. •[104]$pO_2 = 101.3$ to 101,325 Pa. •[105]Annealed in air, polycrystal, 99.995% (at.). •[106]Film. •[107]Annealed in air, polycrystal, 99.999% (by mass). •[108]Single crystal; coefficient of self-diffusion proportional to pressure p^{β}, where $\beta = 0.16$ at 1245°C and 0.20 at 1380°C; annealed in oxygen. •[109]Same, annealed in CO_2. •[110]Single crystal 99.999% purity. •[111]For single crystal with 0.012 fractional content of chromium ions. •[112]Same, with 0.032 fractional content of chromium ions. •[113]Chemical diffusion into single crystal (purity 99.92%) through [100] plane, with 1% (mol.) Cr_2O_3. •[114]Same, with 4.1% (mol.) Cr_2O_3. •[115]Same, with 4.5% (mol.) Cr_2O_3. •[116]Same, with 5.2% (mol.) Cr_2O_3. •[117]Same, with 8.1% (mol.) Cr_2O_3. •[118]Annealed in oxygen at $p = 101.325$ kPa; single crystal with impurities Co 220 ppm, Ag 80 ppm, Fe 60 ppm. •[119]Combined diffusion into single crystal. •[120]$pO_2 = 18.239$ kPa. •[121]$D^{\bullet(\alpha)}_{Cu}$ according to Moore and Zalikson. •[122]$D \sim (pO_2)^{1/2}$. •[123]Chemical diffusion. •[124]Density 95%, annealed in air. •[125]$pO_2 = 101.325$ kPa. •[126]Annealed in saturated zinc vapor; effective diffusion coefficient. •[127]Fused, glassy. •[128]Diffusion into amorphous and very slightly devitrified oxide, below transition temperature. •[129]Annealed in oxygen at $p = 2.026$ Pa; single crystal. •[130]Annealed in oxygen at 93.325 kPa. •[131]Monoclinic ZrO_2, impurities 0.08%; self-diffusion is independent of pO_2 at pressures from 21,278 to $1.1325 \cdot 10^{-14}$ Pa. •[132]α-Nb_2O_5, annealed in $H_2 + H_2O$; isotope exchange method, $pO_2 = 1.3332$ kPa. •[133]Same, $pO_2 = 1.013 \cdot 10^{-12}$ Pa. •[134]$pO_2 = 21.278$-101.325 kPa. •[135]$pO_2 = 101.325$ kPa. •[136]Annealed in argon; radioactive method (removal of layers); values of D at 400 and 450°C and activation energies are slightly low. •[137]Single crystal, 99.99% purity. •[138]Single crystal $D \sim (pO_2)^{-1/5}$. •[139]Through vacant sites. •[140]Through interstices (volume diffusion). •[141]Surface diffusion. •[142]Annealed in oxygen; diffusion coefficient is independent of oxygen partial pressures in interval from 101,325 to 2.0265 Pa; impurities in single crystal: 0.15% SrO, 0.08% CaO, 0.004% MgO, 0.003% Fe_2O_3, 0.002% Al_2O_3, 0.002% Cr_2O_3. •[143]Optical method, impurities ~0.1% (at.). •[144]Optical method; impurities 0.3% Gd (at.). •[145]Isotope exchange method; annealed in oxygen at $pO_2 = 3.920$ kPa. •[146]Same, at 28.598 kPa. •[147]Isotope exchange method, for near-stoichiometric oxide. •[148]Same, for oxide with composition far from stoichiometric. •[149]Single crystal, volume self-diffusion. •[150]High-purity single crystal. •[151]$O/U = 2.0-2.1$. •[152]$O/U = 2.0-2.34$. •[153]Annealed in $Ar + O_2$; UO_{2+x}, $x \leq 0.0015$. •[154]Same, $x = 0.006$. •[155]Same, $x = 0.020$. •[156]Same, $x = 0.10$. •[157]Same, $x = 0.12$. •[158]Same, $x = 0.16$. •[159]Sintered; reduction from $UO_{2.08}$. •[160]Sintered. •[161]Sintered, 93%. •[162]Sintered, 97%. •[163]Unsintered powder. •[164]Fused. •[165]Fused, $O/U = 1.98$. •[166]Sintered 90%. •[167]UO_{2-y}, $0.00 < y \leq 0.066$. •[168]UO_{2+x}, $0.00 < x \leq 0.250$. •[169]$O/U = 2.08$, annealed in air. •[170]Single crystal, annealed in $CO + CO_2$; lattice diffusion. •[171]Microcrystalline $^{238}PuO_2$, isotope exchange method.

24. VAPORIZATION RATE [1; 139; 140; 156; 183; 187; 202; 640]

Oxide	Temperature, °K	Vaporization rate G, kg/(m²·sec)	Vapor composition
1	2	3	4
BeO	1500—3000	—	BeO, Be, O, O_2, $(BeO)n$, $n=2, 3, 4, 5, 6$
	1773	$0,2\cdot10^{-8}$	—*1, *2
	2023	$0,55\cdot10^{-6}$	—
	2273	$0,33\cdot10^{-4}$	—
	2423	$0,24\cdot10^{-3}$	—
NO_2	313	—	31% NO_2*3
N_2O_4	—	—	69% N_2O_4
	373	—	88% NO_2
	—	—	12% N_2O_4
	~413	—	100% NO_2
MgO	1500—3000	—	MgO, Mg, O, O_2*1, *4
	2273	$3,5\cdot10^{-4}$	—
Al_2O_3	—	—	—*1, *5
	2273	$1,6\cdot10^{-6}$	—
SiO_2	—	—	—*1, *6
CaO	1500—3000	—	CaO, Ca, O, O_2*1, *7
	1623—1750	$\lg G=9,16-27400/T$	Ca, O
Sc_2O_3	2073—2173	—	ScO, O
	2173—2473	$\lg G=11,1-37\,000/T$	—*8
	2553—2673	$\lg G=10,39-35\,000/T$	ScO, Sc, O, Sc_2O
Ti_2O_3	3300 ± 300	—	TiO, TiO_2
VO_2	1573—1923	—	VO_2
Ga_2O_3	1523—1723	—	O_2, Ga, GaO, Ga_2O
GeO_2	1223—1323	—	GeO, O_2, (GeO_2), (GeO_3)
As_2O_3	< 2073	—	$(As_2O_3)_2$
	418—553	$\lg(G'\sqrt{T})=(11,85\pm\pm0,11)-5109/T$	—*9
	418—553	$\lg(G'\sqrt{T})=(17,10\pm\pm0,10)-8946/T$	—*10
SrO	1500—3000	—	SrO, Sr, O, O_2*1, *11
	700—1600	$\lg G=7,91-23\,660/T$	Sr, O
Y_2O_3	2273	$1,2\cdot10^{-6}$	—*12
	2443—2703	$\lg G=9,35-32\,500/T$	YO, O, Y_2O, Y_2O_2
ZrO_2	—	—	—*1, *13
NbO	1673—2173	—	NbO_2, NbO, (Nb)*1, *14
NbO_2	1673—1973	—	NbO_2
Nb_2O_5	1653	$2\cdot10^{-6}$	—
	1723	$4,1\cdot10^{-6}$	—
MoO_2	—	—	MoO_2, MoO_3, O_2
MoO_3	1223—1273	—	$(MoO_3)_3$
	811	$0,568\cdot10^{-6}$	—

1	2	3	4
	922	$4,867\cdot10^{-6}$	—
	977	$39,5\cdot10^{-6}$	—
	1033	$6,183\cdot10^{-4}$	—
In_2O	1073—1273	—	In, In_2O, O_2
In_2O_3	1423—1723	—	O_2, In, In_2O
Sb_2O_3	773—953	$\lg(G'\sqrt{T}) = (12,12\pm\pm0,07)-8880/T$	—[*15]
	773—953	$\lg(G'\sqrt{T}) = (15,10\pm\pm0,17)-12\,248/T$	—[*16]
BaO	1173—1473	—	BaO
	—	—	BaO, Ba_2O_2, Ba_2O_3[*17]
	1100—1500	$\lg G = 8,7 - 20\,000/T$	—
	1200—1800	$\lg G = 7,56 - 18\,900/T$	BaO, Ba_2O, BaO_2, Ba_2O_2[*18]
La_2O_3	1773—2073	—	LaO, O
	2233—2503	$\lg G = 9,42-28\,500/T$	LaO, O, La, La_2O, La_2O_2, La_2O_3
Ce_2O_3	—	—	CeO, CeO_2, O
CeO_2	1673—1823	—	CeO, CeO_2, O
Pr_2O_3	1723—2023	—	PrO, PrO_2, O, Pr
Pr_6O_{11}	—	—	O, PrO, PrO_2, Pr
Nd_2O_3	1773—2073	—	NdO, O
	2253—2433	$\lg G = 7,76 - 24\,800/T$	NdO, O
Sm_2O_3	1873—2123	—	Sm, SmO, O
	2233—2653	$\lg G = 10,40-31\,800/T$	Sm, SmO, O
Eu_2O_3	1773—2023	—	EuO, Eu, O
	1973—2193	$\lg G = 11,68-30\,400/T$	EuO, Eu, O
Gd_2O_3	1973—2173	—	GdO, Gd, O
	2343—2603	$\lg G = 11,17-35\,200/T$	GdO, Gd, O
Tb_2O_3	1973—2173	—	TbO, Tb, TbO_2
Tb_4O_7	—	—	O, TbO, TbO_2
Dy_2O_3	1973—2173	—	DyO, Dy, O
	2433—2643	$\lg G=9,35-30\,900/T$	DyO, Dy, O
Ho_2O_3	2073—2273	—	HoO, Ho, O
	2493—2693	$\lg G = 16,0-48\,700/T$	HoO, Ho, O
Er_2O_3	2073—2223	—	ErO, Er, O
	2493—2693	$\lg G=9,74-33\,100/T$	ErO, ErO, O
Tu_2O_3	2443—2643	$\lg G=7,82-27,500/T$	TuO, Tu, O
Yb_2O_3	1873—2073	—	Yb, O
	2363—2623	$\lg G=8,85-29\,600/T$	Yb, O, YbO
Lu_2O_3	2073—2223	—	LuO, Lu, O
	2553—2703	—	LuO, Lu, O
HfO_2	2173—2443	—	HfO, O
Tl_2O_3	723—1173	—	Tl_2O, Tl, O_2

24 – Continued

1	2	3	4
ThO_2	2173—2373	$\lg G = 10,84 - 37\,100/T$	ThO, ThO_2, O
	—	—	—*1, *19
UO_2	2023—2173	—	UO_2

*1Temperature for appreciable volatization. *2In vacuo at 2100°C, in air at 2400°C. *3Pressure 101.325 kPa. *4In vacuo at 1600°C, in air at 1900°C. *5In air at 1750°C. *6In air at 1800°C. *7In air at 1700°C. *8Sintered, relative density 0.96–0.97, zero open porosity. *9Rate of condensation, arsenolite. *10Same, claudetite. *11In air at 1600°C. *12Sintered ceramic, 99.85% Y_2O_3, relative density 96.5%, zero apparent porosity. *13In vacuo >2300°C. *14In vacuo at 1700°C. *15Rate of condensation, senarmontite. *16Same, valentinite. *17At high temperatures. *18Primarily BaO. *19In vacuo >2300°C.

25. VAPOR PRESSURE [1; 27; 29; 33; 42; 43; 66; 77; 81; 113; 124; 136; 140; 156; 174; 180; 187; 195]

Temperature or temperature interval, °K	Vapor pressure P, Pa
1	2
	H_2O
—	$\lg p = 8,882065 - 2445,5645/T + 8,2312 \lg T - 0,0167006\, T + 1,205 \cdot 10^{-5}\, T^2$ *1
$T_m - T_b$	$\lg p = 3,332 - 2940/T + 3,86 \lg T - 3,41 \cdot 10^{-3}\, T + 4,9 \cdot 10^{-8}\, T^2$ *2
	D_2O
298	Total 2673,1*2
373	" 95938,5*2
	Li_2O
1233—1573	$\lg p = 15,31 - 16\,200/T$*1
1000—1600	Metal $\lg p = 11,71 - 18\,760/T$
298	Total $1,439 \cdot 10^{-66}$ *3
800	Oxide $7,787 \cdot 10^{-16}$ *3
2800	" $3,236 \cdot 10^{4}$
4800	" $5,948 \cdot 10^{6}$
	BeO
1500	Oxide $3,50 \cdot 10^{-10}$ *3
2000	" $1,59 \cdot 10^{-5}$
2500	" $7,04 \cdot 10^{-2}$
3000	" 31,9
1500	Metal $6,03 \cdot 10^{-8}$*3

1	2
2000	Metal $1{,}28\cdot10^{-3}$
2500	" $1{,}41$
3000	" $1{,}99\cdot10^{2}$
1500	Atomic oxygen $5{,}64\cdot10^{-8}$ *3
2000	" " $1{,}21\cdot10^{-3}$
2500	" " $1{,}25$
3000	" " $1{,}61\cdot10^{2}$
1500	Molecular oxygen $1{,}96\cdot10^{-9}$ *3
2000	" " $3{,}25\cdot10^{-5}$
2500	" " $7{,}42\cdot10^{-2}$
3000	" " $20{,}3$
1500	Total $1{,}19\cdot10^{-7}$ *3
2000	" $2{,}97\cdot10^{-3}$
2500	" $2{,}80$
3000	" $4{,}13\cdot10^{2}$
2103—2573	$\lg p=13{,}162-33240/T$ *1, *4
2250—2413	$\lg p=20{,}63-34230/T-2\lg T$ *4
—	Metal $\lg p=12{,}10-29800/T$
293	Oxide $2{,}656\cdot10^{-115}$ *3
800	" $1{,}528\cdot10^{-33}$
2800	" $2{,}489$
4800	" $1{,}179\cdot10^{5}$
	B_2O_3
1330—1808	$\lg p=11{,}748-16960/T$
293	Oxide $6{,}464\cdot10^{-58}$ *3
800	" $1{,}327\cdot10^{-12}$
2800	" $1{,}272\cdot10^{6}$
	CO
—	$\lg p=11{,}01-1410/T-0{,}87\lg T-3{,}89\cdot10^{-3}\,T$ *5
	N_2O
129—183	$\lg p=11{,}704-1232/T$ *1
285—309	$\lg p=9{,}6313-858{,}63/T$ *2
	NO
73—92	$\lg p=12{,}173-851{,}8/T$ *1
109,3—125	$\lg p=10{,}565-681{,}1/T$ *2
	N_2O_3
248—273	$\lg p=12{,}42-2058/T$ *2

25 — Continued

1	2
	NO_2 (N_2O_4)
202,3—262	$\lg p = 14,77 - 2750/T$ *1
262—375	$\lg p = 10,94 - 1746/T$ *2
283	Total $58,75 \cdot 10^3$ *6
313	" $24,11 \cdot 10^4$
343	" $74,46 \cdot 10^4$
	N_2O_5
118,8—305	$\lg p = 14,89 - 3017/T$ *1
	O_3
90—243	$\lg p = 10,37803 - 0.001966943\,T - 814,941587/T$
	F_2O
—	$\lg p = 9,5141 - 578,64/T$
	Na_2O
298—1190	$\lg p = 13,74 - 24044/T$ *1
	MgO
1500	Oxide $2,60 \cdot 10^{-6}$ *3
2000	" $2,02 \cdot 10^{-1}$
2500	" $1,59 \cdot 10^2$
3000	" $1,29 \cdot 10^4$
1500	Metal $2,06 \cdot 10^{-5}$ *3
2000	" $3,80 \cdot 10^{-1}$
2500	" $1,29 \cdot 10^2$
3000	" $6,01 \cdot 10^3$
1500	Atomic oxygen $3,72 \cdot 10^{-6}$ *3
2000	" " $8,21 \cdot 10^{-2}$
2500	" " $31,9$
3000	" " $1,67 \cdot 10^3$
1500	Molecular oxygen $8,40 \cdot 10^{-6}$ *3
2000	" " $1,48 \cdot 10^{-1}$
2500	" " $48,6$
3000	" " $2,19 \cdot 10^3$
1500	Total $3,54 \cdot 10^{-5}$ *3
2000	" $8,11 \cdot 10^{-1}$
2500	" $3,69 \cdot 10^2$
3000	" $2,27 \cdot 10^4$
—	Metal $\lg p = 12,37 - 26100/T$
293	Oxide $1,145 \cdot 10^{-88}$ *3
800	" $9,152 \cdot 10^{-24}$
2800	" $2,387 \cdot 10^3$
4800	" $1,158 \cdot 10^7$

1	2
	Al_2O_3
2000	Total $5,77 \cdot 10^{-1}$
2323	" 46,263
2600—2900	$\lg p = 13,42 - 27320/T$ [*2]
	SiO
1345	Oxide 0,871 [*7]
1345	" 3,95 [*8]
1345	" 0,507 [*9]
1336—1450	" $\lg p = 13,078 - 16790/T$
	SiO_2
1800	Total $162,65 \cdot 10^{-5}$ [*10]
1900	» $1013,25 \cdot 10^{-5}$
2133—2503	$\lg p = 15,55 - 26430/T$ [*2]
293	Oxide $2,818 \cdot 10^{-87}$ [*3]
800	" $3,081 \cdot 10^{-23}$
2800	" $1,790 \cdot 10^{3}$
4800	" $9,068 \cdot 10^{6}$
	P_2O_3
343	Total $\sim 29,33 \cdot 10^{2}$
	P_2O_5
842	Total $73,33 \cdot 10^{2}$
298—842	$\lg p = 14,67 - 8250/T$ [*11]
298—631	$\lg p = 12,82 - 4940/T$ [*12]
942—964	$\lg p = 10,23 - 4520/T$ [*2]
	SO_2
293	Total $32,92 \cdot 10^{4}$ [*2]
323	" $85,09 \cdot 10^{4}$
197—264	$\lg p = 14,200 - 1868/T - 15,9 \cdot 10^{-3} T + 15,6 \cdot 10^{-6} T^{2}$
	SO_3
297	Total 32064 [*2]
312,95	" 78927
371,35	" $93,20 \cdot 10^{4}$
407,65	" $18,94 \cdot 10^{5}$
$0 - T_m$	$\lg p = 13,56 - 2680/T$ [*13]
$0 - T_m$	$\lg p = 14,09 - 2860/T$ [*14]
$0 - T_m$	$\lg p = 16,12 - 3610/T$ [*15]
$T_m - T_b$	$\lg p = 12,02 - 2230/T$ [*2]

25 – Continued

1	2
	Cl_2O
—	$\lg p = 10{,}002 - 1375{,}7/T$
273	Total 93 190
	ClO_2
—	$\lg p = 10{,}48 - 1548/T$
273	Total 65 327
	Cl_2O_6
—	$\lg p = 9{,}2 - 2070/T$ *1
273	Total 41,33
—	$\lg p = 11{,}4 - 2690/T$ *2
	Cl_2O_7
270—303	$\lg p = 10{,}29 - 1861{,}8/T$
273	Total 2986,4
	K_2O
298—2500	$\lg p = 13{,}74 - 24262/T$
	CaO
1500	Oxide $6{,}05\cdot10^{-7}$ *3
2000	" $3{,}99\cdot10^{-2}$
2500	" 27,8
3000	" $1{,}73\cdot10^{3}$
1500	Metal $4{,}78\cdot10^{-7}$ *3
2000	" $1{,}77\cdot10^{-2}$
2500	" 9,27
3000	" $5{,}28\cdot10^{2}$
1500	Atomic oxygen $3{,}38\cdot10^{-7}$ *3
2000	" " $1{,}16\cdot10^{-2}$
2500	" " 5,92
3000	" " $3{,}43\cdot10^{2}$
1500	Molecular oxygen $6{,}94\cdot10^{-8}$ *3
2000	" " $3{,}04\cdot10^{-3}$
2500	" " 1,67
3000	" " 91,9
1500	Total $1{,}49\cdot10^{-6}$ *3
2000	" $7{,}19\cdot10^{-2}$
2500	" 44,6
3000	" $2{,}69\cdot10^{3}$
2000	" $2{,}53\cdot10^{-2}$

1	2
1617—1728	$\lg p = 12{,}26 - 27\,400/T$ [*16]
293	Oxide $6{,}598 \cdot 10^{-89}$ [*3]
800	" $3{,}755 \cdot 10^{-24}$
2800	" $4{,}413 \cdot 10^{2}$
4800	" $1{,}105 \cdot 10^{6}$
1000—2860	Total $\lg p = 12{,}85 - 28\,020/T$
2860—T_b	" $\lg p = 11{,}22 - 23\,370/T$
	Sc_2O_3
2400—2573	ScO $\lg p = 12{,}59 - 32\,800/T$
2173—2473	ScO $\lg p = 14{,}42 - 37\,400/T$
2551	0,582
2567	0,796
2551	Atomic oxygen 0,148
2567	" " 0,203
	TiO
1847—1968	$\lg p = 15{,}44 - 29\,421/T - 0{,}583 \cdot 10^{-3}\,T$ [*17]
1000—2020	Total $\lg p = 14{,}24 - 29\,560/T$ [*18]
2020—....	" $\lg p = 12{,}70 - 26\,450/T$
1000—1980	" $\lg p = 13{,}42 - 26\,320/T$ [*19]
1980—2020	" $\lg p = 11{,}74 - 23\,000/T$
2020—3330	" $\lg p = 12{,}19 - 23\,910/T$
	Ti_2O_3
1971—2151	$\lg (p_{TiO}\, p_{TiO_2}) = 31{,}68 - 64\,700/T - 1{,}26 \cdot 10^{-3}\,T$ [*17]
	TiO_2
1000—2020	Total $\lg p = 14{,}37 - 30\,060/T$ [*1]
1849—2010	$\lg p = 16{,}20 - 30\,361/T - 0{,}492 \cdot 10^{-3}\,T$ [*20]
298—2113	$\lg p = 13{,}618 - 29\,945/T$ [*21]
	VO
298—1950	$\lg p = 13{,}028 - 26820/T$
	V_2O_5
T_m —1500	$\lg p = 7{,}17 - 7100/T$ [*2]
1215—1530	$\lg p = 4{,}64 - 5905/T$ [*22]
750—850	$\lg p\,(O_2) = 18{,}20 - 16\,300/T$ [*1]
835—940	$\lg p\,(V_4O_{10}) = 7{,}32 - 9800/T$ [*1]
870—950	$\lg p\,(V_6O_{14}) = 6{,}50 - 9500/T$ [*1]

25 – Continued

1	2
	Cr_2O_3
1690—1930	$\lg p(\mathrm{Cr}) = 13{,}77 - 27\,350/T$
1820—2020	$\lg p(\mathrm{CrO}) = 10{,}55 - 23\,256/T$
1840—2010	$\lg p(\mathrm{CrO_2}) = 14{,}14 - 30\,769/T$
	MnO
—	Oxide $\lg p = 14{,}540 - 26\,400/T - 5{,}72\cdot10^{-4}\,T$ *23
1000—2058	Total $\lg p = 11{,}62 - 21\,880/T$ *1
2058—3000	" $\lg p = 10{,}861 - 20\,320/T$ *2
	FeO
843—3000	Oxide $\lg p = 14{,}56 - 27\,016/T - 0{,}55\cdot10^{-3}T$
—	$\lg p = 14{,}70 - 24\,200/T$ *1, *3
—	$\lg p = 13{,}72 - 22\,580/T$ *2, *3
	Fe_3O_4
—	$\lg p = 14{,}24 - 22\,780/T$ *1, *3
—	$\lg p = 10{,}38 - 15\,560/T$ *2, *3
	CoO
1000—2000	Oxide $\lg p = 14{,}75 - 26\,670/T - 4{,}37\cdot10^{-4}\,T$ *3
	NiO
1438—1566	Oxide $\lg p = 5{,}01 - 25\,586/T - 7{,}67\cdot10^{-4}\,T + 7{,}21\cdot10^{-8}\,T^2$ *24
298—1600	$\lg p = 15{,}20 - 25\,500/T - 0{,}767\cdot10^{-3}\,T + 7{,}21\times10^{-8}\,T^2$
	CuO
873	Total $178{,}65\cdot10^{-7}$
1073	" $153{,}32\cdot10^{-4}$
1223	" $906{,}59\cdot10^{-4}$
	ZnO
1573	Total 200 (160) *25
1673	" 400 (413)
1773	" 1387
	Ga_2O_3
1796—1955	$\lg p = 15{,}464 - 27\,098/T$ *1

1	2
	GeO
915—978	$\lg p = 17{,}65 - 13\,770/T$ [*1]
1000	Total $\lg p = -12{,}575$
2000	" $\lg p = -1{,}25$
	GeO_2
1000—1389	Total $\lg p = 14{,}11 - 17\,240/T$ [*1]
1389—2000	" $\lg p = 12{,}26 - 14\,540/T$ [*2]
1000—1389	$(GeO_2)_n$ $\lg p = 14{,}99 - 22\,460/T$ [*1]
1389—T_b	$(GeO_2)_n$ $\lg p = 14{,}81 - 22\,340/T$ [*2]
—	GeO $\lg p = 13{,}42 - 11\,350/T$ [*3, *26]
—	GeO $\lg p = 14{,}20 - 11\,980/T$ [*26, *27]
1150—1350	Total 0,133 — 1,33
1153—1253	$\lg p = 12{,}28 - 15\,620/T$ [*1, *28]
1313—1373	$\lg p = 18{,}370 - 25\,517/T$ [*2, *28]
	As_2O_3
298—586	$\lg p = 12{,}28 - 4880/T$ [*29]
298—506	$\lg p = 15{,}82 - 6680/T$ [*30]
T_m—T_b	$\lg p = 2{,}12 - 3130/T$ [*2]
413—553	Total $\lg p = (13{,}48 \pm 0{,}06) - 5402/T$ [*1, *31]
413—573	" $\lg p = (12{,}93 \pm 0{,}07) - 5246/T$ [*1, *32]
583—673	" $\lg p = (8{,}64 \pm 0{,}06) - 2788/T$ [*2]
	SeO_2
298—T_{subl}	$\lg p = 23{,}52 - 6170/T - 3{,}02 \lg T$
523—663	$\lg p = 14{,}63 - 5740/T$ [*1]
533—663	$\lg p = 12{,}63 - 4420/T$ [*2]
	SeO_3
400—453	$\lg p = 9{,}327 - 2562/T$ [*2]
	SrO
1500	Oxide $1{,}08 \cdot 10^{-5}$ [*3]
2000	" 0,274
2500	" $1{,}06 \cdot 10^{2}$
3000	" $3{,}96 \cdot 10^{3}$
1500	Metal $4{,}56 \cdot 10$ c [*3]
2000	" $8{,}18 \cdot 10^{-2}$
2500	" 27,5
3000	" $1{,}07 \cdot 10^{3}$
1500	Atomic oxygen $1{,}57 \cdot 10^{-6}$ [*3]
2000	" " $3{,}29 \cdot 10^{-2}$

25 – Continued

1	2
2500	Atomic oxygen 12.5
3000	» » $5,72\cdot10^2$
1500	Molecular oxygen $1.50\cdot10^{-6}$ *3
2000	" " $2,43\cdot10^{-2}$
2500	" " 7,52
3000	" " $2,55\cdot10^2$
1500	Total $1,84\cdot10^{-5}$ *3
2000	" 0,412
2500	" $1,53\cdot10^2$
3000	" $5,86\cdot10^3$
1495—1635	$\lg p = 15,48 - 30\,700/T$ *16
1290—1650	$\lg p = 9,264 - 23\,660/T + 0,5 \lg T$ *33
293	Oxide $4,009\cdot10^{-81}$ *3
800	" $1,791\cdot10^{-21}$
2800	" $1,219\cdot10^3$
4800	" $1,399\cdot10^6$
1000—2730	Total $\lg p = 13,62 - 26\,130/T$
2730—T_b	» $\lg p = 11,127 - 22\,070/T$
	Y_2O_3
2500	YO 0,274
2600	YO 0,874
2700	YO 2,432
2500	Atomic oxygen 0.0537
2600	" " 0,1712
2700	" " 0,4357
2493—2703	Total $\lg p = 13,82 - 36150/T$
	ZrO_2
2000	Total $3,28\cdot10^{-9}$
2273	" 8,00
3273	" $2,13\cdot10^2$
2014—2290	$\lg p = 16,99 - 34\,383/T - 7,98\cdot10^{-4}\,T$
2200—2500	$\lg p = 13,312 - 37\,421/T$
1000—2950	Total $\lg p = 13,94 - 37\,150/T$
2950—T_b	» $\lg p = 12,60 - 33\,170/T$
1000—1478	ZrO $\lg p = 14,15 - 32\,860/T$
1478—2128	ZrO $\lg p = 13,74 - 32\,200/T$
2128—2950	ZrO $\lg p = 13,321 - 31320/T$
2950—T_b	ZrO $\lg p = 12,21 - 28\,020/T$
293	Oxide $4,785\cdot10^{-118}$ *3
800	" $2,025\cdot10^{-34}$
2800	" 1,259
4800	" $5,550\cdot10^4$
	NbO_2
1938	Total 0,0803
2031	" 0,425

1	2
2081	Total 1,006
2122	" 1,493
2213—2393	$\lg p = 14{,}54 - 30\,300/T$
	Nb_2O_5
1653	Total $11{,}47 \cdot 10^{-4}$
1723	" $2{,}40 \cdot 10^{-3}$
	MoO_3
298—1066	$\lg p = 32{,}81 - 16\,140/T - 5{,}53 \lg T$
—	$\lg p = 14{,}077 - 15\,110/T + 1{,}46 \lg T - 1{,}32 \times \times 10^{-3}\, T$ *1, *34
$T_m - T_b$	$\lg p = 36{,}19 - 14\,560/T - 7{,}04 \lg T$ *2
298—T_m	$\lg p = 34{,}20 - 16\,150/T - 5{,}53 \lg T$ *4, *35
$T_m - T_b$	$\lg p = 36{,}66 - 14110/T - 7{,}08 \lg T$
873	$(MoO_3)_n$ 0,616
898	$(MoO_3)_n$ 1,49
923	$(MoO_3)_n$ 6,91
948	$(MoO_3)_n$ 20,5
973	$(MoO_3)_n$ 47,5
—	$4{,}576 \lg p = (85{,}2 \pm 0{,}2) - 75\,400/T$ *1
—	$4{,}576 \lg p = 47{,}48 - 35\,200/T$ *2
	Tc_2O_7
—	$\lg p = 20{,}404 - 7205/T \pm 8\%$ *1
—	$\lg p = 11{,}124 - 3571/T \pm 1\%$ *2
	CdO
1773	Total $10{,}13 \cdot 10^4$
298—T_{subl}	$\lg p = 18{,}95 - 14\,590/T - 1{,}76 \lg T$
	In_2O_3
1563—1763	$\lg p = 16{,}478 - 27\,791/T$
	SnO
—	$\lg p = 12{,}900 - 13\,160/T$ *36
	SnO_2
1623—1748	$\lg p = 18{,}6 - 28\,800/T$ *37
	Sb_2O_3
742—914	$\lg p = 13{,}437 - 9625/T$ *38
742—839	$\lg p = 14{,}320 - 10\,360/T$ *39
929—1073	$\lg p = 7{,}262 - 3900/T$ *2

25 – Continued

1	2
773—929	Total $\lg p = (13{,}86 \pm 0{,}07) - 9568/T$ *40
773—T_m	" $\lg p = (13{,}47 \pm 0{,}08) - 9535/T$ *41
	TeO_2
298—T_m	$\lg p = 25{,}63 - 13\,940/T - 3{,}52 \lg T$ *42
	Cs_2O
298—1800	$\lg p = 13{,}74 - 33\,880/T$
	BaO
1500	Oxide $1{,}71 \cdot 10^{-2}$ *3
2000	" 46,3
2500	" $3{,}17 \cdot 10^{3}$
3000	" $3{,}85 \cdot 10^{4}$
1500	Metal $4{,}62 \cdot 10^{-6}$ *3
2000	" $6{,}67 \cdot 10^{-2}$
2500	" 16,0
3000	" $5{,}45 \cdot 10^{2}$
1500	Atomic oxygen $1{,}58 \cdot 10^{-6}$ *3
2000	" " $1{,}90 \cdot 10^{-2}$
2500	" " 8,72
3000	" " $3{,}52 \cdot 10^{2}$
1500	Molecular oxygen $1{,}52 \cdot 10^{-6}$ *3
2000	" " $1{,}87 \cdot 10^{-2}$
2500	" " 3,64
3000	" " 96,3
1500	Total $1{,}71 \cdot 10^{-2}$ *3
2000	" 46,4
2500	" $3{,}19 \cdot 10^{3}$
3000	" $3{,}95 \cdot 10^{4}$
—	Oxide $\lg p = 12{,}12 - 21\,900/T$
—	Metal $\lg p = 9{,}936 - 9718/T$
1200—1700	$\lg p = 10{,}99 - 19\,700/T$
1200—1555	$\lg p = 15{,}57 - 26\,000/T$
1200—1800	$\lg p = 10{,}76 - 19\,400/T$
1100—1500	$\lg p = 11{,}72 - 20\,000/T$
293	Oxide $6{,}889 \cdot 10^{-63}$ *3
800	" $3{,}667 \cdot 10^{-15}$
2800	" $1{,}361 \cdot 10^{4}$
4800	" $2{,}454 \cdot 10^{6}$
1000—2196	Total $\lg p = 12{,}10 - 21110/T$
2196—T_b	" $\lg p = 10{,}130 - 16\,800/T$
	La_2O_3
—	Metal $\lg p = 10{,}36 - 20\,850/T$ *43
—	LaO $\lg p = 12{,}54 - 23\,360/T$ *43

1	2
—	LaO $\lg p = 12{,}33 - 28\,500/T$
2234	LaO 0,4908
2307	LaO 1,062
2353	LaO 2,382
2412	LaO 3,842
2441	LaO 5,619
2234	Atomic oxygen 0,0787
2307	" " 0,1702
2353	" " 0,3820
2412	" " 0,6162
2441	" " 0,9013
2233—2443	Total $\lg p = 13{,}03 - 29\,370/T$
4473	" $1{,}013 \cdot 10^5$
	CeO_2
1800	Atomic oxygen 1.013
	Pr_2O_3
2000	PrO 0,0101
2000	$PrO_2 < 5{,}07 \cdot 10^{-4}$
2000	Metal $1{,}013 \cdot 10^{-3}$
	Nd_2O_3
2000	NdO $< 5{,}07 \cdot 10^{-3}$
2000	Metal $< 5{,}07 \cdot 10^{-4}$
—	NdO $\lg p = 10{,}67 - 24\,800/T$
2255	" 0,502
2281	" 0,664
2306	" 0,858
2332	" 1,117
2383	" 1,692
2434	" 2,676
2255	Atomic oxygen 0,080
2281	" " 0,108
2306	" " 0,137
2332	" " 0,178
2383	" " 0,270
2434	" " 0,427
2253—2433	Total $\lg p = 12{,}89 - 29\,370/T$
	Sm_2O_3
2000	SmO, Sm $\approx 3{,}04 \cdot 10^{-4}$
—	SmO $\lg p = 13{,}30 - 31\,800/T$
	EuO
1334—1758	Metal $\lg p = (11{,}27 \pm 0{,}13) - (16\,590 \pm 205)/T$

25 – Continued

1	2
	Eu_3O_4
1604—2016	Metal $\lg p = (11,17 \pm 0,18) - (18\,840 \pm 305)/T$
	Eu_2O_3
2000	Metal $\approx 7,09 \cdot 10^{-3}$
2000	EuO $\approx 9,12 \cdot 10^{-4}$
—	EuO $\lg p = 14,58 - 30\,400/T$
	Gd_2O_3
—	GdO $\lg p = 14,06 - 35\,200/T$
	Er_2O_3
—	ErO $\lg p = 12,62 - 33\,100/T$
	Tu_2O_3
2300	Metal 0,1013
	Yb_2O_3
2300	Metal 0,1013
—	$\lg p = 11,72 - 29\,600/T$
	Lu_2O_3
2300	LuO 0,0101
	HfO_2
2940	Total 3,039*44
	WO_3
$1000 - T_m$	$\lg p = 17,75 - 24\,600/T$
	ReO_2
923—1058	$\lg p = 10,89 - 14\,347/T$*1
	ReO_3
598—693	$\lg p = 14,40 - 10\,882/T$*1
573—713	$\lg p = 9,870 - 4966/T$
	Re_2O_7
—	$\lg p = 14,90 - 7320/T$*1
523—598	$\lg p = 15,14 - 6416,9/T$*45
$273 - T_m$	$\lg p = 17,12 - 7300/T$
$T_m - T_b$	$\lg p = 11,22 - 3950/T$

1	2
	ReO_4
473—593	$\lg p = 7{,}6100 - 1738{,}7/T$
	OsO_4
273—315	$\lg p = 12{,}82 - 2580/T$*46
273—329	$\lg p = 11{,}76 - 2955/T$*47
$T_m - T_b$	$\lg p = 10{,}13 - 2065/T$
	Tl_2O
453—588	$\lg p = 13{,}63 - 6612/T$
	PbO
887—1151	$\lg p = 13{,}71 - 13\,900/T$*48
298—T_m	$\lg p = 16{,}48 - 13\,480/T - 0{,}92 \lg T - 0{,}35 \times \times 10^{-3}\, T$*49
$T_m - T_b$	$\lg p = 16{,}97 - 13\,300/T - 0{,}81 \lg T - 0{,}43 \cdot 10^{-3}\, T$
$T_m - T_b$	$\lg p = 21{,}59 - 13\,310/T - 2{,}77 \lg T$*34
273	Oxide $2{,}510 \cdot 10^{-34}$ *3
800	" $2{,}034 \cdot 10^{-4}$
2800	" $10{,}116 \cdot 10^{6}$
	ThO_2
1200	Total $133 \cdot 10^{-15}$
2050—2250	$\lg p = 13{,}65 - 37\,100/T$
2500—2900	$\lg p = 12{,}991 - 34\,890/T$
2000—3273	$\lg p = 12{,}21 - 31\,600/T$
1000—3273	Total $\lg p = 14{,}51 - 37610/T$
1000—3278	" $\lg p = 13{,}26 - 30\,830/T$*50
	UO
2000	Total $1{,}013$*3

25 – Continued

1	2
	UO_2
1600—2800	Total $\lg p = 15{,}423 - 37\,195/T + 3{,}5162 \cdot 10^6/T^2 + 2{,}6178 \cdot 10^9/T^3$
1603—1803	$\lg p = 12{,}181 - 38\,220/T$*51
	UO_3
1600	Total 10,13*3
	PuO_2
>1793	$\lg p = 13{,}135 - 27\,900/T$
—	$\lg p = 13{,}078 - 29\,240/T$
	Am_2O_3
4540	Total $1{,}013 \cdot 10^5$
1000—1340	Atomic oxygen $\lg p = 11{,}835 - 31\,690/T$*3
1340—2225	" " $\lg p = 11{,}592 - 31\,280/T$
1000—1340	Molecular oxygen $\lg p = 11{,}761 - 36\,870/T$*3
1340—2225	" " $\lg p = 11{,}149 - 36\,050/T$
1000—1340	Metal $\lg p = 11{,}358 - 31\,560/T$ *3
1340—2225	" $\lg p = 11{,}149 - 31\,280/T$
1000—1340	AmO $\lg p = 12{,}027 - 32\,000/T$*3
1340—2225	AmO $\lg p = 11{,}512 - 31\,310/T$
1000—1340	Total $\lg p = 12{,}280 - 31\,790/T$ *3
1340—2225	" $\lg p = 11{,}899 - 31\,280/T$

*1 Over solid phase. *2 Over liquid phase. *3 Calculated values. *4 Knudsen method. *5 Over solid phase at liquid H_2 temperature. *6 Mixture of NO_2 and N_2O_4. *7 According to data of Gyunter, Gel'd, and Kochnev. *8 According to data of Schäfer and Hornle. *9 According to data of Porter and Chupek. *10 Cristobalite. *11 Vapor of α-P_4O_{10}. *12 Metastable P_2O_5, vapor P_4O_{10}. *13 α-SO_3. *14 β-SO_3. *15 γ-SO_3. *16 Method of Langmuir and Claasen. *17 Method of Knudsen and Grove. *18 Over condensed TiO. *19 Over system Ti(s)/TiO(s). *20 According to data of Grove. *21 Anatase. *22 V_4O_{10}. *23 Upon sublimation. *24 Method of Langmuir. *25 Faizer (Feiser); data require verification. *26 Over system Ge/GeO_2. *27 Experimental data. *28 Insoluble form. *29 Claudetite I, vapor As_4O_6. *30 Arsenolite, vapor As_4O_6. *31 Arsenolite. *32 Claudetite. *33 Method of Langmuir and Moore. *34 Kelley. *35 Method of Knudsen, Kubaschewski, and Evans. *36 According to data of Platteev and Meier. *37 According to data of Faizer (Feiser) and Veselovskii. *38 Orthorhombic, vapor Sb_4O_6. *39 Cubic, Sb_4O_6 vapor. *40 Senarmontite; sublimation pressure. *41 Valentinite; sublimation pressure. *42 Tellurite. *43 Along low-oxygen boundary of field of γ-La_2O_3. *44 Tungsten effusion cell. *45 Powder, 99.99% Re. *46 Tetragonal. *47 Cubic. *48 β-PbO. *49 Yellow. *50 Over system Th/ThO_2. *51 Uraninite, vapor U_2O_4.

Note. When there is no indication of the oxide component to which the pressure refers, it should be considered that the data are for the total pressure of the oxide vapors.

26. DISSOCIATION PRESSURE [1; 62; 174; 182; 184; 195; 202]

Temperature or interval, °K	Dissociation pressure p or equation for dissociation pressure, Pa
1	2
	Li_2O_2
468	$1{,}013 \cdot 10^5$
	BeO
2000	$6{,}69 \cdot 10^{-4}$
298—1556	$\lg p_{O_2} = 17{,}27 - 62\,710/T - 0{,}726 \lg T$*1
1556—2673	$\lg p_{O_2} = 17{,}98 - 63\,710/T - 0{,}726 \lg T$
298—1556	$\lg p_{O} = 14{,}62 - 44\,540/T - 0{,}363 \lg T$
1556—2673	$\lg p_{O} = 14{,}97 - 45\,080/T - 0{,}363 \lg T$
1000	$\lg p_{O_2} = -24{,}78$*2
2000	$\lg p_{O_2} = -4{,}494$
3000	$\lg p_{O_2} = +1{,}01$
1000	$\lg p_{O} = -28{,}19$
2000	$\lg p_{O} = -2{,}920$
3000	$\lg p_{O} = +2{,}06$
	MgO
2000	$0{,}506$
923—1380	$\lg p_{O_2} = 16{,}79 - 63\,540/T - 0{,}11 \lg T$*3
923—1380	$\lg p_{O} = 14{,}37 - 45\,000/T - 0{,}05 \lg T$
1000	$\lg p_{O_2} = -13{,}648$*2
2000	$\lg p_{O_2} = -0{,}787$
3075	$\lg p_{O_2} = +3{,}560$
1000	$\lg p_{O} = -14{,}126$
2000	$\lg p_{O} = -1{,}068$
3075	$\lg p_{O} = +3{,}442$
	Al_2O_3
2000	$2{,}63 \cdot 10^{-3}$
1000	$\lg p_{O_2} = -23{,}135$*2

26 – Continued

1	2
2000	$\lg p_{O_2} = -4,581$
3000	$\lg p_{O_2} = +1,318$
1000	$\lg p_{O} = -18,871$
2000	$\lg p_{O} = -2,965$
3000	$\lg p_{O} = +2,213$
	SiO
298—1690	$\lg p_{O_2} = 13,05—44\,240/T$*4
1690—2075	$\lg p_{O_2} = 16,17—49\,520/T$
1000	$\lg p_{O_2} = -28,03$*2
2000	$\lg p_{O_2} = -7,70$
3000	$\lg p_{O_2} = -0,964$
1000	$\lg p_{O} = -21,319$
2000	$\lg p_{O} = -4,52$
3000	$\lg p_{O} = +1,08$
	SiO_2
1000	$\lg p_{O_2} = -13,150$*2
2000	$\lg p_{O_2} = +0,258$
3000	$\lg p_{O_2} = +4,492$
1000	$\lg p_{O} = -13,877$
2000	$\lg p_{O} = -0,543$
3000	$\lg p_{O} = +3,799$
	CaO
2000	$2,13 \cdot 10^{-2}$
298—1122	$\lg p_{O_2} = 15,35—66\,150/T$*5
1122—1760	$\lg p_{O_2} = 16,22—67\,130/T$
298—1122	$\lg p_{O} = 13,65—46\,260/T$
1122—1760	$\lg p_{O} = 14,09—46\,750/T$
1000	$\lg p_{O_2} = -16,60$*2
2000	$\lg p_{O_2} = -2,519$
3500	$\lg p_{O_2} = +3,060$

1	2
1000	$\lg p_{O} = -15,60$ [*2]
2000	$\lg p_{O} = -1,93$
3500	$\lg p_{O} = +3,713$
	TiO
298—1980	$\lg p_{O_2} = 19,95-59\,280/T - 1,57 \lg T$ [*6, *7]
1980—2020	$\lg p_{O_2} = 20,96-61\,250/T - 1,57 \lg T$
2020—3240	$\lg p_{O_2} = 17,93-55\,130/T - 1,57 \lg T$
1000	$\lg p_{O_2\,(max)} = -39,09$ [*8]
1000	$\lg p_{O_2\,(min)} = -36,49$
1000	$\lg p_{O_2} = -31,514$ [*2]
2020	$\lg p_{O_2} = -9,526$
3240	$\lg p_{O_2} = -1,666$
1000	$\lg p_{O} = -23,059$ [*2]
2020	$\lg p_{O} = -5,374$
3240	$\lg p_{O} = +1,151$
	Ti_2O_3
1000	$\lg p_{O_2\,(max)} = -36,49$ [*8]
1000	$\lg p_{O_2\,(min)} = -25,09$
1000	$\lg p_{O_2} = -33,15$ [*6, *9]
1873	$\lg p_{O_2} = -12,43$
1000	$\lg p_{O} = -23,87$ [*6, *9]
1873	$\lg p_{O} = -7,35$
1000	$\lg p_{O_2} = -25,478$ [*2]
2020	$\lg p_{O_2} = -6,16$
2400	$\lg p_{O_2} = -2,84$
1000	$\lg p_{O} = -20,041$
2020	$\lg p_{O} = -3,69$
2400	$\lg p_{O} = -0,98$
	Ti_3O_5
1000	$\lg p_{O_2} = -25,07$ [*2, *10]

26 – Continued

1	2
2020	$\lg p_{O_2} = -6{,}24$
1000	$\lg p_{O} = -19{,}839$ *2, *10
2020	$\lg p_{O} = -3{,}73$
1100—1323	$\lg p_{O_2} = 14{,}56 - 36\,800/T$ *11
	TiO_2
1000	$\lg p_{O_2\,(max)} = -20{,}69$ *8
2000	$1{,}52 \cdot 10^{-5}$
298—2183	$\lg p_{O_2} = 15{,}06 - 31\,910/T$ *2
1000	$\lg p_{O} = -15{,}73$ *2
2020	$\lg p_{O} = -0{,}98$
	TiO_x
1400	$5{,}77 \cdot 10^{-11}$ *2, *12; $6{,}15 \cdot 10^{-11}$ *2, *13
"	$7{,}19 \cdot 10^{-11}$ *2, *14; $7{,}90 \cdot 10^{-11}$ *2, *15
"	$7{,}70 \cdot 10^{-11}$ *2, *16; $9{,}12 \cdot 10^{-11}$ *2, *17
	VO
1000	$\lg p_{O_2\,(max)} = -29{,}5$ *8 $\lg p_{O_2\,(min)} = -28{,}2$
	V_2O_3
—	$\lg p_{O_2} = 15{,}61 - 40\,600/T$ *18
	V_3O_5
1400—1700	$\lg p_{O_2} = 11{,}91 - 19\,400/T$ *19
	V_4O_7
1400	$\lg p_{O_2} = -3{,}62$ *20
1500	$\lg p_{O_2} = -2{,}58$ *21
1600	$\lg p_{O_2} = -1{,}59$ *22
1700	$\lg p_{O_2} = -0{,}67$ *23

1	2
	VO_2
1000—1120	$\lg p_{O_2} = 5,51 - 11200/T$*24
1120	$\lg p_{O_2} = 6,06 - 11020/T$*24
	MnO
1000	$\lg p_{O_2 (max)} = -29,7$*8
1000	$\lg p_{O_2 (min)} = -5,7$
1517—2058	$\lg p_{O_2} = 13,63 - 41700/T$*6,*25
2058—2368	$\lg p_{O_2} = 10,54 - 35410/T$
298—1517	$\lg p_{O} = 12,29 - 33280/T$*6,*25
1517—2058	$\lg p_{O} = 12,79 - 34030/T$
2058—2368	$\lg p_{O} = 11,24 - 30910/T$
1000	$\lg p_{O_2} = -10,74$*2
2000	$\lg p_{O_2} = +0,09$
3000	$\lg p_{O_2} = +3,300$
1000	$\lg p_{O} = -12,57$*2
2000	$\lg p_{O} = -0,57$
3000	$\lg p_{O} = +3,238$
	Mn_3O_4
1073	1373,2*26
1173	1106,6
1273	906,6
1373	733,3
1473	573,3
800—1445	$\lg p_{O_2} = 18,574 - 23750/T$*8
1445—1800	$\lg p_{O_2} = 16,073 - 20170/T$
800—1445	$\lg p_{O_2} = 19,02 - 26270/T$*6,*27
1445—1773	$\lg p_{O_2} = 19,60 - 27110/T$
	Mn_2O_3
849—1029	$\lg p_{O_2} = 13,06 - 10100/T$*28

26 – Continued

1	2
1152—1257	$\lg p_{O_2} = 12{,}59 - 9520/T$
—	$\lg p_{O_2} = 13{,}629 - 10\,910/T$
	MnO_2
—	$\lg p_{O_2} = 13{,}132 - 6890/T$
	FeO
1023	$213{,}58 \cdot 10^{-17}$
1223	$198{,}65 \cdot 10^{-11}$
2723	$21{,}27 \cdot 10^{-3}$
1000	$\lg p_{O_2\,(max)} = -15{,}5$[*8]; $\lg p_{O_2\,(min)} = -14{,}2$
813—1473	$\lg p_{O_2} = (11{,}62 \pm 0{,}02) - (27\,300 \pm 20)/T$[*8, *29]
923—1623	$\lg p_{O_2} = (11{,}68 \pm 0{,}02) - (27\,360 \pm 50)/T$
1000	$\lg p_{O_2} = -15{,}78$[*6, *30]
2000	$\lg p_{O_2} = -2{,}33$
3000	$\lg p_{O_2} = +1{,}83$
830—1670	$\lg p_{O_2} = 11{,}73 - 27\,500/T$
1000	$\lg p_{O} = -15{,}19$[*6, *30]
2000	$\lg p_{O} = -1{,}84$
3000	$\lg p_{O} = +2{,}47$
1000	$\lg p_{O_2} = -11{,}01$[*31]
2000	$\lg p_{O_2} = +0{,}22$
3000	$\lg p_{O_2} = +3{,}63$
660—1670	$\lg p_{O_2} = 11{,}76 - 22\,740/T$
1000	$\lg p_{O} = -12{,}69$[*31]
2000	$\lg p_{O} = -0{,}50$
3000	$\lg p_{O} = +3{,}401$
	Fe_3O_4
996—1496	$\lg p_{O_2} = 18{,}01 - 32\,500/T$[*8, *32]
1173—1473	$\lg p_{O_2} = (18{,}36 \pm 0{,}06) - (33\,210 \pm 90)/T$[*8, *32]
1000	$\lg p_{O_2\,(max)} = -14{,}2$[*8]; $\lg p_{O_2\,(min)} = -9{,}0$

26 – Continued

1	2
1452	$\lg p_{O_2} = +1,06$[*8, *33]; $\lg p_{O_2} = +2,08$[*8, *34]
1581	$\lg p_{O_2} = +1,61$[*8, *35]; $\lg p_{O_2} = +2,14$[*8, *36]
	$\lg p_{O_2} = +3,599$[*8, *37]
1666	$\lg p_{O_2} = +1,93$[*8, *38]; $\lg p_{O_2} = +2,24$[*3, *39]
	$\lg p_{O_2} = +3,60$[*8, *40]; $\lg p_{O_2} = +4,314$[*8, *41]
1731	$\lg p_{O_2} = +2,15$[*8, *42]; $\lg p_{O_2} = +2,35$[*8, *43]
	$\lg p_{O_2} = +3,61$[*8, *44]; $\lg p_{O_2} = +4,324$[*8, *45]
	$\lg p_{O_2} = +5,01$[*8, *46]
998	$5,666 \cdot 10^{-15}$
1200	$6,11 \cdot 10^{-10}$
1400	$5,10 \cdot 10^{-6}$
1490—1600	$\lg p_{O_2} = 19,568 - 34\,400/T$
298—1642	$\lg p_{O_2} = 18,09 - 32\,650/T$[*6, *47]
1642—1870	$\lg p_{O_2} = 23,98 - 42\,330/T$[*6, *47]
700	$\lg p_{O_2} = -21,59$[*31]
1000	$\lg p_{O_2} = -11,32$
1400	$\lg p_{O_2} = -4,70$
1650	$\lg p_{O_2} = -2,02$
670—1670	$\lg p_{O_2} = 12,40 - 23\,790/T$
700	$\lg p_{O} = -23,74$[*31]
1000	$\lg p_{O} = -12,96$
1400	$\lg p_{O} = -5,87$
1650	$\lg p_{O} = -3,10$
	Fe_2O_3
700	$29,878 \cdot 10^{-25}$
1030—1200	$\lg p_{O_2} = 19,11 - 25\,200/T$
900—1000	$\lg p_{O_2} = 11,14 - 13\,750/T$[*48]
843—1873	$\lg p_{O_2} = 19,71 - 26\,090/T$[*8]

1	2
1373—1573	$\lg p_{O_2} = 19,91-25\,900/T$ [*49]
1193—1373	$\lg p_{O_2} = 11,01-14\,882/T$
1373—1703	$\lg p_{O_2} = 9,01-32\,406/T$
	CoO
298—1768	$\lg p_{O_2} = 12,42-24\,440/T$ [*6, *50]
1768—2077	$\lg p_{O_2} = 13,32-26\,080/T$
2077—	$\lg p_{O_2} = 10,60-20\,390/T$
298—1768	$\lg p_{O} = 12,18-25\,280/T$ [*6, *50]
1768—2083	$\lg p_{O} = 12,64-26\,220/T$
2083—	$\lg p_{O} = 11,28-23\,380/T$
1000–1500	$\lg p_{O_2} = 13,16-25\,200/T$ [*8, *51]
1000	$\lg p_{O_2\,(max)} = -9,5$ [*8]
1000	$\lg p_{O_2\,(min)} = +2,5$
1000—1240	$\lg p_{O_2} = 17,61-15\,600/T$ [*8, *52]
	Co_3O_4
1073	1333
1243	$102\cdot 10^3$
	NiO
1000	$\lg p_{O_2} = -10,24$ [*6, *53]
2000	$\lg p_{O_2} = +2,71$
2500	$\lg p_{O_2} = 5,15$
1000	$\lg p_{O} = -12,42$ [*6, *50]
2000	$\lg p_{O} = +0,68$
2500	$\lg p_{O} = 3,24$
1000	$\lg p_{O_2\,(max)} = -11,5$ [*8]
1000—1500	$\lg p_{O_2} = 13,91-24\,100/T$ [*54]

1	2
	Ni_2O_3
673	10 266[*8, *55]
773	14 799
873	20 132
	Cu_2O
1000	$\lg p_{O_2\,(max)} = -4,6$[*8]; $\lg p_{O_2\,(min)} = -2,0$[*8]
1000—1300	$\lg p_{O_2} = 12,57-17\,390/T$[*56];
"	$\lg p_{O_2} = 14,03-12\,780/T$[*57]
"	$\lg p_{O_2} = 10,57-10\,800/T$[*58]
298	$3\cdot 10^{-48}$
473	$3\cdot 10^{-47}$
673	$1\cdot 10^{-23}$
1337	0,019
1363	0,049
1423	0,187
1523	4,00
1623	30,66
—	$\lg p_{O_2} = 14,423-17\,940/T-0,242\,\lg T-0,000641\cdot T$
	GeO
1000	$\lg p_{O_2} = -14,30$[*2]
2000	$\lg p_{O_2} = -0,88$
1000	$\lg p_{O} = -14,45$[*2]
2000	$\lg p_{O} = -1,12$
	GeO_2
700—1210	$\lg p_{O_2} = 14,38-28\,870/T$[*6, *59]
1210—1389	$\lg p_{O_2} = 15,73-30\,530/T$[*6, *59]
1389—2000	$\lg p_{O_2} = 13,23-27\,190/T$
1000	$\lg p_{O_2} = -3,61$[*2]

26 – Continued

1	2
2000	$\lg p_{O_2} = +4{,}49$
1000	$\lg p_O = -9{,}10$ [*2]
2000	$\lg p_O = +1{,}57$
1100	$5{,}07 \cdot 10^{-26}$ [*60]
1200	$3{,}20 \cdot 10^{-24}$
1300	$13{,}20 \cdot 10^{-23}$
1350	$6{,}53 \cdot 10^{-22}$
	SrO
2000	$6{,}99 \cdot 10^{-2}$
1000	$\lg p_{O}{}^{2} = -14{,}381$ [*2]
2000	$\lg p_{O_2} = -5{,}614$
3500	$\lg p_{O_2} = +3{,}402$
1000	$\lg p_O = -14{,}494$ [*2]
2000	$\lg p_O = -1{,}481$
3500	$\lg p_O = +3{,}894$
	Y_2O_3
2000	$1{,}62 \cdot 10^{-6}$
	ZrO_2
2000	$3{,}34 \cdot 10^{-9}$
1673—2673	$\lg p_{O_2} = 8{,}01 - 22\,800/T$
1000	$\lg p_{O_2} = -32{,}54$ [*2]
2000	$\lg p_{O_2} = -8{,}623$
2950	$\lg p_{O_2} = -1{,}132$
4000	$\lg p_{O_2} = +2{,}453$
1000	$\lg p_O = -23{,}575$ [*2]
2000	$\lg p_O = -4{,}986$
2950	$\lg p_O = +0{,}911$
4000	$\lg p_O = +3{,}899$

1	2
	NbO
1000	$4{,}05 \cdot 10^{-30}$ *62
	NbO_2
1050—1300	$\lg p_{O_2} = 12{,}61 - 37\,700/T$ *63
	$\lg p_{O_2} = 13{,}21 - 32\,800/T$ *64
	MoO_2
1025—1325	$\lg p_{O_2} = 13{,}74 - 30\,050/T$ *65
1000	$1{,}013 \cdot 10^{-5}$ *66
	PdO
1148	$1{,}013 \cdot 10^{5}$
	Ag_2O
298	50,65
473	$1{,}77 \cdot 10^{5}$
575	$20{,}77 \cdot 10^{5}$
673	$110{,}62 \cdot 10^{5}$
773	$393{,}35 \cdot 10^{5}$
873	$1039{,}34 \cdot 10^{5}$
1073	$4233 \cdot 10^{5}$
—	$\lg p_{O_2} = 11{,}2909 - 2859/T$
—	$\lg p_{O_2} = 11{,}390 - 2942/T$
	SnO_2
873	$1{,}47 \cdot 10^{-35}$
1273	$5{,}13 \cdot 10^{-22}$
1473	$6{,}69 \cdot 10^{-10}$
	BaO
2000	0,101
983—2090	$\lg p_{O_2} = 16{,}64 - 59\,450/T$ *8

26 – Continued

1	2
1000	$\lg p_{O_2} = -14{,}77$[*2]
2000	$\lg p_{O_2} = -2{,}063$
3000	$\lg p_{O_2} = +1{,}740$
1000	$\lg p_{O} = -14{,}690$[*2]
2000	$\lg p_{O} = -1{,}706$
3000	$\lg p_{O} = +2{,}424$
	La_2O_3
2000	$4{,}36\cdot 10^{-7}$
	CeO_2
2000	$2{,}33\cdot 10^{-4}$
	Pr_2O_3
1073—1473	$\lg p_{O_2} = 16{,}21 - 17\,390/T$[*67]
	HfO_2
2000	$1{,}01\cdot 10^{-10}$
	Ta_2O_5
1050—1300	$\lg p_{O_2} = 13{,}61 - 42\,000/T$[*68]
	$W_{18}O_{49}$
973—1273	$\lg p_{O_2} = 15{,}61 - 29\,600/T$[*69]
	PtO_2
1373—1823	$\lg p = (5{,}210 \pm 0{,}047) - (8585 \mp 74)/T$[*70]
	Pb_3O_4
—	$\lg p_{O_2} = 4{,}9 - 7089/T + 1{,}75 \lg T + 0{,}000216\cdot T$
723	1399,88
748	3199,73
773	6932,74
798	14 798,7
823	29 730,8
848	56 261,9
873	113 323,7

1	2
	ThO_2
2000	$1{,}82 \cdot 10^{-10}$
298—1968	$\lg p_{O_2} = 14{,}41 - 64\,040/T$ [*6, *71]
1968—3273	$\lg p_{O_2} = 14{,}820 - 64\,860/T$
1000	$\lg p_{O_2} = -33{,}82$ [*2]
2000	$\lg p_{O_2} = -9{,}66$
3000	$\lg p_{O_2} = -1{,}758$
1000	$\lg p_{O} = -24{,}21$ [*2]
2000	$\lg p_{O} = -5{,}509$
3000	$\lg p_{O} = +0{,}674$
	UO_2
1723	$<3{,}039 \cdot 10^{-2}$
2000	$4{,}66 \cdot 10^{-8}$
—	$\lg p_{O_2} = 18{,}49 - 24\,100/T$ [*72]; $\lg p_{O_2} = 13{,}75 - 56\,400/T$ [*73]
—	$\lg p_{O_2} = -6{,}944$ [*74]; $\lg p_{O_2} = 6{,}644$ [*75]
1273	$\lg p_{O_2} = -6{,}284$ [*76]; $\lg p_{O_2} = -6{,}034$ [*77]
"	$\lg p_{O_2} = -5{,}914$ [*78]; $\lg p_{O_2} = -5{,}804$ [*79]
"	$\lg p_{O_2} = -5{,}674$ [*80]; $\lg p_{O_2} = -5{,}584$ [*81]
"	$\lg p_{O_2} = -5{,}529$ [*82]; $\lg p_{O_2} = -5{,}419$ [*83]
"	$\lg p_{O_2} = -5{,}324$ [*84]; $\lg p_{O_2} = -5{,}244$ [*85]
"	$\lg p_{O_2} = -5{,}184$ [*86]; $\lg p_{O_2} = -5{,}114$ [*87]
"	$\lg p_{O_2} = -4{,}999$ [*88]; $\lg p_{O_2} = -4{,}904$ [*89]
"	$\lg p_{O_2} = -4{,}829$ [*90]; $\lg p_{O_2} = -4{,}494$ [*91]
"	$\lg p_{O_2} = -4{,}254$ [*92]; $\lg p_{O_2} = -4{,}084$ [*93]
"	$\lg p_{O_2} = -3{,}924$ [*94]
	U_3O_8
1018	3866
1123	14 399
	AmO_2
1397	266,6 [*95]
1450	400,0
1464	533,3

26 — Continued

1	2
1508	666,6
1572	933,3
1606	1333,2
1662	2000,0
1770	2666,4
1933	4000,0

*[1]Equilibrium partial pressure in system $BeO(s)/Be(s)/O_2/O$. *[2]Partial pressure upon dissociation. *[3]Equilibrium partial pressure in system MgO/Mg. *[4]Equilibrium pressure in condensed system (Si/SiO). *[5]Equilibrium partial pressure in system CaO/Ca. *[6]Equilibrium pressure in system. *[7]In condensed system (Ti/TiO). *[8]Equilibrium pressure. *[9]TiO/Ti_2O_3. *[10]The same equilibrium pressures in system Ti_3O_5/Ti_2O_3. *[11]Equilibrium pressure along high-oxygen boundary of Ti_3O_5 field. *[12]x = 1.9297. *[13]x = 1.9326. *[14]x = 1.9363. *[15]x = 1.9396. *[16]x = 1.9428. *[17]x = 1.954. *[18]Pressure along low-oxygen boundary of V_2O_3 field. *[19]Partial pressure of oxygen along high-oxygen boundary of V_3O_5 field. *[20]With $CO_2/H_2 = 51.2$ and x = 1.666 in VO_x. *[21]With CO_2/H_2 = 33.7 and x = 1.667. *[22]With $CO_2/H_2 = 26.7$ and x = 1.666. *[23]With $CO_2/H_2 = 21$ and x = 1.667. *[24]Equilibrium pressure of oxygen along low-oxygen boundary of VO_2 field. *[25]$MnO(s)/Mn(s)/O_2/O$. *[26]Over mechanical mixture of hausmannite and manganosite. *[27]$Mn_3O_4/MnO/O_2$. *[28]Over stable oxide of manganese in equilibrium with hausmannite. *[29]Wustite in equilibrium with iron. *[30]Wustite/iron/oxygen. *[31]Partial pressure with gaseous dissociation. *[32]Wustite in equilibrium with magnetite. *[33]With $\delta = 0.0149$ in formula $Fe_{3-\delta}O_4$. *[34]Same, with $\delta = 0.0411$. *[35]Same, with $\delta = 0.0036$. *[36]Same, with $\delta = 0.0116$. *[37]Same, with $\delta = 0.0608$. *[38]Same, with $\delta = 0.0018$. *[39]Same, with $\delta = 0.0043$. *[40]Same, with $\delta = 0.0420$. *[41]Same, with $\delta = 0.0812$. *[42]Same, with δ = (0). *[43]Same, with $\delta = 0.0028$. *[44]Same, with $\delta = 0.0276$. *[45]Same, with $\delta = 0.060$. *[46]Same, with $\delta = 0.1051$. *[47]Wustite/magnetite. *[48]For hematite with near-stoichiometric composition. *[49]Along low-oxygen boundary of hematite field. *[50]$CoO(s)/Co(s)/O_2/O$. *[51]Along low-oxygen boundary of CoO field. *[52]Along high-oxygen boundary. *[53]NiO/Ni. *[54]Equilibrium pressure on low-oxygen boundary of NiO field. *[55]Over mechanical mixture of nickel(II) oxide and higher oxide. *[56]Equilibrium pressure over cuprous oxide in equilibrium with metallic copper. *[57]Same, with cupric oxide. *[58]Equilibrium pressure for stoichiometric cuprous oxide. *[59]Ge/GeO_2. *[60]Very low dissociation pressures indicate virtual absence of any dissociation to Ge and O_2; dissociation to GeO and O_2 is improbable, in view of the instability of GeO at high temperatures. *[61]Equilibrium pressure along low-oxygen boundary of ZrO_2 field. *[62]Partial pressure of oxygen for NbO in equilibrium with metallic phase. *[63]Equilibrium pressure along low-oxygen boundary of NbO_2 field. *[64]Same, along high-oxygen boundary. *[65]Equilibrium pressure along low-oxygen boundary of $MoO_{2-\gamma}$ field. *[66]Point of pressure of high-oxygen boundary corresponding to equilibrium of dioxide with Mo_4O_{11} phase. *[67]Equilibrium pressure over low-oxygen boundary of non-stoichiometric phase $PrO_y(Pr_2O_3 + Pr_5O_8)$. *[68]Equilibrium pressure along low-oxygen boundary of $Ta_2O_{5-\gamma}$ field. *[69]Along high-oxygen boundary of $WO_{2.72}$ field. *[70]Dissociation tension. *[71]ThO_2/Th. *[72]Equilibrium pressure along high-oxygen boundary of $UO_2(UO_2/U_3O_8)$ field. *[73]Same, along low-oxygen boundary of field. *[74]Equilibrium pressure with $\gamma = 0.0025$ in $UO_{2+\gamma}$. *[75]Same, with $\gamma = 0.0027$. *[76]Same, with $\gamma = 0.0030$. *[77]Same, with $\gamma = 0.0033$. *[78]Same, with $\gamma = 0.0035$. *[79]Same, with $\gamma = 0.0037$. *[80]Same, with $\gamma = 0.0040$. *[81]Same, with $\gamma = 0.0043$. *[82]Same, with $\gamma = 0.0045$. *[83]Same, with $\gamma = 0.0050$. *[84]Same, with $\gamma = 0.0055$. *[85]Same, with $\gamma = 0.0060$. *[86]Same, with $\gamma = 0.0065$. *[87]Same, with $\gamma = 0.0070$. *[88]Same, with $\gamma = 0.0080$. *[89]Same, with $\gamma = 0.0090$. *[90]Same, with $\gamma = 0.0100$. *[91]Same, with $\gamma = 0.0150$. *[92]Same, with $\gamma = 0.0200$. *[93]Same, with $\gamma = 0.0250$. *[94]Same, with $\gamma = 0.0300$. *[95]From graph.

MOLECULAR PROPERTIES

1. SURFACE TENSION OF LIQUID OXIDES [224; 257; 258; 305]

Temperature, °C	Surface tension σ, mJ/m²
1	2
H_2O*1 [260]	
—10	77,1
0	75,6
5	74,9
10	74,2
15	73,5
20	72,8
25	72,0
30	71,2
35	70,4
40	69,5
45	68,7
50	67,9
60	66,2
70	64,4
80	62,6
90	60,7
100	58,8
110	56,9
120	54,9
130	52,8
D_2O*1 [260]	
15	73,4
20	72,6
25	71,9
30	71,1
35	70,3
99	58,5
110,8	56,0
B_2O_3*1 [224, 257]	
450	63,5
700	72,35
740	73,77
780	75,18
800	75,89
820	76,6
860	78,01
900	79,43
940	80,85
980	82,26
1000	82,97
1060	85,09
1100	86,51
1140	87,93
1200	90,05
1260	92,17
1300	93,59
1340	95,01
1400	97,13
Al_2O_3 [224]	
2050*2	690
2320*3, *4	690
2350*3, *4	670
2400*3, *4	650
2500	610
2550	580
2600	560
2700	520
2800	470
2900	420
2950	400
3000	370
3100	350
SiO_2*4 [224]	
1470	295,8
1500	298,2
1520	298,8
1550	299,4
1600	301,3
1650	302,9
1700	304,7
1750	305,9
1800	307,5
PO*1 [224]	
23,8	13,2
P_2O_5*3 [224]	
100	60,0
150	59,0
200	57,9
250	56,9
300	55,8
569	50,7
TiO_2*3 [258, 305]	
1850	335
2350	270
1850	380
V_2O_5*3 [305]	
700	94
Fe_2O_3*3 [258]	
1415—1423	585
GeO_2*3 [224]	
1116	248
1200	252,8
1250	255,6
1300	258,4
1350	261,2
1400	264,0
Nb_2O_5*3 [305]	
1460	220
MoO_3*3 [305]	
825	70
Ta_2O_5*3 [305]	
1910	280

1 – Continued

1	2	1	2	1	2
WO_3^{*3} [305]		PbO^{*3} [305, 259]		Bi_2O_3 [128]	
1490	100	900 1000	132 134,8	824 830—900	209,7 σ=232,3— —0,0274 *t*

*1In air. *2In helium. *3In vacuo. *4In argon.

2. VISCOSITY

A. Dynamic Viscosity η of Water as a Function of Temperature and Pressure [1; 638]

Temperature, °K	Viscosity $\eta \cdot 10^6$, Pa· sec, at indicated pressure, 10^5 Pa (or kgf/cm²)				
	1,02 (1,0)	51,0 (50)	102,0 (100)	204,0 (200)	306,0 (300)
273	1792	1781	1770	1748	1726
283	1307	1301	1296	1289	1281
293	1002	1001	1000	998	995
303	797	797	798	798	800
313	653	653	654	656	658
323	546	547	549	552	555
333	466	468	469	472	476
343	404	406	408	411	416
353	355	358	361	366	372
363	315	317	324	330	337
373	282	287	293	301	309
473	—	139	141	145	149
573	—	—	94	96	99
673	—	—	—	—	43

B. Dynamic Viscosity η of Water Vapor as a Function of Temperature and Pressure [637]

Temperature, °C	Saturation	Viscosity $\eta \cdot 10^7$, Pa·sec, at indicated pressure, 10^5 Pa (or kgf/cm²)							
		1,02 (1,0)	20,4 (20)	40,8 (40)	61,2 (60)	81,6 (80)	107,0 (100)	153,0 (150)	204,0 (200)
100	124	124	—	—	—	—	—	—	—
120	133	132	—	—	—	—	—	—	—
140	142	140	—	—	—	—	—	—	—
160	152	148	—	—	—	—	—	—	—

1	2	3	4	5	6	7	8	9	10
180	163	155	—	—	—	—	—	—	—
200	175	163	—	—	—	—	—	—	—
220	188	171	184	—	—	—	—	—	—
240	201	179	190	—	—	—	—	—	—
260	217	186	196	211	—	—	—	—	—
280	235	194	203	217	232	—	—	—	—
300	257	202	210	223	236	250	—	—	—
320	282	209	217	230	242	255	268	—	—
340	318	217	225	236	248	261	272	318	—
360	365	225	232	243	254	266	278	308	—
380	—	232	240	250	261	272	283	311	361
400	—	240	247	257	268	279	289	316	354
420	—	247	254	264	275	284	295	322	354
440	—	254	262	271	282	291	302	329	358
460	—	262	269	279	288	298	308	335	363
480	—	269	277	285	294	304	315	342	370
500	—	277	284	292	301	311	322	349	380

C. Dynamic Viscosity η of Certain Gases as a Function of Temperature and Pressure [637]

Carbon monoxide CO

Pressure, 10^5 Pa (or kgf/ cm^2)	Viscosity $\eta \cdot 10^8$, Pa·sec, at indicated temperature, °C						
	0	25	50	100	150	200	250
2,04(2)	1660	1765	1870	2075	2270	2450	2625
20,4(20)	1690	1795	1900	2105	2290	2470	2645
51,0(50)	1750	1850	1945	2145	2320	2500	2670
102,0(100)	1895	1990	2050	2225	2385	2560	2715
153,0(150)	2080	2140	2175	2320	2465	2630	2770
204,0(200)	2300	2305	2315	2430	2550	2710	2830
306,0(300)	2740	2680	2640	2680	2765	2870	2970
408,0(400)	3175	3065	2985	2950	2970	3035	—
510,0(500)	3600	3450	3330	3225	3185	3200	—
612,0(600)	—	3825	3660	3490	3405	3400	—
816,0(800)	—	4550	4300	4010	3820	3760	—

Carbon dioxide CO_2

Pressure, 10^5 Pa (or kgf/cm^2)	Temperature, °C								
	20	30	40	60	90	100	150	200	250
1,02 (1)	1 463	1 510	1 560	1 650	1 790	1 825	2045	2254	2456
20,4 (20)	1 560	1 590	1 630	—	1 820	1 855	2075	2280	2485
51,0 (50)	1 850	1 830	1 805	—	1 900	1 950	2140	2340	2540
71,4 (70)	2 320	2 500	2 120	—	2 030	2 040	2190	2385	2590
81,4 (80)	2 680	5 310	2 470	—	2 110	2 100	2220	2420	2610
91,8 (90)	8 000	5 980	3 350	—	2 200	2 160	2250	2450	2640
102,0 (100)	8 280	6 500	4 875	—	2 300	2 240	2290	2480	2665
122,4 (120)	8 820	7 300	6 030	—	2 590	2 440	2390	2540	2730
153,0 (150)	9 520	8 080	6 950	4 650	3 120	2 850	2560	2660	2810
204,0 (200)	10 400	9 090	8 000	6 020	4 190	3 800	2980	2900	2980
255,0 (250)	11 170	9 930	8 850	6 980	5 130	4 700	3480	3200	3180
306,0 (300)	11 820	10 610	9 590	7 810	5 960	5 470	3980	3540	3380
408,0 (400)	12 960	11 780	10 800	9 130	7 300	6 780	4980	4260	3890
510,0 (500)	14 000	12 860	11 850	10 230	8 400	7 880	5920	4960	4440
612,0 (600)	15 000	13 850	12 820	11 200	9 370	8 800	6760	5640	5020
714,0 (700)	—	14 840	13 720	12 110	10 200	9 620	7530	6280	5570
816,0 (800)	—	15 600	14 600	12 950	11 000	10 400	8250	6900	6100

Note. Dynamic viscosity ($\eta \cdot 10^8$, Pa · sec) of CO_2 at 15°C, at low pressures:

Pressure, Pa	Viscosity
$1 \cdot 10^5$	1490
$0,51 \cdot 10^5$	1490
$0,27 \cdot 10^4$	1480
$0,27 \cdot 10^3$	1470
79	1380

Oxygen O_2

Pressure, 10^5 Pa (or kgf/cm²)	Temperature, °C		
	15	50	100
1,02(1,0)	1993	2182	2437
22,4(22,0)	2013	2211	2460
42,9(42,0)	2055	2249	2490
63,2(62,0)	2114	2295	2525
83,7(82,0)	2188	2345	2564
104,0(102,0)	2269	2401	2608
122,4(120,0)	2351	2463	2655

Pressure, 10^5 Pa (or kgf/cm²)	Temperature, °C		Pressure, 10^5 Pa (or kgf/cm²)	Temperature, °C	
	15,8	54,8		15,8	54,8
1,02(1,0)	1997,5	2207	306,0(300)	3220	—
20,4(20)	2024,0	—	408,0(400)	3769	—
51,0(50)	2093,5	2288	510,0(500)	4330	—
102,0(100)	2250,0	2434	612,0(600)	4862,0	—
153,0(150)	2451,5	2610	714,0(700)	5392,0	—
204,0(200)	2688,5	—			

D. Dynamic Viscosity of Oxides

Temperature, °K	η, Pa·sec
1	2
H_2O [638]	
259*1	$85\cdot10^{9}$
273*1	$1\cdot10^{10}$
273*2	$17921\cdot10^{-7}$
283	$13077\cdot10^{-7}$
293	$10050\cdot10^{-7}$
303	$8007\cdot10^{-7}$
313	$6560\cdot10^{-7}$
323	$5494\cdot10^{-7}$
333	$4688\cdot10^{-7}$
343	$4061\cdot10^{-7}$
353	$3565\cdot10^{-7}$
363	$3165\cdot10^{-7}$
373	$2838\cdot10^{-7}$
273*3	$883\cdot10^{-8}$
298	$975\cdot10^{-8}$
323	$1065\cdot10^{-8}$
348	$1157\cdot10^{-8}$
373	$1250\cdot10^{-8}$
423	$1435\cdot10^{-8}$
473	$1615\cdot10^{-8}$
523	$1800\cdot10^{-8}$
573	$1985\cdot10^{-8}$
673	$2350\cdot10^{-8}$
773	$2720\cdot10^{-8}$
873	$3090\cdot10^{-8}$
973	$3460\cdot10^{-8}$
1073	$3820\cdot10^{-8}$
1273	$4560\cdot10^{-8}$
D_2O[638]	
303	$969\cdot10^{-6}$
318	$713\cdot10^{-6}$
333	$552\cdot10^{-6}$
348	$445\cdot10^{-6}$
363	$365\cdot10^{-6}$
373	$323\cdot10^{-6}$
398	$252\cdot10^{-6}$
423	$208\cdot10^{-6}$
448	$175\cdot10^{-6}$
473	$151\cdot10^{-6}$
498	$135\cdot10^{-6}$
523	$124\cdot10^{-6}$
BeO [1]	
1473	$185\cdot10^{9}$
1573	$68\cdot10^{9}$
1673	$38\cdot10^{9}$
1773	$127\cdot10^{8}$
1873	$93\cdot10^{8}$
1973	$72\cdot10^{8}$
B_2O_3 [1] *4	
573	$4193\cdot10^{5}$
773	3990
973	1000
1 273	6,31
B_2O_3 [1]	
1023	43,6
1073	26,0
1173	11,8
1273	7,0
1373	4,0

1	2
CO [637]*5	
68,55	$287\cdot10^{-6}$
73,2	$224\cdot10^{-6}$
75,2	$203\cdot10^{-6}$
77,8	$186\cdot10^{-6}$
82,8	$165\cdot10^{-6}$
90,1	$146\cdot10^{-6}$
99,6	$116\cdot10^{-6}$
111,6	$100\cdot10^{-6}$
129,6	$66\cdot10^{-6}$
CO [637]*6	
80*7	$533\cdot10^{-8}$
100*7	$668\cdot10^{-8}$
120*7	$796\cdot10^{-8}$
153	$1030\cdot10^{-8}$
163	$1090\cdot10^{-8}$
173	$1130\cdot10^{-8}$
198	$1275\cdot10^{-8}$
223	$1400\cdot10^{-8}$
248	$1528\cdot10^{-8}$
273	$1662\cdot10^{-8}$
293	$1749\cdot10^{-8}$
298	$1766\cdot10^{-8}$
323	$1872\cdot10^{-8}$
348	$1980\cdot10^{-8}$
373	$2076\cdot10^{-8}$
423	$2271\cdot10^{-8}$
473	$2452\cdot10^{-8}$
523	$2622\cdot10^{-8}$
573	$2788\cdot10^{-8}$
673	$3090\cdot10^{-8}$
773	$3370\cdot10^{-8}$
873	$3630\cdot10^{-8}$
973	$3870\cdot10^{-8}$
1073	$4100\cdot10^{-8}$
1273	$4530\cdot10^{-8}$
CO_2 [637]*6	
173	$886\cdot10^{-8}$
198	$1007\cdot10^{-8}$
223	$1126\cdot10^{-8}$
248	$1247\cdot10^{-8}$
273	$1367\cdot10^{-8}$
293	$1463\cdot10^{-8}$
298	$1486\cdot10^{-8}$
323	$1607\cdot10^{-8}$
348	$1716\cdot10^{-8}$
373	$1827\cdot10^{-8}$
423	$2045\cdot10^{-8}$
473	$2254\cdot10^{-8}$
523	$2456\cdot10^{-8}$
573	$2646\cdot10^{-8}$
673	$2994\cdot10^{-8}$
773	$3309\cdot10^{-8}$
873	$3605\cdot10^{-8}$
973	$3876\cdot10^{-8}$
1073	$4140\cdot10^{-8}$
1273	$4658\cdot10^{-8}$
N_2O*3	
198	$990\cdot10^{-8}$
223	$1115\cdot10^{-8}$
248	$1240\cdot10^{-8}$
273	$1360\cdot10^{-8}$
293	$1460\cdot10^{-8}$
298	$1482\cdot10^{-8}$
323	$1595\cdot10^{-8}$
348	$1712\cdot10^{-8}$
373	$1822\cdot10^{-8}$
423	$2040\cdot10^{-8}$
473	$2245\cdot10^{-8}$
523	$2450\cdot10^{-8}$
573	$2649\cdot10^{-8}$
673	$3030\cdot10^{-8}$
773	$3375\cdot10^{-8}$
NO*3	
123	$860\cdot10^{-8}$
133	$930\cdot10^{-8}$
143	$1000\cdot10^{-8}$
153	$1064\cdot10^{-8}$
163	$1130\cdot10^{-8}$
173	$1198\cdot10^{-8}$
198	$1357\cdot10^{-8}$
223	$1510\cdot10^{-8}$
248	$1656\cdot10^{-8}$
273	$1800\cdot10^{-8}$
293	$1899\cdot10^{-8}$
298	$1920\cdot10^{-8}$
323	$2035\cdot10^{-8}$
348	$2156\cdot10^{-8}$
373	$2272\cdot10^{-8}$
423	$2475\cdot10^{-8}$
473	$2682\cdot10^{-8}$
523	$2870\cdot10^{-8}$
573	$3055\cdot10^{-8}$
673	$3400\cdot10^{-8}$
773	$3700\cdot10^{-8}$
873	$4010\cdot10^{-8}$
973	$4275\cdot10^{-8}$
1073	$4535\cdot10^{-8}$
1273	$5075\cdot10^{-8}$
NO_2 [1]	
293	$189\cdot10^{-7}$
N_2O_4 [637]*6	
273,87	$5220\cdot10^{-7}$
278,24	$4954\cdot10^{-7}$
282,30	$4720\cdot10^{-7}$
288,51	$4401\cdot10^{-7}$
O_2 [637]*5	
54,4	$873\cdot10^{-6}$
54,9	$772\cdot10^{-6}$
56,4	$717\cdot10^{-6}$
57,1	$638\cdot10^{-6}$
59,7	$631\cdot10^{-6}$
61,7	$521\cdot10^{-6}$
63,5	$476\cdot10^{-6}$
68,9	$377\cdot10^{-6}$
72,3	$323\cdot10^{-6}$
77,4	$273\cdot10^{-6}$
80,0	$250\cdot10^{-6}$
90,1	$190\cdot10^{-6}$
O_2 [637]*6	
83	$635\cdot10^{-8}$
93	$710\cdot10^{-8}$
103	$785\cdot10^{-8}$
113	$860\cdot10^{-8}$
123	$935\cdot10^{-8}$
133	$1010\cdot10^{-8}$
143	$1085\cdot10^{-8}$
153	$1150\cdot10^{-8}$
163	$1218\cdot10^{-8}$
173	$1286\cdot10^{-8}$
198	$1452\cdot10^{-8}$
223	$1612\cdot10^{-8}$
248	$1753\cdot10^{-8}$
273	$1910\cdot10^{-8}$
293	$2026\cdot10^{-8}$
298	$2052\cdot10^{-8}$
323	$2182\cdot10^{-8}$

1	2
348	$2310\cdot10^{-8}$
373	$2437\cdot10^{-8}$
423	$2674\cdot10^{-8}$
473	$2867\cdot10^{-8}$
523	$3103\cdot10^{-8}$
573	$3310\cdot10^{-8}$
673	$3686\cdot10^{-8}$
773	$4030\cdot10^{-8}$
873	$4350\cdot10^{-8}$
973	$4652\cdot10^{-8}$
1073	$4940\cdot10^{-8}$
1273	$5470\cdot10^{-8}$

O_3 [81, Vol. 3] *8

1	2
78	$420\pm1\cdot10^{-5}$
90	$155\pm2\cdot10^{-5}$

Al_2O_3 [1] *9

1	2
1573	$12\cdot10^{10}$
1723	$4\cdot10^{10}$
1873	$9\cdot10^{9}$
1973	$78\cdot10^{8}$

Al_2O_3 [236] *10

1	2
1873	$45\cdot10^{8}$
1973	$196\cdot10^{7}$
2073	$92\cdot10^{7}$

SiO_2 [1] *4

1	2
1673	$563\cdot10^{7}$
1973	$3162\cdot10^{3}$
2273	$399\cdot10^{2}$
2573	3990

SO_2 [1] *8

1	2
239,5	$551\cdot10^{-6}$
262,5	$428\cdot10^{-6}$
272,9	$394\cdot10^{-6}$

SO_2*3

1	2
198	$855\cdot10^{-8}$
223	$955\cdot10^{-8}$
248	$1053\cdot10^{-8}$
273	$1158\cdot10^{-8}$
293	$1250\cdot10^{-8}$
298	$1270\cdot10^{-8}$
323	$1390\cdot10^{-8}$
348	$1500\cdot10^{-8}$
373	$1611\cdot10^{-8}$
423	$1830\cdot10^{-8}$
473	$2040\cdot10^{-8}$
523	$2255\cdot10^{-8}$
573	$2455\cdot10^{-8}$
673	$2820\cdot10^{-8}$
773	$3155\cdot10^{-8}$
873	$3465\cdot10^{-8}$
973	$3760\cdot10^{-8}$
1073	$4040\cdot10^{-8}$
1273	$4540\cdot10^{-8}$

TiO_2 [236] *10

1	2
1593	$382\cdot10^{7}$
1673	$196\cdot10^{7}$
1753	$100\cdot10^{7}$
2123–2623	$\eta=5\cdot10^{-5}\exp(33000/RT)$*11

V_2O_5 [236]*10

1	2
713	$250\cdot10^{8}$
753	$505\cdot10^{7}$
793	$124\cdot10^{7}$

Cr_2O_3 [236]*10

1	2
1773	$625\cdot10^{7}$
1853	$330\cdot10^{7}$
1933	$184\cdot10^{7}$

Fe_2O_3 [236]*10

1	2
1373	$635\cdot10^{7}$
1443	$321\cdot10^{7}$
1523	$157\cdot10^{7}$

CoO [236]*10

1	2
1173	$123\cdot10^{8}$
1253	$45\cdot10^{8}$
1353	$196\cdot10^{7}$
1443	$103\cdot10^{7}$

NiO [236]*10

1	2
1423	$575\cdot10^{7}$
1523	$346\cdot10^{7}$
1623	$225\cdot10^{7}$

ZnO [236]*10

1	2
1173	$122\cdot10^{8}$
1353	$43\cdot10^{8}$
1453	$28\cdot10^{8}$
1533	$21\cdot10^{8}$

GeO_2 [1]

1	2
1773*12	200
1823	120
1873	80
1923	50
1973	32
1573*13	2800
1673	750
1773	200
1573*14	9000
1673	1100
1773	380

ZrO_2 [236]*10

1	2
2123	$134\cdot10^{7}$
2273	$86\cdot10^{7}$
2423	$59\cdot10^{7}$

Nb_2O_5 [236]*10

1	2
1493	$161\cdot10^{8}$
1553	$75\cdot10^{8}$
1633	$30\cdot10^{8}$

MoO_3 [236]*10

1	2
723	$465\cdot10^{7}$
773	$270\cdot10^{7}$
823	$163\cdot10^{7}$
873	$103\cdot10^{7}$

D – Continued

1	2	1	2	1	2
HfO_2 [236]*10		Ta_2O_5 [236]*10		WO_3 [236]*10	
				1143	1280·10⁷
2100	214·10⁷	1583	1430·10⁷	1243	440·10⁷
2473	166·10⁷	1673	500·10⁷	1303	250·10⁷
2573	124·10⁷	1773	164·10⁷	1373	137·10⁷

*1Ice. *2Water. *3Viscosity of gases and vapors at pressure of $1.013 \cdot 10^5$ Pa [637]. *4From graph. *5Liquefied gas. *6Gas at pressure of 101,325 Pa. *7[1]. *8Liquid. *9From graph; 99.9% purity; hot-pressed. *10Viscosity determined in process of hot-pressing at $176.52 \cdot 10^5$ Pa. *11TiO_2, 99.5% purity; method of damped vibrations under vacuum of $133.322 \cdot 10^{-4}$ Pa and in argon; calculational error ± 8%; activation energy for viscous flow 135.65 kJ/mole, entropy of viscous flow 16.747 J/(mole · °K). *12From graph; logarithmic scale. *13Data of Kurkjian and Douglas. *14Data of Mackenzie.

3. CONSTANTS IN VAN DER WAALS EQUATION*1

$$\left(p+\frac{a}{v^2}\right)(v-b)=RT$$

Oxide	$a \cdot 10^{-5}$ $N \cdot m^4/kmole^2$	b, m^3/kmole	Oxide	$a \cdot 10^{-5}$, $N \cdot m^4/kmole^2$	b, m^3/kmole
H_2O	5,445	0,03053	N_2O_4	5,265	0,04427
CO	1,480	0,03988	O_2	1,355	0,03185
CO_2	3,580	0,04270	O_3	1,840*3	0,03544*2
N_2O	3,770	0,04418	SO_2	6,690	0,05640
NO	1,335	0,02791			

*1In converting from values of b listed in the Chemist's Handbook (2nd ed., Vol. 1, Khimiya, Moscow–Leningrad, 1966) to the values indicated above, the factor 22.415 was used, and for values of a the factor 0.0050. *2An erroneous value is given in [81, Vol. 3, p. 655]. *3Calculated from relationship $a = p_{cr} 27b^2$.

4. CRITICAL PARAMETERS [1; 9]

Oxide	Critical temperature T_{cr}, °K	Critical pressure $p_{cr} \cdot 10^{-5}$, Pa	Critical volume V_{cr}, 10^3, m^3/kmole	Critical density ρ_{cr}, kg/m^3
H_2O	647,30±0,05	221,1±0,1	56,3	320±10
H_2O_2	730	216,8	—	—
D_2O	644,05±0,10	218,5±0,3	55,1	363
Li_2O	>6000	—	—	—

1	2	3	4	5
BeO	>6000	—	—	—
B_2O_3	>6000	—	—	—
CO	132,92±0,03	34,978±0,05	93,06±1,0	301±3
CO_2	304,15±0,05	73,85±0,05	94,04	468±1
N_2O	309,58±0,01	72,53±0,01	97.27	453±1
NO	180,3	65,46±0,2	58	520
$N_2O_4(NO_2)$	431,2	100,29±1,01	82	560
O_2	154,78±0,05	50,79±0,10	78	410±20
O_3	261,05	55,4±0,1	89,4	537
F_2O	215,15±0,1	49,5	97,6	553
MgO	>6000	—	—	—
Al_2O_3	>6000	—	—	—
SiO_2	>6000	—	—	—
SO_2	430,65±0,2	78,8±0,1	122	524±5
SO_3	491,15±0,5	82,05±0,5	126	633
CaO	>6000	—	—	—
SrO	>6000	—	—	—
PbO	5300	3292	—	—

5. MOLECULAR PARAMETERS [1]

Oxide	Arithmetic average velocity of molecules at 0°C, m/sec	Average number of collisions of molecule per second at 0°C and 101,325 Pa, $\bar{z} \cdot 10^{-9}$	Average length of molecular free path at 0°C and 101,325 Pa, $\bar{\lambda} \cdot 10^{10}$, m
H_2O	566,5	14,02	404
CO	471,3	7,99	590
CO_2	375,99	9,67	389
SO_2	300,4	10,3	290

6. AVERAGE LENGTH OF FREE PATH OF PHONONS

Temperature, °K	$\bar{\lambda}$, nm	Temperature, °K	$\bar{\lambda}$, nm
1	2	1	2
BeO [19] *1		773	3,4
		873	2,8
373	13,8	973	2,0
473	8,0	1073	1,7
573	5,1	1173	1,4
673	4,3	1273	1,2

6 — Continued

1	2
MgO [19] *1	
373	7,7
473	4,5
573	3,0
673	2,5
773	2,0
873	1,7
973	1,5
1073	1,2
1173	1,1
1273	1,0
Al_2O_3 [19] *1	
373	2,8
473	2,1
573	1,7
673	1,2
773	1,0
873	0,8
973	0,7
1073	0,6
1173	0,5
1273	0,4
TiO [45] *2	
77	191,0
200	35,0
273	19,0
300	13,0
400	10,0
500	10,0
600	10,0
700	12,0
800	14,0
900	16,0
1000	18,0
1100	20,0

1	2
TiO_2 [19] *1	
423	1,0
473	0,8
673	0,6
873	0,4
1073	0,3
VO [50] *2	
77	83,2
200	13,1
273	6,6
300	6,1
400	4,2
500	3,0
600	3,1
700	3,6
800	3,8
900	4,1
1000	4,3
1100	4,3
UO_2 [116] *3	
140—300	$1/\bar{\lambda}=2,21\cdot10^8+9,42\cdot10^5T$
U_4O_9 [116] *3	
100—300	$1/\bar{\lambda}=9,10\cdot10^8+5,01\cdot10^6 T$

*1 From graph. *2 Calculation. *3 m^{-1}.

MECHANICAL PROPERTIES

1. MODULUS OF NORMAL ELASTICITY

Modulus of normal elasticity E, GPa	Temperature, °K
1	2
H_2O [1]	
2,8	271
BeO [1]	
304,02—127,49*1	293—1473
49,04*2	293
61,78*3	293
102,97*4	293
157,89*5	293
227,52*6	293
294,21*7	293
356*97*8	293
308,92—127,49*9	293—1473
MgO [1]	
209,87—27,46*10	296—1573
208,9*11	293
295,20—219,68*12	298—1573
246,7*13	293
309,6*14	293
345,8*15	293
307,2*16	298
284*0*17	298
Al_2O_3 [1]	
374,63—147,1	293—1773
370,70*18	293
368,74*19	293
344,2*20	—
342,0*21	298
360,0*22	298
377,0*23	298
395*24	298
403,9*25	298
SiO_2 [1]	
111*26	293
103*27	293
98*28	293
90*29	293
85*30	293
79*31	293
76*32	293
71*33	293
52*34	293
CaO [211]	
181*35	0—273
Sc_2O_3 [140]	
199*36	293
177*36	473
175*36	673
173*36	873
170*36	1073
TiO_2 [1]	
88,263*37	—
244,0*38	—
254,5*39	—
273,0*40	—
284,2*41	—
289,4*42	—
FeO [1]	
1,079*43	1073
1,177*43	1073
ZnO [238]	
57,0*44	298
72,0*45	298
93,0*46	298
114,0*47	298
123,5*48	298
123,6*49	298
ZrO_2 [1]	
168,68—94,15*50	293—1633
186,83*51	
168,68—100,03*52	293—1573
186,3—88,26*53	293—1773
141,2—100*54	293—1473
166,72—81,4*55	293—1573
122,3*56	—
Gd_2O_3 [212]	
150,0*57	293
128,0*58	293
107,5*59	293
88,0*60	293
67,0*61	293
45,0*62	293
142,0*63	273
135,0*63	673
128,0*63	1073
122,0*63	1473
120,0*63	1673
90,0*64	273
83,0*64	673
77,0*64	1073
74*64	1473
70*64	1673
54*65	273
50*65	1073
40*65	1473

1 – Continued

1	2	1	2	1	2
Dy_2O [240]		132*79	473	240,27—192,2*84	293—1473
51,6*66	—	130*79	773	236,35*85	573
89,6*67	—	124*79	1273	216,7*85	873
124,0*68	—	104*80	273	205,95*85	1173
170,0*69	—	102*80	473	137,3*86	293
Yb_2O_3 [218]		100*80	773	125,5*86	873
188*70	293	95*80	1273	107,88*86	1273
178*71	293	86*81	273	83,36*86	1473
162*72	293	84*81	473	144,16*87	293
150*73	293	82*81	773	**UO_2** [1]	
135*74	293	80*81	1273	171,62*88	—
120*75	293	**ThO_2** [1]		181,43—162,8	293—1073
105*76	293	137,30—97,1*82	293—1473	**U_3O_8** [1]	
90*77	293	223,6*83	293	172,2	293
77*78	293				
135*79	273				

*1Bending method; density 2700 –2800 kg/m³. *2Density 1800 kg/m³. *32000 kg/m³. *42200 kg/m³. *52400 kg/m³. *62600 kg/m³. *72800 kg/m³. *83000 kg/m³. *9Porosity 3–7%. *10Porosity 11%, density 3150 kg/m³. *11Porosity 3–7%. *12Porosity 2%, density 3510 kg/m³. *13, *14, *15Single crystal, static loading, for directions E_{100}, E_{110}, E_{111}. *16Density 3580 kg/m³ according to [215]. *17Density 3460 kg/m³, porosity 3.3% according to [216, p. 46]. *18Hot-pressed. *19Fired at 2023°K. *20Density 3580 kg/m³, porosity 3.5% according to [216]. *21 – *25From graph; density 3800, 3850, 3900, 3950, and 3990 kg/m³ respectively, according to [238]. *26 – *34Quartz fibers, diameter 3, 4, 5, 7, 10, 15, 76, 71, and 52 μm, respectively. *35At 0°K, by extrapolation; specimens polycrystalline. *36Static method; specimen in form of disk with d = 32 mm and h = 1 mm. *37Density 4100 kg/m³ according to [1]. *38 – *41From graph: density 4000, 4100, 4200, and 4250 kg/m³, respectively, according to [238]. *42Single crystal, according to [238]. *43Tested in tension; specimens polycrystalline, porosity 1%, sintered under a pressure p = 22 MPa at T = 1353°K. *44 – *48Polycrystalline; from graph; density 4000, 4500, 5000, 5500, and 5680 kg/m³, respectively. *49Single crystal. *50Sintered; bending test on thin rods. *51Density 6100 kg/m³. *52Stabilized with 4.2% MgO by mass. *53Stabilized with 2% MgO by mass. *54Stabilized with 5% CaO by mass; density 4930 kg/m³. *55Stabilized with 4% CaO by mass; total porosity 4%. *56Density 5430 kg/m³, porosity 5.6%, according to [216]. *57 – *62Polycrystalline, monoclinic; from graph; volume porosity 0.00, 0.08, 0.16, 0.24, 0.32, and 0.40%, respectively. *63Porosity 3.47%. *64Porosity 22.97%. *65Porosity 36.6%. *66 – *68Polycrystalline; from graph; density 3,300, 5,000, and 10,000 kg/m³, respectively. *69Nonporous. *70 – *78Polycrystalline; volume porosity 0.00, 0.10, 0.06, 0.10, 0.14, 0.18, 0.22, 0.26, and 0.30, respectively. *79Porosity 14.3%. *80Porosity 21.4%. *81Porosity 27.1%. *82Specimen sintered at 2173°K. *83Hot-pressed, density 9570 kg/m³. *84Density 9630 kg/m³ with addition of 0.5% CaO. *85Composition 99.5% ThO_2 + 0.5% CaO. *86Specimens fired at 2173°K. *87Specimens sintered; porosity 3 – 7%. *88Porosity 3 – 10%.

Shear modulus G, GPa	Temperature, °K
1	2
H_2O [1]	
2,746—2,942*1	—
BeO [1]	
100,03*2	273—1073
96,11*2	1273
48,05*2	1373
MgO [1]	
76,49—22,56*3	573—1573
113,76—35,31*4	573—1673
129,3*5	298
Al_2O_3 [1]	
124,55—105,92*6	298—1323
89,24*6	1473
77,48*6	1623
60,00*6	1773
186,33*7	293
347,36*8	293
347,36*8	293
291,76*8	773
274,60*8	1273
225,86*8	1373
202,51*8	1473
119,45*8	1573
15,79*8	1673
12,36*8	1773
163,4*9	298
85*10	—
115*11	—
135*12	—
160*13	—
SiO_2 [1]	
31	293
66*14	293
61*15	293
58*16	293
53*17	293
48*18	293
42*19	293
39*20	293
35*21	293
CaO [211]	
73,5*22	0
74,0*22	273
TiO_2 [238]	
96,0*23	298
107,0*24	298
111,5*25	298
113,54*26	298
ZnO [238]	
21,0*27	298
27,0*28	298
34,0*29	298
42,5*30	298
45,6*31	298
45,5*32	298
ZrO_2 [1]	
60,80	293
67,67*33	298
97,09*33	1573
96,79*34	1573
CeO_2 [1]	
62,47	303
34,32	1373

2 — Continued

1	2	1	2	1	2
15,79	1573	35*43	273	79,44*48	1073
69,04*35	—	30*43	1073	58,84*48	1373
Gd_2O_3 [212]		28*43	1473	38,25*48	1573
58,0*36	293	56—53*44	273—1473	93,17—83,36*49	303—973
51,0*37	293	**Yb_2O_3 [218]**		**UO_2 [1]**	
43,0*38	293	52*45	273	0,98—1,96	298
35,0*39	293	50*45	773		
27,0*40	293	48*45	1273		
17,5*41	293	41—40*46	273—1273		
21*42	273	34—32*47	273—1273		
18*42	1073	**ThO_2 [1]**			
16*42	1473	98,07*48	303		

*1Ice. *2Density 2700–2800 kg/m^3. *3Specimens obtained by slipcasting; porosity 12%. *4Hydrostatic pressing; porosity 2%. *5Polycrystalline; reasonance method according to [215]. *6From pure α-Al_2O_3; density 3840 kg/m^3. *7Single crystal. *8Nonporous. *9Single crystal according to [238]. *10From graph; density 3400 kg/m^3. *11Density 3600 kg/m^3. *123800 kg/m^3. *134000 kg/m^3. *14–*21Fused quartz fibers; diameter 3, 4, 5, 7, 10, 15, 20, and 30 μm, respectively. *22Single crystal; value at 0°K extrapolated. *23–*25Polycrystalline; from graph; density 4050, 4200, and 4250 kg/m^3, respectively. *26Single crystal. *27–*31Polycrystalline; from graph; density 4000, 4500, 5000, 5400, and 5700 kg/m^3, respectively. *32Single crystal. *33Stabilized with 2.5% MgO by mass. *34Density 5200 kg/m^3, total porosity 4%, open porosity 0%; stabilized with 4% CaO by mass. *35Fired at 1773°K; obtained by pressing at p = 490 MPa. *36–*41Polycrystalline, monoclinic; from graph; volume porosity 0.00, 0.08, 0.16, 0.24, 0.32, and 0.40, respectively. *42From graph; porosity 36.6%. *43Porosity 22.97%; from graph. *44Porosity 3.47%; from graph. *45Porosity 14.3%; from graph. *46Porosity 21.4%. *47Porosity 27.1%. *48Porosity 1.7%, density 9530 kg/m^3; sintered at 2103°K. *49Density 9700 kg/m^3; composition 99.5% ThO_2 + 0.5% CaO.

3. POISSON'S RATIO

Poisson's ratio	Temperature, °K
1	2
BeO [1]	
0,36—0,38[*1]	293
0,34—0,35	293—1273
0,40	1373
MgO [1]	
0,36—0,33	298—1273
0,47	1373
Al_2O_3 [1]	
0,27—0,36[*2]	293—1273
0,47[*2]	1373
0,26—0,45[*3]	298—1473
0,32—0,53[*4]	298—1473
0,29—0,49[*5]	298—1473
0,25—0,44[*6]	298—1473
0,13—0,30[*7]	298—1473
SiO_2 [47]	
0,17[*8]	—
CaO [211]	
0,23[*9]	0
0,22[*9]	273
ZrO_2 [1]	
0,29[*10]	293
0,36[*11]	293
CeO_2 [1]	
0,515[*12]	293
Gd_2O_3 [212]	
0,266—0,268[*13]	293
Yb_2O_3 [218]	
0,264[*14]	273—1273
0,295—0,288[*15]	293—1273
ThO_2 [1]	
0,17[*16]	293
0,28—0,29[*17]	873—1173
0,280—0,285[*18]	303—1173

[*1]Density 2700–2800 kg/m^3. [*2]Sintered alumina, nonporous. [*3]Porosity 0.1%. [*4]Porosity 0.2%. [*5]Porosity 0.3%. [*6]Porosity 0.4%. [*7]Porosity 0.5%. [*8]Fused quartz fiber. [*9]Polycrystalline; value at 0°K obtained by extrapolation. [*10]Density 5200 kg/m^3; stabilized with 4% CaO by mass. [*11]Stabilized with 2.5% MgO. [*12]Fired at 1773°K; pressed at p = 0.5 MPa. [*13]Polycrystalline; volume porosity 0.00–0.40. [*14]Porosity 27.1%. [*15]Porosity 14.3%. [*16]Sintered, T = 2193°K; high-purity. [*17]99.5% ThO_2 + 0.5% CaO. [*18]Density 9700 kg/m^3.

Tensile strength σ, MPa	Temperature, °K
1	2
H_2O [1]	
0,98*1	258—261
BeO [1]	
144,55	293
137,2—68,6*2	293—1273
102,9—27,44*3	293—1473
MgO [1]	
96,04——41,16*4	293—1573
110,25—98,00*5	873—1073
55,17—41,3*5	473—1573
96,0—82,3*6	293—1273
64,19—21,95*7	303—1673
29,4*8	—
392*9	—
980*10	—
490*11	—
88*12	—
68,6*13	—
58,6*14	—

Tensile strength σ, MPa	Temperature, °K
1	2
Al_2O_3 [1]	
258,62—29,50*15	293—1673
259—127,40*16	293—1473
44,10—10,78*16	1583—1773
258,72*17	293
43,90—12,03*17	1573—1773
SiO_2 [1]	
9000*18	293
8000*19	293
6500*20	293
5500*21	293
4800*22	293
3900*23	293
3000*24	293
2300*25	293
1700*26	293
14500*27	293
TiO_2 [1]	
54,88—41,16*28	293—1273
58,8—68,6*29	293

Tensile strength σ, MPa	Temperature, °K
1	2
FeO [1]	
36,26*30	1073
29,40*31	1073
Zr_2O_2 [1]	
145,53—12,74*32	293—1813
137,20*33	293
119,56—69,97*34	673—1573
137,20—69,97*35	293—1573
145,53—82,52*36	473—1473
12,74*36	1813
ThO_2 [1]	
82,32—48,02*37	293—1273
UO_2 [1]	
82,30—123,48*38	293—1273

*1 At low loading rates. *2 Porosity 3–7%. *3 Density 2800 kg/m^3. *4 Porosity 11%. *5 Rectangular specimens obtained by casting. *6 Porosity 3–7%. *7 Porosity 11%. *8 Single crystal without any special treatment. *9 Single crystal after chemical polishing and annealing. *10 – *14 Fiberlike single crystals grown by condensation; diameter 2–6, 10, 20, 30, and 40 μm, respectively. *15 Dense pieces made of high-purity Al_2O_3. *16 Sintered alumina. *17 Density 3900 kg/m^3. *18 – *27 Quartz fibers, diameter 1.5, 2.0, 3.0, 4.0, 5.0, 7.0, 10.0, 15.0, 20.0, and 30.0 μm, respectively. *28 Porosity 3–7%. *29 Density 4100 kg/m^3. *30 Polycrystalline, porosity 1%; sintered at T = 1353°K. *31 Single crystal. *32 Stabilized with 4.2% MgO by mass. *33 Density 5600 kg/m^3. *34 Porosity 6100 kg/m^3. *35 Porosity 1–4%. *36 Specimens fired at 2173°K. *37 Porosity 3–7%. *38 Porosity 3–10%. *39 Density 8,300–10,020 kg/m^3.

Compressive strength σ_c, MPa	Temperature, °K
1	2
H_2O [1]	
0,98—1,96*1	273
5,586—6,86*1	261—258
0,49—0,98*2	273
1,47—4,704*2	261—258
BeO [1]	
782,04	293
243,53—48,02	1273—1873
784,0*3	293
490,0—49*3	773—1873
48,02*4	293
89,18*5	293
137,20*6	293
205,80*7	293
411,60*8	293
754,60*9	293
1176,0*10	293
1372,0*11	293
782,04*12	293
439,04—48,02*12	1073—1873
784,98*13	298
489,02—49*13	1073—1873
MgO [1]	
1372,00—33,32*14	293—2073
Al_2O_3 [1]	
2940*15	293
1740—49*15	673—1873
2548—96*16	273—1773
SiO_2 [228]	
176—970*17	293—1573
CaO [227]	
181—220*18	1773
140—159*19	2273
TiO_2 [1]	
245*20	293
ZrO_2 [1]	
2058—19,6*21	293—1773
1173,06—19*22	1273—1773
2058*23	
1960*24	
980*25	293
490*26	293
98*27	293
49*28	293
39,2*29	293
ThO_2 [1]	
1372*30	293
1078*30	673
588*30	873
490*30	1073
353*30	1273
196—39*30	1473—1673
1509,2*31	293
1558,2*32	293
950,6*33	293
499,8*34	293
216,58*35	293
1468*36	293
487*36	1073
350*36	1273
196,00*36	1473
10,29*36	1773
UO_2 [1]	
960,4*37	293
480,2*38	293
411,6*39	293

*1 At low loading rates. *2 At high loading rates. *3 Density 2800–2900 kg/m³, porosity 6%. *4–*9 Density 1800–2800 kg/m³, in stops of 200 kg/m³. *10 Density 2900 kg/m³. *11 Density 2950 kg/m³. *12 Density 3030 kg/m³. *13 Density 2700–2800 kg/m³, according to [14, p. 347]. *14 Sintered; density 3480 kg/m³. 15 Sintered alumina. 16 Density 3900 kg/m³, according to [226]. *17 Quartz ceramic, 99.99% SiO_2. *18 Density 3140–3180 kg/m³; in vacuo; grain size 50–90 μm. *19 Density 3160 kg/m³; argon medium; grain size 120–250 μm. *20 Density 3100 kg/m³. *21 Density 5600 kg/m³. *22 Density 6100 kg/m³. *23–*29 Grade kh. ch. [chemically pure], stabilized with 4.2% MgO by mass; porosity 5, 9, 16, 26, 48, 58, and 62%, respectively. *30 Sintered at 2173°K. *31–*35 Sintered at 2073°K; 99.9% ThO_2; values dependent on porosity (%) and grain diameter (μm) as follows: *31 Porosity 6.7%, diameter 25.1 μm. *32 8.6, 16.4. *33 23.2, 6.1. *34 31.3, 9.4. *35 31.3, 42.9. *36–*39 Density 9,690, 10,020, 9,080, and 8,300 kg/m³, respectively.

Bending strength, MPa	Temperature, °K	Notes
1	2	3
	BeO [1]	
178.4–96.0	293–1473	–
17.84–99.96	293	Density 1960–2380 kg/m^3
250.9–61.7	293	Grain size 7–100 μm
34.3	293	Grain size 200 μm
	MgO [1]	
89.18	293	Density 3400 kg/m^3
43.61–8.82	1623–1903	Density 3400 kg/m^3
245	–	Fiberlike single crystals [219]
441	–	Theoretical, for single crystals
	Al_2O_3 [1]	
294–393	293	Density 3900 kg/m^3
156.8	293	Fired at 2023°K
26.5–27.4	293	Sapphire
185	293	Density 3850 kg/m^3, porosity 3.5% [216]
	SiO_2 [229]	
		Density, from graph:
35.8–73.5	273–1273	1930–1970 kg/m^3
58.8–110.0	273–1273	1980–2020 kg/m^3
69.5–137.0	273–1273	2100–2150 kg/m^3
106–171	873–1273	2170–2200 kg/m^3
88–177	673–1353	Quartz ceramic [228]
	Sc_2O_3 [140]	
176–186	293–973	Fired at 1900–2100°C
176–1873	1273–1873	v = 5 N/sec (loading rate)
	TiO_2 [1]	
134.26	293	Density 4100 kg/m^3

1	2	3
		SmO_2 [1]
196–245	293	Density 6100 kg/m^3
9.75–0.61	300–2300	Porosity 18.6%, stabilized with tech. CaO
132	–	Density 5430 kg/m^3, porosity 5.6%
		Sm_2O_3 [1]
12.74	293	Pressed at 19.6 MPa; oxygen/ acetylene medium; fired at 1573°K
13.72	293	Fired at 1773°K
		TaO [235]
345	293	Film, d = 20 μm
		Gd_2O_3 [219]
18.82	293	Fired at 1573°K
19.60	293	Fired at 1773°K
		ThO_2 [1]
		Sintered at 1923–2123°K
		Porosity, %:
73.5–19.1	293	7.2–33.1
98.98–27.44	293	6.1–33.1
109.76–31.56	293	6.7–33.1
115.64–16.46	1273	6.7–33.1
107.80–38.42	293	8.6–33.0
		UO_2 [1]
		Sintered at 2273°K
82.3–59.1	298	Density 8,500–10,100 kg/m^3
94.08–79.18	673–973	Density 9,560–10,060 kg/m^3
123.5–56.84	1273	Density 8,390–10,020 kg/m^3

7. HARDNESS ON MINERALOGICAL SCALE [1, pp. 256–258]

Oxide	Hardness (Mohs scale)	Notes
1	2	3
H_2O	1,5—6,0	268—198 K
BeO	9	Bromellite
	7—9	—
MgO	5—6,5	—
	5,50—5,75	Periclase
Al_2O_3	9	Corundum, sapphire, ruby
SiO_2	7	Quartz, tridymite
	6	Agate, flint, jasper
	6,5	Cristobalite
	8,5	Mineral
CaO	4,0—4,5	—
	4,0	Calcium oxide
	3,5	Lime
TiO_2	5—6,5	Anatase, rutile
V_2O_3	8,5	Karelianite
VO_2	3,5	
Cr_2O_3	8,5	Eskolaite
MnO	5,6—4,75	Manganosite
Mn_2O_3	5,5—7,0	Braunite, bixbyite
Mn_3O_4	4	Hausmannite
MnO_2	2,0—6,0	Pyrolusite
	7,0	Polianite
FeO	5,0	Iozite (wustite)

Oxide	Hardness (Mohs scale)	Notes
1	2	3
Fe_3O_4	5,5—6,5	Magnetite
Fe_2O_3	6,75	Hematite
NiO	5,0—5,5	Bunsenite
Cu_2O	3,5—4,0	Cuprite
CuO	3,5	—
	3,5	Tenorite
ZnO	4,0—5,7	—
	4,0	Zincite
Ga_2O_3	9,0	—
GeO_2	7,5	—
As_2O_3	1,5	Arsenolite
	2,5	Claudetite
	2,5	Claudetite
As_4O_6	1,5—2,0	Arsenolite
SeO_2	2,0	Selenolite
SrO	3,5	—
Y_2O_3	5,5	—
ZrO_2	6,5—7,0	—
	7,0—8,0	Sintered
	6,5	Baddeleyite
Nb_2O_3	6,5	—
MoO_3	1,5	
	1,0—2,0	Molybdite
CdO	3,0	—
In_2O_3	7,0	—

1	2	3	1	2	3
SnO_2	6,0—7,0	Cassiterite	Ta_2O_5	6,0	—
	6,5	—	WO_2	5,5—6,0	—
Sb_2O_3	2,0—2,5	Senarmonite	HgO	2,0	Montroydite
	2,5—3,0	Valentinite	Tl_2O_3	6,5	Avicennite
Sb_4O_6	2,0—2,5	Senarmontite	PbO	2,0	Litharge, massicot
Sb_2O_4	4,0—5,0	Cervantite	Pb_3O_4	2,0—3,0	Red lead
TeO_2	2,0	Tellurite	PbO_2	5,5	Plattnerite
BaO	3,0	—	Bi_2O_3	4,5	Bismite, sillénite
Ce_2O_3	5,0—5,5	—			—
CeO_2	6,0	—	ThO_2	6,5	Thorianite
	6,0	Cerianite	UO_2	3,5	Density 10 750 kg/m^3
Yb_2O_3	6,0	—		5,5—6.0	Uraninite
			U_3O_8	3,5	—

8. MICROHARDNESS

Microhardness, MPa	Load $P \cdot 10^{-5}$, N	Notes
1	2	3
BeO [1]		
14906,64		Time under load 10 sec
12 290	98 070	—
9950	49 035	—
MgO [1]		
9081,3—9277,4		Periclase single crystal
7453,3—1961,4	49035—98070	293—923 K
11 250	49 035	Single crystal

8 — Continued

1	2	3
	Al_2O_3 [1]	
20153,4—21487,1	196 140	Corundum, mineral
20447,6—20202,4	49 035	Synthetic corundum
19614,0—21575,4	196 140	Synthetic corundum
26675,04	49 035	Synthetic sapphire
25498,2	196 140	" "
24674,4	98 070	Synthetic leucosapphire
24321,36	98 070	Light-red ruby
22458,03	98 070	Dark red ruby
		Ruby; position of plane of observation:
23370,08	49 035	parallel to axis
25253,02	49 035	perpendicular to axis
		Synthetic coroundum; psoition of axis of tip relative to main axis of crystal:
21575,40	98 070	perpendicular
19025,60	98 070	parallel
22703,2	98 070	at a 60° angle
24600±1120*1	196 140	
	SiO_2 [1]	
11033,56—12121,45	196 140	Quartz
7571,0—8394,8	50 016	Smoky quartz, (0001) face
9708,9	50 016	Clear quartz, (0001) face
12062,6*2	49 035	Quartzite, (1010) face
		Quartz; position of axis of tip relative to main axis of crystal:
10983,8—10817,1*2	49 035	perpendicular
12356,82*3	49 035	parallel
10983,84*3	49 035	(1010) face
11081,9*3	49 035	(1011) face
10591—10737,7*3	49 035	Onyx
	CaO [1]	
6045	49 035	Grade "ch" [pure], sintered at T = 2023°K in air and argon medium; porosity 12%; time under load 10 sec

1	2	3
	Tl_6O [1]	
		Annealing temperature, °K:
5099,64	196 140	673
5393,85	196 140	873
5688,06	196 140	1073
	Tl_3O [1]	
		Annealing temperature, °K:
5197,71	196 140	673
5295,78	196 140	873
5589,99	196 140	1073
	TlO [1]	
19614,00—6374,55	49035—98070	293—1023 K
	TiO_2 [1]	
7845,66—1961,40	49035—98070	398—923 K
6001,88	98 070	Rutile
10 750	49 035	Hot-pressed at T = 1723°K, fired at T = 1473°K; time under load 10 sec
19400±274*4	98 070	
	V_2O_5 [236]	
1380±264	19 614	
	VO_2 [233]	
8560—6410	29 421	Single crystal, 278–364°K
	Cr_2O_3 [1]	
29 150	98 070	Grade ch.d.a. [pure, for analysis]; argon medium, sintered at T = 2273°K, porosity 21%; time under load 10 sec
27300±1190*4	196 140	Hausmannite
	MnO [1]	
5727,29	98 070	Manganite
1961,40—411,89	49035—98070	353—1073 K, from graph
	Mn_3O_4 [1]	
4207,20	98 070	Hausmannite

8 – Continued

1	3	3
	FeO [1]	
5393,85—98,07	49035—98070	293—1273 K, from graph
	Fe_3O_4 [1]	
4677,94—7884,83	49 035	Magnetite
4677,94—7257,18	49035—98070	Magnetite (1101) plane
	Fe_2O_3 [1]	
10983,84—6864,9	49 035	Hematite
9022,44—10415,03	98070—196140	Hematite, large crystals
12400—705*[1]	98 070	
	CoO [1]	
3775,7—196,1	49035—98070	293—1163 K, from graph
3610±235*[1]	49 035	
	NiO [1]	
3922,80—1961,40	49035—98070	393—723 K, from graph
4310±402*[1]	49 035	
	Cu_2O [1]	
2010,43—2030,05	49 035	Cuprite
	CuO [1]	
2049,66—2490,98	49 035	Tenorite
	ZnO [1]	
1471,05—3118,63	49 035	Zincite
3150±362*[1]	49 035	
	Ga_2O_3 [237]	
		Single crystal
6370±245	—	(100) plane
8420±245	—	(001) plane
	Y_2O_3 [242]	
7250—1765	—	293—1373 K

1	2	3
ZrO_2 [1]		
9807—2942	49 035	543—1073 K, stabilized with CaO
12 080	98 070	Stabilized with CaO, annealed at T = 1173°K, hot-pressed at T = 2253°K
		$T = 2253$ K
1153±28*1	98 070	
Nb_2O_5 [1]		
7260	39 228	Hot-pressed at T = 1473°K, annealed at T = 1273°K; porosity 16%; time under load 10 sec
7340*1	98 070	
MoO_3 [236]		
5700±461	98 070	
SnO_2 [1]		
9885,46—12043	196 140	Cassiterite
14318,22	98 070	»
13533,66	49 035	Cassiterite
Hf_6O [230]		
5390	—	After annealing at T = 1073°K
Hf_3O [230]		
6860—7840	—	Annealed at T = 1073 – 1923°K
9310		Cast specimens
HfO_2 [236]		
10250±822	98 070	
7650—2540*4	—	293—1373 K
Ta_2O_5 [236]		
10 400±254	98 070	
WO_3 [236]		
1500±147	49 035	

8 — Continued

1	2	3
		ThO_2 [1]
9689,32—10934,81	98070—196140	Thorianite
		UO_2 [1]
7669,07—8228,07	98070—196140	Uraninite
6600,1—7875,02	98070—196140	Pitchblende

*1 According to [236]. *2 According to [231]. *3 According to [232]. *4 According to [242].

9. COMPRESSIBILITY [1]

Oxide	Compressibility coefficient β · 10^{-11}, m^2/N	Pressure p · 10^{11} m^2/N Pa	Temperature, °K	Notes
H_2O	12.0	300	266	Ice I
	4.7	7500	283	Ice II
MgO	0,72	125	273	
Al_2O_3	0,40	125	273	Corundum
SiO_2	2,75	1	298	Quartz
CaO	—	—	—	Polycrystalline specimens
	0,92	—	273	According to [211]
	0,89	—	0	Calculated by extrapolation
TiO_2	0,59	125	273	Rutile
Fe_2O_3	0,60	125	273	Hematite
Fe_3O_4	0,55	1	273	Magnetite
Cu_2O	0,93	—	293	According to [217]
ZnO	0,78	125	273	Zincite
SnO_2	0,49	125	273	Cassiterite

10. ELASTIC CONSTANTS C_{ij}, GPa

C_{11}	C_{12}	C_{13}	C_{33}	C_{44}	C_{66}	Temperature, °K
1	2	3	4	5	6	7
			H_2O [225]			
15,00	7,75	6,75	16,25	3,25	—	150
14,25	7,25	6,50	15,50	3,00	—	200
13,25	6,75	6,00	14,75	2,75	—	250

1	2	3	4	5	6	7
			D_2O [225]			
16,00	8,00	5,25	18,50	4,00	—	150
15,00	7,50	4,75	17,75	3,75	—	200
14,25	7,25	4,50	16,75	3,70	—	
			MgO^{*1} [250]			
289,3	87,7	—	—	154,8	—	300
306,17	93,8	—	—	157,6	—	4—77
296,47	95,1	—	—	155,9	—	296
			$Al_2O_3^{*2}$ [1]			
492,0	168,4	116,4	490,2	146,8	—	—
496	109	48	502	206	193,5	300
			SiO_2 [1]			
77,8				31,58	—	
			CaO^{*3} [211]			
205	62	—	—	76	—	200
207	63	—	—	75	—	150
209	64	—	—	75	—	100
223	59	—	—	81	—	Room temp.
			TiO_2^{*4} [244]			
288,6	197,0	—	—	—	227,2	4
284,2	192,4	—	—	—	216,2	100
274,0	181,0	—	—	—	197,5	250
271,4	178,0	149,6	484,0	124,4	194,8	298
270,1	176,4	148,6	481,8	124,0	192,5	323
267,4	173,4	146,9	477,0	123,0	188,2	373
262,3	167,7	143,2	468,3	121,3	180,2	473
257,1	162,0	140,8	459,2	119,4	173,1	573
			MnO^{*5} [245]			
223	120	—	—	79	—	298
			Cu_2O^{*6} [217]			
116,5	100,3	—	—	12,1	—	
121,1	105,4	—	—	10,9	—	4,2

10 – Continued

1	2	3	4	5	6	7
			ZnO*7 [1]			
209,7	121,1	105,1	210,9	42,5	44,3	
			SrO*8 [246]			
177,6 173,0	41,3 45,0	— —	— —	59,5 56,0	— —	140 Room temp.
			ZrO_2*9 [247]			
204 223	87 97,3	— —	— —	158 154	— —	77 77
			TeO_2*10 [248]			
53,2 56	48,6 51,6	21,2 27,2	108,5 105,1	24,4 27,0	55,2 66,8	

*1First line of values from [215], density 3580 kg/m³. *2Second line of values from [250], single crystal, density 3986 kg/m³. *3Fourth line of values from [243], single crystal. *4Single crystal, rutile. *5Single crystal. *6By pulse-echo method. *7Single crystal, phase method. *8Second line of values from [243], single crystal. *9Stabilized with 8 and 12% (mol.) Y_2O_3, respectively. *10Second line of values from [249].

10 – Continued. $S_{ij} \cdot 10^{14}$, m²/N [215; 238]

Oxide	S_{11}	S_{12}	S_{13}	S_{14}	S_{33}	S_{44}	S_{66}	Temperature, °K	Density, kg/m³
MgO	4,024	0,936	—	—	—	6,461	—	—	3580
Al_2O_3	2,18	0,50	0,16	—0,49	2,02	5,04	5,36	300	3980*1
	2,353	0,716	0,364	0,489	2,170	6,940	6,138	298	
	2,404	0,747	0,382	0,503	2,216	7,038	6,302	300	
	2,408	0,756	0,756	—0,504	2,218	6,328	6,328	300	

Oxide	S_{11}	S_{12}	S_{13}	S_{14}	S_{33}	S_{44}	S_{66}	Temperature, °K	Density, kg/m^3
TiO_2	11,80 6,55	9,03 3,76	0,86 0,86	— —	2,74 2,59	8,33 8,00	6,25 5,16	298 298	4260*2
ZnO	7,855	3,431	2,205	—	6,939	23,546	22,573	298	5680*3

*1 Single crystal, data from various sources. *2 Single crystal. *3 Single crystal, C_{144} = 113, C_{166} = −659, C_{456} = 147.

11. SONIC VELOCITY [1; 45; 50; 135]

Oxide	Temperature, °K	Sonic velocity, m/sec	Notes
H_2O	273	3980	Ice
H_2O	293	1410	—
	298	1496	—
	313	1530	—
D_2O	283	13,48—13,83	—
	313	14,30	—
	333	14,53	—
	353	14,57	—
	373	14,46	—
BeO	—	v_L=11 910; v_S=7280	Porosity 0,006
	—	v_L=11 450; v_S=7020	" 0,047
	—	v_L=10 940; v_S=6770	" 0,064
	—	v_L=10 250; v_S=6370	" 0,116
	—	v_L=9060; v_S=5630	" 0,189
CO_2	293	260	—
O_2	293	316	—
TiO	—	3330	Calculated
VO	—	4510	"

ELECTRICAL AND MAGNETIC PROPERTIES

1. ELECTRICAL RESISTANCE [1]

Resistivity, ohms · m	Temperature, °K
1	2
H_2O	
$33,2\cdot10^{4}$	273
$53\cdot10^{3}$	323
BeO	
$9\cdot10^{9}$	773
$3\cdot10^{7}$ *1	1073
$1,6\cdot10^{6}$	1273
$3\cdot10^{4}$	1573
$2\cdot10^{3}$	1773
350 *1	1873
65 *1	2073
16	2273
Na_2O_2	
$2,5\cdot10^{2}$	293
60	392
$1\cdot10$	557
MgO	
$10^{12}—10^{13}$	573
$9\cdot10^{11}$	773
$1\cdot10^{7}$	973
$6\cdot10^{6}$	1273
$7\cdot10^{4}$	1573
$2\cdot10^{3}$	1773
5	2273
Al_2O_3	
$1\cdot10^{14}$	287
$3\cdot10^{12}$	473
$3\cdot10^{10}$	673
$4\cdot10^{8}$	873
$3,5\cdot10^{6}$	1073
$5\cdot10^{4}$	1273
$1\cdot10^{4}$	1373
$8,1\cdot10^{2}$	1873
$4,2\cdot10^{2}$	1998
$2,9\cdot10^{2}$	2083
$2,2\cdot10^{2}$	2148
SiO_2	
$1\cdot10^{12}$	293
$1\cdot10^{4}$	973
$2\cdot10^{3}$	1173
$3\cdot10^{2}$	1473
$2\cdot10^{2}$	1673
SO_2	
$66\cdot10^{3}$	308
CaO	
$70\cdot10^{4}$	1036
$41,7\cdot10^{3}$	1203
$10,4\cdot10^{2}$	1508
20,4	1643
Sc_2O_3	
$4,4\cdot10^{5}$	1000
TiO *2	
$2,6\cdot10^{2}$	200
$3,0\cdot10^{2}$	300
$3,1\cdot10^{2}$	400
TiO_2	
$3\cdot10^{5}$	773
$1,2\cdot10^{2}$	1073
$8,5\cdot10^{2}$	1473
V_2O_3	
$5,5\cdot10^{-5}$	293
$1,75\cdot10^{-5}$	2233
VO_2	
10,7	295
2,7	369
0,4	600
V_2O_5	
$4,9\cdot10^{2}$	148
$8,5\cdot10$	169
13,7	195
5,6	213
1,9	233
$8,1\cdot10^{-1}$	158
$4,1\cdot10^{-1}$	273
$3,0\cdot10^{-1}$	292
$2,7\cdot10^{-1}$	298
$7,9\cdot10^{-2}$	375
$4,9\cdot10^{-2}$	440
$1\cdot10^{-2}$	1273
Cr_2O_3	
$1,3\cdot10^{3}$	623
$2,3\cdot10^{1}$	1473
$12,65\cdot10^{2}$	618
$7,8\cdot10^{1}$	1023
$4,0\cdot10^{1}$	1273
$2,13\cdot10^{1}$	1488
CrO_2	
$5\cdot10^{-5}$	513
MnO	
$1\cdot10^{8}$	293
Mn_3O_4	
$2,0\cdot10^{3}$	833
$1,43\cdot10^{1}$	1273
$7,42\cdot10^{-1}$	1553

1	2
FeO	
$\sim 15,6\cdot10^{-4}$	873
$15\cdot10^{-5}$	1173
$2\cdot10^{-3}$	1473
$8,6\cdot10^{-4}$	1573
Fe_3O_4	
$1\cdot10^{-4}$	293
$1,3\cdot10^{-1}$	873
$2,2\cdot10^{-2}$	1273
$7,7\cdot10^{-3}$	1593
Fe_2O_3	
62,4	843
10,4	973
$8,2\cdot10^{-1}$	1273
$6,8\cdot10^{-1}$	1285
CoO	
$1\cdot10^{6}$	293
1	573
$1\cdot10^{-3}$	1273
Co_3O_4	
$1\cdot10^{2}$	—
NiO	
10^{11}	293
6,7	863
1,4	1273
0,24	1518
Cu_2O	
10^{6}—10^{7}	293
70	400
CuO	
1—10	293
$5\cdot10^{-2}$	973
$1\cdot10^{-3}$	1273
ZnO	
94,4	433
0,59	878
0,026	1273
0,02	1693
Ga_2O	
0,25—0,02	40—4,2
SrO	
40	—
SrO_2	
$8\cdot10^{2}$	1473
Y_2O_3	
$5,4\cdot10^{4}$	1000
ZrO_2	
$3\cdot10^{4}$	573
$8\cdot10^{3}$	673
$7\cdot10^{2}$	873
80	1073
10	1273
4	1473
$6\cdot10^{-1}$	1673
$1\cdot10^{-1}$	1873
$5\cdot10^{-2}$	2073
$1\cdot10^{-2}$	2273
Nb_2O_5	
$8\cdot10^{-3}$	1473
$2,8\cdot10^{-3}$	1773
RuO_2 [*3]	
$35,2\cdot10^{-6}$	300
PdO [*4]	
0,01—10	Room temp.
—	—
CdO	
67,3	1068
7,13	1273
0,32	1473
0,05	1603
$(0,01—0,05)\times$ $\times10^{-3}$	—
In_2O_3	
10^{4}—10^{7} [*5]	—
10^{-3}	293
SnO_2	
$4\cdot10^{-4}$	293
65,6	1058
2,56	1273
0,6	1473
0,1	1593
BaO	
$1\cdot10^{4}$	573
0,22	773
La_2O_3	
10^{6}	833
100	1373
CeO [*6]	
$4\cdot10^{10}$	293
Ce_2O_3 [*7]	
$2\cdot10^{7}$	293
CeO_2	
10^{4}	770
650	1073
3,4	1473
22,4	1103
2,13	1273
$1,87\cdot10^{-1}$	1483
Pr_2O_3	
1950	1000
PrO_2	
0,5	1000
Pr_6O_{11}	
$8,3\cdot10^{-3}$	1000

1 – Continued

1	2
Nd_2O_3	
110	1000
Sm_2O_3	
420	1000
EuO	
10^2	20*8
10^6	298*9
Eu_2O_3	
10	1380
Gd_2O_3	
$1\cdot10^4$	1000
Tb_2O_3	
10	667
~1	1000
Dy_2O_3	
$1{,}67\cdot10^2$	1000
Ho_2O_3	
10^7	690
Er_2O_3	
$\sim10^5$	1000
Tu_2O_3	
$\sim10^6$	833

1	2
Yb_2O_3	
$3{,}5\cdot10^3$	1000
Lu_2O_3	
$\sim10^6$	1000
HfO_2	
$5\cdot10^7$	673
10	1773
Ta_2O_5	
10^3	293
WO_3	
$2\cdot10^3$	—
ReO_2	
$8\cdot10^{-6}$	293
$2\cdot10^{-6}$	Liquid air
ReO_3	
$2\cdot10^{-5}$	293
IrO_2*3	
$(49\pm2)\cdot10^{-8}$	300
PbO	
10^7	293*10
$2{,}6\cdot10^5$	657
$2{,}7\cdot10^3$	745
12,2	1060
1	1123
PbO_2	
$9{,}08\cdot10^{-5}$	273
10^{-4}	—

1	2
Bi_2O_3	
10^6—10^8	293
$2{,}34\cdot10^6$	498
$1{,}44\cdot10^3$	697
60,1	918
1,0	973
ThO_2	
$4\cdot10^{11}$	293
$1{,}2\cdot10^{10}$	773
$2{,}6\cdot10^5$	823
$8\cdot10^3$	1073
$1{,}1\cdot10^3$	1273
150	1478
7	1573
2	1773
0,01	2000
UO_2	
$14\cdot10^3$	293
$5\cdot10^2$	373
50	473
5	673
0,33	973
0,11	1273
$3\cdot10^{-3}$	1375
U_3O_8	
$1\cdot10^5$	293
$4\cdot10^4$	323
$1{,}1\cdot10^4$	373
$3{,}3\cdot10^3$	423
$6\cdot10^2$	473
$2\cdot10^2$	523
$8\cdot10^2$	573
33	623
14,3	673
7	723
4	773

*1 Sintered oxide, bulk density 2250 kg/m^3. *2 From graph. *3 [322], single crystal. *4 [314]. *5 [323], polycrystalline films. *6 [326]. *7 [325]. *8 [327]. *9 [328]. *10 [290].

Temperature, °K	Thermo-emf coefficient, μV/°K
1	2
H_2O [1]	
—	1000
Ti_6O [283]	
400	—4
600	—6,5
800	—8,2
1000	—6,5
1200	—5,5
1400	—5,5
Ti_5O [283]	
400	—3,1
600	—7,5
800	—1
1000	—10
1200	—8
1400	—7,8
Ti_3O [283]	
400	—3,2
600	—6
800	—9
1000	—11,5
1200	—11,25
1400	—11
TiO* [1]	
100	—6
200	—4
293	—5
400	—7
TiO	
600	—9
800	—11,75
1000	—13
1400	—12,5
TiO_2 [283]	
400	0,75
600	—2,75
800	—6
1000	—9
1200	—12

Temperature, °K	Thermo-emf coefficient, μV/°K
1	2
V_4O_7 [267]	
250	—(10—12)
V_5O_9 [267]	
—	~(10—20)
V_6O_{11} [330]	
< 177	—10
V_7O_{13} [330]	
< 120	—1
VO_2 [285]	
—	—500
V_2O_5 [1]	
293	1320
873	700
973	400—420
1073	290—300
1273	200
1373	170—200
Cr_2O_3 [1]	
273	900
MnO [272]	
400	1800
500	1500
660	1200
1000	500
Mn_2O_3 [1]	
—	385
FeO [1]	

Temperature, °K	Thermo-emf coefficient, μV/°K
1	2
Fe_3O_4 [1]	
—	—430
Fe_2O_3 [1]	
95	380
360	75
CoO [1]	
1000	—400
1400	—250
NiO [1]	
—	240
Cu_2O [1]	
293	700
CuO*2 [289]	
500	1050
554	930
667	760
714	800
853	950
ZnO [1]	
313	220
423	290
513	330
Y_2O_3 [282]	
1450	270±30*3
1450	100±10*4
Nb_2O_5 [1]	
1573	800—300
1673	500—510

2 — Continued

1	2
1400	−11,5
CdO [1]	
—	−30
—	−40
In_2O_3 [1]	
500	3000
563	1400
SnO [1]	
—	−140
Pr_2O_3 [1]	
300—1000	1—15
Nd_2O_3 [282]	
1450	300

1	2
—	−500
Nd_2O_3 [282]	
1450	600
WO_3 [1]	
—	−740
PbO [290]	
573	910*5
673	800*5
763	900*5
773	830*5
973	800*5
573	700*6
673	550*6

1	2
1773	190—210
Bi_2O_3 [1]	
773	1000
813	960
853	1000
973	−30
1013	0
1043	200
1213	300
1273	400—500
Th_2O [1]	
1400—1800	2200—4000
U_3O_8 [291]	
30—850	Positive
≥850	Changes to negative

*1 From graph. *2 From graph, pO_2 = 67 Pa. *3 Polycrystal, pO_2 = 50 Pa. *4 Polycrystal, pO_2 = 1 Pa. *5 Heating rate 100°C/h. *6 Heating rate 15–20°C/h.

3. THERMAL EMISSION PROPERTIES [1, pp. 284-287]

Oxide	Work function		Oxide	Work function	
	$A\cdot10^{19}$, J	A, eV		$A\cdot10^{19}$, J	A, eV
1	2	3	1	2	3
BeO	5,45—7,50	3,4—4,7	TiO_2*4, *1	9,97	6,21
B_2O_3	7,5	4,7	TiO_2*5, *1	6,2	3,87
MgO	4,95—7,05	3,1—4,4	FeO	6,15	3,85
Al_2O_3	7,5	4,7	NiO	8,9	5,55
SiO_2	8,0	5,0	Cu_2O	7,85	4,9
CaO*1, *2	2,82	1,76	Cu_2O*6	8,25	5,15
	3,8	2,4	CuO*6	8,55	5,34
			CuO	6,95	4,35
Sc_2O_3*3	6,45	4,04	SrO*1, *7	2,04	1,27
TiO	4,78—4,98	2,96—3,1	Y_2O_3*8, *1	3,2	2,0

1	2	3	1	2	3
ZrO_2*9, *1	5,00	3,12	Tb_2O_3*11	5,3	3,3
MoO_3	6,80	4,25	Dy_2O_3*12, *1	4,4	2,75
Cs_2O	1,58–1,87	0,99–1,17	Ho_2O_3*12, *1	5,12	3,2
BaO*1, *2	1,58	0,99	Er_2O_3*12, *1	4,8	3,0
La_2O_3*10, *1	4,95	3,1	Yb_2O_3*12, *1	5,3	3,3
Ce_2O_3*11	5,12	3,21	Lu_2O_3*12, *1	6,18	3,86
Pr_2O_3*10, *1	5,55	3,47	HfO_2*13, *1	6,00	3,76
Nd_2O_3*10, *1	5,28	3,3	HfO_2*14, *1	4,5	2,81
Sm_2O_3*10, *1	5,12	3,21	Ta_2O_5	7,42	4,65
Eu_2O_3*10, *1	5,4	3,38	WO_2	7,9	4,95
Gd_2O_3*10, *1	5,25	3,29	ThO_2*15, *1	4,34	2,71

*1 Richardson constant R = 10^4 A/(m^2 · kg · °C^2). *2 R = 10^{-2}. *3 A_1 = 400 and A_2 = 9.3, obtained from Richardson lines above and below the inflection point. *4 Not activated; R = 4.08 · 10^5. *5 Activated; measurements at 1750–2000°K; R = 0.458. *6 Photoemission. *7 R = 10^{-3}. *8 Activation temperature 1723–1773°K; R = 0.55. *9 Activated; measurements at 1700–2000°K; R = 0.363. *10 Effective work function at 1700°K; R = 120 at 1600°K. *11 Effective work function at 1700°K. *12 Effective work function at 1600°K; R = 120. *13 Activated; measurements at 1300–1500°K; R = 3.81. *14 Activated; measurements at 1500–1900°K; R = 0.49. *15 R = 21.

4. GALVANOMAGNETIC PROPERTIES [1] (H indicates mobility of holes, and E indicates mobility of electrons)

Oxide	Hall constant, m^3/C	Temperature, °C	Mobility, m^2/(V·sec)
1	2	3	4
SiO_2	—	—	E 10^{-11}
TiO	$5,4\cdot10^{-10}$	80—400	E $1\cdot10^{-4}$
	—	293	E $0,36\cdot10^{-4}$
	$4\cdot10^{-6}$*1	300	—
TiO_2	$2\cdot10^{-6}$	500	—
	—	1375—1573	E $1\cdot10^{-5}$
VO	—	1200	E $7\cdot10^{-6}$
V_2O_3	—	—	E $5\cdot10^{-7}$
VO_2	$(0,1-0,15)10^{-6}$	—	E $0,1\cdot10^{-4}$
V_2O_5	$288,00\cdot10^{-6}$	241	—
	$61,32\cdot10^{-6}$	259	—
	$22,40\cdot10^{-6}$	277	—
	$16,20\cdot10^{-6}$	287	—
	$2,93\cdot10^{-6}$	302	—

4 – Continued

1	2	3	4
	$-11{,}73\cdot10^{-6}$	318	—
	$-50{,}4\cdot10^{-6}$	347	—
	$-93{,}7\cdot10^{-6}$	359	—
	$-129{,}1\cdot10^{-6}$	374	—
V_2O_5*2	$(7{,}5-17{,}6)10^{-6}$	293	E 10^{-6}—10^{-5}
MnO*3	—	—	H $9{,}57\cdot10^{-4}$
MnO_2*4	$-53\cdot10^{-6}$	—	—
Fe_3O_4	$(8-15)10^{-4}$	293	E $1\cdot10^{-3}$
Fe_2O_3	$\sim2{,}2\cdot10^{-6}$	—	—
CoO	—	700	H $7\cdot10^{-2}$
NiO	—	—	E $5\cdot10^{-2}$
	—	293	H $7\cdot10^{-6}$
	—	300	H $4\cdot10^{-7}$
	$1\cdot10^{3}$	500	—
Cu_2O	0,26	293	H $(6-8)\cdot10^{-3}$
	0,10	400	—
	0,01	500	—
CuO	—	293	E $0{,}26\cdot10^{-4}$
ZnO	—	293	E $2\cdot10^{-2}$
ZnO*5	—	—	E $5\cdot10^{-3}$
ZnO	$-2{,}05\cdot10^{-6}$	300	—
β-Ga_2O_3	—	160	E $11\cdot10^{-3}$
Nb_2O_5	—	High	E $0{,}4\cdot10^{-4}$
CdO	$-(0{,}24-1{,}12)\cdot10^{-6}$	—	E $65\cdot10^{-4}$
CdO*6	—	—	E $50\cdot10^{-2}$
In_2O_3*7	—	—	E $(55-60)10^{-4}$
In_2O_3	—	293	E $16\cdot10^{-3}$
SnO_2	—	293	E $(2-3)\ 10^{-3}$
	—	1023	E $(5-6)\ 10^{-3}$
BaO*8	—	400—820	E $(3-5)\cdot10^{-4}$
$PrO_{<1,5}$	—	—	E 0,1—1
PbO*9	—	300	E $(5-100)\cdot10^{-4}$
PbO_2	—	293	E $1\cdot10^{-4}$
UO_2	—	293	H $1\cdot10^{-3}$
U_3O_8	—	—	E $0{,}79\cdot10^{-4}$

*1 Specimen was held in hydrogen for 5 min at 973°K. *2 [286]. *3 With up to 3% of other oxides as impurities. *4 Egyptian pyrolusite; contains 80% MnO_2. *5 Sintered specimens. *6 [293]. *7 [282]. *8 Uncolored crystal. *9 [294].

5. MAGNETIC PROPERTIES. A. Magnetic Susceptibility [1; 261; 262; 268; 276; 277; 281; 334]. (In the table, the magnetic susceptibility is designated as MS)

Oxide	$MS \cdot 10^9$, m^3/kg	Temperature, °K
1	2	3
H_2O	—0,7019*1	273
	—0,7177*2	273
	—0,727	413
Li_2O	—0,57	293
BeO	0	—
B_2O_3	—0,55	293
CO	—0,4	293
CO_2	—0,42	—
N_2O	—0,3*3	293
NO	48,66	295
N_2O_2	3,59	100
N_2O_3	—0,206*4	291
N_2O_4	—0,276*3	257
NO_2	—0,429*4	285
	3,26*5	428
N_2O_5	—0,332*3	289
O_2	5*3	20
	310*4	54
	107,8*5	289
MgO	—0,25*6	—
Al_2O_3	—0,36	298
SiO_2	—0,45	—
P_2O_5	—0,46	291
SO_2	—0,285*	298
ClO_3^-	—30,2	—
KO_2	45	298
CaO	—0,27	—
Sc_2O_3	—0,117	—
TiO	2,3±0,3	100—40
Ti_3O_5	8,1	298
TiO_2	—0,066	—
VO	50,1	288
V_2O_3	~7,5*7	180
VO_2	3,75	286
Cr_2O_3	25,5	—
CrO_2	20,1	87
	12,6	1608
CrO_3	60	—
MnO	89,9	84
	68,3	293
	51,8	692
Mn_3O_4	126	98
	1,7059±	77*11
	55,8	293
	25,5	1364
Mn_2O_3	69,0	294
	29,5	1201
MnO_2	68,5	84
	27	298
Fe_3O_4	18,5	973
	11,1	1173
Fe_2O_3	10,5	973
	10,0	1173
CoO	74,5	—
Co_2O_3	34,3	293
NiO	54	293
Cu_2O	—(0,213± ±0,003)*8	298
CuO	3,8	298
ZnO	—0,36	—
GeO	—0,33	293
As_2O_3	—0,27	—
SeO_2	—0,24	293
BrO_3	—39,8	—
SrO	—0,06	—
Y_2O_3	0,5	293
ZrO_2	—0,112	—
NbO_2	0,31	90
	0,18	293
Nb_2O_5	—0,10	293
Mo_3O	0,30	289
Mo_2O_3	—0,35	289
MoO_2	0,33	293
Mo_5O_{14}	0,53—0,80	350*9
	0,57—0,85	598*9
AgO	0,16	298
SnO	—0,14	293
SnO_2	—0,26	293
Sb_2O_3	—0,19	287
TeO_2	—0,14	291
IO_3^-	—51,4*10	—
Ba_2O	—0,24	—
BaO	—0,13	—
La_2O_3	—0,4	293
CeO_2	30	—
Lu_2O_3	—0,3	293

5 – Continued

1	2	3	1	2	3
	±0,0038		HfO_2	–0,110	—
	0,5381±	305	TaO_2	0,32	293
	±0,0014		Ta_2O_5	–0,07	—
Pr_2O_3	26,4	296	WO_2	0,216	298
Pr_4O_7	16	288	WO_3	–0,065	298
Pr_6O_{11}	14,6	296	Re_2O_7	90	—
PrO_2	10,7	296	ReO_2	157±5*12	80
Nd_2O_3	29,1	293		148±5*12	300
	16,6	556	ReO_3	0,086	77–300
Pm_2O_3	7,85	293	OsO_2	0,528	55
Sm_2O_3	5,8	293		0,536	77
Eu_2O_3	30	293	OsO_4	–0,045	77
Gd_2O_3	140	293		–0,048	300
	57,8	673	$PtO_{1,38}$	0,783	—
	44,8	873	Pt_2O_3	–0,085	298
	32,4	1273	HgO	–0,216*13	298
Tb_2O_3	195,5	296		–0,221*14	298
Tb_4O_7	160,7	296	PbO	0,21*13	293
Tb_6O_{11}	145,0	296		0,20*14	293
Dy_2O_3	219	293		–0,13	291
	89,8	723	Pb_3O_4	–0,24	291
Ho_2O_3	229	292	Bi_2O_3	–0,17	298
	50,4	1273	ThO_2	–0,06	298
Er_2O_3	189	293	UO_2	7,5	—
	102	553	U_2O_5	1,80	293
Yb_2O_3	38	293	U_3O_8	0,95	—

B. Effective Magnetic Moment [1]

Oxide	Effective moment P_{ef}, Bohr magnetons	Temperature, °K	Oxide	Effective moment p_{ef}, Bohr magnetons	Temperature, °K
1	2	3	1	2	3
NO	1,535	112,5	PrO_2	2,61	296
	1,732	194,7	Nd_2O_3	3,79	293
	1,841	289,2	EuO	7,3	—
O_2	2*15	—	Gd_2O_3	7,95	293
NaO_2	1,77*16	225	Tb_2O_3	9,62	296
KO_2	1,20*16	150	Tb_4O_7	8,70	296
	1,03	150	Tb_6O_{11}	8,49	296
Ti_3O_5	0,2	—	Dy_2O_3	9,67*17	423
CrO_2	2,95	293		10,6	573*18
Fe_3O_4	4,2	858	Ho_2O_3	10,09	673
RbO_2	1,89	90	Tu_2O_3	7,2	—
	1,89	293	UO_2	2,92	—
CsO_2	1,91	90	U_2O_5	1,77±0,03	293
	1,89	293	U_3O_8	1,59±0,04	293
Pr_2O_3	3,55	293	UO_3	2,06	289
Pr_6O_{11}	2,77	296			

C. Temperature of Magnetic Transition [1; 279; 336] (C is the Curie point; N is the Néel point)

Oxide	Temperature of magnetic transition, °K	Oxide	Temperature of magnetic transition, °K
1	2	1	2
Ti_2O_3	475 (C)	Nb_2O	450 (C)
V_2O_3	338 (C)	Pr_2O_3	57 (C)
V_4O_7	250*19 (C)	Pr_6O_{11}	85 (C)
V_5O_9	130 (C)	PrO_2	127 (C)
V_2O_4	343 (C)	Nd_2O_3	32 (C)
CrO	393—403 (C)	EuO	70,3 (C)
CrO_2	397 (C)		71,5 (C)
	100*20 (N)	Eu_3O_4	7,8 (C)
α-Cr_2O_3	32±1 (N)	Gd_2O_3	18 (C)
MnO	116 (N)	TbO	3 (N)
MnO_2	92 (N)	Tb_2O_3	24 (C)
	610*8 (C)	Tb_4O_7	19 (C)
FeO	198 (N)	Tb_6O_{11}	32 (C)
Fe_3O_4	858 (C)	Dy_2O_3	21 (C)
Fe_2O_3	950 (N)	Ho_2O_3	14 (C)
CoO	291 (N)	Er_2O_3	13 (C)
	289*8 (N)	Tu_2O_3	42 (N)
Co_3O_4	40*21 (N)	Yb_2O_3	104 (C)
Co_2O_3	33±1 (N)	UO_2	30 (C)
NiO	550 (N)	U_3O_8	170 (C)
CuO	458 (C)		

*1Ice. *2Water. *3Solid. *4Liquid. *5Gas. *6Sintered polycrystal. *7From graph. *8Single crystal. *9Metastable. *10Ionic diamagnetic susceptibility. *11With 0.4% added Gd; measurements performed in nitrogen atmosphere by Faraday method. *12Monoclinic. *13Red. *14Yellow. *15Wide range of temperatures; gas. *16Powder; impurities <2%. *17Experimental. *18Theoretical. *19Single crystal obtained by transport reaction method. *20Added sulfur and technetium >0.17%. *21Undergoes ferromagnetic transition below temperature of 40°K.

6. DIELECTRIC PROPERTIES [1]

Oxide	Frequency, Hz	Temperature, °K	Dielectric constant
1	2	3	4
H_2O	—	255	3,2
	—	273	87,83
	—	293	81
	—	303	76,47
	—	323	69,73
	10^6	383	1,0126
	—	413—423	1,00785
	10^8	273	88
D_2O	—	298	78,54

6 – Continued

1	2	3	4
BeO	10^5	293	6,3
	—	—	7,35
	$(0,3-10)\ 10^6$	298	1,86—0,20
	$(45-90)\ 10^4$	473	1,3—1,4
	$(45-90)\ 10^4$	673	2,0—2,1
B_2O_3	$(1-50)\ 10^3$	773—1073	3,1—3,2
	$4,5\cdot10^8$	—	3—8
CO	—	273	1,00069
	10^6	298	1,00634
CO_2	10^6	273	1,00099
NO	10^6	298	1,00059
N_2O	10^6	298	1,00103
Na_2O	$4,5\cdot10^8$	—	15,0
MgO	$(0,3-10)\ 10^6$	298	3,20—0,20*1
	$(45-90)\ 10^4$	473	3,2—3,1
	$(45-90)\ 10^4$	673	3,5—3,6
	—	—	9,8*2
Al_2O_3	—	293	10,5—12*3
	—	293	12,3*4
SiO	—	—	2,6—4*5
SiO_2	—	—	3,5—4,1*6
	$3\cdot10^7$	290—295	4,34*7
	$3\cdot10^7$	290—295	4,27*8
SO_2	10^6	265	1,0100
	—	293	14*7
SO_3	10^6	350	1,001270
K_2O	$4,5\cdot10^8$	—	16,0(20,3)*8
CaO	$(0,3-10)\ 10^6$	298	3,00±0,10
	$(45-90)\ 10^4$	473	3,00
	$(45-90)\ 10^4$	673	3,8—3,7
	—	—	11,8
TiO_2	—	—	40,0—80,0
	10^8	290—295	86*9
	—	290—295	170*10
	10^4	298	160*11
	$3\cdot10^7$	303	100*12
V_2O_5	—	—	13,84
Cr_2O_3	$4,5\cdot10^8$	—	9,2
MnO	$4,5\cdot10^8$	—	13,8
	$10^2—10^4$	303	18,1*13
FeO	$4,5\cdot10^8$	—	16,0
CoO	$10^2—10^{10}$	298	12,9
	$4.5\cdot10^8$	—	13,4
Co_2O_3	—	—	9,59
NiO	—	—	5—10
	10^5	298	11,9
Ni_2O_3	—	—	8,88*14
	—	—	11,24*15
	—	—	12,29*16
	—	—	12,87*17

1	2	3	4
Cu_2O	—	—	8,58[*18]
	—	—	10,26[*19]
	10^2—10^5	—	7,50±0,15
	Low frequencies	—	30[*20]
CuO	—	—	9,77[*21]
	—	—	10,68[*22]
ZnO	$(0,3-10)\cdot 10^6$	298	18,00—1,00
	$4,5\cdot 10^8$	—	14,4
	—	—	8,22[*23]
β-Ga_2O_3	$(5-500)10^3$	297	10,2±0,3[*24]
SrO	$(45-90)10^4$	373	3,2—3,25
	$(45-90)10^4$	473	3,8—3,85
	$(45-90)10^4$	623	4,9—4,85
Y_2O_3	$(0,5-1,5)10^6$	293	14,0[*25]
Nb_2O_5	—	—	50±5[26]
	—	—	30±5[27]
	—	—	35±5[28]
CdO	$4,5\cdot 10^8$	—	17,2[*25]
SnO_2	—	—	24[*19]
	—	298	9,0±0,5[*29]
		298	14±2[*30]
BaO	$(0,3-10)10^6$	298	4,00—0,10[*1]
La_2O_3	$(0,5-1,5)10^6$	293	20,8[*25]
CeO_2	$(0,5-1,5)10^6$	293	21,2[*25]
Pr_2O_{11}	$(0,5-1,5)10^6$	293	69,9
Nd_2O_3	$(0,5-1,5)\cdot 10^6$	293	19,7[*25]
Sm_2O_3	10^3	—	21,5[*1]
	$(0,5-1,5)\cdot 10^6$	293	18,4[*25]
Eu_2O_3	$(0,5-1,5)10^6$	293	10,2[*25]
Gd_2O_3	$(0,5-1,5)10^6$	293	11,4[*25]
Tb_4O_7	$(0,5-1,5)10^6$	293	32,1[*25]
Dy_2O_3	$(0,5-1,5)10^6$	293	12,1[*25]
Ho_2O_3	$(0,5-1,5)10^6$	293	12,3[*25]
Er_2O_3	$(0,5-1,5)10^6$	293	12,5[*25]
Tu_2O_3	$(0,5-1,5)10^6$	293	12,6[*25]
Yb_2O_3	$(0,5-1,5)10^6$	293	12,6[*25]
Lu_2O_3	$(0,5-1,5)10^6$	293	12,9[*25]
Ta_2O_5	—	—	15[*31]
WO_3	—	—	20,2[*32]
	—	—	29,08[*33]
HgO	10^5	298	12,6[*34]
	10^5	298	9,4[*35]
PbO	$4,5\cdot 10^3$	—	22,0
	—	—	31,34[*36]
PbO_2	—	—	26[*19]
UO_2	7		21,7—0,5[*37]

6 — Continued

1	2	3	4
U_4O_9	9,4·10^9	Low	70[*38]
U_3O_8	—	—	41,77
UO_3	—	—	1,86—11

[*1]Sintered specimens. [*2]Crystal. [*3]Sintered alumina. [*4]Independent of frequency. [*5][262]; high-frequency; dependent on deposition conditions. [*6]Fused quartz. [*7]Liquid; for very long waves ($\lambda = \infty$). [*8]Value of 20.3 is anomalous. [*9]Perpendicular to optical axis. [*10]Parallel to optical axis. [*11]Perpendicular to c-axis. [*12]Along c-axis. [*13][297]. [*14]Density 2430 kg/m^3. [*15]Density 2460 kg/m^3. [*16]Density 2700 kg/m^3. [*17]Density 2810 kg/m^3. [*18]Pressed. [*19]Polycrystal. [*20][298]; highly dependent on temperature and frequency. [*21]Density 3100 kg/m^3. [*22]Density 4690 kg/m^3. [*23]Density 1800 kg/m^3. [*24][298]. [*25]Accuracy 1.5%. [*26][299]; in direction of a-axis. [*27]Same, in direction of b-axis. [*28]Same, c-axis. [*29][300]; parallel to tetragonal axis; wavelength 10–50 μm. [*30]Same, perpendicular to c-axis. [*31][301]. [*32]Density 2900 kg/m^3. [*33]Density 4200 kg/m^3. [*34]Red. [*35]Yellow. [*36]Density 7060 kg/m^3. [*37][333]. [*38][302].

7. WIDTH OF FORBIDDEN BAND AND ACTIVATION ENERGY

A. Width of Forbidden Band [1]

Oxide	Width of forbidden band		Oxide	Width of forbidden band	
	10^{19}, J	eV		10^{19}, J	eV
1	2	3	1	2	3
BeO	8,32	5,20	PdO	0,06	0,04[*30]
B_2O_3	7,84	4,89	AgO	2,52	1,57[*31]
MgO	11,68	7,30	SnO_2	5,65	3,54
Al_2O_3	4,0	2,5	Sb_2O_3	6,72	4,20
	5,75	3,58	TeO_2	>4,8	>3,0
CaO	>8,96	>5,60	BaO	2,88—3,32	1,85—2,08
TiO_2	5,9	3,7		6,4—6,9	4,0—4,3
	4,87	3,05[*10]	La_2O_3	8,65	5,40
V_2O_5	0,8	0,5	CeO_2	5,45	3,41
Cr_2O_3	7,7	4,8[*13]	Nd_2O_3	7,02	4,40
MnO_2	2,0	1,3	Sm_2O_3	8,0	5,0
Fe_2O_3	3,36	2,10	EuO	0,68	0,42[*31]
CoO	0,96	0,60	Eu_2O_3	7,2	4,5
	1,12	0,70	Gd_2O_3	8,5	5,3
NiO	2,73—3,04	1,72—1,90	Tb_2O_3	4,85	3,0
Cu_2O	2,91—3,33	1,83—2,08	Dy_2O_3	8	5
ZnO	4,18	2,62[*2]	Ho_2O_3	8,65	5,42
	5,1	3,2	Er_2O_3	8,65	5,42
SrO	9,1	5,7[*13]	Tu_2O_3	7,24	4,53
ZrO_2	3,2	2,0	Yb_2O_3	8,35	5,22
Nb_2O_5	3,55	2,21[*27]	Lu_2O_3	8,7	5,5
	6,4	4,0[*28]	WO_3	3,4	2,2[*13]
MoO_3	6,04	3,75[29]	UO_2	2,08	1,30

Oxide	Activation energy		Oxide	Activation energy	
	10^{19}, J	eV		10^{19}, J	eV
1	2	3	1	2	3
H_2O	0,92	0,58*1	NiO	1,49	0,93
	1,67	0,10*1	Cu_2O	0,40	0,25*22
BeO	3,42	2,14*2		0,26	0,16*23
	0,77	0,48*3	CuO	1,12	0,70
MgO	1,86	1,16	SrO	3,00—3,22	1,87—2,05
	4,8	3*4	Y_2O_3	2,7	1,7*9
	6,4	4*5	ZrO_2	0,08	0,05*24
Al_2O_3	2,08	1,3*2		2,75	1,72*25
SiO_2	2,12	1,32*6		1,34	0,84*26
	14,5	0,88*7	CdO	1,20	0,75*32
	0,16	0,1*8		2,68	1,67*33
	0,56	0,35*8	In_2O_3*34	$6\cdot10^{-2}$	$3,7\cdot10^{-2}$
Sc_2O_3	3,52	2,2*9		$6\cdot10^{-5}$	$3,7\cdot10^{-5}$
TiO_2	1,76	1,1	SnO_2	1,12	0,70
V_2O_3	0,51	0,32*11	La_2O_3	4,58	2,86
V_3O_5	0,77	0,48*11	CeO_2	4,3	2,7
V_4O_7	0,69	0,43*11	Pr_2O_3	1,12	0,70
V_5O_9	0,48	0,30*11	Pr_6O_{11}	1,92	1,20
V_6O_{11}	0,11	0,07*11	PrO_2	1,41	0,88
	0,19	0,12*11	Nd_2O_3	3,60	2,24
V_7O_{13}	0,49	0,31*12	Sm_2O_3	3,64	2,27
V_6O_{13}	0,38	0,21*12	Eu_2O_3	2,95	1,84
V_2O_5	0,32	0,20*13	Gd_2O_3	4,65	2,9
	0,72	0,45*14	Tb_4O_7	1,28	0,8
	0,75	0,47	Dy_2O_3	4,8	3,0
	1,6	1,0	Ho_2O_3	4,55	2,8
Cr_2O_3	0,32	0,20	Er_2O_3	5,22	3,26
	0,64	0,40*15	Tu_2O_3	5,05	3,17
	2,56	1,60	Yb_2O_3	4,78	2,99
CrO_2	0,32	0,20	Lu_2O_3	6,3	3,9
MnO	3,60	2,24*16	WO_3	1,0	0,6*13
	2,44	1,92*17	PbO	1,6	1,0
	6,05	3,76*18		4,0	2,5
Mn_3O_4	0,83	0,58	PbO_2	2,4	1,5
MnO_2	0,48	0,30	Bi_2O_3	2,24	1,44*36
FeO	0,55	0,34*19	ThO_2	5,6	3,5*37
Fe_3O_4	0,48	0,30*19	U_2O_4	1,20	0,75*38
	0,12	0,08*20		0,96	0,6*38
CoO	1,09	0,67*21	U_3O_8	0,77	0,48*39

*1 [303]; pure ice. *2 At high temperatures. *3 At low temperatures. *4 [304]; at low oxygen pressures. *5 Same, at high oxygen pressures. *6 Quartz, perpendicular to axis. *7 Parallel to axis. *8 [323, p. 86]. *9 [282]; T = 1100°K. *10 T > 800°K. *11 [328]. *12 [306]. *13 [305]. *14 At 243°K. *15 At 279 – 314°K. *16 [307]. *17 At 673°K. *18 At 1373°K. *19 [308]. *20 [309]. *21 [310]; at 300°K. *22 [311]. *23 [312]. *24 At 373°K. *25 At 1173°K. *26 At 1273°K. *27 At 1473°K. *28 At 1773°K. *29 [313]. *30 [314]. *31 [315]; at 4.2°K. *32 [316]. *33 [293]. *34 At T > 40°K and T < 40°K. *35 For oxides of rare-earth elements, the temperature interval for the activation energy is approximately 500 – 1500°K. *36 [262]. *37 For high vacuum. *38 [319]; at 93°K and 349°K. *39 [291].

OPTICAL PROPERTIES

1. COLOR [1; 19; 45; 52; 67; 80; 81; 171]

Oxide	State	Color
1	2	3
H_2O	*l*	Colorless*[1]
	cr	Colorless*[2]
Li_2O	cr	Colorless
Li_2O_2		Yellow
LiO_3		Red
BeO	cr	Colorless
	cr	White*[3]
	p	White
B_2O_2		Colorless
B_2O_3	gl	Colorless
B_4O_5		Colorless
CO	g	Colorless
CO_2	g	Colorless
N_2O	g	Colorless
NO	g	Colorless
N_2O_3	g	Reddish brown
	l	Blue
	l	Dark sky-blue
	cr	Snow-white
	cr	Pale sky-blue
$N_2O_4(NO_2)$	g	Brown
	g	Reddish brown
	l	Light yellow
	l	Reddish brown
	cr	Colorless
N_2O_5	cr	Colorless
O_2	g	Colorless
	l	Pale blue
	s	Pale blue
	cr	Blue
O_3	g	Sky-blue tint
		Blue

1	2	3
O_3	*l*	Dark blue
	s	Almost black
	cr	Dark violet
F_2O	g	Colorless
	l	Bright yellow
F_2O_2	g	Brown
	l	Cherry-red
		Orange
Na_2O	cr	Light yellowish
	cr	Colorless
Na_2O_2	p	White*[4]
NaO_2	p	Yellow
NaO_3		Red
MgO	cr	Colorless*[5]
	p	White
		Colorless to yellowish or green
MgO_2		Colorless*[6]
Al_2O_3		Colorless
α-Al_2O_3	cr	Colorless*[7]
	cr	Blue*[8]
	cr	Red*[9]
	cr	Black*[10]
β-Al_2O_3	cr	Colorless*[11]
γ-Al_2O_3	ap	Colorless*[12]
SiO	sf	Black or brownish black
	rs	Yellowish cinnamon-brown
Si_2O_3	sf	Golden yellow
SiO_2	gl	Colorless
	cr	Colorless*[13]
P_2O_3	cr	White
P_4O_6	cr	White
P_2O_4	cr	Colorless, lustrous
P_2O_5	cr	Colorless
	p	White
P_2O_6		Violet
SO	g	Colorless
	s	Orange
		Orange-red
S_2O_2	g	Colorless
S_2O_3	cr	Greenish blue

1 — Continued

1	2	3
	cr	Sky-blue
	cr	Bluish
SO_2	g	Colorless
	l	Colorless
	s	White
SO_3	s	Colorless
	gl	Colorless
SO_4		White
	s	Colorless
Cl_2O	g	Yellowish cinnamon-brown
	g	Brownish yellow
	l	Reddish cinnamon-brown
	l	Reddish brown
ClO_2	g	Greenish yellow
	l	Red
	cr	Orange-red
Cl_2O_6	oi	Dark red
	s	Bright orange-red*[14]
Cl_2O_7	*l*	Colorless
K_2O	cr	Yellowish
	cr	White
	cr	Colorless
K_2O_3	cr	Colorless
KO_3	cr	Red
CaO	cr	Colorless
	p	White
CaO_2	p	White
CaO_4	p	Yellow
Sc_2O_3	p	White
TiO	cm	Golden yellow*[15]
	p	Golden yellow
	p	Dark cinnamon-brown*[16]
Ti_3O_4		Black
Ti_2O_3	p	Dark violet
Ti_3O_5	p	Sky-blue
		Blue
α-TiO_2	cr	Red or cinnamon-brown, but thin lamellae are colorless*[17]
β-TiO_2	cr	Cinnamon-brown or yellow; sky-blue or green*[18]

1	2	3
γ-TiO_2	cr	Cinnamon-brown, yellowish, reddish, or black*[19]
TiO_2	p	White
VO	cr	Light grey with metallic luster
	p	Grey
	p	Black
V_2O_3	cr	Black
	p	Black
VO_2	cr	Blue-black
	cr	Blue or sky-blue
	p	Black or brown
$V_{12}O_{26}$	cr	Bright green with metallic luster
V_2O_5	cr	Red or reddish yellow
	crp	Red*[20]
	p	Orange-yellow
	p	Brown
CrO	p	Brick-red
		Red
		Black
Cr_3O_4		Reddish orange*[21]
Cr_2O_3	cr	Green
		Dark green
	p	Green
Cr_5O_{12}	cr	Black or dark cinnamon-brown
Cr_6O_{15}		Black
Cr_8O_{21}		Black
CrO_3	p	Dark red
CrO_5		Blue
MnO	cr	Emerald-green on fresh fracture but becomes black; streak cinnamon-brown*[22]
	cr	Grey-green
	cr	Brown
	p	Green
		Emerald-green*[23]
Mn_2O_3	cr	Brown
		Black
Mn_3O_4	cr	Black to cinnamon-brown
MnO_2	cr	Black
	p	Black

1 – Continued

1	2	3
		Steel-grey*[24]
Mn_2O_7	*l*	Greenish black
FeO	cr	Black*[25]
	p	Black
		Dark
Fe_3O_4	cr	Dark*[26]
		Dark red
$Fe_3O_4 \cdot 4H_2O$		Black
α-Fe_2O_3	cr	From steel-grey to iron-black with diamond luster; dark red in lump form*[27]
		Brownish red
γ-Fe_2O_3		Cinnamon-brown
CoO	p	Olive-green
		Light to dark cinnamon-brown
Co_3O_4	cr	Black
Co_2O_3	p	Cinnamon-brown or black*[28]
		Dark cinnamon-brown
NiO	cr	Dark green or cinnamon to black*[29]
Ni_2O_3	p	Black
		Greyish black
NiO_2	am	Black
Cu_2O	cr	Bright red*[30]
		Red or liver-red
CuO	cr	From iron-grey to black*[31]
		Black
Cu_2O_3	p	Red
CuO_2		Cinnamon-brown to black
ZnO	cr	Colorless to dark red*[32]
	p	White
ZnO_2		Colorless*[33]
Ga_2O	p	Dark cinnamon-brown
GaO		Grey
Ga_2O_3	p	White
GeO	cr	Yellow
	cr	Lemon-yellow with slight greenish tint*[34]
	p	Dark grey
	p	Yellowish cinnamon-brown
GeO_2	cr	Colorless
	cr	White

1	2	3
α-As_2O_3	cr	Colorless to white*[35]
β-As_2O_3	cr	White*[36]
As_2O_3	p	White
As_2O_5	am	Colorless
SeO_2	g	Yellowish green
	cr	White, lustrous
SeO_3	cr	Colorless
Br_2O	*l*	Brown
		Cinnamon-brown
BrO_2	s	Cinnamon-brown
	s	Light yellow
Br_3O_8	cr	Colorless
Rb_2O		Yellowish
Rb_2O_2		Colorless
RbO_2		Yellow
RbO_3		Orange
SrO	cr	Colorless
	p	White
SrO_2	p	White
SrO_4	p	Yellow
Y_2O_3	cr	Colorless
	p	White
	p	Green*[37]
		Pale cinnamon-brown
ZrO_2	cr	Colorless to cinnamon-brown*[38]
	p	White
NbO	p	Grey
	p	Black
		Black to cinnamon-brown
Nb_2O_3		Blue-black
NbO_2		Black
		Blue-black
Nb_2O_5		Colorless
		White*[39]
	ml	Greyish black
Mo_2O_3		Black
MoO_2	cr	Dark cinnamon-brown with lilac tint
Mo_2O_5		Violet-black
γ-Mo_4O_{11}	cr	Dark violet
η-Mo_4O_{11}	cr	Cherry-red

1 — Continued

1	2	3
β-Mo_8O_{23}	cr	Sky-blue
ξ-Mo_9O_{26}	cr	Black
β'-Mo_9O_{26}	cr	Dark blue
MoO_3	cr	Colorless with greenish tint
	cr	Straw-yellow*[40]
	cr	Almost black*[41]
Tc_2O_7	cr	Light yellow
RuO_2	p	Black-grey with metallic luster*[42]
RuO_4	cr	Golden yellow, cinnamon-brown, or yellow
Rh_2O_3		Grey-black
		Grey
PdO	p	Black
Pd_2O_3	p	Chocolate-black*[43]
PdO_2		Dark red
Ag_2O	cr	Brown
		Dark brown
AgO	cr	Dark grey with metallic luster
		Black
CdO	cr	Cinnamon-brown or red to black
	cr	From light cinnamon-brown to dark brown*[44]
		Brown*[45]
		Dark cinnamon-brown with slight metallic luster*[46]
		Greenish yellow*[47]
		Dark blue*[48]
CdO_2		Colorless*[49]
InO	cr	Black
In_2O_3	cr	Yellow
SnO		Different varieties may be black, bluish black, grey, red, green, or greenish brown
SnO_2	cr	Yellow to cinnamon-brown, rarely red, grey, or white*[50]
		White
α-Sb_2O_3	cr	Colorless or grey-white*[51]
β-Sb_2O_3	cr	Colorless to white or colored*[52]
Sb_2O_3	cr	Colorless
	p	White

1 – Continued

1	2	3
Sb_4O_6		From white to yellow
Sb_2O_4	cr	Pale yellow to yellowish or reddish*[53]
Sb_2O_4	cr	Snow-white*[54]
	p	White
Sb_2O_5	cr	Light yellow
TeO_2	cr	White*[55]
	cr	Colorless
TeO_3	p	Yellow
I_2O_4	p	Yellow
I_4O_9	p	Yellowish
I_2O_5	cr	White
Cs_2O	cr	Yellow
		Orange
Cs_2O_2		Colorless
CsO_2		Yellow
CsO_3		Red
BaO	cr	Colorless
	p	White
BaO_2	cr	Colorless*[56]
	p	White
BaO_4		Yellow
La_2O_3	cr	Colorless
	p	White
Ce_2O_3		Colorless
	p	Greenish-yellow
CeO_2	p	Yellowish white
		White*[57]
Pr_2O_3	p	Greenish yellow
Pr_6O_{11}	p	From dark brown to black
		Cinnamon-brown to black
Nd_2O_3	p	Light blue to light purple*[58]
		Lilac
Sm_2O_3	p	Slightly yellow
		From white to slightly yellow
		Rust-brown*[59]
EuO		Garnet-red to black
$Eu_{16}O_{21}$	cr	Orange-yellow*[60]
Eu_3O_4	p	Dark red
Eu_2O_3	p	White with reddish tint (pale rose)
		Cream

1 — Continued

1	2	3
Gd_2O_3	p	White
		Colorless
		Dark cinnamon-brown* [61]
Tb_2O_3	p	White
		Colorless
		Yellow
Tb_4O_7	p	Dark brown
		Dark cinnamon-brown
Dy_2O_3	p	Pale cream, almost white
		Colorless
Ho_2O_3	p	Yellow
		Pale yellow
Er_2O_3	p	Pale rose-red
		Rose
Tu_2O_3	p	White with slight greenish tint
Yb_2O_3	p	White
		Colorless
Lu_2O_3	p	White
		Colorless
HfO_2		White
Ta_4O		Blue-black
TaO_2		Black
TaO_x		Black* [62]
Ta_2O_5	p	White
W_3O		Cinnamon-brown
	cr	Cinnamon-brown
	p	
WO_2		Cinnamon-brown
$WO_{2.03}$		Cinnamon-brown
$WO_{2.5}$		Reddish violet
$WO_{2.65}$		Reddish violet
$WO_{2.82}$		Reddish violet
$WO_{2.88}$		Reddish violet with blue tint
$WO_{2.95}$	cr	Blue
WO_3	cr	Golden yellow to green* [63]
	cr	Green* [64]
	p	Lemon- or orange-yellow
		Light yellow or cinnamon-brown
		Blue
ReO_2	cr	Black

1	2	3
ReO_3	cr	Red with metallic luster
	p	Red with metallic luster
Re_2O_7	cr	Light yellow
		Yellow
OsO	p	Black
OsO_2	p	Cinnamon-brown or black
OsO_4	cr	Pale yellow
	cr	Almost colorless
Ir_2O_3		Black
IrO_2		Blue-black
PtO	p	Black
	p	Greyish violet
Pt_2O_3	p	Cinnamon-brown
PtO_2	cr	Black
	p	Dark cinnamon-brown
PtO_3		Reddish cinnamon-brown
Au_2O	p	Greyish violet
AuO		Dark green
Au_2O_3	p	Black-brown
	p	Dark cinnamon-brown
Hg_2O		Black
HgO		Orange-red*[65]
	p	Yellow*[66]
HgO_2	p	Yellow*[67]
Tl_2O	p	Black
Tl_2O_3		Cinnamon-brown
α-PbO	cr	Yellowish red to red*[68]
β-PbO	cr	Yellow*[69]
Pb_3O_4	cr	Red*[70]
PbO_2	cr	Tar-black*[71]
		Dark cinnamon-brown
		Dark cinnamon-brown
α-Bi_2O_3	cr	Grey-green to bright yellow*[72]
γ-Bi_2O_3	cr	Olive-grey to green*[73]
Bi_2O_3	p	Yellow*[74]
Bi_2O_4	p	Cinnamon-brown
Bi_2O_5	p	Brownish red
PoO_2	cr	Red*[75]
		Yellow*[76]
ThO_2	cr	Dark grey to black*[77]

1 – Continued

1	2	3
	cr	White
		Colorless
PaO	p	Black
PaO_2	p	Black
Pa_2O_5	p	White
UO		Grey with metallic luster
UO_2	cr	Cinnamon-brown* [78]
		From brown to black
		Cinnamon-brown to black
	cr	Violet-black (changes to yellow at edges)* [79]
	p	From cinnamon-brown to black
U_4O_9		Black
β-$UO_{2.25}$		Black
γ-$UO_{2.34}$		Black
U_3O_7		Bluish black
U_2O_5		Black
U_3O_8		Dark green
		From olive-green to black
		Dark green or olive-green
		Dark green or black
UO_3	cr	Yellow* [80]
	cr	Amber-yellow to brownish yellow* [81]
	p	Orange-yellow
		Orange
		Dark orange
		Greyish yellow
U_2O_7		Orange
$UO_4 \cdot 2H_2O$		Yellow
NpO_2	p	Green* [82]
		Cinnamon-brown
		Apple-green
Np_3O_8	p	Cinnamon-brown
		Chocolate-brown
PuO	p	Black
		Black with metallic luster
Pu_2O_3		Silvery
		Silvery with semimetallic luster
PuO_2	cr	Yellow

1	2	3
	p	From yellowish green to cinnamon-brown
	p	Yellow, yellow-green, or dark cinnamon-brown
AmO		Black
Am_2O_3		Cinnamon-brown*[83]
		Reddish orange
		Reddish cinnamon-brown*[84]
AmO_2	p	Black
Cm_2O_3		White
CmO_2		Black
Bk_2O_3		Yellowish green*[85]

*[1]Bluish green in thick layers; sky-blue in layers thicker than 2 m. *[2]Pale sky blue in large mass. *[3]Bromellite. *[4]Technical product has a slight yellowish tinge due to NaO_2 present as an impurity. *[5]Periclase. *[6]$MgO_2 \cdot H_2O$. *[7]Corundum. *[8]Sapphire. *[9]Ruby. *[10]Emerald. *[11]Unstable. *[12]Unstable at normal temperatures. *[13]α- and β-cristobalite, α- and β-tridymite, and α- and β-quartz. *[14]At –78°C. *[15]Pressed mixture of TiO_2 and Ti heated to 1700°C in vacuo. *[16]TiO_2 reduced with magnesium. *[17]Rutile. *[18]Anatase or octahedrite. *[19]Brookite. *[20]In finely divided state, orange or yellow. *[21]In polarized light. *[22]Manganosite. *[23]Rapidly darkens in air. *[24]Pyrolusite. *[25]Wüstite. *[26]Magnetite. *[27]Hematite. *[28]$Co_2O_3 \cdot H_2O$. *[29]Bunsenite. *[30]Cuprite. *[31]Tenorite. *[32]Zincite. *[33]$ZnO_2 \cdot xH_2O$. *[34]In transmitted light, greenish yellow with clearly evident pleochroism from rich yellow-green to very pale green. *[35]Claudetite. *[36]Arsenolite. *[37]With nonstoichiometric composition. *[38]Baddeleyite. *[39]Upon heating, acquires a yellow tinge that disappears when cooled. *[40]Along a- and b-axes. *[41]Along c-axis. *[42]Upon heating in vacuo to 958°C, changes over to a blue crystalline form. *[43]$Pd_2O_3 \cdot xH_2O$. *[44]Depends on method of preparation. *[45]Obtained at low temperatures. *[46]After calcining at 800 – 900°C *[47]Obtained from hydrate by calcining at 350°C. *[48]After calcining hydrate at 800°C. *[49]$GdO_2 \cdot xH_2O$. *[50]Cassiterite. *[51]Senarmontite. *[52]Valentinite. *[53]$Sb_2O_4 \cdot nH_2O$ (stibiconite?). *[54]Becomes yellow upon heating. *[55]Tellurite. *[56]$BaO_2 \cdot 8H_2O$. *[57]Yellow in heated state. *[58]Calcined at 900°C. *[59]After firing at 1300 – 1500°C. *[60]"Ortho-I" phase. *[61]Fired at 1300 – 1500°C. *[62]$2 < x < 2.5$. *[63]$WO_3 \cdot H_2O$ (tungstite). *[64]$WO_3 \cdot 2H_2O$ (hydrotungstite). *[65]Montroydite; red oxide, becomes black upon heating, but recovers its bright orange-red color upon cooling. *[66]Fine crystalline powder (grain size ~2 μm). *[67]Red in solution. *[68]Litharge. *[69]Massicot. *[70]Red lead. *[71]Plattnerite. *[72]Bismite. *[73]Sillénite. *[74]Becomes orange or brown upon heating (recovers original color after cooling). *[75]High-temperature modification. *[76]Low-temperature modification. *[77]Thorianite. *[78]Uraninite. *[79]$2UO_2 \cdot 7H_2O$ (ianthinite). *[80]$4UO_3 \cdot 9H_2O$ (schoepite). *[81]$7UO_3 \cdot 11H_2O$ (becquerelite). *[82]Lustrous black crystals were obtained. *[83]Hexagonal modification. *[84]Cubic modification. *[85]Monoclinic modification.

2. REFRACTIVE INDEX [1; 19; 52; 67; 77; 353; 636]

Oxide	Refractive index			Notes
	n_g	n_m	n_p	
1	2	3	4	5
H_2O	1,3104	1,309	1,309	Ice
	—	1,33299	—	At 20°C, 101.3 kPa, water

2 — Continued

1	2	3	4	5
	—	1,3289	—	At 50°C
	—	1,3178	—	At 100°C
H_2O_2	—	1,4067(*D*)	—	At 25°C
D_2O	—	1,32795(*D*)	—	Heavy water; at 25°C
Li_2O	—	1,644	—	
BeO	1,733	—	1,719	Crystalline (bromellite)
	—	1,590	—	Uncalcined powder
	—	1,692	—	Fired at 900°C
	—	1,716	—	" " 1300° C
	—	1,718	—	" " 1700° C
B_2O_3	—	1,459	—	Glass
MgO	—	1,737	—	Periclase
Al_2O_3	1,760	—	1,768	α-Modification (corundum)
	—	1,736	—	γ-Modification
	—	1,690—1,695	—	γ-Al_2O_3, finely dispersed cubic form
	—	1,65	—	Gel after calcining
SiO	2,15	—	2,06	
	—	2,15—1,95	—	For $\lambda = 0,4 \div 0,7$
Si_2O_3	1,658	—	1,654	
Si_3O_4	1,90	—	1,80	
SiO_2	1,5405		1,5329	α-Quartz; for Na light
	1,553	—	1,544	β-Quartz
	1,484	—	1,487	α-Cristoballite
	—	1,486	—	β-Cristobalite
	—	1,486—1,492	—	Metacristobalite
	1,473	1,470	1,469	α-Tridymite
	1,481	1,479	1,479	γ-Tridymite
	—	1,462	—	Quartz glass
	1,538	—	1,532	Chalcedony
	1,826± ±0,002	—	1,799± ±0,002	Stishovite
	1,597	—	1,594	Coesite
	1,513	—	1,522	Keatite
	—	1,425±0,002	—	Melanophlogite
P_2O_5	1,624	—	1,599	—
CaO	—	1,837	—	Lime
Sc_2O_3	—	1,990	—	—
	—	1,91	—	—
TiO_2	2,908	—	2,621	α-Modification, rutile
	2,488	—	2,561	β-Modification, anatase
	2,7004	2,5843	2,5831	γ-Modification, brookite
	2,9467	—	2,6506	Rutile
	2,5688	—	2,6584	Anatase
	2,809	—	2,677	Brookite
Cr_2O_3	—	2,5(Li)	—	
MnO	—	2,16(Li)	—	Manganosite
FeO	Almost transparent (*n*=2,32)			Wustite

1	2	3	4	5
Fe_2O_3	2,759	—	2,988(Li)	Hematite
	2,78	—	3,01(Li)	α-Modification
Fe_3O_4	Almost transparent ($n=2,42$ Na)			Magnetite
NiO	—	2,27(Li)	—	—
	—	2,23(Na)	—	Bunsenite
Cu_2O	—	2,816—2,534	—	—
	—	2,849(D)	—	Cuprite
CuO	—	2,63(Li)	—	Tenorite
	—	3,18—2,63	—	—
ZnO	2,020	2,004(D)	2,004	Zincite
GeO_2	1,724± ±0,003	—	1,697± ±0,001	Quartz-like modification
	1,653	—	1,633	Chalcedony-like modification
	—	>1,606	—	Amorphous
	—	1,606	—	Fused
As_2O_3	2,01	1,92	1,87	α-Modification, claudetite
	—	1,755(Na)	—	β-Modification, arsenolite
	—	1,748(Li)	—	In red part of spectrum
SeO_2	—	>1,76	—	—
SrO	—	1,87	—	—
Y_2O_3	—	1,910—1,915	—	—
ZrO_2	2,20	2,19	2,13	Baddeleyite
Nb_2O_5	—	2,33	—	—
CdO	—	2,49(Li)	—	—
In_2O_3	—	1,95(1,25 μm)	—	In IR region of spectrum
SnO_2	2,0929	—	1,9968	Cassiterite
Sb_2O_3	—	2,087(Na)	—	Senarmontite
	—	2,073(Li)	—	Senarmontite, in red part of spectrum
	2,358	2,35	2,18	Valentinite
Sb_2O_4	2,04	—	1,83	Cervantite
TeO_2	2,35	2,18	2,00(Li)	Tellurite
BaO	—	1,98(Na)	—	—
La_2O_3	—	1,85—1,95	—	—
CeO_2	—	2,00	—	—
	—	2,40	—	—
Pr_6O_{11}	—	1,92—2,05	—	—
Nd_2O_3	—	1,85	—	A—Nd_2O_3, 500° C
	—	1,88	—	" 600° C
	—	1,97	—	" 700° C
	—	2,00	—	" 800° C
	—	2,05	—	—
Sm_2O_3	<2,14	—	2,08	
	—	1,82—1,86	—	Average for uncalcined material

2 – Continued

1	2	3	4	5
	—	2,04—2,06	—	After calcining at 1600—1700° C
	2,095	2,093	2,070	—
Eu_2O_3	—	1,87(20° C)—2,03(1400° C)	—	—
	2,10—2,14	—	2,08—2,1	—
$Eu_{18}O_{21}$	—	1,89—1,92	—	—
Gd_2O_3	2,10—2,14	—	2,05—2,08	—
	—	1,82	—	—
Tb_2O_3	2,08—2,051	—	2,04	—
Dy_2O_3	⩾2,051	—	⩾2,03	—
	—	1,88	—	—
Ho_2O_3	—	1,960	—	—
Er_2O_3	—	1,955	—	—
Tu_2O_3	—	1,95	—	—
Yb_2O_3	—	1,865	—	—
	—	1,940	—	—
	—	1,947	—	—
Lu_2O_3	—	1,930	—	—
HfO_2	—	1,98—2,02	—	—
HgO	2,65	2,50	2,37	Montroydite
PbO	2,535(Li)	—	2,665	Litharge
	2,71(Li)	2,61	2,51	Massicot
PbO_2	<2,30	—	2,30	Plattnerite
Bi_2O_3	2,63	—	2,63	α-Modification, bismite
	—	2,42	—	γ-Modification, sillénite
ThO_2	—	2,09—2,15	—	Thorianite
UO_2	Usually not transparent ($R \approx 12-15$ %)			In visible region of spectrum
	—	2,35	—	—
Pu_2O_3	—	2,44±±0,02(Li)	—	—
PuO_2	—	2,402(Na)	—	—
	—	2,35±±0,005(Li)	—	—

3. EMISSIVITY. A. Integral Normal Emissivity

Integral normal emissivity $\varepsilon_{\pm}$ in smooth surface	Integral normal emissivity $\varepsilon_{\pm}$ in powder	Temperature, °K	Integral normal emissivity $\varepsilon_{\pm}$ in smooth surface	Integral normal emissivity $\varepsilon_{\pm}$ in powder	Temperature, °K
1	2	3	1	2	3
BeO[*1] [19, 93]			—	0,706	1300
—	0,665	1200	—	0,746	1400

A – Continued

1	2	3
—	0,785	1500
—	0,819	1600
—	0,843	1700
—	0,867	1800
—	0,894	1900
—	0,931	2000
BeO*2 [19, 93]		
0,336	—	1200
0,361	—	1300
0,392	—	1400
0,420	—	1500
0,439	—	1600
0,453	—	1700
0,463	—	1800
0,470	—	1900
0,474	—	2000
0,475	—	2100
0,475	—	2150
BeO*3 [19]		
—	0,66	1173
—	0,70	1273
—	0,74	1373
—	0,77	1473
—	0,81	1573
—	0,83	1673
—	0,86	1773
—	0,88	1873
—	0,92	1973
—	0,96	2073
BeO*4 [19]		
—	0,33	1173
—	0,35	1273
—	0,38	1373
—	0,41	1473
—	0,43	1573
—	0,45	1673
—	0,47	1773
—	0,475	1873
—	0,48	1973
—	0,48	2073
MgO*5 [19, 93]		
0,74	—	73
0,74	—	173
0,73	—	273
0,715	—	373
0,68	—	473
0,64	—	573
0,60	—	673
0,56	—	773
0,49	—	873
0,44	—	973
0,41	—	1073
0,38	—	1173
0,37	—	1273
0,33	—	1373
0,325	—	1473
0,32	—	1573
0,315	—	1673
0,31	—	1773
MgO*6 [93]		
0,73	—	100
0,73	—	200
0,72	—	300
0,70	—	400
0,65	—	500
0,62	—	600
0,57	—	700
0,52	—	800
0,47	—	900
0,42	—	1000
0,38	—	1100
0,35	—	1200
0,33	—	1300
0,30	—	1400
0,29	—	1500
0,28	—	1600
0,28	—	1700
0,29	—	1800
0,32	—	1900
0,36	—	2000
0,41	—	2100
0,49	—	2200
0,58	—	2300
Al_2O_3*7 [19, 93]		
0,73	—	73
0,74	—	173
0,74	—	273
0,735	—	373

A – Continued

1	2	3
0,73	—	473
0,72	—	573
0,71	—	673
0,67	—	773
0,65	—	873
0,62	—	973
0,58	—	1073
0,55	—	1173
0,53	—	1273
0,49	—	1373
0,47	—	1473
0,45	—	1573
0,44	—	1673
0,43	—	1773
$Al_2O_3^{*8}$ [93]		
0,79	—	400
0,71	—	600
0,60	—	800
0,52	—	1000
0,46	—	1200
0,42	—	1400
0,40	—	1600
0,39	—	1800
SiO_2^{*9} [19]		
0,72	—	573
0,69	—	673
0,65	—	773
0,61	—	873
0,57	—	973
0,52	—	1073
0,48	—	1173
SiO_2^{*10} [93]		
0,38—0,42	—	1100—1700
CaO^{*11} [93]		
0,27	—	1123—1550
TiO_2^{*12} [93]		
0,82	—	400
0,83	—	500
0,84	—	600
0,85	—	700
0,86	—	800
0,87	—	900
0,875	—	1000
0,88	—	1100
0,89	—	1200
0,90	—	1300
$Cr_2O_3^{*13}$ [93]		
—	0,74	1123-1523
$Cr_2O_3^{*14}$ [93]		
—	0,86	1123
—	0,89	1200
—	0,91	1300
—	0,93	1400
—	0,95	1500
—	0,95	1523
$Fe_2O_3^{*15}$ [93]		
—	0,57	1100
—	0,63	1200
—	0,65	1300
—	0,74	1400
NiO^{*16} [93]		
0,36	—	600
0,38	—	700
0,40	—	800
0,43	—	900
0,47	—	1000
0,50	—	1100
0,55	—	1200

1	2	3
ZnO^{*17} [93]		
—	0,24	1160
—	0,33	1200
—	0,49	1300
—	0,58	1400
—	0,63	1500
ZnO^{*18} [93]		
0,91	—	1140
0,81	—	1240
0,82	—	1330
$Y_2O_3^{*19}$ [93]		
—	0,33	1273
ZrO_2^{*20} [93]		
—	0,26	1100
—	0,25	1200
—	0,27	1300
—	0,28	1400
—	0,33	1500
—	0,37	1600
—	0,45	1700
—	0,52	1800
—	0,59	1900
—	0,65	2000
—	0,71	2100
—	0,74	2200
—	0,77	2300
—	0,78	2400
—	0,79	2500
—	0,80	2600
—	0,80	2700
—	0,80	2800
ZrO_2^{*21} [93]		
—	0,81	20
—	0,75	400
—	0,65	600
—	0,53	800
—	0,42	1000
—	0,37	1200
—	0,37	1400
—	0,39	1600
—	0,46	1800
—	0,55	2000
—	0,62	2200
—	0,66	2400
—	0,69	2600
CeO_2^{*22} [93]		
—	0,35	1100
—	0,73	1200
—	0,91	1400
—	0,94	1600
—	0,95	1800
—	0,94	2000
—	0,93	2200
—	0,92	2300
$Sm_2O_3^{*23}$ [19]		
—	0,47	623
—	0,48	673
—	0,49	773
—	0,50	873
—	0,50	973
—	0,50	1073
—	0,48	1173
—	0,46	1273
—	0,47	1373
—	0,50	1473
—	0,53	1573
—	0,56	1673
—	0,56	1773
—	0,53	1873
$Gd_2O_3^{*24}$ [19]		
—	0,33	773
—	0,38	873
—	0,38	973
—	0,35	1073
—	0,29	1173

A – Continued

1	2	3
—	0,20	1273
—	0,15	1373
—	0,12	1473
—	0,10	1573
—	0,10	1673
—	0,11	1773
—	0,13	1873
Er_2O_3[*25] [19]		
—	0,05	673
—	0,13	773
—	0,27	873
—	0,35	973
—	0,34	1073
—	0,28	1173
—	0,22	1273
—	0,18	1373
—	0,10	1473
—	0,08	1573
—	0,07	1673
—	0,07	1773
HfO_2[*26] [93]		
—	0,80	1100
—	0,82	1200
—	0,83	1400
—	0,83	1600
—	0,81	1800
—	0,82	2000
—	0,89	2200
—	0,88	2400

1	2	3
Ta_2O_5[*27] [93]		
—	29	1273
ThO_2[*28] [93]		
—	0,64	573
—	0,60	673
—	0,57	773
—	0,52	873
—	0,48	973
—	0,43	1073
—	0,39	1173
ThO_2[*29] [19, 93]		
—	0,62	600
—	0,52	800
—	0,43	1000
—	0,38	1200
—	0,37	1400
—	0,41	1600
—	0,50	1800
—	0,61	2000
—	0,68	2200
—	0,72	2400

[*1] Hot-pressed in graphite mold, fired in air up to 1300°C; uniformly dark; density 2.85 g/cm³. [*2] Hot-pressed, fired in air up to 1300°C; white; density 2.778 g/cm³. [*3] Specimens darkened, directly from mold; data from graph. [*4] Specimens decolorized when calcined in air; data from graph. [*5] Fused; data from graph. [*6] Recommended values; from graph. [*7] Measurements in air; from graph. [*8] Recommended values; from graph. [*9] Pure oxide; from graph. [*10] Crystalline quartz. [*11] Film. [*12] Hemispherical radiation in vacuo from coating with thickness of 63 μm; from graph. [*13] Grain size 0.15 – 1.5 μm; from graph. [*14] Grain size 1.5 – 8.0 μm; from graph. [*15] From graph. [*16] Hemispherical radiation in air from film with thickness of 2 μm; from graph. [*17] Particle size 0.5 μm; from graph. [*18] Hemispherical radiation in vacuo from single crystal. [*19] Calculated value on the basis of $\varepsilon_{\lambda n}$. [*20] 97.1% ZrO_2 and 2.17% CaO, density 4.65 g/cm³; from graph. [*21] Recommended values. [*22] Pressed and sintered, density 6.87 g/cm³; from graph. [*23] Fired at 1850°C, density 7.62 g/cm³; from graph. [*24] From graph. [*25] From graph. [*26] Pressed and sintered, density 9.55 g/cm³; from graph. [*27] Calculated from graph of $\varepsilon_{\lambda n}$. [*28] Spectrally pure; from graph. [*29] Recommended values; from graph.

B. Monochromatic Normal Emissivity

Monochromatic normal emissivity $\varepsilon_{\lambda n}$ smooth surface	Monochromatic normal emissivity $\varepsilon_{\lambda n}$ powder	Wavelength λ, μm	Temperature, °K
1	2	3	4
BeO*1 [19]			
—	0,542	0,665	1200
—	0,543	0,665	1300
—	0,546	0,665	1400
—	0,552	0,665	1500
—	0,559	0,665	1600
—	0,568	0,665	1700
—	0,577	0,665	1800
—	0,587	0,665	1900
BeO*2 [19]			
0,212	—	0,665	1200
0,209	—	0,665	1300
0,210	—	0,665	1400
0,213	—	0,665	1500
0,217	—	0,665	1600
0,222	—	0,665	1700
0,228	—	0,665	1800
0,235	—	0,665	1900
BeO*3 [19]			
—	0,55	0,665	1073
—	0,54	0,665	1173
—	0,54	0,665	1273
—	0,54	0,665	1373
—	0,55	0,665	1473
—	0,56	0,665	1573
—	0,57	0,665	1673
—	0,58	0,665	1773
—	0,59	0,665	1873
—	0,595	0,665	1973
—	0,60	0,665	2073
BeO*4 [19]			
—	0,23	0,665	1900

Monochromatic normal emissivity $\varepsilon_{\lambda n}$ smooth surface	Monochromatic normal emissivity $\varepsilon_{\lambda n}$ powder	Wavelength λ, μm	Temperature, °K
1	2	3	4
BeO*5 [93]			
—	0,21	1,0	1223
—	0,08	2,0	1223
—	0,13	3,0	1223
—	0,38	4,0	1223
—	0,72	5,0	1223
—	0,87	6,0	1223
—	0,91	7,0	1223
—	0,89	8,0	1223
—	0,82	9,0	1223
—	0,49	10,0	1223
—	0,38	11,0	1223
—	0,36	12,0	1223
—	0,34	13,0	1223
—	0,33	14,0	1223
—	0,32	15,0	1223
MgO*6 [19, 93]			
0,17	—	0,665	1073
0,18	—	0,665	1173
0,20	—	0,665	1273
0,23	—	0,665	1373
0,27	—	0,665	1473
0,31	—	0,665	1573
0,35	—	0,665	1673
0,40	—	0,665	1773
0,44	—	0,665	1873
MgO*7 [1]			
0,28	—	1,0	293
0,25	—	2,0	293
0,25	—	3,0	293
0,30	—	4,0	293
0,33	—	5,0	293
0,49	—	6,0	293
0,72	—	7,0	293
0,85	—	8,0	293
0,93	—	9,0	293

B – Continued

$Al_2O_3^{*8}$ [19, 93]

1	2	3	4
—	0,22	0,665	1073
—	0,25	0,665	1173
—	0,29	0,665	1273
—	0,34	0,665	1373
—	0,38	0,665	1473
—	0,42	0,665	1573
—	0,47	0,665	1673
—	0,51	0,665	1773
—	0,56	0,665	1873

$Al_2O_3^{*9}$ [19]

1	2	3	4
—	0,29	0,665	973
—	0,25	0,665	1073
—	0,24	0,665	1173
—	0,24	0,665	1273
—	0,25	0,665	1373

$Al_2O_3^{*10}$ [93]

1	2	3	4
—	0,12	1,0—3,0	1273
—	0,22	4,0	1273
—	0,69	5,0	1273
—	0,85	6,0	1273
—	0,97	7,0	1273
—	0,98	8,0	1273
—	0,96	9,0	1273
—	0,98	10,0	1273
—	0,55	11,0	1273
—	0,50	12,0	1273
—	0,47	13,0	1273
—	0,46	14,0	1273
—	0,45	15,0	1273

$Al_2O_3^{*11}$ [93]

1	2	3	4
0,14	—	16,0	77
0,78	—	20,0	77
0,30	—	25,0	77
0,40	—	30,0	77
0,28	—	35,0	77
0,20	—	40,6	77
0,13	—	44,0	77

SiO^{*12} [93]

1	2	3	4
0,32	—	2,0	873
0,38	—	3,0	873
0,33	—	4,0	873
0,28	—	5,0	873
0,25	—	6,0	873
0,23	—	7,0	873
0,24	—	8,0	873
0,37	—	9,0	873
0,38	—	10,0	873
0,30	—	11,0	873
0,23	—	12,0	873
0,21	—	13,0	873
0,20	—	14,0	873

SiO_2^{*13} [93]

1	2	3	4
—	0,06	1,0—3,0	1273
—	0,60	4,0	1273
—	0,92	6,0	1273
—	0,96	8,0	1273
—	0,84	10,0	1273
—	0,96	12,0	1273

SiO_2^{*14} [1]

1	2	3	4
—	0,25	1,0	293
—	0,22	2,0	293
—	0,30	3,0	293
—	0,62	4,0	293
—	0,92	5,0	293
—	0,97	6,0	293
—	0,98	7,0	293
—	0,96	8,0	293
—	0,85	9,0	293

TiO_2^{*15} [93]

1	2	3	4
—	0,27	1,0	1223
—	0,15	2,0	1223
—	0,20	3,0	1223
—	0,30	4,0	1223
—	0,32	5,0	1223
—	0,50	6,0	1223
—	0,67	7,0	1223

1	2	3	4
—	0,76	8,0	1223
—	0,80	9,0	1223
—	0,84	10,0	1223
—	0,85	11,0	1223
—	0,86	12,0	1223
—	0,87	13,0	1223
—	0,88	14,0	1223
—	0,89	15,0	1223
	Cr_2O_3*16 [93]		
—	0,68	1,0—5,0	1273
—	0,69	6,0	1273
—	0,73	7,0	1273
—	0,78	8,0	1273
—	0,81	9,0	1273
—	0,84	10,0	1273
—	0,88	11,0	1273
—	0,91	12,0	1273
—	0,88	13,0	1273
—	0,85	14,0	1273
—	0,82	15,0	1273
	NiO*17 [93]		
—	0,78	1,0	1273
—	0,76	2,0	1273
—	0,74	3,0	1273
—	0,73	4,0	1273
—	0,74	5,0	1273
—	0,76	6,0	1273
—	0,77	7,0	1273
—	0,78	8,0	1273
—	0,80	9,0	1273
—	0,82	10,0	1273
—	0,88	11,0	1273
—	0,91	12,0	1273
—	0,92	13,0	1273
—	0,88	14,0	1273
—	0,86	15,0	1273
	Y_2O_3*18 [93]		
—	0,27	1,0	1273
—	0,27	2,0	1273
—	0,26	3,0	1273
—	0,24	4,0	1273
—	0,30	5,0	1273
—	0,43	6,0	1273
—	0,60	7,0	1273
—	0,67	8,0	1273
—	0,72	9,0	1273
—	0,77	10,0	1273
—	0,79	11,0	1273
—	0,80	12,0	1273
—	0,79	13,0	1273
—	0,78	14,0	1273
—	0,77	15,0	1273
	ZrO_2*19 [93, 19]		
—	0,42	0,665	1200
—	0,46	0,665	1400
—	0,49	0,665	1600
—	0,54	0,665	1800
	ZrO_2*20 [93]		
—	0,61	0,665	2000
—	0,69	0,665	2200
—	0,77	0,665	2400
—	0,80	0,665	2600
—	0,82	0,665	2800
	TeO_2*21 [93]		
—	0,09	2,5	533; 643
—	0,31	3,0	533; 643
—	0,21	4,0	533; 643
—	0,47	5,0	533; 643
—	0,88	6,0	533; 643
	CeO_2*22 [93]		
0,29	—	0,665	1300
0,28	—	0,665	1400

B – Continued

1	2	3	4
0,27	—	0,665	1500
0,26	—	0,665	1600
0,25	—	0,665	1700
0,24	—	0,665	1750
Sm_2O_3*23 [19, 1]			
—	0,43	0,64	973
—	0,48	0,64	1073
—	0,49	0,64	1173
—	0,49	0,64	1273
—	0,46	0,64	1373
—	0,39	0,64	1473
—	0,36	0,64	1573
—	0,33	0,64	1673
—	0,32	0,64	1773
—	0,30	0,64	1873
Sm_2O_3*24 [19, 1]			
—	0,78	1,0	973
—	0,71	1,0	1073
—	0,66	1,0	1173
—	0,61	1,0	1273
—	0,57	1,0	1373
—	0,50	1,0	1473
—	0,46	1,0	1573
—	0,39	1,0	1673
—	0,30	1,0	1773
—	0,25	1,0	1873
Gd_2O_3*25 [19]			
—	0,39	0,64	1073
—	0,26	0,64	1173
—	0,18	0,64	1273
—	0,11	0,64	1373
—	0,08	0,64	1473
—	0,07	0,64	1573
—	0,06	0,64	1673
—	0,03	0,64	1773
—	0,04	0,64	1873
Gd_2O_3*26 [19]			
—	0,80	1,0	1073
—	0,62	1,0	1173
—	0,47	1,0	1273
—	0,35	1,0	1373
—	0,27	1,0	1473
—	0,21	1,0	1573
—	0,16	1,0	1673
—	0,13	1,0	1773
—	0,11	1,0	1873
Er_2O_3*27 [19, 1]			
—	0,34	0,64	973
—	0,26	0,64	1073
—	0,19	0,64	1173
—	0,15	0,64	1273
—	0,12	0,64	1373
—	0,10	0,64	1473
—	0,08	0,64	1573
—	0,07	0,64	1673
—	0,06	0,64	1773
Er_2O_3*28 [19]			
—	0,57	1,0	973
—	0,54	1,0	1073
—	0,53	1,0	1173
—	0,52	1,0	1273
—	0,49	1,0	1373
—	0,41	1,0	1473
—	0,29	1,0	1573
—	0,23	1,0	1673
—	0,20	1,0	1773
Lu_2O_3*29 [93]			
—	0,15	0,68	500
—	0,14	0,68	600
—	0,14	0,68	700
—	0,28	0,68	800
—	0,47	0,68	900
—	0,54	0,68	1000
—	0,56	0,68	1100
—	0,57	0,68	1200

1	2	3	4
—	0,67	0,68	1300
—	0,57	0,68	1400
HfO_2*30 [93]			
0,70	—	0,65	1800
0,70	—	0,65	2000
0,72	—	0,65	2200
0,77	—	0,65	2400
0,81	—	0,65	2600
Ta_2O_5*31 [93]			
—	0,37	1,0	1223
—	0,08	2,0	1223
—	0,25	4,0	1223
—	0,68	6,0	1223
—	0,87	8,0	1223
—	0,88	10,0	1223

1	2	3	4
—	0,87	12,0	1223
—	0,81	14,0	1223
ThO_2 [19]			
—	0,35	0,65	1473—1773
UO_2*32 [93]			
0,40± 0,02	—	0,65	2073—2373
UO_2 [93]			
—	0,51± ±0,03	0,65	2073—2373
UO_2 [93]			
—	0,416	0,65	3033—3133

*1 Hot-pressed in graphite mold, fired in air up to 1300°C; uniformly dark; density 2.85 g/cm³. *2 Hot-pressed, polished, and fired in air up to 1300°C; white; density 2.778 g/cm³. *3 Specimens darkened, directly from mold; data from graph. *4 Specimens decolorized when calcined in air; from graph. *5 Specimen in form of plate with thickness of 1.65 mm, sintered for 2 h at 1700°C; density 1.84 g/cm³; from graph. *6 Fused; from graph. *7 For fused surface, polished with very fine emery; from graph. *8 From graph. *9 High-density Al_2O_3; from graph. *10 Hot-pressed at 2123°K; density 3.35 g/cm³; thickness 0.81 mm; from graph. *11 Sapphire crystal, thickness 0.79 mm; from graph. *12 Film with thickness 0.1 μm; from graph. *13 Cold-pressed with subsequent sintering for 1 h at 1823°K; density 1.53 g/cm³; from graph. *14 From graph. *15 Specimen thickness 1.75 mm, prepared by sintering TiO_2 for 2 h at 1673°K; density 3.87 g/cm³; from graph. *16 Cold-pressed, sintered 2 h at 1223°K; density 3.29 g/cm³; from graph. *17 Cold-pressed, sintered 2 h at 1673°K; density 5.32 g/cm³; from graph. *18 Sintered for 2 h at 2023°K; density 4 g/cm³; from graph. *19 Stabilized with CaO; from graph. *20 Radiation at angle of 44°; stabilized with CaO; from graph. *21 Specimen thickness 7.65 mm; from graph. *22 Calculated from Wien formula for film with thickness of 50 μm; from graph. *23 Fired at 1850°C; density 7.62 g/cm³; from graph. *24 Fired at 1850°C; density 7.62 g/cm³; from graph. *25 From graph. *26 From graph. *27 From graph. *28 From graph. *29 Pressed and subsequently fired for 24 h at 1773°K; thickness 3 mm; from graph. *30 Coating stabilized with Y_2O_3, thickness 0.06–0.3 mm; from graph. *31 Sintered for 2 h at 1673°K; thickness 1.2 mm; density 6.51 g/cm³; from graph. *32 Polished dense specimen.

4. SPECTRA OF OPTICAL CONSTANTS [76; 93; 611; 613; 614; 616–619]

Reflection

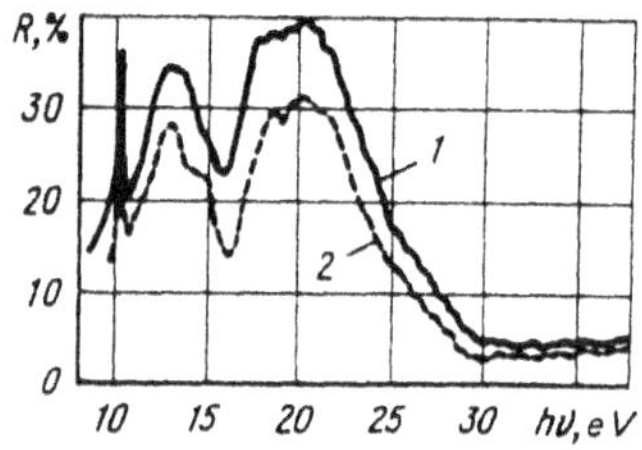

Fig. 1. Coefficient of reflection R of BeO single crystal in UV region of spectrum: 1) $\vec{E} \perp \vec{c}$; 2) $\vec{E} \parallel \vec{c}$ at 100°K.

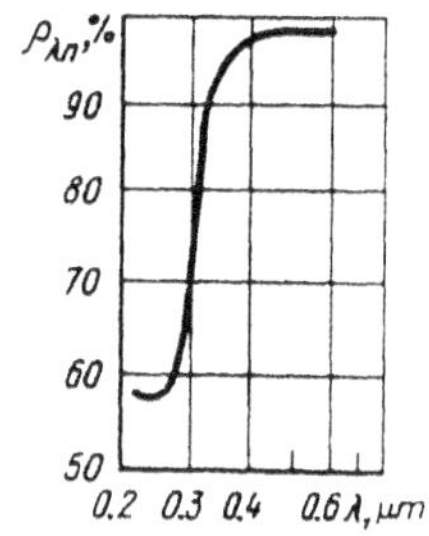

Fig. 2. Monochromatic normal reflectance $\rho_{\lambda n}$ of BeO (sintered powder, density 1.84 g/cm³) in UV and visible regions of spectrum, at 298°K.

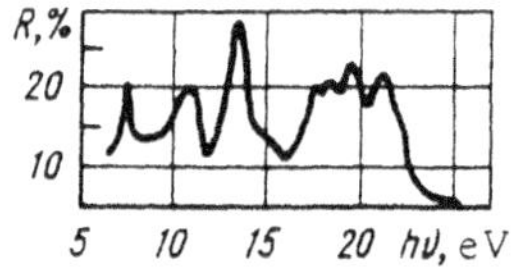

Fig. 3. Coefficient of reflection R of MgO single crystal in UV region of spectrum, at 300°K.

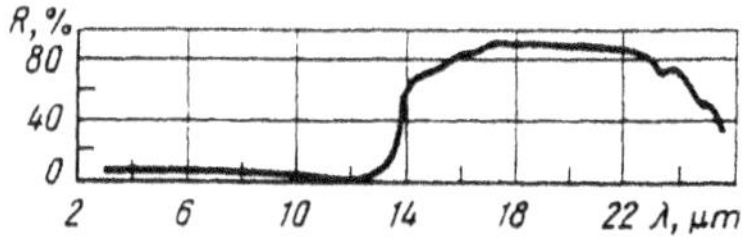

Fig. 4. Coefficient of reflection R of MgO single crystal in IR region of spectrum, at 298°K.

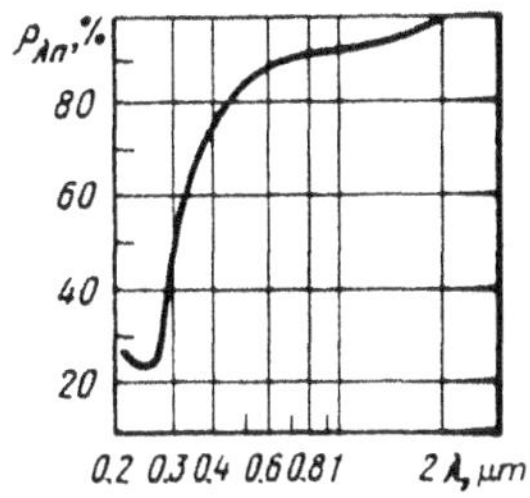

Fig. 5. Monochromatic normal reflectance $\rho_{\lambda n}$ of Al_2O_3 (sintered powder, density 3.45 g/cm^3) in UV, visible, and near-IR regions of spectrum, at 298°K.

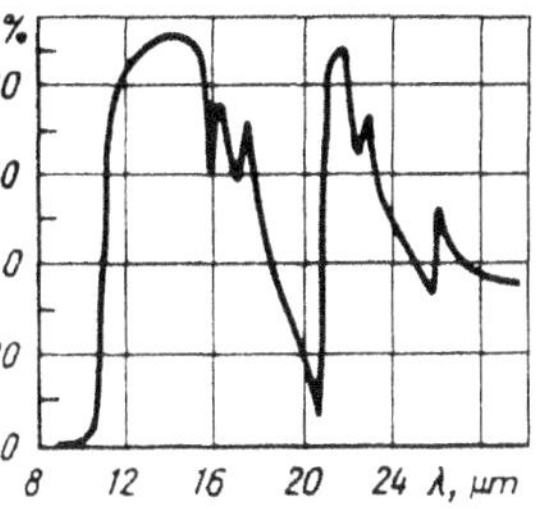

Fig. 6. Coefficient of reflection R of Al_2O_3 single crystal (sapphire) for ordinary beam in IR region of spectrum, at 298°K.

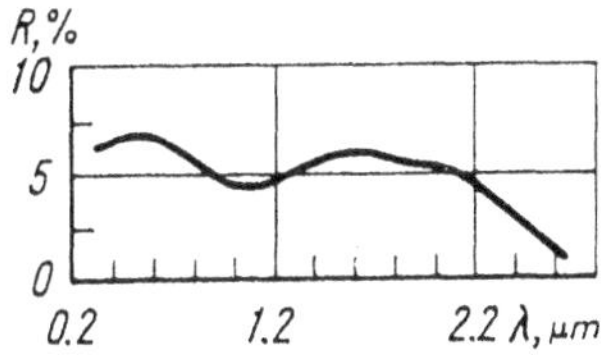

Fig. 7. Coefficient of reflection R of SiO_2 crystal (fused quartz) in visible and near-IR regions of spectrum, at 298°K.

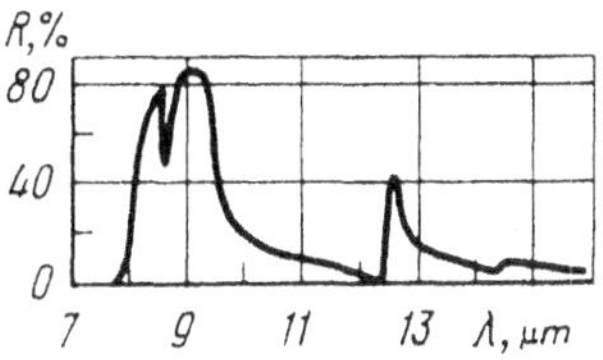

Fig. 8. Coefficient of reflection R of SiO_2 (crystalline quartz) for ordinary ray in IR region of spectrum, at 298°K.

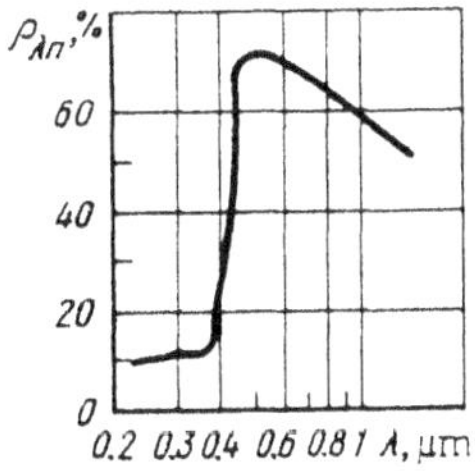

Fig. 9. Monochromatic normal reflecance $\rho_{\lambda n}$ of TiO_2 (pressed powder, grain size ~ 58 μm) in UV, visible, and near-IR regions of spectrum, at 298°K.

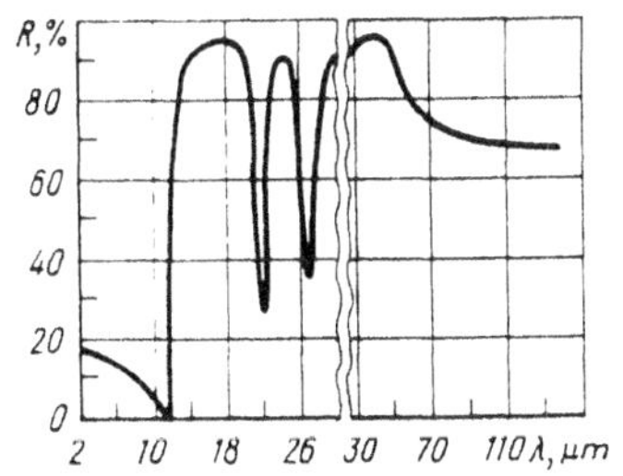

Fig. 10. Coefficient of reflection R of TiO_2 single crystal (rutile) for ordinary ray in IR region of spectrum, at 298°K.

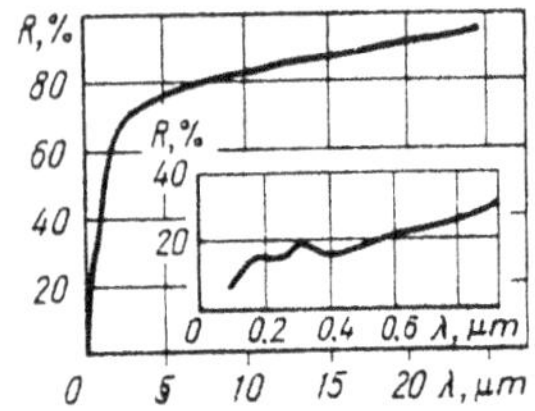

Fig. 11. Coefficient of reflection R of V_2O_3 single crystal in UV, visible, and IR regions of spectrum, at 298°K.

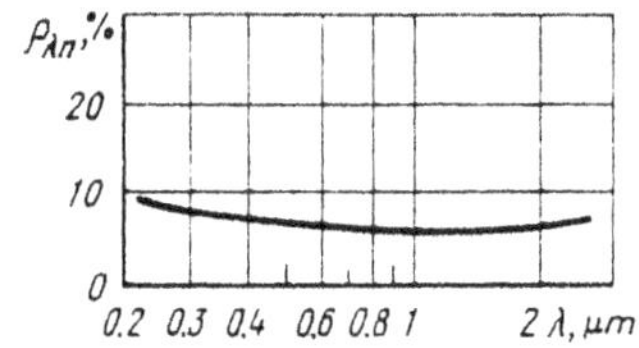

Fig. 12. Monochromatic normal reflectance $\rho_{\lambda n}$ of Cr_2O_3 (sintered powder, density 3.15 g/cm^3) in UV, visible and IR regions of spectrum, at 298°K.

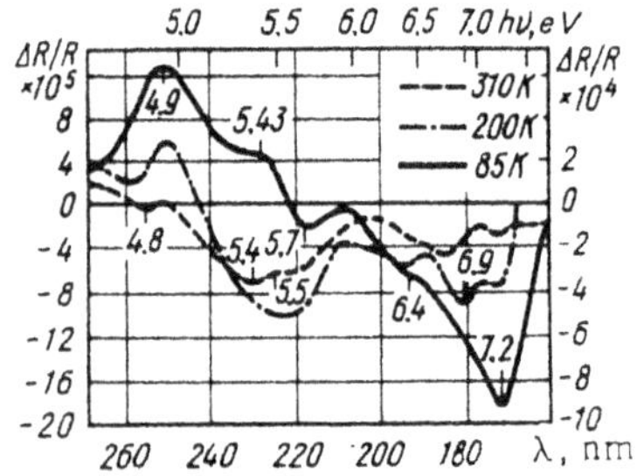

Fig. 13. Thermal reflection spectra of MnO single crystal in UV region at 310 and 200°K (left-hand scale) and 85°K (right-hand scale).

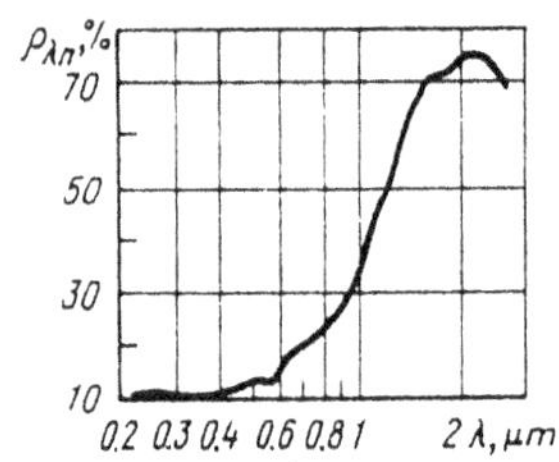

Fig. 14. Monochromatic normal reflectance $\rho_{\lambda n}$ of MnO (pressed powder, grain size ~ 58 μm) in UV, visible, and near-IR regions of spectrum, at 298°K.

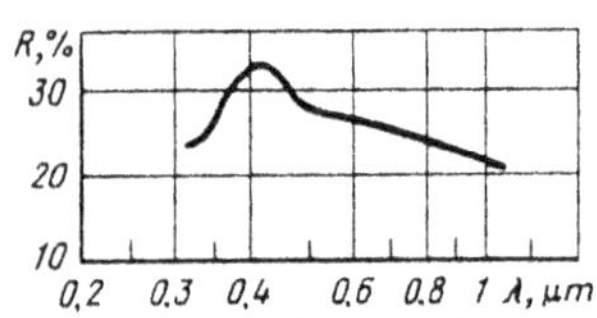

Fig. 15. Coefficient of reflection R of α-Fe_2O_3 single crystal in visible and near-IR regions of spectrum, at 298°K.

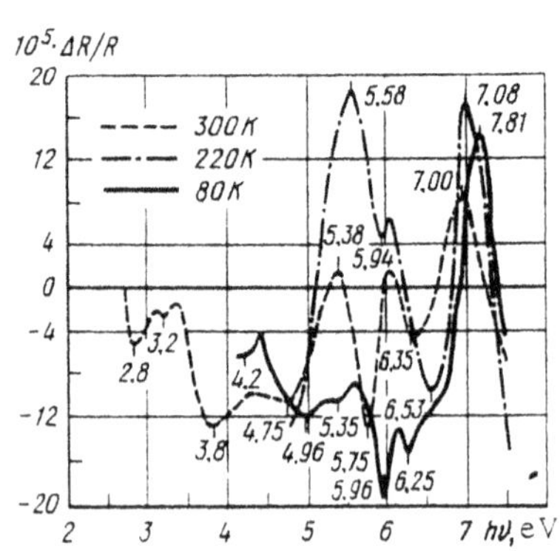

Fig. 16. Thermal reflection spectra of CoO single crystal in UV region at 300, 220, and 80°K.

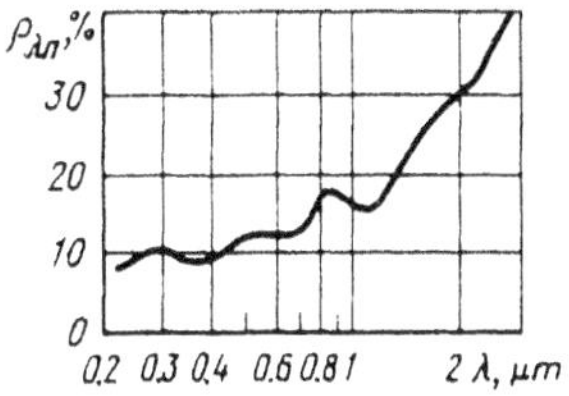

Fig. 17. Monochromatic normal reflectance $\rho_{\lambda n}$ of NiO (sintered powder, density 4.81 g/cm^3) in UV, visible, and near-IR regions of spectrum, at 298°K.

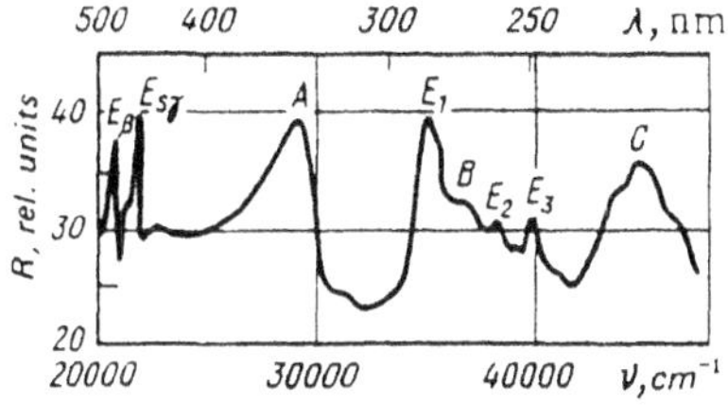

Fig. 18. Coefficient of reflection R of Cu_2O single crystal in visible and UV regions of spectrum, at 77°K.

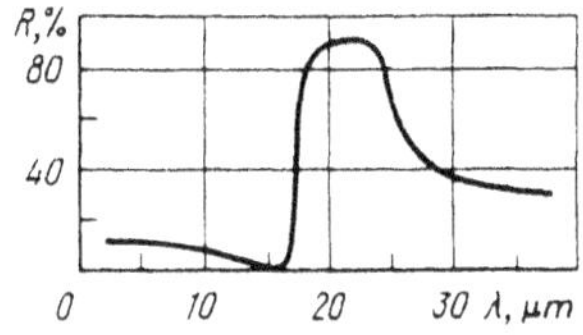

Fig. 19. Coefficient of reflection R of ZnO single crystal in IR region of spectrum, at 298°K.

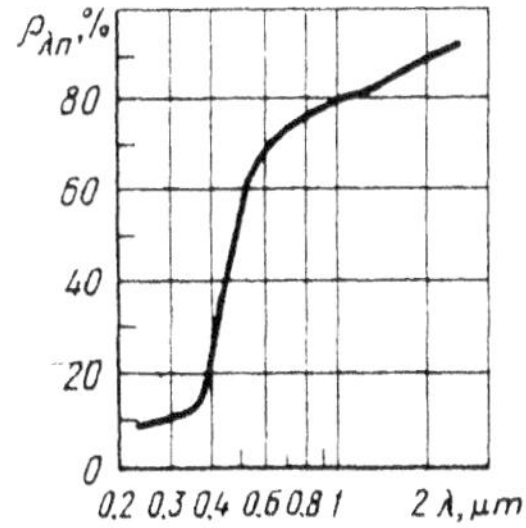

Fig. 20. Monochromatic normal reflectance $\rho_{\lambda n}$ of ZrO_2 (pressed powder, grain size ~ 58 μm) in UV, visible, and near-IR regions of spectrum, at 298°K.

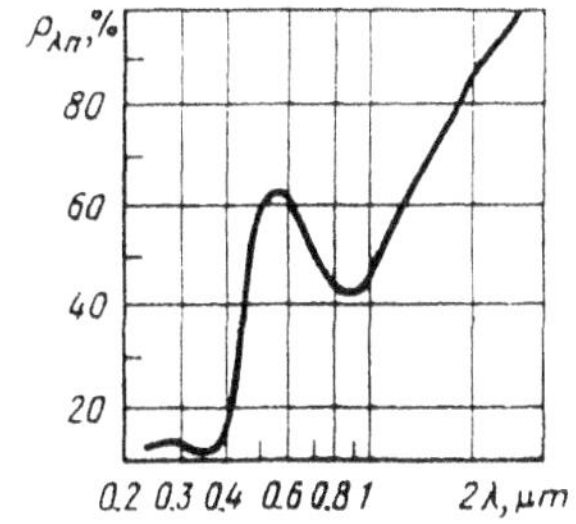

Fig. 21. Monochromatic normal reflectance $\rho_{\lambda n}$ of MoO_3 (sintered powder) in UV, visible, and near-IR regions of spectrum, at 298°K.

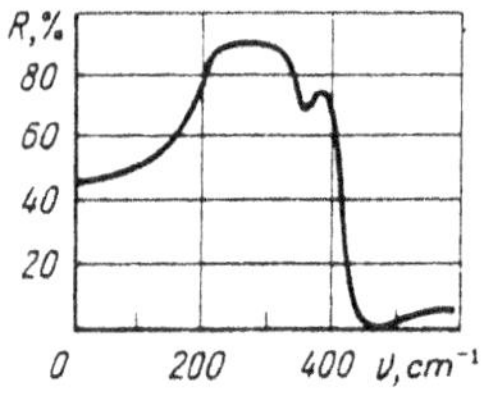

Fig. 22. Coefficient of reflection R of EuO single crystal in IR region of spectrum, at 300°K.

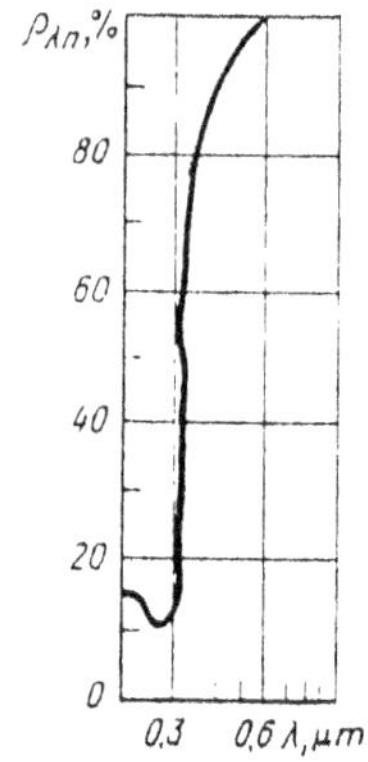

Fig. 23. Monochromatic normal reflectance $\rho_{\lambda n}$ of Ta_2O_5 (sintered powder, density 6.51 g/cm^3) in UV and visible regions of spectrum, at 298°K.

Refraction

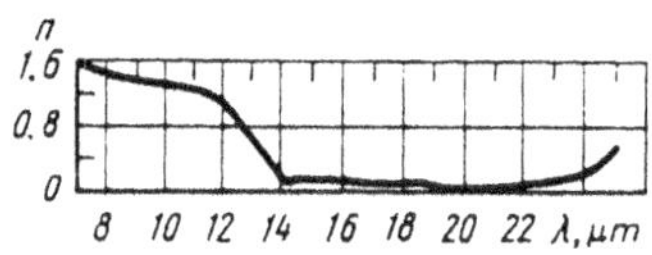

Fig. 24. Refractive index n of MgO single crystal, at 298°K.

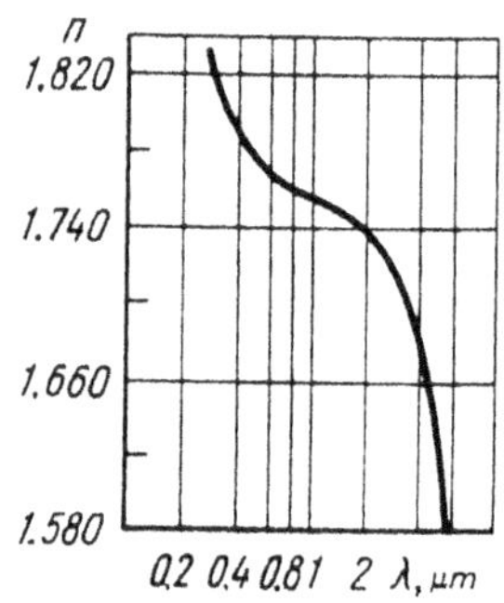

Fig. 25. Refractive index n of Al_2O_3 single crystal (sapphire) for ordinary ray in UV, visible, and IR regions of spectrum, at 297°K.

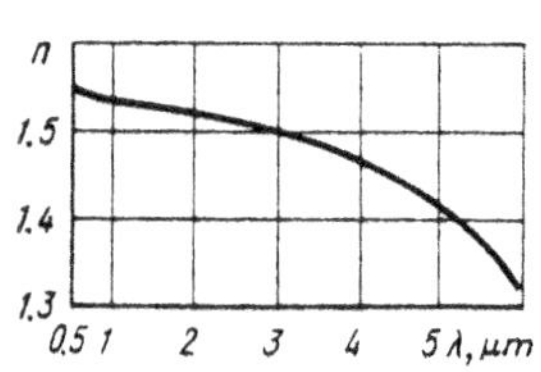

Fig. 26. Refractive index n of crystalline quartz SiO_2 for ordinary ray in IR region of spectrum, at 298°K.

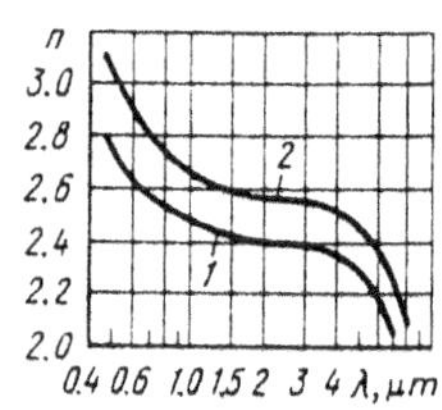

Fig. 27. Refractive index n of TiO_2 single crystal (rutile) for ordinary (1) and extraordinary (2) rays in visible and IR regions of spectrum, at 298°K.

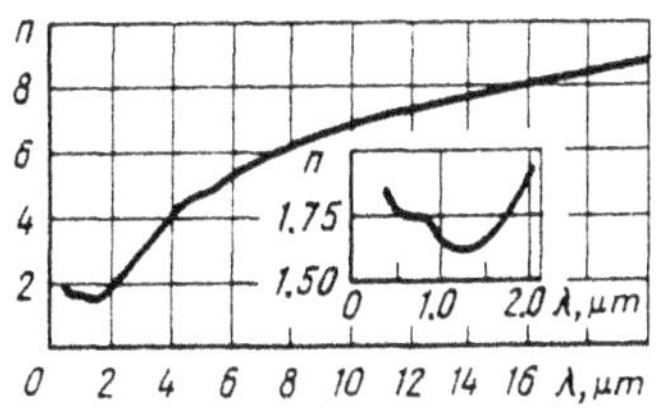

Fig. 28. Refractive index n of V_2O_3 single crystal in visible and IR regions of spectrum, at 298°K.

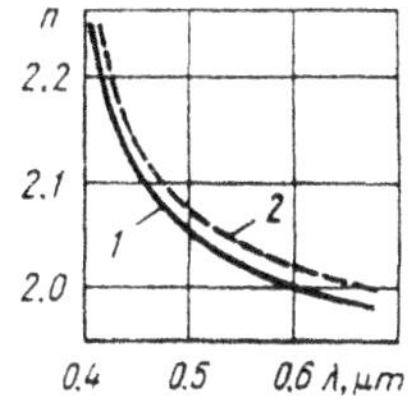

Fig. 29. Refractive index n of ZnO single crystal for ordinary (1) and extraordinary (2) rays in visible region of spectrum, at 298°K.

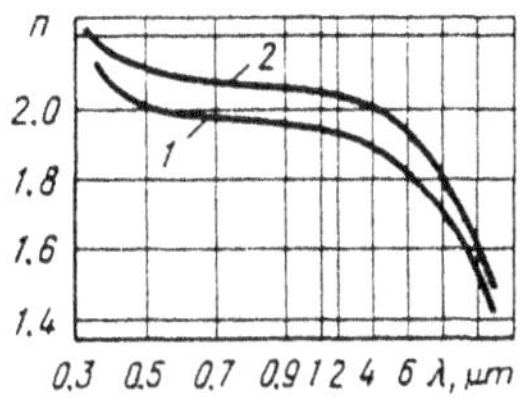

Fig. 30. Refractive index n of SnO_2 single crystal for ordinary (1) and extraordinary (2) rays in visible and IR regions of spectrum, at 298°K.

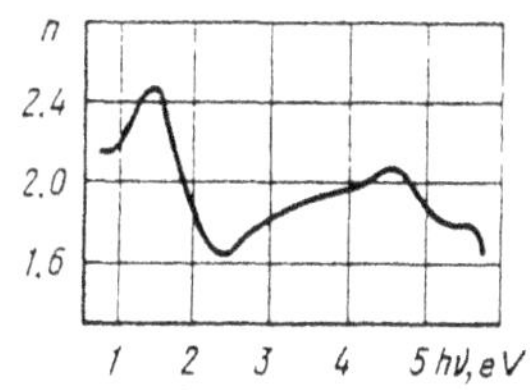

Fig. 31. Refractive index n of EuO single crystal in UV and visible regions of spectrum, at 298°K.

Absorption

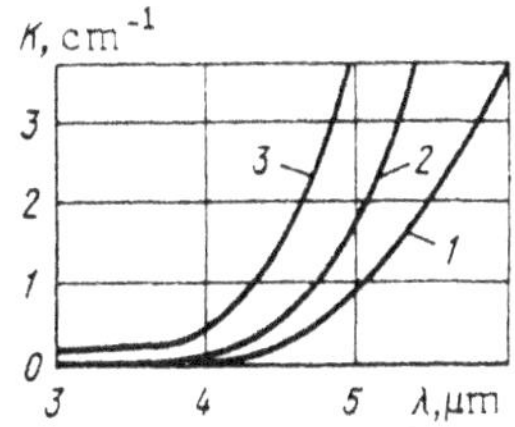

Fig. 32. Absorption coefficient K of Al_2O_3 single crystal (sapphire) in IR region of spectrum, at 293 (1), 773 (2), and 1273°K (3).

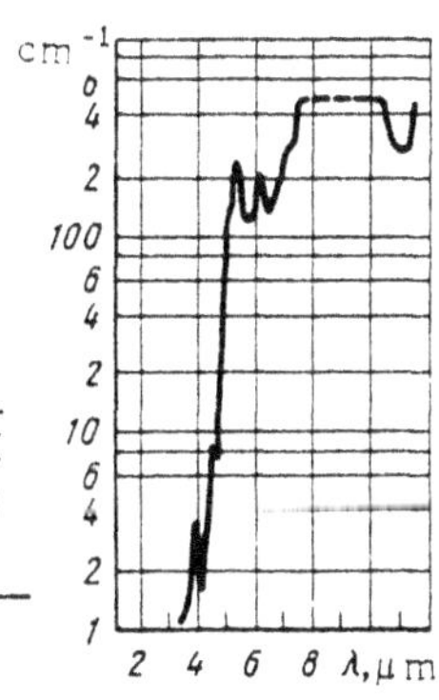

Fig. 33. Absorption coefficient K of SiO_2 crystal (fused quartz) in IR region of spectrum, at 298°K.

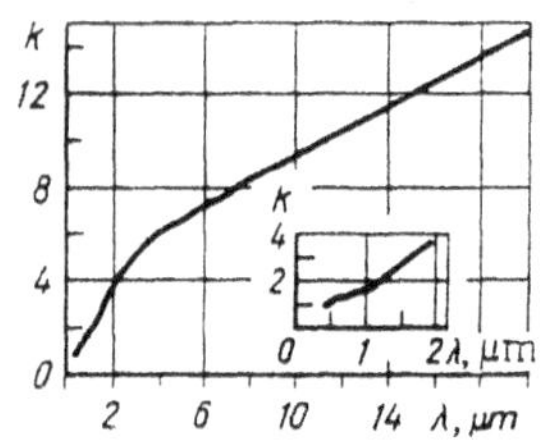

Fig. 34. Absorbance k of V_2O_3 single crystal in visible and IR regions of spectrum, at 298°K.

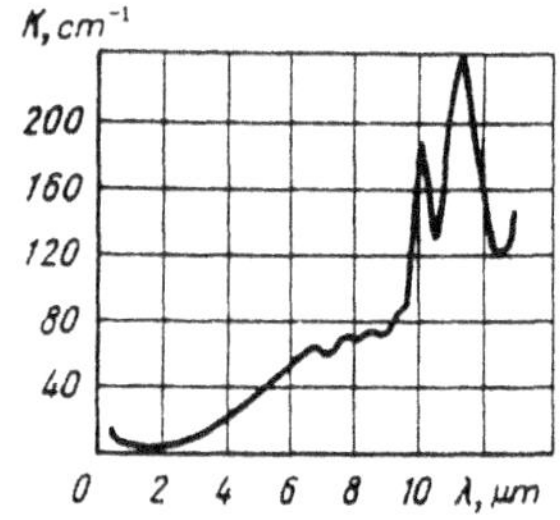

Fig. 35. Absorption coefficient K of ZnO single crystal in IR region of spectrum, at 298°K.

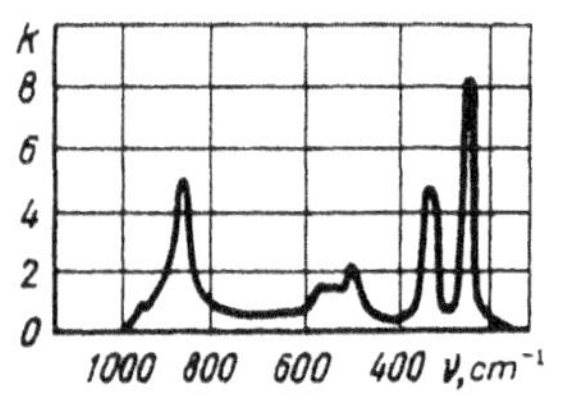

Fig. 36. Absorbance k of GeO_2 crystal (hexagonal modification) in IR region of spectrum, at 298°K.

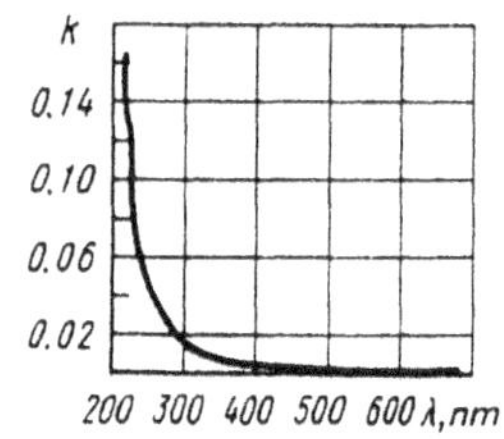

Fig. 37. Absorbance k of ZrO (crystalline film, thickness 90.2 nm) in UV and visible regions of spectrum, at 298°K.

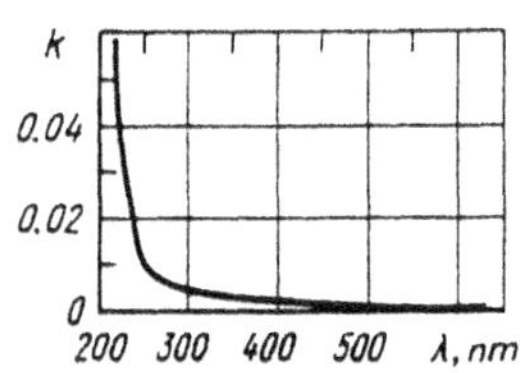

Fig. 38. Absorbance k of NdO (crystalline film, thickness 92.6 nm) in UV and visible regions of spectrum, at 298°K.

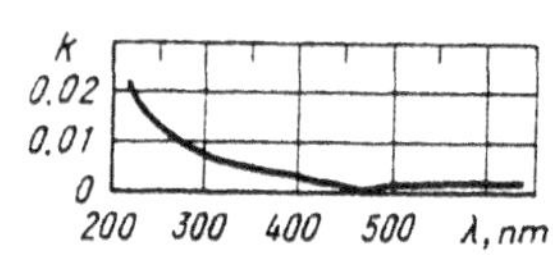

Fig. 39. Absorbance k of ThO_2 (crystalline film, thickness 95.2 nm) in UV and visible regions of spectrum, at 298°K.

Features of optical properties	Notes
1	2
H_2O	
$n_e - n_0 = 0.0014$ (F) [67]	Ice at −3°C
In range of dynamic pressures 100−1000 MPa, refractive index varies linearly with density [338]	Pure water
In UV region of spectra, stimulated luminescence of γ-irradiated polycrystals is observed [343]	Pure ice
BeO	
$n_e - n_0 = 0.014$ [67]	Bromellite
In spectral region 250−1000 nm, gives almost 100% light transmission [339]	Polycrystalline layers, $\rho \approx 10^{13}$ ohm · cm
Red edge of photoeffect 374.5 nm [19]	Crystal
B_2O_3	
$n_0 - n_e = 0.033$ [67]	Hexagonal modification
CO	
Fluoresces in 380−800 nm region [340, 341]	Upon photodissociation of CO_2 molecules by lines 76.4, 78.9, 83.5, 87.9, 90.1, 92.3 nm
CO_2	
Fluoresces in 280−480 nm region [342]	Upon excitation of CO_2 molecules by photons with $\lambda \approx 46.2$, 52.5, 55.5, 58.7, 61.0, 62.9, 63.7, 68.6, 70.3, 71.5 nm
NO	
Absorption bands in IR spectrum: $\nu \approx 1400$, 1683, 1840, 1875 cm^{-1} [378]	Gas
Absorption bands in IR spectrum ($t = -190°C$) $\nu \approx 950$, 1000, 1075, 1400, 1725 cm^{-1} [378]	Solid condensate, white

5 – Continued

1	2
NO_2	
Absorption bands in IR spectrum: ($t = -190°C$) $\nu \approx 970$, 1085, 1122, 1150, 1380, 1683 cm^{-1} [378]	Solid condensate
F_2O	
In spectral region for $\lambda \leq 540.0$ nm, continuous absorption is observed with maxima at 421.0, 358.0, and 294.0 nm [81, Vol. 5]	Gas
MgO	
Optically isotropic [67]	Periclase
Beginning of edge of fundamental absorption in UV region of spectrum $> 25{,}000$ cm^{-1} at $T = 298°K$ [350]	Single crystal
Position of exciton peaks in UV reflection spectrum at 25°K: $h\nu \approx 7.689$, 7.715, 7.752, 7.768 eV [362]	"
Fine structure of UV thermal reflection spectrum at 85°K: $h\nu \approx 7.67$, 7.70, 7.74, 7.76, 7.84 eV (minima in thermal reflection spectrum) [379]	"
Freshly prepared specimens are highly transparent in 0.22–8.0 μm region, but gradually become cloudy in air. With increasing temperature of the substrate, the refractive index of the layer increases [345]	Films obtained by thermal vaporization
Absorption bands in IR transmission spectrum: $\nu \approx 400$, 560, and 680 cm^{-1} [347]	Microcrystals

1	2
At 80°K, in UV absorption spectrum close to $h\nu \approx 5$ eV, F-center band is observed [89]	Single crystal
Absorption band of V^--center is in region ~2.3 eV [348]	Crystal
At 80–90 and 120–130°C, in thermoluminescence and thermostimulated conduction spectra, maxima are observed, related to decay of V^- and V^0 centers [344]	Pure single crystals, x-ray irradiated
At 298.77°K and with laser excitation, photoluminesces in region with $\nu \approx$ 14,350 cm^{-1} [377]	Annealed powders and single crystals, doped with He, Ar, Fe, and Cr ions
Al_2O_3	
$n_0 - n_e = 0.0082$ (D); pleochroic [67]	Corundum
Transparent in near-UV, visible and IR regions of spectrum up to $\lambda \approx 5$ μm; long-wave region of transmission 100–1000 μm [76]	Sapphire
Thermoluminesces upon excitation in x-ray region [352]	Sapphire
Maximum of absorption band in thermoluminescence spectrum at $\lambda = 540$ nm [349]	Film
SiO	
Optically anisotropic, but x-ray amorphous; $n_g - n_p = 0.09$ [52]	Crystal
Characteristic absorption bands in IR spectrum: $\lambda \approx 8.2$, 8.4, 9.3, 10.4, 12.7 μm [52]	
Si_2O_3	
Optically anisotropic [52]	Scaly flakes
Characteristic absorption bands in IR spectrum: $\lambda \approx 9.6$ and ≈11.5 μm [52]	

5 – Continued

1	2
SiO_2	
Optically anisotropic (isotropic within limits of 200–275°C) [52, 67]	α-Quartz
(+) 2V° = 0; $n_e - n_0 = 0.0091$;	
(+) 2V° = 0; $n_e - n_0 = 0.0066$ [52, 67]	β-Quartz
(+) 2V° = 0; 35; $n_g - n_p = 0.004$ [52, 67]	α-Tridymite
$n_e > n_0$ [52]	β-Tridymite
(+) 2V° = 50; $n_g > n_p$ [52]	γ-Tridymite
$n_0 - n_e = 0.003$ [67]	α-Cristoballite
(–) 2V° = 40; $n_0 > n_e$ [52]	β-Cristoballite
Transparent in spectral region 0.22–4.5 μm [76]	Fused quartz
Long-wave transmission region 100–500 μm [66]	Polycrystalline and fused quartz
Shows electrooptical effect with transmission limit ~ 4 μm [76]	Crystalline quartz
In spectral range of 3–5 μm, absorption coefficient varies linearly with temperature (1270–1870°K) [351]	Fused quartz
Position of luminescence maxima: hν ≈ 3.1, 4.4, 6.2, 7.2 eV [346]	Quartz crystals with photo-excitation in UV region
Natural, unirradiated quartz shows thermoluminescence [64]	–
P_2O_3	
In the process of oxidation, when the pressure is decreased, emission is observed [81, Vol. 5]	White crystalline mass
P_2O_5	
$n_e - n_0 = 0.025$ [67]	Tetragonal phase
(–) 2V° = 65; $n_g - n_p = 0.044$ [67]	Orthorhombic (metastable) phase

1	2
$n_e - n_0 = 0.002$ [67]	Hexagonal phase
Under the influence of natural light, shows a green luminescence, the intensity increasing with decreasing temperature [81, Vol. 5]	
CaO	
Optically isotropic [67]	Lime
At 85°K, in UV thermal reflection spectrum, fine structure is observed with minima at $h\nu \approx 6.93$, 6.97, 7.00, 7.04, 7.098 eV [379]	Single crystal
At 77°K, in 200–300 nm spectral region, significant light absorption is observed with maxima at 215 and 270 nm [372]	Crystal, deformed along ⟨100⟩
Band of F-center is observed in absorption spectrum at 5°K near $h\nu \approx 3.6$ eV [89]	Single crystal
In absorption spectrum at 6°K, F^+-center band is observed ($\lambda \approx 374$ nm), corresponding to radiation band at $\lambda \approx 398$ nm [380]	Single crystal, doped with Mg
Maximum in sky-blue luminescence ($\lambda_{exc} = 266$ nm, $T = 295$°K) is observed at $h\nu \approx 2.6$ eV (band half-width ~ 0.7 eV) [372]	Crystal, deformed along ⟨100⟩
Phosphoresces; upon adsorption of oxygen, chemiluminescence is observed [65]	Specimens pure or doped with with Bi, Sb, or Tl
At 298°K with x-ray excitation, luminesces in 300–650 nm region [381]	Film thickness 1 μm, doped with Sm and Tu, subsequently annealed in interval $t = 100–800$°C

5 — Continued

1	2
Sc_2O_3	
Transmittance in visible and near-IR ($\lambda < 2$ μm) as high as 99.4% [382]	Films
$TiO_{0.83-1.30}$	
Absorption bands in IR spectrum: $\nu \approx 475, 515, 800, 1080$ cm^{-1} at 298°K [176]	Powder pressed in CsI matrix
TiO	
Absorption bands in IR spectrum: $\nu \approx 410, 465, 515, 590, 1080$ cm^{-1} at 298°K [359]	Powder pressed in KBr matrix
Ti_2O_3	
Absorption bands in IR spectrum: $\nu \approx 410, 480, 515, 1080$ cm^{-1} at 298°K [359]	Powder pressed in KBr matrix
TiO_2	
At 25°C, $n_e - n_0 = 0.2869$; pleochroic [67]	Rutile
At 25°C, $n_0 - n_e = 0.0732$; weakly pleochroic [67]	Anatase
At 25°C, $n_g - n_p = 0.1173$ (Na); weakly pleochroic [67]	Brookite
Birefringent; at $\lambda = 1.0$ μm, $n_e - n_0 = 0.26$ [76]	Brookite
Temperature dependence of refractive index for $\lambda = 0.436$ μm: $(dn_0/dT)10^5 = -4.36$ and $(dn_e/dT)10^5 = -8.6$ [76]	Rutile
In electroreflection spectrum at 84°K, fine structure (sensitive to direction of light polarization) is observed in region $h\nu \approx 3.0-3.7$ eV [358]	Single crystal
In IR spectrum at 298°K, absorption peak is observed with $h\nu \approx 1.4$ eV (light absorption by small-radiance polarons) [363]	Rutile

1	2
At 300°K in IR spectrum of reflection coefficient, minima are observed with $\nu \approx 380$, 460, and 870 cm^{-1} [365]	Rutile
Absorption bands in IR spectrum: $\nu \approx 415$, 550, 650, and 1080 cm^{-1} at 298°K [359]	Powder, pressed in KBr matrix
At 77°K, maximum of broad asymmetric luminescence band is observed at $\lambda = 850$ nm [368]	Single crystal
Photoluminescence band at 77°K is characterized by large Stokes shift ~1.22 eV [70]	Polycrystal
In photoconduction spectrum, near the edge of natural absorption band ($\lambda \approx 390$ nm), fine structure is observed [70]	Amorphous films
$VO_{0.79-1.29}$	
Absorption bands in IR spectrum: $\nu \approx 475$, 515, 800, 1080 cm^{-1} at 298°K [176]	Powder, pressed in CsI matrix
V_2O_3	
At 93°K, in edge absorption spectrum, two maxima are observed: $h\nu \approx 0.2$ eV ($\vec{E} \parallel \vec{C}$) and $h\nu \approx 0.32$ eV ($\vec{E} \perp \vec{C}$) [374]	Crystals, annealed at 1500°C
$VO_{1.84-1.87}$	
Very fine crystals with strong light reflection [52]	–

1	2
VO_2	
At 340°K (critical temperature for semiconductor/metal transition) a weak minimum is observed in the light reflection spectra at 1.65 eV, and a more distinct minimum near 5 eV [82]	Single crystal
V_2O_5	
Transmission limit in UV region ~2.2 eV; in spectrum of reflection coefficient at 298°K, a maximum is observed in the region $h\nu$ =2.8-3.1 eV $(\vec{E} \| \vec{a})$ [364]	Single crystals with various stoichiometric compositions
Light absorption by small-radius polarons at $h\nu \approx 0.84$ and 1.24 eV (T = 298°K) [94]	Single crystal
Red limit of photoeffect ~224 eV [71]	Single crystal
Cr_3O_4	
Optically anisotropic [52]	Crystal
Cr_2O_3	
Position of maximum in UV absorption spectrum at $\lambda \approx 400$ nm [357]	Film
Red limit of photoeffect ~3.4 eV at 298°K [354]	Powder
MnO	
Optically isotropic [67]	Manganosite
Red limit of photoeffect ~3–4 eV [354]	
FeO	
Optically isotropic [67]	Wustite
Fe_2O_3	
$n_0 - n_e = 0.23$ (Li); pleochroic [67]	Hematite

1	2
Red limit of photoeffect ~ 3.3 eV at 298°K [354]	Powder
Fe_3O_4	
Optically isotropic [67]	Magnetite
At 298°K, in 0.15–0.8 eV spectral region, refractive index n decreases from 3 to 2, and absorbance k from 2 to 1 [383]	Single crystal of natural magnetite
At 119°K (temperature of metal/nonmetal transition), on curve for temperature dependence of optical conductance, there are two maxima for $h\nu = 0.13$ and 0.6 eV [383]	Single crystal of natural magnetite
CoO	
Blue limit of transmission: $\nu \approx 21{,}000$ cm^{-1} at 298°K [350]	Single crystal
Red limit of photoeffect ~ 3–4 eV at 298°K [354]	
NiO	
Optically isotropic [67]	Bunsenite
Red limit of photoeffect ~ 3.7 eV at 298°K [354]	Powder
Cu_2O	
Optically isotropic [67]	Cuprite
Rotates polarization plane in magnetic field [52]	Platelets of cuprite
In photoluminescence spectra ($\lambda_{exc} = 488$ nm) at liquid helium temperatures, radiation lines are observed with $\lambda \approx 0.72$ and 0.82 μm (vacancies of oxygen), and also with $\lambda \approx 0.91$ and 1.01 μm (vacancies of copper) [83]	Single crystal

1	2
In photoluminescence spectra at 4–77°K, together with direct radiation of exciton from state IS, O (crossover transition), a number of bands are observed from the annihilation of excitons (IS, K) with simultaneous emission or absorption of phonons (noncrossover transitions) [61]	Crystal
Shows photoeffect, equal in absolute magnitude to effect for metallic copper [52]	Cuprite
In spectrum of longitudinal photoconduction (at 298°K), two maxima are observed with $h\nu \approx 1.45$ and 2.35 eV [87]	Quenched crystals
CuO	
$n_g > n_p$; pleochroic [67]	Tenorite
Shows photoeffect [52]	Tenorite
ZnO	
$n_e - n_0 = 0.016$ (Na); at 298°K, normally fluoresces in UV light [67]	Natural zincite (99.63% ZnO)
Birefringent; $n_e - n_0 = 0.015$ ($\lambda = 0.589$ μm) [76]	Single crystal
Becomes yellow upon calcining; assumes original color when cooled [81, Vol. 5]	Crystal
Position of maxima in UV spectrum of reflection coefficient ($\vec{E} \perp \vec{c}$) at 298°K: $h\nu \approx 3.30$, 3.35, 7.0, 9.2, 12.6, 14.0, 15.2, 17.1, 19.3, 20.8 eV [376]	Single crystal

1	2
Position of maxima in diffuse reflection spectrum (340–380 nm) at 77°K: $\lambda \approx 354.6$, 358.8, 362.8, 364.8, 367.8, 369.2 nm; positions of bands in photoluminescence and cathodoluminescence spectra in 360–400 nm region are identical, but their intensities are different [384]	Powder
Position of maximum absorption coefficient in UV region: $h\nu \approx 3.355$ eV at 300°K and ~ 3.42 eV at 77°K [361]	Single crystal
Position of photoluminescence maxima ($\lambda_{exc} = 365$ nm, T = 83°K) $\lambda \approx 367.9$, 374.7, 383.4, 392.0 nm; position of maxima of green-yellow series of thermoluminescence ($\lambda_{exc} \geq$ 360 nm, T = 110–180°K) $\lambda \approx$ 403, 513, 590 nm [355, 72, 385]	Single crystals
With nitrogen laser excitation ($\lambda_{exc} = 3371.5$ Å, T = 4.2 °K) an emission band is observed in the region of 370.0–372.0 nm [360]	Single crystal
Upon excitation by electron pulse beam ($h\nu_e \approx$ 40 eV, T = 10°K), as a result of exciton recombination, a line of stimulated radiation is observed with $\lambda \approx$ 372.6 nm (threshold of generation 3.5 Å/cm^2) [386]	Single crystal

5 – Continued

1	2
Ga_2O_3	
Position of photoluminescence maxima (λ_{exc} = 254 nm) after annealing in oxygen (nitrogen): $\lambda \approx 420$ nm at 300°K and $\lambda \approx$ 365 nm at 77°K [356]	Single crystal, n-type
GeO_2	
$n_e - n_0 = 0.006$ to 0.011 [67]	Tetragonal phase
$n_e - n_0 = 0.04$ [67]	Hexagonal phase
As_2O_3	
(+) 2V° = 58; $n_g - n_p = 0.14$ [67]	Claudetite
Optically isotropic; possibly anomalous birefringence [67]	Arsenolite
SeO_2	
$n_e > n_0$ [67]	Crystal
SrO	
Optically isotropic [67]	Crystal
Position of exciton peaks in reflection spectrum at 5°K: $h\nu \approx$ 5.711, 5.784, 6.082, 6.126 eV [362]	Single crystal
Absorption band of F^+-center observed at λ = 400 nm, and photoluminescence maxima (T = 77°K) at $\lambda \approx$ 458, 500, 554 nm [86]	Crystals irradiated with fast neutrons (protons) with energy of 3.5 MeV (dose 10^{17} particles/cm^2)
At 300°K, maxima are observed in IR absorption spectrum at $\nu \approx$ 90, 101, 108, 116, 126, 134, 142, 156, 164, 206, 224, 255 cm^{-1} [367]	Powder; grain size ~ 8 μm
Y_2O_3	
Optically isotropic [67]	Crystal
Transmittance in visible and near-IR (λ < 2 μm) as high as 99.4% [382]	Films

1	2
Has photoelectric properties; in interval 21–37°K in strong fields, the current/voltage curves (on coordinates of log I vs. $V^{1/2}$) are straight lines [387]	Polycrystalline films with thickness 0.15–0.267 μm
ZrO_2	
(–) 2 V° = 30; $n_g - n_p = 0.07$, pleochroic [67]	Baddeleyite
Position of absorption band maxima in UV spectrum: λ ≈ 229 (cubic modification) and λ ≈ 248.4 nm (monoclinic modification) [357]	Films
Position of absorption bands in IR spectrum: ν ≈ 425, 525, 685, 760 cm^{-1} [359]	Powder, pressed in KBr matrix
Spectral position of photoluminescence bands (λ_{exc} = 313, 365, nm) at 77°K: hν ≈ 2.74, 2.42, 2.20, 1.98, 1.76 eV [402]	Powder
Nb_2O_5	
Acquires a yellow color upon heating to 400–500°C, color disappearing after cooling, similar to ZnO [80]	–
Intense absorption of light in 260–330 μm region, with maximum at λ ≈ 278 nm [52, 357]	Thin film
MoO_2	
Shows photoelectric effect [52]	Crystal
MoO_3	
$n_g > n_p$; (+) 2 V° = 13; 2 H° = 117°15′ (for red light) and 2 H° = 127° (for blue light); pleochroic [52]	Crystal

5 – Continued

1	2
Limit of transmission in UV beginning at $\lambda \approx 0.4$ μm; in 0.4–2 μm region, refractive index n changes from 3 to 2; an absorption band related to a color center is observed at $\lambda = 850$ nm [368, 313]	Films with thickness of 0.08–1.75 μm
Ag_2O	
Luminesces at low temperatures of 0.9–2 μm region [370, 61]	Single crystals
Has exciton photoconductance at 77°K [370]	Single crystals
CdO	
Optically isotropic [67]	Crystal
At 8–13°K, a minimum is observed in the IR absorption spectra at $\lambda \approx 20.5$ μm (long-wave absorption of light by small-radius polarons) [373]	Films, n-type, obtained by cathode sputtering
In_2O_3	
Position of edge of fundamental absorption in UV region at 298°K: $h\nu \approx 3.75$ eV [634]	Films
SnO_2	
At 289°K, $n_e - n_0 = 0.0973$ (for $\lambda = 578$ nm); at 808°K, $n_e - n_0 = 0.0866$ (for $\lambda = 578$ nm) [67]	Cassiterite
Transparent in visible and near-IR regions; position of absorption bands in IR spectrum at 298°K: $\lambda \approx 3.1$, 8.5–9.0, 16.4 μm [388]	Thin films
Sb_2O_3	
Optically isotropic; sometimes has anomalous birefringence [67]	Senarmonite

1	2
(–) 2 V° ≈ 0; $n_g - n_p = 0.17$ [67]	Valentinite
Has photoconductance in region of fundamental absorption edge (hν ≈ 3.0–3.5 eV [389, 52]	Crystal
TeO_2	
(–) 2 V° large; $n_g - n_p = 0.35$ (Li) [67]	Tellurite
Has nonlinear refractive index: $\delta n = (56 \pm 6) \cdot 10^{-8}$ cgse units [85]	Crystal
BaO	
Optically isotropic [67]	Crystal
Position of exciton peaks in reflection spectrum at 5°K: hν ≈ 3.91, 4.03, 4.08, 4.24 eV [362]	Single crystal
Absorption bands of F- and F^+-centers close to 2.3 and 2.0 eV (respectively) are observed in absorption spectra at 77–423°K [88]	Crystals
In interval 103–423°K with photoexcitation ($h\nu_{exc} = 3.65$–3.9 eV), a recombination luminescence band is observed (hν ≈ 2.92 eV) [390]	Powder
La_2O_3	
Strong light absorption in UV, beginning at hν ≈ 5.6 eV [623]	Thin films (C-form)
Ce_2O_3	
Depending on degree of reduction (T = 623–1273°K), color changes from grey-turquoise to black-blue [52]	Crystalline powder
CeO_2	
Optically isotropic [52]	Crystalline powder

5 — Continued

1	2
Beginning of long-wave absorption edge in UV region at 298°K: $\nu \approx 25{,}500\ cm^{-1}$ [391]	Single crystal
Position of fundamental absorption edge in UV at 298°K: $h\nu \approx 5.1$ eV [623]	Thin films, thickness 60.0–400.0 nm (C-form)
Pr_2O_3	
Position of fundamental absorption edge in UV at 298°K: $h\nu \approx 5.56$ eV [623]	Thin films, thickness 60.0–400.0 nm (C-form)
Pr_6O_{11}	
Position of fundamental absorption edge in UV at 298°K: $h\nu \approx 5.35$ eV [623]	Thin films, thickness 60.0–400.0 nm
At $\lambda = 589.3$ nm (T = 298°K), refractive index and absorbance vary with thickness, as follows (respectively): $1.791 < n < 1.889$ and $0.001 < k < 0.032$ [392]	Thin films, thickness below 3/4 λ
Nd_2O_3	
Position of fundamental absorption edge in UV at 298°K: $h\nu \approx 4.0$ eV [623]	Thin films, thickness 60.0–400.0 nm (C-form)
Sm_2O_3	
Optically anisotropic; 2 V° = 10–20 [52]	Crystal
Position of fundamental absorption edge in UV at 298°K: $h\nu \approx 4.86$ eV (C-form) and 5.04 eV (B-form) [623]	Thin films, thickness 60.0–400.0 nm
EuO	
Upon excitation by He–Cd laser radiation or electron beam, a luminescence band is observed close to 3.36 eV [403]	Single crystal

1	2
At $\lambda \approx 1.0\ \mu m$ (T = 298°K), a maximum is observed in the photoconductance spectrum; the maximum photo-emf is located at $\lambda \approx 0.6\ \mu m$ [375]	Single crystal, $\rho \approx 10^8$ ohms · cm
$Eu_{16}O_{21}$	
(+) 2 V° = 25; $n_g > n_p$ [52]	Ortho-1 phase
Eu_2O_3	
Position of fundamental absorption edge in UV at 298°K: $h\nu \approx 4.48$ eV (C-form) and 4.90 eV (B-form) [623]	Thin films, thickness 60.0 – 400.0 nm
At 77°K, in transmission spectrum, light absorption lines are observed with $\nu \approx$ 18,719; 18,747; 18,947; 18,964; 19,013; 19,102 cm^{-1} [371]	Fine powder (C-form)
Position of absorption maxima in tramission spectrum in visible region at 4.2°K: $\lambda \approx$ 573.5, 576.0, 578.4, 582.0, 582.2 nm [371]	Fine powder (B-form)
Gd_2O_3	
Position of fundamental absorption edge in UV at 298°K: $h\nu \approx 5.45$ eV [623]	Thin films, thickness 60.0 – 400.0 nm (C-form)
Maximum frequency of IR absorption spectrum: $\nu_l \approx 535$ cm^{-1} [404]	Powder, pressed in CsI matrix
Tb_2O_3	
Position of fundamental absorption edge in UV at 298°K: $h\nu \approx 4.77$ eV [623]	Thin films, thickness 60.0 – 400.0 nm (C-form)
Tb_4O_7	
Position of fundamental absorption edge in UV at 298°K: $h\nu \approx 4.95$ eV [623]	Thin films, thickness 60.0 – 400.0 nm

1	2
Dy_2O_3	
Position of fundamental absorption edge in UV at 298°K: $h\nu \approx 4.86$ eV [623]	Thin films, thickness 60.0–400.0 nm (C-form)
Maximum frequency of IR absorption spectrum: $\nu_l \approx 550$ cm^{-1} [404]	Powder, pressed in CsI matrix
Ho_2O_3	
Position of fundamental absorption edge in UV at 298°K: $h\nu \approx 5.27$ eV [623]	Thin films, thickness 60.0–400.0 nm (C-form)
Maximum frequency of IR absorption spectrum: $\nu_l \approx 562$ cm^{-1} [404]	Powder, pressed in CsI matrix
Er_2O_3	
Position of fundamental absorption edge in UV at 298°K: $h\nu \approx 5.21$ eV [623]	Thin films, thickness 60.0–400.0 nm (C-form)
Maximum frequency of IR absorption spectrum: $\nu_l \approx 565$ cm^{-1} [404]	Powder, pressed in CsI matrix
Tu_2O_3	
Position of fundamental absorption edge in UV at 298°K: $h\nu \approx 5.25$ eV [623]	Thin films, thickness 60.0–400.0 nm (C-form)
Maximum frequency of IR absorption spectrum: $\nu_l \approx 568$ cm^{-1} [404]	Powder, pressed in CsI matrix
Yb_2O_3	
Position of fundamental absorption edge in UV at 298°K: $h\nu \approx 5.30$ eV [623]	Thin films, thickness 60.0–400.0 nm (C-form)
Maximum frequency of IR absorption spectrum: $\nu_l \approx 573$ cm^{-1} [404]	Powder, pressed in CsI matrix

1	2
Lu_2O_3	
Position of fundamental absorption edge in UV at 298°K: $h\nu \approx 5.52$ eV [623]	Thin films, thickness 60.0–400.0 nm (C-form)
Maximum frequency of IR absorption spectrum: $\nu_l \approx 580$ cm^{-1} [404]	Powder, pressed in CsI matrix
HfO_2	
Position of maximum absorption band in UV spectrum at $\lambda \approx 213$ nm [357]	Film
Position of absorption bands in IR spectrum: $\nu \approx 425$, 535, 685, 770 cm^{-1} (at 298°K) [359]	Powder, pressed in KBr matrix
Ta_2O_5	
When heated to red heat, retains its white color [80]	
Beginning of absorption band in UV ~ 300 nm, with maximum at $\lambda \approx 209$ nm [52]	Thin film
Electroluminesces [393]	Powder
Red limit of photoeffect ~ 1.5 eV [394]	Film obtained by anodic oxidation
WO_3	
Limit of light transmission in UV $\lambda \approx 380$ nm (amorphous state) and ~ 450 nm (after crystallization; red limit of photoeffect ~ 325 nm (amorphous state) and ~ 550 nm (after crystallization) [395]	Thin films
$WO_3 \cdot H_2O$	
(–) 2 V° = 27; $n_g - n_p = 0.22$ [67]	Tungstite
$WO_3 \cdot 2H_2O$	
(–) 2 V° = 52; $n_g - n_p = 0.34$; pleochroic [67]	Hydrotungstite

5 – Continued

1	2
ReO_3	
Position of reflection minimum in visible region of spectrum: $h\nu \approx 2.1$ eV (edge of plasma reflection) [396]	Single crystal
HgO	
$n_g - n_p = 0.28 \pm 0.04$ (Li); (+) 2 V° is large; red oxide blackens when heated, but recovers original color upon cooling; yellow oxide becomes red upon heating [67; 81, Vol. 4]	Montroydite
In IR absorption spectrum at 293°K, two maxima are observed with $\nu \approx 491$ and 595 cm^{-1}, these bands broadening when the oxide changes from the red-orange to the pale yellow modification [397]	Powders
Phosphoresces in spectral region 2.0–4.5 eV. Maximum intensity of emission ($h\nu$ = 2.39 ± 0.03 eV) is observed at 133°K (red modification of oxide) [63]	Powders
Tl_2O_3	
Position of fundamental absorption edge in visible region of spectrum at 298°K: $h\nu \approx 2.1$ eV [635]	Films
PbO	
$n_0 - n_e = 0.13$ (Li) [67]	Litharge
2 V° ≈ 90; $n_g - n_p = 0.20$ (Li); pleochroic [67]	Massicot
Position of fundamental absorption edge at 298°K: $h\nu \approx 1.7$–3.4 eV [399]	Films and powders with enanthiotropic conversions

1	2
Position of maximum in absorption band in UV at 293°K is $\lambda \approx 350$ nm; maximal value of refractive index $n = 2.7-2.8$ is observed in the visible region (red modification of oxide) [398]	Thin films
In absorption and photoconductance spectra, fine structure is observed in the region of 1.4 and 3.3 eV at 298°K [52, 399]	Single crystals
Pb_3O_4	
$n_0 \geq n_e$; strongly pleochroic [67]	Minium
At 291°K, spectral position of absorption edge coincides with photoconductance maximum: $h\nu \approx 2.1$ eV [366]	Powder (red lead), grain size ~ 8 μm
PbO_2	
$n_0 > n_e$; weak birefringence [67]	Plattnerite
Bi_2O_3	
Optically isotropic [67]	Sillénite
$n_g \geq n_p$ [67]	Bismite
Limit of light transmission in visible region ~ 1.7 eV [400]	Annealed films
Has photosensitivity [52]	Crystal
ThO_2	
Optically isotropic [67]	Thorianite
Limit of transmission in UV ~ 213 nm [401]	Crystal, annealed in oxygen at 1673°K
Position of maxima in absorption spectra at $\lambda \approx 246$ and 300 nm	Crystals, reduced at 1673°K, and also doped with Ca^{2+} and Y^{3+} ions
Fluoresces in spectral region 390–470 nm [401]	

5 – Continued

1	2
UO_2	
Nontransparent; reflection coefficient in visible region R = 12.5–15.0% [67]	Uraninite
At 3°K, in far-IR region, absorption maxima are observed with $\nu \approx 17.6$, 19.2, 79.0, 100.0 cm^{-1} [153]	Single crystal
$2UO_2 \cdot 7H_2O$	
(–) 2 V° small; $n_g - n_p = 0.246$; pleochroic [67]	Ianthinite
$4UO_3 \cdot 5H_2O$	
(–) 2 V° = 48; $n_g - n_p = 0.12$; strongly pleochroic [67]	Crystal (orthorhombic modification)
$4UO_3 \cdot 9H_2O$	
(–) 2 V° = 89; $n_g - n_p = 0.045$ [67]	Schoepite
$7UO_3 \cdot 11H_2O$	
(–) 2 V° = 31; $n_g - n_p = 0.095$; pleochroic [67]	Becquerelite

NUCLEAR PROPERTIES AND RADIATION EFFECTS

1. MICROSCOPIC AND MACROSCOPIC ABSORPTION AND SCATTERING CROSS SECTIONS [405, 406]

Oxide	Molecular density, $\times 10^{28}$ molecules/m^3	σ_a	σ_s	Σ_a	Σ_s
		$\times 10^{-28}$ m^2/molecule		m^{-1}	
H_2O	3,35	0,66	103	2,2	345
D_2O	3,31	0,001	13,6	0,0033	44,9
BeO	7,28	0,01	6,8	0,073	50,1
MgO*	5,44	0,063	7,8	0,344	42,5
Al_2O_3	2,35	0,46	15,4	1,08	36,2
SiO_2	2,66	0,13	10,1	0,345	26,8
CaO	3,64	0,43	7,2	1,57	26,2
TiO	3,20	5,8	12,4	18,5	39,7
Cr_2O_3	2,06	5,8	18,6	11,9	38,3
ZnO	4,16	1,06	7,8	4,4	32,4
SrO	2,70	1,16	14,2	3,13	38,4
Y_2O_3	1,29	2,6	18,6	3,36	24,0
ZrO_2	3,12	0,18	16,4	0,56	51,2
CdO	3,26	3315	11,2	10 850	36,5
Sm_2O_3	1,28	16 500	22,6	21 100	28,9
Eu_2O_3	1,27	8740	30,2	11 100	38,3
Dy_2O_3	1,26	2200	214	2 770	270
ThO_2	2,25	7	20,9	15,8	47,0
UO_2	2,23	7,6	16,7	16,9	37,2

*Sections of MgO and other oxides calculated according to formulas: $\sigma_i = \sigma_{ie} + \sigma_{iO}$ (i = a, s), where σ_{ie} is the microscopic section of the element forming the oxide and σ_{iO} is the microscopic section of oxygen; $\Sigma_i = \rho_e\sigma_{ie} - \rho_O\sigma_{iO}$, where ρ_e is the atomic density of the element in the oxide (number of atoms of the oxide-forming element in unit volume) and ρ_O is the atomic density of oxygen in the oxide (number of oxygen atoms in the oxide composition per unit volume).

2. NUCLEAR PROPERTIES OF OXIDE MODERATORS [406]

Oxide	Density, kg/m^3	Molecular mass	Molecular density, $\times 10^{28}/m^3$	Absorption cross section for thermal neutrons, $\times 10^{-28}$ m^2	Scattering cross section for thermal neutrons, $\times 10^{-28}$ m^2	Scattering cross section for epithermal neutrons, $\times 10^{-28}$ m^2	Log-mean energy loss of neutron in one collision	Path length of neutrons with energies $4 \cdot 10^{-21}$ J before absorption, m
H_2O	1000	18	3,35	0,66	110	46	0,948	0,45
D_2O	1100	20	3,32	0,00092	15	10,5	0,570	328
BeO	2800	25	6,75	0,0092	11,1	9,8	0,173	16,1

2 — Continued

Oxide	Transport length of neutron path, m	Retardation time, sec	Diffusion time, sec	Albedo (infinite)	Diffusion length, m	Fermi age, m^2	Fermi age (to energy equal to $2.3 \cdot 10^{-19}$ J), m^2
H_2O	0,00246	$1.0 \cdot 10^{-5}$	0,00021	0,82	0,0288	0,0033	0,0030
D_2O	0,024	$4,6 \cdot 10^{-5}$	0,15	0,97	1,00	0,0120	0,0100
BeO	0,0165	$7,8 \cdot 10^{-5}$	0,0068	0,93	0,30	0,0143	—

3. THRESHOLD ENERGIES OF REACTIONS LEADING TO FORMATION OF NEW ELEMENTS IN OXIDES [407]

Oxide	Isotope	Threshold energy, $\times 10^{-12}$ J (or MeV)			
		(γ, n)	(n, 2 n)	(n, α)	(n, p)
All	^{16}O	2,61(16,3)	2,77(17,3)	0,58(3,6)	1,63(10,2)
BeO	^{9}Be	0,27(1,67)	0,3(1,85)	0,048(0,3)	—
MgO	^{24}Mg	2,62(16,4)	2,73(17,1)	—	0,78(4,9)
	^{25}Mg	1,17(7,3)	1,20(7,5)	—	0,66(4,1)
	^{26}Mg	1,79(11,2)	1,85(11,6)	—	—
Al_2O_3	^{27}Al	2,05(12,8)	2,11(13,2)	0,39(2,44)	0,31(1,96)
SiO_2	^{28}Si	2,7(16,8)	2,78(17,4)	—	0,64(4,0)
	^{29}Si	1,36(8,5)	1,39(8,7)	—	0,53(3,3)
CaO	^{40}CaO	2,54(15,9)	2,61(16,3)	—	—
TiO_2	^{46}Ti	2,13(13,3)	2,18(13,6)	—	—
	^{49}Ti	1,39(8,7)	1,42(8,9)	—	—
ZnO	^{64}Zn	—	1,91(11,9)	—	—
	^{66}Zn	—	1,81(11,3)	—	—
	^{67}Zn	—	1,14(7,1)	—	—
	^{68}Zn	—	1,65(10,3)	—	—
SrO	^{86}Sr	—	1,54(9,6)	—	—
	^{87}Sr	—	1,36(8,5)	—	—
	^{88}Sr	—	1,79(11,2)	—	—
ZrO_2	^{90}Zr	—	1,97(12,3)	—	—
	^{91}Zr	—	1,17(7,3)	—	—

4. CERTAIN CHARACTERISTICS OF ISOTOPES FORMED IN OXIDES BY RADIATION [407]

Oxide	Isotope	Type of decay	Half-life sec	Maximal radiation energy, $\times 10^{-13}$ J		Product of decay
				of particle	of γ-rays	
1	2	3	4	5	6	7
All	^{1}H	Stable	—	—	—	—
	^{4}He	Stable	—	—	—	—
	^{13}C	Stable	—	—	—	—
	^{16}N	$\beta-$	7,4	16,5	11,4	^{16}O
	^{15}O	$\beta+$	126	2,7		^{15}N
BeO	$^{3}H^{*}$	$\beta-$	$3{,}87\cdot10^{-8}$	0,0288		^{3}He
	^{6}He	$\beta-$	0,83	5,6		^{6}Li
	^{8}Be	2α	10^{-16}	0,058		^{4}He
MgO	^{21}Ne	Stable	—	—		—

4 – Continued

1	2	3	4	5	6	7
	^{22}Ne	Stable	—	—	—	—
	^{23}Ne	β−	40,2	7,0	3,34	^{23}Na
	^{24}Na	β−	$5,4\cdot10^{4}$	2,22	2,21	^{24}Mg
	^{25}Na	β−	62	6,4	2,56	^{25}Mg
	^{23}Mg	β+	11	4,72		^{23}Na
Al_2O_3	^{24}Na	β−	$5,4\cdot10^{4}$	2,22	2,21	^{24}Mg
	^{27}Mg	β−	570	2,8	1,62	^{27}Al
	^{26}Al	β+	6,7	5,12		^{26}Mg
SiO_2	^{25}Mg	Stable	—	—	—	—
	^{26}Mg	Stable	—	—	—	—
	^{28}Al	β−	138	4,58	2,86	^{28}Si
	^{29}Al	β−	394	4,0	3,9	^{29}Si
	^{27}Si	β+	4,0	6,02		^{27}Al
CaO	^{37}Ar	E.C.	$2,94\cdot10^{5}$			^{37}Cl
	^{40}K	β−, E.C.	$4,1\cdot10^{16}$	2,11	2,34	^{40}Ca, ^{40}Ar
	^{39}Ca	β+	0,9	8,17		^{39}K
ZrO	^{87}Sr	Stable	—	—	—	—
	^{88}Sr	Stable	—	—	—	—
	^{90}Y	β−	$2,31\cdot10^{5}$	3,58	2,8	^{90}Zr
	^{89}Zr	β+	$2,84\cdot10^{5}$	1,44	1,46	^{89}Y

* Tritium is formed in accordance with reactions: $^{6}Li + n = ^{4}He + ^{3}H$.

5. RADIATION-INDUCED CHANGES IN VOLUME

Φ_{fn}, *1 × 10^{24} m^{-2}	T_{irr}, °K	$\Delta V/V_0$, %	Φ_{fn}, *1 × 10^{24} m^{-2}	T_{irr}, °K	$\Delta V/V_0$, %
1	2	3	1	2	3
BeO, single crystals [408, 409]			**BeO**, polycrystals [409, 410, 411, 428]		
1	373	+0,3	1,7	348—373	+0,7*2
2	373	+0,6	2,6	348—373	+2,0*2
3	373	+1,2	5,6	348—373	+4,1*2
5	373	+2,3	7,4	348—373	+6,3*2
8	373	+3,3	10	348—373	+8,3*2
10	373	+3,7	1,2	783	+0,2*2
25	373	+4,5	5,6	793	+1,7*2
18	383	+1,7	4,5	863	+1,5*2
11	923	+0,9	1,5	943	+0,15*2
41	923	+3,8	6,6	943	+1,3*2

5 — Continued

Φ_{fn}, [*1] $\times 10^{24}$ m^{-2}	T_{irr}, °K	$\Delta V/V_0$, %
1	2	3
12	963	+2,5[*2]
2,7	348—373	+0,8[*3]
5,2	348—373	+2,8[*3]
10	348—373	+5,9[*3]
14	348—373	+7,1[*3]
9	808	+2,4[*3]
11	858	+1,9[*3]
12	963	+1,7[*3]
2,8	348—373	+0,5[*4]
5,3	348—373	+1,9[*4]
10	348—373	+4,5[*4]
14	348—373	+6,8[*4]
9	808	+1,4[*4]
11	858	+1,4[*4]
12	963	+1,3[*4]
1,9	373	+1,2[*5]
2,4	373	+1,6[*5]
3,1	373	+2,2[*5]
4,0	373	+2,4[*5]
5,0	373	+2,9[*5]
5,3	373	+3,4[*5]
2,0	373	+1,4[*6]
2,5	373	+0,9[*6]
3,0	373	+2,8[*6]
4,3	373	+4,0[*6]
5,0	373	+4,6[*6]
5,5	373	+4,9[*7]
1,7	373	+0,5[*7]
2,1	373	+0,9[*7]
2,9	373	+1,3[*7]
3,0	373	+1,4[*7]
3,5	373	+1,6[*7]
3,6	373	+1,9[*7]
4,0	373	+1,6[*7]
4,2	373	+2,4[*7]
4,6	373	+2,4[*7]
4,9	373	+2,6[*7]
5,1	373	+2,9[*7]
5,2	373	+2,6[*7]
5,5	373	+3,1[*7]
5,8	373	+2,6[*7]
8	923	+2,1[*8]
16	923	+4,2[*8]
24	923	+6,3[*8]
10	1373	+0,9[*9]
20	1373	+1,4[*9]
10	1373	+2,1[*10]
20	1373	+2,1[*10]

5 – Continued

Φ_{fn}, *1 $\times 10^{24}$ m^{-2}	T_{irr}, °K	$\Delta V/V_0$, %
1	2	3
40	1373	+2,2 *10
10	1373	+1,5 *11
20	1373	+2,6 *11
10	1373	+2,7 *12
20	1373	+3,0 *12
40	1373	+3,4 *12
10	1273	+0,5 *13
20	1273	+1,1 *13
30	1273	+1,7 *13
50	1273	+2,8 *13
90	1273	+5,1 *13
10	1273	+0,6 *14
20	1273	+1,3 *14
30	1273	+2,1 *14
50	1273	+2,6 *14
90	1273	+3,0 *14
MgO, single crystals [412]		
0,15	348—373	+0,1
0,6	348—373	+0,4
1,4	348—373	+0,6
2,2	348—373	+0,7
3,0	348—373	+0,8
4,8	348—373	+1,0
6,0	348—373	+1,1
Al_2O_3, polycrystals [413, 414]		
1	523	+0,22
2	523	+0,45
3	523	+0,70
4	523	+0,85
6	523	+1,15
8	523	+1,30
1	748	+0,25
2	748	+0,40
3	748	+0,55
4	748	+0,65
6	748	+0,80
8	748	+0,85
52	673—973	+5

*1 Φ_{fn} is the influence of neutrons with energies $E_n > 1.6 \cdot 10^{-13}$ J; T_{irr} is the specimen temperature during irradiation; $\Delta V/V_0$ is the relative change in volume; ρ_0 is the density before irradiation (ρ_{rel} is the relative density before irradiation, %); d_g is the grain diameter. *2 Hot-pressed, $\rho_{rel} = 97-98$%, $d_g = 12-15$ μm. *3 Cold-pressed, $\rho_{rel} = 96-98$%, $d_g = 8-12$ μm. *4 Cold-pressed, $\rho_{rel} = 95-97$%, $d_g = 2-3$ μm. *5 BeO, UOX grade, + 1% Be, $\rho_0 = 2980$ kg/m^3, $d_g = 9$ μm. *6 BeO (UOX) + 2% Be, $\rho_0 = 2960$ kg/m^3, $d_g = 12$ μm. *7 BeO (UOX) + 0.2% Al_2O_3 + 0.1% SiO_2, $\rho_0 = 2920$ kg/m^3, $d_g = 10$ μm. *8 Cold-pressed, $d_g = 17-25$ μm. *9 Cold-pressed, $\rho_0 = 2700$ kg/m^3, $d_g = 17-24$ μm; irradiation time $\tau_{irr} = 7.95 \cdot 10^6$ sec. *10 $\rho_0 = 2700$ kg/m^3, $d_g = 17-24$ μm; $\tau_{irr} = 1.59 \cdot 10^7$ sec. *11 $\rho_0 = 2900$ kg/m^3, $d_g = 23-25$ μm; $\tau_{irr} = 7.95 \cdot 10^6$ sec. *12 $\rho_0 = 2900$ kg/m^3, $d_g = 23-25$ μm; $\tau_{irr} = 1.59 \cdot 10^7$ sec. *13 Cold-pressed, $\rho_0 = 2900$ kg/m^3, $d_g = 5$ μm. *14 Cold-pressed, $\rho_0 = 2600$ kg/m^3, $d_g = 5$ μm.

Φ_{fn}, $\times 10^{24}$ m^{-2}	T_{irr}, °K	$\Delta\rho/\rho_0$, %	ρ_0, kg/m^3
1	2	3	4
BeO, sintered [415, 409]			
0,086	373	−0,1[*1]	2900
1,2	373	−0,7[*1]	2900
1,6	373	−1,0[*1]	2900
1,8	373	−1,2[*1]	2900
2,0	373	−1,3[*1]	2900
2,1	373	−1,5[*1]	2900
3,7	373	−3,4[*1]	2900
3,9	373	−3,8[*1]	2900
4,0	373	−4,2[*1]	2900
0,082	373	−0,1[*2]	2900
3,6	373	−3,6[*2]	2900
4,5	373	−4,6[*2]	2900
0,061	373	−0,1[*3]	2900
3,8	373	−4,2[*3]	2900
6,0	373	−5,8[*3]	2900
0,088	373	−0,1[*4]	2900
1,3	373	−0,7[*4]	2900
1,8	373	−1,1[*4]	2900
2,3	373	−1,3[*4]	2900
4,0	373	−3,9[*4]	2900
5,4	373	−5,4[*5]	2940
2,0	1133	−3,19[*6]	2839
2,0	1133	−2,61[*6]	2833
2,1	1133	−1,26[*6]	2806
2,1	1133	−2,33[*6]	2800
MgO[*7] [408]			
0,15	348—373	−0,1	
0,6	348—373	−0,4	
1,4	348—373	−0,6	
2,2	348—373	−0,7	
3,0	348—373	−0,8	
4,8	348—373	−1,0	
6,0	348—373	−1,1	
Al_2O_3[*8] [408, 416—418]			
0,2	353	−0,1	
0,5	353	−0,23	
1,0	353	−0,38	
2,0	353	−0,6	
3,0	353	−0,8	
4,0	353	−0,95	
5,0	353	−1,03	
0,6[*9]	323	−0,35[*10]	3983
6,0[*9]	323	−0,98[*10]	3983
0,3	323	−0,17[*11]	3559
0,5		−0,39[*11]	
SiO_2, crystals [418]			
0,66	373	−3,5[*12]	
0,013	313—373	−0,07[*12]	2648
0,049	313—373	−4,7[*12]	
11	523—573	−17,7[*12]	
2		−14,7[*12]	2600
2		−13,0[*13]	2600
2		−13,0[*14]	2600
SiO_2, amorphous [417, 418]			
0,07	323	+0,9	2210
2	323	+1,8	2210
0,013	313—373	+0,2	2209
0,059	313—373	+2,8	2268
11	523—573	+2,1	2251
0,2	323	+1,9	2196
0,5	323	+2,05	2196
0,7	323	+2,3	2204
4	323	+1,2	2204
2	323	+3	2210
2,2		+2,8	
TiO_2 [416, 419]			
0,6	323	−0,5	4010
3	323	−0,75	4010
$MgO \cdot Al_2O_3$ [419]			
4		0	3600
$3BeO \cdot Al_2O_3 \cdot 6SiO_2$ [418]			
3,6		−6	
$2MgO \cdot SiO_2$ [418]			
0,6		−0,85	3056
$ZrO_2 \cdot SiO_2$ [418, 419]			
0,5		−0,67	3730
3,6	323	−4	3730
Steatite [418]			
0,7	323	−1,3	2796
Porcelain [419]			
0,6		−0,3	3410
4		−0,6	3410
Mica [419]			
0,4[*15]		−3,8	2845
2[*15]		−14	2845
0,8[*16]		−2,4	
1[*16]		−2,8	

*1 Grade AOX, d_g = 15 – 20 μm. *2 Grade HPA, d_g = 15 – 20 μm. *3 Grade UOX, d_g = 15 – 20 μm. *4 UOX + 0.5% MgO, d_g = 15 – 20 μm. *5 UOX + 3% ZrO_2, d_g = 15 – 20 μm. *6 Cold-pressed, +0.6% CaO. *7 Radiation change in density of MgO equivalent to change in volume. *8 Single crystals and polycrystals. *9 E_n > 1.6 · 10^{-17} J. *10 Sapphire. *11 Sintered. *12 Quartz. *13 Cristobalite. *14 Tridymite. *15 Irradiated in water-moderated reactor. *16 Irradiated in graphite-moderated reactor.

Φ_{fn}, $\times 10^{24}$ m^{-2}	T_{irr}, °K	$\Delta a/a_0$, %	$\Delta c/c_0$, %
1	2	3	4
	BeO [409, 410, 420, 421]		
1	373	+0,03	+0,22*1
2	373	+0,05	+0,50*1
4	373	+0,10	+1,15*1
6	373	+0,12	+1,85*1
8	373	+0,13	+2,55*1
10	373	+0,13	+3,0*1
5,5	783—813	+0,02	+0,6*2
9	793—823	+0,02	+1,42*2
4,5	853—873	+0,02	+0,51*2
11	843—873	+0,02	+1,2*2
12	943—973	+0,02	+1,0*2
5	348—373	+0,105	+1,6*3
5	348—373	+0,14	+0,5*4
5	348—373	+0,10	+1,4*5
1,9	1173	0	+0,03*6
1,9	1173	0	0,02—0,04*7
4	383	+0,10	+1,00*8
7	383	+0,12	+2,56*9
10	383	+0,13	+2,98*8
10	383	+0,13	+3,26*9
8	923	0	+1,50*8
17	923	0	+2,26*8
43	923	+0,04	+2,09*8
8	1373	0	0*8
19	1373	0	+0,19*8
34	1373	0	0*8
9	923	0	+1,58*9
17	923	+0,01	+1,40*9
29	923	0	+2,12*9
45	923	+0,05	+2,04*9
7	1373	0	0*9
19	1373	0	+0,34*9
34	1373	0	0*9
	MgO*10 [412, 415]		
0,2	373	+0,04	—
0,5	373	+0,12	—
0,8	373	+0,17	—
1	373	+0,18	—
1,5	373	+0,195	—
1,8	373	+0,20	
2,0	373	+0,19	
2,5	373	+0,16	
3,0	373	+0,125	
3,5	373	+0,095	

1	2	3	4
4,0	373	+0,07	
4,5	373	+0,05	
2,2	673	+0,085	
	Al_2O_3[*11] [413]		
0,5	373	+0,08	0,08
1,0	373	+0,13	+0,13
1,5	373	+0,18	+0,18
2,0	373	+0,22	+0,22
2,5	373	+0,26	+0,26
3,0	373	+0,28	+0,28
3,5	373	+0,29	+0,32
4,0	373	+0,30	+0,35
5,0	373	+0,32	+0,40
2,8	823	+0,11	+0,11
3,2	873	+0,12	+0,12
2,5	973	+0,07	+0,07
	SiO_2[*12] [418]		
0,09		+0,1	+0,03
0,66	373	+2,2	+0,32
	TiO_2 [418]		
1,1		0	
	ZnO [422]		
4,0	348—373	+0,04	+0,4
	ZrO_2 [418, 419]		
0,015	373	—0,39	—1,94[*13]
1,0		0[*14]	—
2,0		+0,28[*14]	—
	UO_2 [419]		
0,4		+0,04	—
	$MgO\cdot Al_2O_3$ [419]		
1,0		+0,12	—
	$BeO\cdot Al_2O_3$ [418, 419]		
3,6		+1	—
	$2BeO\cdot SiO_2$ [418, 419]		
3,6		+0,7	—

[*1]Single crystals and powders. [*2]Powders obtained by grinding sintered irradiated specimens. [*3]Powder. [*4]Compact. [*5]Ground into powder after irradiation. [*6]Plus 0.6% CaO; compact. [*7]Same as in [*6], but ground after irradiation. [*8]$d_g = 71-74$ μm. [*9]$d_g = 17-24$ μm. [*10]Single crystals and polycrystals. [*11]Single crystals and polycrystals. [*12]Quartz, $a_0 = 0.4903$ nm, $c_0 = 0.5393$ nm; with large fluxes, the crystalline SiO_2 changes to amorphous. [*13]Monoclinic, $\Delta b/b_0 = -1.2\%$; $\Delta\beta/\beta_0 = +0.68\%$. [*14]Cubic, stabilized.

[408, 409, 411, 413, 415, 416, 418, 419, 423–425]

Φ_{fn}, $\times 10^{24}$ m^{-2}	T_{irr}, °K	$\Delta\lambda/\lambda_0$, %	λ_0, W/(m·°C)
BeO			
0.25	373	−14[*1]	168
0.6	373	−69[*1]	168
0.25	373	−32[*2]	197
0.6	373	−80[*2]	197
0.15	343–353	−18[*3]	230
1.2	353–373	−70[*3]	230
4	343–353	−92[*3]	230
1.2	783	−44[*3]	230
1.1	893	−19[*3]	230
1.5	933	−22[*3]	230
2.1	1123	−60[*4]	—
0.4	348	−80[*5]	520
1.1	348	−89[*6]	550
0.4	348	−60[*5]	500
0.86	348	−73[*5]	500
1.1	348	−80[*6]	500
—	573	−58[*7]	—
—	873	−33.7[*7]	—
—	1073	−26.3[*7]	—
MgO			
0.3	323	−40	—
0.8	343	−44	—
Al_2O_3			
0.0015	—	−55[*8]	6000
0.0089	—	−80[*8]	6000
0.02	—	−87[*8]	6000
0.05	—	−97[*8]	6000
0.6	323	−50[*9]	25
6.0	323	−67[*9]	25
0.3	323	−42[*10]	16.8
4.0	323	−78[*10]	16.8
1.0	343	−25[*11]	—
2.0	523	−67[*12]	—
5.0	523	−78[*12]	—
8.0	523	−80[*12]	—
2.0	748	−52[*12]	—
5.0	748	−69[*12]	—
2.0	973	−33[*12]	—
5.0	973	−52[*12]	—
SiO_2			
0.00054	—	−40[*13]	1000
0.018	—	−95[*13]	1000
0.043	—	−98[*13]	1000
0.34	—	−99.8[*13]	1000
0.17	—	+17[*14]	0.52
0.41	—	+35[*14]	0.52
0.6	—	+100[*14]	0.52
TiO_2			
0.6[*15]	323	−33	6.09
3.0[*15]	323	−60	6.09
ZrO_2			
2.0[*15]		−21	5.05
$MgO \cdot Al_2O_3$			
0.7	—	−48[*16]	10.5
4.0	—	−95[*16]	10.5
$2MgO \cdot SiO_2$			
0.6	323	−70[*17]	10.5
1.0	323	−76[*17]	7.5
2.0	323	−76[*17]	7.5
$ZrO_2 \cdot SiO_2$ [418, 425]			
0.5	323	−81[*18]	5
Cordierite			
0.5	323	−72[*19]	3.06
3	323	−72[*19]	3.06
1	323	−70[*19]	3.23
2	323	−73[*19]	3.23
Steatite			
0.7	323	−63[*20]	3.19
1	323	−53[*20]	2.51
2	323	−53[*20]	2.51
Porcelain			
0.6	—	−52[*21]	—
4	—	−68[*21]	—
Mica			
0.4	—	−29[*22]	0.49
2	—	+47[*22]	0.49
Quartz glass			
2	323	0	—

[*1] Hot-pressed, ρ_0 = 2740 kg/m³, T_{meas} = 373°K. [*2] ρ_0 = 3006 kg/m³, T_{meas} = 373°K. [*3] ρ_0 = 2950–2980 kg/m³, T_{meas} = 293°K. [*4] Cold-pressed + 0.6% CaO, ρ_0 = 2800 kg/m³, d_g = 32 μm. [*5] ρ_{rel} = 98%, d_g = 12 μm. [*6] d_g = 35 μm. [*7] Change in thermal diffusivity $\Delta D/D_0$, with $D_0 = 1.5 \cdot 10^{-5}$ m²/sec; ρ_0 = 2900 kg/m³, d_g = 5 μm; irradiation continued to $\Delta V/V_0$ = +0.7%; T_{meas} = 573°K. [*8] Sapphire, T_{meas} = 40°K. [*9] ρ_0 = 3983 kg/m³. [*10] Sintered, ρ_0 = 3559 kg/m³. [*11] T_{meas} = 573°K. [*12] Sintered, porosity 8, 25, and 31%. [*13] α-quartz, T_{meas} = 10°K. [*14] Fused quartz, T_{meas} = 5°K. [*15] $E_n > 1.6 \cdot 10^{-17}$ J. [*16] ρ_0 = 3600 kg/m³. [*17] ρ_0 = 3056 kg/m³, T_{meas} = 313°K. [*18] ρ_0 = 3730 kg/m³, T_{meas} = 313°K. [*19] T_{meas} = 313°K. [*20] ρ_0 = 2796 kg/m³, T_{meas} = 313°K. [*21] ρ_0 = 3410 kg/m³. [*22] ρ_0 = 2845 kg/m³.

Φ_{fn}, $\times 10^{24}$ m^{-2}	T_{irr}, °K	$\Delta E/E_0$, %
1	2	3
BeO [419, 426, 440]		
0,25	373	—19*1
0,25	373	—40*2
0,6	373	—50*3
0,6	373	—64*4
4,2	373	—71,8*5
4,2	373	—54,8*6
MgO [408]		
0,8	343	—5
Al_2O_3 [413; 418]		
0,6*7	323	< —10*8
1,6*7	323	< —10*8
1,5	343	—3*9
1—10	523—973	— (5—10)*9
SiO_2 [416, 418, 425, 427]		
0,04	323	— (3,2—16,5)*10
0,07	328	+1,42*11
0,6	—	< +5*12
1,6	—	< +45*12
0,04	—	+ (0,66—3,8)*13
2	—	+ (3—4)*13
0,07	328	—3,8*14
TiO_2 [418]		
0,6*7	323	< —5
ZrO_2 [418]		
0,6*7	323	< —10
1,6*7	323	< —10
$2MgO \cdot SiO_2$ [418]		
0,6	—	—21
0,8	—	—10
1,6	—	—11
Cordierite [418]		
0,6		—46
0,8		—22
1,6		—27
Steatite [418]		
0,6		< —5
1,6		< —5
$ZrO_2 \cdot SiO_2$ [418]		
0,6		—34
0,8		—42
1,6		—45

*1 $\rho_0 = 2620$ kg/m^3, $E_0^d = 243$ GPa (E_0^d is the dynamic modulus of elasticity before irradiation). *2 $\rho_0 = 2740$ kg/m^3. *3 $\rho_0 = 2704$ kg/m^3, $E_0^d = 264$ GPa. *4 $\rho_0 = 2902$ kg/m^3, $E_0^d = 340$ GPa. *5 $\rho_{rel} = 95\%$, $d_g = 5$ μm, $E_0^{st} = 355$ GPa. *6 $\rho_0 = 95\%$, $d_g = 5$ μm, $E_0^{st} = 365$ GPa. *7 $E_n > 1.6 \cdot 10^{-17}$ J. *8 Polycrystals and single crystals. *9 Polycrystals. *10 Quartz. *11 Amorphous. *12 Quartz glass. *13 Fused. *14 Shear modulus; amorphous.

Φ_{fn}, $\times 10^{24}$ m^{-2}	T_{irr}, °K	$\Delta\sigma/\sigma_0$, %	σ_0, MPa
1	2	3	4
BeO [411, 415, 418, 423, 426, 428, 429]			
In tension			
0,06	—	0	65
0,1	—	+8,3	65
0,15	—	+23,7	65
0,6	373	—82	59—83[*1]
0,6	373	—94	123[*2]
2,2	623—673	—87	102[*3]
2,2	623—673	—80	102[*4]
2,2	623—673	—93	147[*5]
2,2	623—673	—88	91,2[*6]
In compression			
0,25	373	—23,5	1962
0,6	373	—77,5	1962
0,9	373	—98	1962
2	373	—95	981[*7]
2	373	—97	981[*8]
2	373	—93	1470[*9]
2	373	—98	1470[*10]
2,2	673	— (63—76)	1950[*11]
2,2	673	—3 — +10	560[*12]
In bending			
0,086	373	+11,5	241[*13]
2	373	—87,9	241[*13]
3,9	373	—97,4	241[*13]
0,082	373	+7,2	172[*14]
3,6	373	—93,2	172[*14]
0,074	373	+3	206[*15]
0,061	373	+7,3	209[*16]
3,8	373	—97,7	209[*16]
0,088	373	+2,2	219[*17]
2,1	373	—62,8	219[*17]
0,074	373	+14,5	209[*18]
5,4	373	—96,7	209[*18]
1,8	373	—84,6	134[*19]
1,9	373	—29	214[*20]
2,4	373	—48	214[*20]
3,1	373	—45	214[*20]
4,0	373	—68	214[*20]
5,0	373	—73	214[*20]
5,4	373	—74	214[*20]
2,0	373	—71	261[*21]
2,5	373	—81	261[*21]
3,0	373	—89	261[*21]
4,3	373	—91	261[*21]
5,0	373	—94	261[*21]
5,5	373	—98	261[*21]
1,7	373	+51	187[*22]
2,1	373	+28	187[*22]
2,9	373	+35	187[*22]
3,0	373	+6	187[*22]
3,5	373	—33	187[*22]
3,6	373	—27	187[*22]
4,0	373	—48	187[*22]
4,2	373	—30	187[*22]
4,6	373	—46	187[*22]
4,9	373	—38	187[*22]
5,1	373	—53	187[*22]
5,2	373	—41	187[*22]
5,5	373	—53	187[*22]
5,8	373	—75	187[*22]
2,3	373	+175[*23]	
90	1273	0[*24]	
90	1273	—30[*25]	
MgO [415]			
In compression			
2,2	673	+101[*26]	466
2,2	673	+120[*27]	
2,2	673	+17[*28]	294
2,2	673	+29[*29]	
Al_2O_3 [427]			
In bending			
1,3	373	—20[*30]	
1,7	373	+30[*30]	
2,1	373	+60[*30]	
3,5	373	+30[*30]	
5,0	373	+60[*30]	
2,1	373	+15[*31]	
4,2	373	+30[*31]	
5,0	373	+25[*31]	
SiO_2 [427]			
0,024	373	+10[*12]	
2,0	373	+6[*32]	

[*1] ρ_0 = 2600–2700 kg/m³. [*2] ρ_0 = 2840–2880 kg/m³. [*3] ρ_0 = 2670 kg/m³. [*4] ρ_0 = 2740 kg/m³. [*5] ρ_0 = 2800 kg/m³. [*6] ρ_0 = 2850 kg/m³. [*7] ρ_0 = 2520 kg/cm³. [*8] ρ_0 = 2530 kg/m³. [*9] ρ_0 = 2740 kg/m³. [*10] ρ_0 = 2800 kg/m³. [*11] ρ_0 = 2960–2990 kg/m³, T_{meas} = 293°K. [*12] ρ_0 = 2960–2990 kg/m³, T_{meas} = 1273°K. [*13] ρ_0 = 2900 kg/m³, dg = 20 μm; AOX grade of BeO powder. [*14] HPA grade. [*15] HPA + 0.5% MgO. [*16] UOX grade. [*17] UOX + 0.5% MgO. [*18] ρ_0 = 2940 kg/m³, UOX + 3% ZrO_2. [*19] ρ_0 = 2900 kg/m³, d_g = 50 μm, AOX grade. [*20] ρ_0 = 2980 kg/m³, d_g = 9 μm; UOX + 1% Be. [*21] ρ_0 = 2960 kg/m³, d_g = 12 μm; UOX + 2% Be. [*22] ρ_0 = 2920 kg/m³, d_g = 10 μm; UOX + 0.2% Al_2O_3 + 0.1% SiO_2. [*23] d_g = 4 μm; UOX + 0.4% Al_2O_3 + 0.1% SiO_2. [*24] ρ_0 = 2900 kg/m³, d_g = 5 μm; $\Delta V/V_0$ = 5%. [*25] ρ_0 = 2600 kg/m³, d_g = 5 μm; $\Delta V/V_0$ = 3%. [*26] ρ_{rel} = 95%, T_{meas} = 293°K. [*27] ρ_{rel} = 95%, T_{meas} = 673°K. [*28] ρ_{rel} = 83%, T_{meas} = 293°K. [*29] ρ_{rel} = 83%, T_{meas} = 673°K. [*30] Sintered, d_g = 4–6 μm. [*31] Sintered, d_g = 40–200 μm. [*32] Quartz glass.

Φ_{fn}, $\times 10^{24}$ m^{-2}	T_{irr}, °K	$\Delta H/H_0$, %	H_0, GPa	Notes
BeO [430, 431]				
1	373	+25	10,8	ρ_0=2900 kg/m^3, d_g=20–50
2	373	+ (27—45)	10,8	Knoop hardness with load of 0.981 N
4	373	+ (32—41)	10,8	
6	373	+ (32—41)	10,8	
0,3	373	+13	11	d_g=18 μm, ρ_{rel}=97%
0,3	373	+19,5	11,2	d_g=30 μm, with load of 0.981 N
0,3	373	+30	11,6	d_g=50 μm
0,8	373	+18	10,8	d_g=8 "
0,8	373	+24,5	10,9	d_g=12 "
0,8	373	+35	11	d_g=20 "
2,0	373	>+25	—	d_g=5 "
2,0	373	+47	10,9	d_g=14 "
MgO [432]				
0,4	473	+10	—	Single crystals
Al_2O_3 [413, 433]				
1,2	343	+17	—	Single crystals
0,19	473	+6	17	Microlite, +0.5% MgO
0,9	473	+2	17	
SiO_2 [418]				
0,5	343	—38	7,85	Quartz, polycrystals
5,0	343	—50	7,85	Knoop hardness with load of 0.981 N
Forsterite [433]				
0,19	473	—3	7,8	—
0,9	473	0	7,8	
2,0	473	—10,5	7,8	

12. RADIATION-INDUCED CHANGES IN ELECTRICAL PROPERTIES

Oxide	T_{irr}, °K	T_{meas}, °K	Before irradiation	During or after irradiation	Notes
1	2	3	4	5	6
			Conductivity, S/m		
BeO	973	1073	$3 \cdot 10^{-7}$	$3 \cdot 10^{-6}$	After irradiation with fluence $\Phi_{fn} = 7 \cdot 10^{24}$ m^{-2}; before measurement, the irradiated specimen was annealed at 1273°K [409]
Al_2O_3	333	333	$5 \cdot 10^{-12}$	$1 \cdot 10^{-9}$	In process irradiation in reactor (power 6 MW/m^3); plasma-sprayed specimens, $\rho_{rel} = 85\%$ [434]
	400	400	$1 \cdot 10^{-10}$	$1{,}5 \cdot 10^{-9}$	
	500	500	$1 \cdot 10^{-9}$	$3 \cdot 10^{-9}$	
	1000	1000	$1 \cdot 10^{-5}$	$1 \cdot 10^{-5}$	
			Resistivity, ohms·m		
BeO	1173	773	—	Decreases by factor of 10^2 to 10^3	After irradiation with fluence $\Phi_{fn} = 2 \cdot 10^{24}$ m^{-2}; BeO + 0.06% CaO, cold-pressed [409]
SiO_2	433	433	10^{12}	$8 \cdot 10^{7}$	With irradiation at flux density $I_{tn} = 8 \cdot 10^{16}$ $m^{-2} \cdot sec^{-1}$
	433	433	10^{12}	$1 \cdot 10^{10}$	Under same conditions, but after accumulation of fluence $\Phi_{tn} = 1 \cdot 10^{22}$ m^{-2}; quartz [435]
ZrO_2	383	383	$4 \cdot 10^{6}$	$1{,}1 \cdot 10^{5}$	In process of irradiation in reactor; neutron flux density $I_n = 1.3 \cdot 10^{16}$ m^{-2} sec^{-1}; ZrO_2 stabilized with 10% Y_2O_3 [436]
	500	500	$5 \cdot 10^{4}$	$1{,}6 \cdot 10^{4}$	
	667	667	$1{,}6 \cdot 10^{2}$	$1{,}6 \cdot 10^{2}$	

Cordierite	383 500 667	383 500 667	$6,3 \cdot 10^{10}$ $9 \cdot 10^{7}$ $2 \cdot 10^{4}$	$1,6 \cdot 10^{9}$ $3,2 \cdot 10^{7}$ $2 \cdot 10^{4}$	In process of irradiation in reactor; neutron flux density $I_n = 1.3 \cdot 10^{16}\ m^{-2} \cdot sec^{-1}$ [436]
Mullite-corundum ceramic	556 667	556 667	$8 \cdot 10^{10}$ $1 \cdot 10^{7}$	$5 \cdot 10^{8}$ $1,8 \cdot 10^{6}$	In process of irradiation in reactor; neutron flux density $I_n = 1.3 \cdot 10^{16}\ m^{-2} \cdot sec^{-1}$ [436]
			Resistance, ohms		
Al_2O_3	523	—	$2 \cdot 10^{14}$	$2 \cdot 10^{12}$	After irradiation with fluence $\Phi_{tn} = 2 \cdot 10^{22}\ m^{-2}$ (Φ_{tn} is the thermal neutron fluence)
	673	—	10^{10}	10^{8}	In process of irradiation with neutrons [418]
	—	—	10^{12}	10^{7}	In process of irradiation with protons, $E_p = 8 \cdot 10^{-17}$ J, current strength $I_p = 0.3$ mA [418]
	303	303	—	20% decrease 20%	After irradiation with electrons, $E_e = 4 \cdot 10^{-13}$ J [418]
MgO	673	—	—	No change	After irradiation with fluence $\Phi_n = 3.5 \cdot 10^{22}\ m^{-2}$ [418]
	—	—	$7 \cdot 10^{10}$	10^{6}	In process of irradiation with protons, $E_p = 8 \cdot 10^{-17}$ J, current strength, $I_p = 0.3$ mA; after irradiation, resistance is recovered [418]
			Dielectric constant		
Al_2O_3	323 323 323 323	293 293 293 293	$9,2 \pm 0,1$ $9,2 \pm 0,1$ $9,2 \pm 0,1$ $9,2 \pm 0,1$	$9,0 \pm 0,1$ $8,9 \pm 0,1$ $8,4 \pm 0,1$ $8,3 + 0,1$	Measurements made on polycrystals irradiated with respective fluences of $2 \cdot 10^{21}$, $6 \cdot 10^{21}$, $2 \cdot 10^{23}$, and $5 \cdot 10^{23}\ m^{-2}$; frequency $f_{meas} = 1$ MHz [417]

12 – Continued

1	2	3	4	5	6
SiO_2	323 323 323	293 293 293	$3{,}8\pm0{,}1$ $3{,}8\pm0{,}1$ $3{,}8\pm0{,}1$	$3{,}7\pm0{,}1$ $3{,}6\pm0{,}1$ $3{,}6\pm0{,}1$	Measurements made on quartz specimens irradiated with respective fluences of $6\cdot10^{20}$, $6\cdot10^{23}$, and $5\cdot10^{23}$ m^{-2}; frequency f_{meas} =1 MHz [417]
			Dielectric loss, tan δ		
Al_2O_3	323 323 323 323	293 293 293 293	$(0{,}3\pm0{,}1)\cdot10^{-4}$ $(0{,}3\pm0{,}1)\cdot10^{-4}$ $(0{,}3\pm0{,}1)\cdot10^{-4}$ $(0{,}3\pm0{,}1)\cdot10^{-4}$	$(0{,}5\pm0{,}1)\cdot10^{-4}$ $(4{,}0\pm0{,}5)\cdot10^{-4}$ $(2{,}0\pm0{,}25)\cdot10^{-4}$ $(1{,}0\pm0{,}25)\cdot10^{-4}$	Measurements made on polycrystals irradiated with respective fluences of $2\cdot10^{21}$, $6\cdot10^{21}$, $2\cdot10^{23}$, and $5\cdot10^{23}$ m^{-2}; frequency f_{meas} = 1 MHz [417]
SiO_2	323 323 323 323	293 293 293 293	$(0{,}2\pm0{,}1)\cdot10^{-4}$ $(0{,}2\pm0{,}1)\cdot10^{-4}$ $(0{,}2\pm0{,}1)\cdot10^{-4}$ $(0{,}2\pm0{,}1)\cdot10^{-4}$	$(0{,}4\pm0{,}1)\cdot10^{-4}$ $(6\pm0{,}5)\cdot10^{-4}$ $(14\pm1)\cdot10^{-4}$ $(18\pm1)\cdot10^{-4}$	Measurements made on quartz specimens irradiated with respective fluences of $2\cdot10^{21}$, $6\cdot10^{21}$, $2\cdot10^{23}$, and $5\cdot10^{23}$ m^{-2}; frequency f_{meas} =1 MHz [417]
			Dielectric strength, V/m		
Al_2O_3	323	293	$5{,}6\cdot10^{8}$	$2{,}3\cdot10^{6}$	After γ-irradiation; absorbed dose D $D_a=10^7$ J/kg; specimen sintered; specimen composition Al_2O_3 +2% SiO_2 [417]
			Activation energy of conduction, J		
BeO	1173	773	$2{,}63\cdot10^{-19}$	$4{,}32\cdot10^{-19}$	After neutron irradiation, fluence Φ_{fn} = $2\cdot10^{24}$ m^{-2}; specimens cold-pressed; specimen composition BeO + 0.6% CaO [409]

Oxide	Φf_n, $\times 10^{24}$ m^{-2}	T_{irr}, °K	Before irradiation		After irradiation		Notes
			Color				
BeO	3,7	373	Colorless		Reddish brown		Single crystal [437]
	0,08	373	"		Light cinnamon brown		Single crystal [409]
	6,5	373	"		Almost black		[409]
MgO	2,2	673	White		Violet-black		ρ_{rel}=95,5% [415]
	2,2	673	"		Grey-black		ρ_{rel}=83% [415]
Al_2O_3	0,15	303	Colorless		Dark cinnamon brown		Sapphire [439]
			Refractive index				
			n_{p_0}	n_{q_0}	n_p	n_q	
BeO	10	373	1,719	1,733	1,698	1,718	Single crystals [438]
	15	393	1,719	1,733	1,700	1,712	
	18	393	1,719	1,733	1,694	1,708	
	28	393	1,719	1,733	1,694	1,708	
	11	713	1,719	1,733	1,709	1,722	
	21	1100	1,719	1,733	1,709	1,722	
	23	1217	1,719	1,733	1,709	1,722	
	21	1238	1,719	1,733	1,709	1,722	
SiO_2	0,5	323	1,545	1,555	1,530		Quartz; after irradiation, $n_p = n_q$ [418]
	1,0	323	1,545	1,555	1,500		
	2,0	323	1,545	1,555	1,469		
	4,0	323	1,545	1,555	1,4675		
	0,1	323	1,460		1,465		Amorphous; $n_p = n_q$ [418]
	0,5	323	1,460		1,470		
	1,0	323	1,460		1,469		
	4,0	323	1,460		1,467		

14. STORED ENERGY

Φ_{fn}, $\times 10^{24}$ m^{-2}	T_{irr}, °K	E_s, kJ/kg
BeO [408]		
0,06	373	13,4 ± 2,9
0,3	373	30,1 ± 9,2
0,4	373	33,4 ± 16,3
0,7	373	53,4
1,4	373	109—142
2,8	373	205—268
4,2	373	293—368
2,2	673	60—113
2,2	673	75,3
4,8	648	85,3 ± 50
7,5	797	60,2 ± 36,8
10	813	219 ± 18,4
MgO [408]		
0,8	343	71,8
Al_2O_3 [408]		
1,2	343	104 ± 8,4

15. INTERNAL FRICTION

Φ_{fn}, $\times 10^{24}$ m^{-2}	T_{irr}, °K	Logarithmic decrement of attenuation	
		before irradiation	after irradiation
BeO [418]			
1,24	373	0,009	0,028*1
1,61	373	0,004	0,006
1,75	373	0,003	0,125
1,96	373	0,006	0,102
2,00	373	0,004	0,090
2,09	373	0,002	0,172
1,30	373	0,002	0,003*2
1,78	373	0,003	0,087
1,82	373	0,002	0,125
2,12	373	0,003	0,120
2,17	373	0,001	0,086
2,25	373	0,002	0,139

*1 Grade AOX powder. *2 Grade UOX powder + 0.5% MgO.

Φ_{fn}, $\times 10^{24}$ m^{-2}	T_{irr}, °K	Characteristics of defects	
1	2	3	
			BeO
0.28	373	$1.0 \cdot 10^{26}$ def/m^3	Density of atomic defects (def) of lattice (vacancies for imbedded atoms), calculated from thermal neutron scattering data on the assumption that all defects are single [441]
0.58	373	$4.8 \cdot 10^{26}$ def/m^3	
1.8	373	$8.0 \cdot 10^{26}$ def/m^3	
6.8	373	$4.1 \cdot 10^{27}$ def/m^3	
0.65	373	$5 \cdot 10^{22}$ acc/m^3	Density of accumulations (acc) of defects, determined by transmission electron microscopy [409]
1.8	373	$5 \cdot 10^{22}$ acc/m^3	
3.0	373	$3 \cdot 10^{22}$ acc/m^3	
0.65	373	2 nm	Average diameter of accumulations, determined by transmission electron microscopy [409]
1.8	373	3 nm	
3.0	373	7 nm	
15	373	7 nm	
1	373	$1.5 \cdot 10^{24}$ acc/m^3	Density of accumulations of point defects, calculated from data on thermal conductivity at low temperatures on the assumption that accumulations are planar [409]
2	373	$1.5 \cdot 10^{24}$ acc/m^3	
10	373	$2.5 \cdot 10^{24}$ acc/m^3	
2	373	7-10 nm	Average diameter of accumulations as determined by transmission electron microscopy and x-ray analysis; at high irradiation temperatures, accumulations are dislocation loops of imbedded type [420]
2	373	20-25 nm	
2	1073	30-40 nm	

16 – Continued

1	2	3
3	1273	Dislocation loops of imbedded type in {0001} basis planes with average diameter 40–50 nm; dislocation loops of imbedded type in prismatic planes $\{11\bar{2}0\}$ with Burgers vector $a/3$ <1120>, with average diameter 50–60 nm; number of basis loops is far greater than number of prismatic loops
1	1273	With increasing degree of radiation up to $1.1 \cdot 10^{25}$ m^{-2}, average diameter of basis dislocation loops increases to 100 nm, diameter of prismatic dislocation loops to 100–200 nm
25	1273	With further increase in degree of irradiation, number of prismatic dislocation loops decreases, and completely disappears at a fluence of $2.5 \cdot 10^{25}$ m^{-2}; tube-shaped voids appear with a regular hexagonal faceting, with length up to 300 nm and diameter up to 100 nm; determined by transmission electron microscopy [442]
		MgO
1	373	Concentration of vacancies (or interstitial atoms) is about 0.19%; one vacancy reduces crystal volume by about 0.4v, where v is the atomic volume; one imbedded atom increases the crystal volume by about 3v; calculated from data on density and crystal lattice constants [412]
0.3	423	Density of accumulations (acc) of point defects is 10^{23} to $4 \cdot 10^{23}$ acc/m^3, average diameter of accumulations 2–5 nm, and integral area of accumulations $2 \cdot 10^6$ m^{-1}; annealing at temperatures above 1073°K increases the size of the accumulations and reduces their density to $5 \cdot 10^{22}$ acc/m^3; here accumulations are allowed as dislocation loops of the imbedded type in the {110} planes with Burgers vector in the <110> direction

1	2	3
4.3	423	With increasing neutron fluence up to $4.3 \cdot 10^{24}$ m^{-2}, size and density of accumulations increase so much that they begin to interact, resulting in the formation of a dislocation network; determined by transmission electron microscopy [443]
4.5	373	Block substructure formed as a result of interaction of dislocation loops; block size approximately the same along <100>, <110>, and <111> directions, approximately 50 nm; determined by harmonic analysis of x-ray diffraction lines [444]
		Al_2O_3
0.2	423	Average diameter of accumulations of point defects 1.5 nm
1–4	423	2-10 nm
1–5	423	5–7.5 m; dislocation loops in $\{10\bar{1}0\}$ planes with diameter 5–60 nm
1	823	Dislocation loops of imbedded type in prismatic $\{10\bar{1}0\}$ and basis $\{0001\}$ planes, with diameter 15–50 nm
4	1273	Dislocation loops with diameter 30–50 nm; pores with diameter of 5–20 nm, density 10^{20}–10^{21} pores/m^3 [408]

17. RADIATION RESISTANCE

Φ_{fn}, $\times 10^{24}$ m^{-2}	T_{irr}, °K	Specimen characteristics and nature of damage
1	2	3
		BeO
		Hot-pressed, ρ_{rel} – 97–98%, d_g – 12–15 μm
2.5	348–373	Severe microcracking along grain boundaries
5	773–973	Severe microcracking along grain boundaries

17 — Continued

1	2	3
		Sharp increase in open porosity [409]
		Cold-pressed, ρ_{rel}=96−98%, d_g=8−12 μm
5	348−373	Appreciable microcracking
9−12	773−973	Appreciable microcracking [409]
		Cold-pressed, ρ_{rel}=95−97%, d_g=2−3 μm
10	348−373	No microcracking observed
14	348−373	Severe microcracking [409]
		Cold-pressed, ρ_0=2700−2920 kg/m^3, d_g= 20−72 μm
10	373	Appreciable macrocracking
20	373	Breakdown and formation of powder [409]
		Cold-pressed, ρ_0=2700 kg/m^3, d_g=20 μm
12	923	Slight breakdown
20	1373	Slight breakdown
20−35	923	Considerable breakdown [409]
		Cold-pressed, ρ_0=2700 kg/m^3, d_g=34−60 μm
17	923	Considerable breakdown
19−38	1373	Considerable breakdown
35	923	Breakdown with formation of powder [409]
		Cold-pressed, ρ_0=2900 kg/m^3, d_g=24 μm
19−26	923	Considerable breakdown
20−33	1373	Considerable breakdown
25	923	Breakdown with formation of powder [409]
		Cold-pressed, ρ_0=2900 kg/m^3, d_g=72 μm
14−30	923	Breakdown with formation of powder

1	2	3
14–16	1373	Slight breakdown
16–36	1373	Considerable breakdown [409]
		Hot-pressed, ρ_0=2970–3000 kg/m^3
0.25	373	25% loss in compressive strength
0.6	373	75% loss in compressive strength
0.9	373	Severe microcracking, 97% loss in strength
2.0	373	Breakdown with formation of powder
2.2	673	60–75% loss of compressive strength [415]
		Cold-pressed, ρ_0=2900 kg/m^3, d_g=20 μm
2.0	373	63–85% loss of bending strength
4.0	373	95% loss of bending strength [415]
		Cold-pressed with 0.6% CaO, ρ_0=2860 kg/m^3, d_g=33–42 μm
2.0	1173	41–91% loss of compressive strength [409]
		Cast, ρ_{rel}=97%, d_g=30–40 μm
1.0	373	Beginning of microcracking
4.0	373	Breakdown with powder formation [445]
		Cold-pressed with 0.25% Al_2O_3, ρ_{rel}=93.5%, d_g=35 μm
5	373	Microcracking along grain boundaries
7	373	Severe breakdown
10	373	Complete breakdown [445]
		Cast with 0.5%, ρ_{rel}=93.5%, d_g=35 μm
5	373	Beginning of microcracking
20	373	Breakdown of specimens into individual parts [445]
		Hot-pressed, ρ_{rel}=95.5–98.5%, d_g=3–5 μm
40	373	Breakdown with powder formation [445]

17 – Continued

1	2	3
		Cast with 3% Al_2O_3, SiO_2, CaO; ρ_{rel}=96%, d_g=80–100 μm
3	373	No microcracking observed [445]
		Hot-pressed with 1% Be, ρ_0=2950 kg/m^3, d_g=9 μm
1	373	Microcracking, no loss of strength
1.9	373	29% loss of strength [428]
		Hot-pressed with 2% Be, ρ_0=2960 kg/m^3, d_g=12 μm
1	373	Microcracking, no loss of strength
2	373	71% loss of strength [428]
		Cold-pressed with 0.2% Al_2O_3 and 0.1% SiO_2, ρ_0=2920 kg/m^3, d_g=10 μm
2	373	Start of microcracking
2.9	373	No loss of strength
3.6	373	17–23% loss of strength [428]
		Cold-pressed, ρ_{rel}=97%
0.3	373	d_g=50 μm; start of microcracking
0.8	373	d_g=25 μm; start of microcracking
2.2	373	d_g=16 μm; start of microcracking [431]
		Cold-pressed, d_g=5 μm
90	1273	ρ_0=2900 kg/m^3; no loss of strength
90	1273	ρ_0=2600 kg/m^3; 20–40% loss of strength [411]
		MgO [408]
		Single crystals and polycrystals
1–10	373	Embrittlement, no signs of breakdown
		Al_2O_3
		Single crystals and polycrystals
5	373	No signs of breakdown

1	2	3
11	523	No signs of breakdown
11	748	No signs of breakdown
		Extruded, porosity 31%; rolled, porosity 25%
1.4	973	Breakdown [408]
		Pressed, porosity 8%
5.6	973	No cracking
10	973	Cracking [408]
		Sintered, ρ_{rel}=99.8%
23	873	Microcracking
47	873	Macrocracking [414]

CHEMICAL AND CATALYTIC PROPERTIES

1. CHEMICAL PROPERTIES

Reagent	Temp., °C	Character of interaction
1	2	3
	H_2O [1]	
Aluminum phosphide	–	Reacts, with evolution of phosphine
Argon, krypton, xenon	–	Reacts with compression to 15 MPa over supercooled water, with formation of crystal hydrates $x \cdot 6H_2O$
Bromine	20	Reacts: $H_2O + Br_2 = HOBr + HBr$
Bismuth chloride	100	Reacts, forming bismuth oxychloride BiOCl
Alkali metal hydrides	20	Reacts: $MeH + H_2O = MeOH + H_2$
Calcium phosphide		Reacts with evolution of phosphine
Oxygen in atomic state	–	Reacts, forming peroxide
Methane	760–800	Reacts on catalyst (Ni or Co): $2H_2O + CH_4 = CO_2 + 4H_2$
Methane	1200–1400	Reacts: $H_2O + CH_4 = CO + 3H_2$
Acidic oxides	–	React, dissolving to form acids
Basic oxides		React, dissolving to form alkalis
Salts	–	React, dissolving to form hydrates with various numbers of water molecules
Antimony trioxide	20	Reacts, forming antimony oxychloride SbOCl
Carbon monoxide	450	Reacts on iron catalyst: $H_2O + CO = CO_2 + H_2$

1 — Continued

1	2	3
Phosphorus	–	Reacts upon heating under pressure in presence of catalyst: $6H_2O + 2P = 2HPO_3 + 5H_2$
Fluorine	20	Reacts, with evolution of atomic oxygen and molecular O_2, O_3, H_2O, and F_2O
Chlorine	20	Reacts: $H_2O + Cl_2 = HOCl + HCl$
Chlorine	100	Reacts: $2H_2O + 2Cl_2 = 4HCl + O_2$
	H_2O_2 [1; 448, Vol. 1, 450, Vol. 1]	
Metal hydroxides	–	Reacts, forming peroxides: $H_2O + Me(OH)_2 = MeO_2 + 2H_2O$
Alkali-metal hydroxides	–	Promote reductive decomposition: $H_2O_2 = 2H + O_2$
Ferrous sulfate	–	React, oxidized to ferrous sulfate
Gold salts (solutions)	–	React, with liberation of metallic gold
Iodine + ammonia (10% solution)	–	Reacts, forming ammonium iodide
Potassium permanganate	–	Reacts in acidic medium: $5H_2O_2 + 2KMnO_4 + 3H_2SO_4 = K_2SO_4 + 2MnSO_4 + 5O_2 + 8H_2O$
Potassium sulfate	–	Dissolves, 96.2 g per 100 g H_2O_2
Potassium chloride		Dissolves, 63.3 g per 100 g H_2O_2
Acids (solutions)	–	React, causing oxidative decomposition: $H_2O_2 = H_2O + 1/2O_2$
Nitrous acid	–	Reacts; oxidized to nitric acid
Hydroiodic acid	–	Reacts, forming free iodine
Arsenous acid	–	Reacts; oxidized to arsenic acid
Sulfurous acid	–	Reacts; oxidized to sulfuric acid

1 — Continued

1	2	3
Chlorosulfonic acid	–	Reacts forming peroxymonosulfuric acid: $H_2O_2 + HSO_3Cl = H_2SO_5 + HCl$ or persulfuric acid $H_2S_2O_8$
Lead sulfide	–	Reacts; oxidized to lead sulfate
Sodium nitrate	–	Dissolves, 30.9 g per 100 g H_2O_2
Sodium sulfate	0	Dissolves, 26.7 g per 100 g H_2O_2
Sodium chloride	0	Dissolves, 20.5 g per 100 g H_2O_2
Ozone	–	Reacts: $H_2O_2 + O_3 = H_2O + 2O_2$
Mercuric oxide	–	Reacts; reduced to metallic mercury
Silver(I) oxide	–	Reacts, causing reductive decomposition: $H_2O_2 + Ag_2O = 2Ag + H_2O + O_2$
Silver(II) oxide	–	Reacts; reduced to metallic silver
Lead sulfide	–	Reacts; oxidized to lead sulfate
	D_2O [1]	
Aluminum sulfide	–	Reacts forming deuterium sulfide
Iodine + phosphorus	–	Reacts: $4D_2O + P + 5/2I_2 = 5DI + D_3PO_4$
Silicon chloride	–	Reacts, forming deuterium chloride
Magnesium chloride	600	Reacts, forming deuterium chloride
Magnesium or sodium	–	Reacts, with evolution of deuterium
Magnesium nitride	400	Reacts, forming deuteroammonia
Sulfur trioxide	–	Reacts, forming deuterosulfuric acid
Phosphorus tribromide	–	Reacts: $3D_2O + PBr_3 = 3DBr + D_3PO_3$

1	2	3
Phosphoric anhydride	–	Reacts, forming deutero-orthophosphoric acid
Li_2O [1; 446, Vol. 1; 448, Vol. 1; 449; 450, Vol. 2]		
Aluminum, silicon, or magnesium	900–1100	Reacts in presence of calcium oxide: $3Li_2O + CaO + 2Al = 6Li + CaO \cdot Al_2O_3$ and $2Li_2O + 2CaO + Si = 4Li + 2CaO \cdot SiO_2$
Air	–	Reacts, forming lithium carbonate
Water	–	Reacts, forming hydroxide
Hydrogen	–	Reacts upon heating to form LiOH and LiH
Carbon or carbon monoxide	–	Reacts; weakly reducing
Metal oxides	–	React, to form eutectic melts
BeO [446, Vol. 1; 448, Vol. 1; 449]		
Boron	–	Reacts at high temperature, forming boride
Bromine or iodine	–	No reaction
Water	–	No reaction
Helium (liquid)	1000	No reaction
Alkali-metal hydroxides (solutions)	–	No reaction
Alkali-metal hydroxides (melts)	–	React, forming beryllates Me_2BeO_2
Potassium bisulfate	–	Reacts: $BeO + 2KHSO_4 = BeSO_4 + K_2SO_4 + H_2O$
Acids (solutions)	–	No reaction
Hydrofluoric acid	–	Reacts, forming fluoride
Silicon	–	Reacts at high temperature, forming silicide
Carbon	1900	Reacts, forming carbide Be_2C
Carbon tetrachloride	–	Reacts, forming anhydrous chloride
Fluorine or fluorides	20	React
Chlorine	800	Reacts in presence of carbon: $BeO + Cl_2 + C = BeCl_2 + CO$

1 – Continued

1	2	3
Be_2O_3 [1; 448, Vol. 1; 450, Vol. 2; 451]		
Nitrogen	1700	Reacts in presence of carbon, forming boron nitride
Ammonia	1200	Reacts in presence of carbon, ammonium chloride, calcium cyanamide, or sodium amide, forming boron nitride
Ammonium chloride	300	Reacts, forming boron nitride
Sulfur trioxide	–	Reacts upon prolonged heating to form compound $SO_3 \cdot B_2O_3$
Water	20	Reacts, dissolving to form boric acid
Potassium, calcium, or sodium cyanide	2000	Reacts, forming boron nitride
Potassium tetrafluoroborate	600	Reacts: $2B_2O_3 + KBF_4 = BF_3 + KF \cdot B_4O_6$
Calcium chloride or calcium bromide	–	Reacts, forming calcium chloroborate or bromoborate
Hydrofluoric acid or metal fluorides	–	React, forming BF_3
Sodium chloride, bromide, or iodide	250	Reacts: $2B_2O_3 + 2NaI + 1/2O_2 = Na_2B_4O_7 + I_2$
Sodium tetrafluoroborate + sulfuric acid	–	Reacts: $B_2O_3 + 6NaBF_4 + 6H_2SO_4 = 8BF_3 + 6NaHSO_4 + 3H_2O$
Oxides of alkali and alkaline-earth metals	–	React, forming corresponding borates
Sulfur or hydrogen sulfide	1000	No reaction
Carbon disulfide	–	Reacts in presence of carbon forming boron sulfide
Carbon	2000	Reacts, forming carbide B_4C
Fluorine	–	Reacts, forming BF_3
Chlorine, bromine, iodine	1000	No reaction

1	2	3
Chlorine, bromine	600	Reacts in presence of carbon, forming BCl_3 and BBr_3
		CO [1; 448, Vol. 1; 450, Vol. 1]
Ammonia	500	Reacts in presence of ThO_2, forming hydrogen cyanide
Bromine trifluoride	–	Reacts, forming carbonyl bromofluoride
Water	20	Does not react; at elevated temperatures and pressures, reacts to form formic acid
Water vapor	Up to 830	Reacts: $H_2O + CO = CO_2 + H_2$
Hydrogen	250	Reacts in presence of nickel: $CO + 3H_2 = CH_4 + H_2O$
Air or oxygen	700	Reacts, oxidizing to carbon dioxide
Alkali metal hydroxides (solutions)	120	React at pressure of 50 kPa, forming formates $Me(CO_2H)$
Iron, nickel, or cobalt	–	Reacts upon heating to form metal carbonyls
Iodine pentafluoride	–	Reacts: $3CO + IF_5 = COIF + 2COF_2$
Iodine pentoxide	–	Reacts: $5CO + I_2O_5 = 5CO_2 + I_2$
Potassium	80	Reacts, forming potassium salt of hexahydroxybenzene $K_6(CO)_6$
Potassium dichromate	–	Reacts in presence of mercurcy salts, oxidizing
Potassium permanganate	–	Reacts in presence of finely divided silver, oxidizing
Calcium carbide	–	Reacts upon heating to form carbonate
Acids	–	No reaction
Alkali metals dissolved in liquid ammonia	–	React, forming alkali metal carbonyls
Cuprous chloride	–	Reacts in hydrochloric acid or ammonia solution, forming $CuCl \cdot CO \cdot 2H_2O$

1	2	3
Metal oxides	–	Reacts upon heating; reduced to metals
Palladium chloride (solution)	20	Reacts: $CO + PdCl_2 + H_2O = CO_2 + Pd + 2HCl$
Sulfur	–	Reacts upon heating to form carbonyl sulfide COS
Selenium	780	Reacts, forming carbonyl selenide $COSe$
Salts of gold or platinum	20	React; reduced to metal
Salts of metals	–	React, forming complex compounds
Fluorine	–	Reacts, forming carbon tetrafluoride CF_4 or carbonyl fluoride COF_2
Chlorine	20	Reacts in direct sunlight, forming phosgene $COCl_2$
Chlorine monofluoride	–	Reacts, forming carbonyl chlorofluoride $COClF$
Alkali-metal methylates	–	React upon heating to form acetates
	CO_2 [1; 448, Vol. 1; 450, Vol. 1]	
Ammonia	20	Reacts, forming ammonium carbamate $CO(NH_2)(ONH_4)$
Water	20	Reacts, forming carbonic acid
Alkali-metal hydroxides	20	Reacts, forming carbonate salts
Potassium, magnesium, or zinc	With heating	Reacts, forming carbonate salts
Carbon	800	Reacts, forming carbon monoxide
	N_2O [1; 447, Vol. 2; 448, Vol. 1; 450, Vol. 1]	
Ammonia	–	Reacts: $3N_2O + 2NH_3 = 4N_2 + 3H_2O$
Water	20	Dissolves 0.629 volumes of gas per volume of water; no chemical reaction

1	2	3
Hydrogen	–	Reacts: $N_2O + H_2 = N_2 + H_2O$, giving flash if in contact with flame
Oxygen	–	No reaction
Sodium amide	–	Reacts, forming sodium azide
	NO [1; 448, Vol. 1; 450, Vol. 1]	
Bromine, chlorine, or fluorine	–	Reacts, forming nitrosyl halides NOI
Ferrous sulfate	–	Reacts, forming nitrosulfate
Water	–	Dissolves very slightly
Nitrogen dioxide	–	Reacts, forming N_2O_3
Hydrogen	–	Reacts: $2NO + 2H_2 = N_2 + 2H_2O$
Sodium	150–170	Reacts, forming sodium nitrosyl NaNO
Sodium chloride (saturated solution)	–	No reaction
Oxidizers (CrO_3, $HMnO_4$, HOCl, etc.)	–	Reacts, oxidizing to HNO_3
Stannous chloride	–	Reacts in hydrochloric acid solution, forming hydroxylamine and ammonia
Sulfur dioxide	–	Reacts, reducing to nitrous oxide in presence of water; under high pressure, reacts: $2NO + 2SO_2 = 2SO_3 + N_2$
Hydrogen sulfide	–	Reacts: $2NO + 2H_2S = N_2 + 2S + 2H_2O$
Tetrafluorohydrazine	–	Reacts, forming $F_2N \cdot NO$
Bivalent chromium salts	–	React, reducing to ammonia in neutral solution and to hydroxylamine in acidic solution
	N_2O_3 [1; 448, Vol. 1; 450, Vol. 1]	
Ammonia	–	Reacts with strong cooling, to form nitrosamide NH_2NO
Water	–	Reacts, forming nitrous acid, which rapidly decomposes to form nitric acid

1 – Continued

1	2	3
Alkali-metal hydroxides	–	React, forming nitrites
Sulfuric acid	–	Reacts, forming nitrosonium bisulfate $NOHSO_4$
Tetrafluoroboric acid	–	Reacts, forming nitrosyl tetrafluoroborate $NOBF_4$
Cupric oxide	–	Reacts when incandescent, with evolution of nitrogen
	NO_2 [1; 448, Vol. 1; 450, Vol. 1]	
Ammonia (liquid)	–	Reacts explosively
Water	20	Reacts: $2NO_2 + H_2O = HNO_3 + HNO_2$
Water in presence of excess oxygen	–	React: $2NO_2 + 2MeOH = 4HNO_3$
Alkali-metal hydroxides (solutions)	20	Reacts: $2NO_2 + 2MeOH = MeNO_3 + MeNO_2 + H_2O$
Fluorine	–	Reacts, forming gaseous nitronyl fluoride NO_2F
	N_2O_5 [1; 448, Vol. 1; 450, Vol. 1]	
Water	20	Reacts, forming nitric acid
Hydrogen peroxide (100%)	80	Reacts, forming an explosive substance, pernitric acid
	OF_2 [1; 448, Vol. 1; 450, Vol. 1; 452]	
Water	–	Does not react; solubility 7 volumes per 100 volumes H_2O
Water vapor	–	Reacts explosively: $OF_2 + H_2O = 2HF + O_2$
Alkali-metal hydroxides (solutions)	–	React, decomposing
	Na_2O [446, Vol. 1; 450, Vol. 2]	
Ammonia (gaseous)	20	Reacts: $Na_2O + NH_3 = NaNH_2 + NaOH$
Water	20	Reacts: $Na_2O + H_2O = 2NaOH$
Hydrogen	180	Reacts: $Na_2O + H_2 = NaH + NaOH$
Oxygen	400	Reacts, forming Na_2O_2
Iodine	–	Reacts upon heating to form NaI

1	2	3
Carbon dioxide	20	Reacts: $Na_2O + CO_2 = Na_2CO_3$
Sulfur	–	Reacts upon heating to form Na_2S
Fluorine or chlorine	–	Reacts upon heating to form NaF or NaCl
Na_2O_2 [1; 446, Vol. 1; 448, Vol. 1; 450, Vol. 2]		
Water	–	Reacts: $Na_2O_2 + H_2O = 2NaOH + H_2O_2$
Bismuth oxide	600	Reacts, forming sodium bismuthate $NaBiO_3$
Iodine	–	Reacts: $3Na_2O_2 + 3I_2 = 5NaI + NaIO_3 + O_3$
Ethanol	0	Reacts: $Na_2O_2 + C_2H_5OH = C_2H_5ONa + NaOOH$
Carbon	–	Reacts: $3Na_2O_2 + 2C = 2Na_2CO_3 + Na$
Carbon dioxide	–	Reacts: $Na_2O_2 + CO_2 = Na_2CO_3 + 1/2O_2$
Carbon monoxide	–	Reacts: $Na_2O_2 + CO = Na_2CO_3$
Sulfur	–	Reacts
Ether, acetic acid, or nitrobenzene	–	Reacts, with flash
MgO [1; 446, Vol. 1]		
Nitrogen + carbon disulfide vapor	800	React, forming sulfide MgS
Air	–	No reaction
Water	–	Reacts slowly, forming hydroxide
Atmospheric moisture and carbon dioxide	–	React, forming basic magnesium carbonate
Acids	–	React, dissolving
Magnesium chloride (30% solution)	–	Reacts, forming oxychloride Mg_2OCl_2
Trivalent metal oxides	–	React upon fusion, forming $Mg(M_2O_4)$
Metals (potassium, sodium), silicon, ferrosilicon, calcium carbide	–	React at high temperature, reducing

1 – Continued

1	2	3
Carbon	–	Reacts, forming carbide
Chlorine + carbon monoxide	750	React, forming magnesium chloride
Al_2O_3 [1; 446, Vol. 1; 448, Vol. 1; 450, Vol. 2]		
Nitrogen	1700	Reacts in presence of carbon, forming aluminum nitride
Boron trifluoride	450	Reacts
Hydrogen fluoride (gaseous)	–	Reacts upon heating to form AlF_3
Water		No reaction
Hydrogen, carbon monoxide, hydrocarbons	Up to 1800	No reaction
Alkali-metal hydroxides and carbonates	–	React upon fusion, forming aluminates
Potassium bisulfate or pyrosulfate	–	React upon fusion, forming sulfate
Acids and alkalis	1800	No reaction
Silicon	1800	Reacts upon heating in vacuo: $Al_2O_3 + 2Si = Al_2O + 2SiO$
Oxides Me_2O and MeO	100	Reacts upon fusion, forming anhydrous aluminates $MeAlO_2$ and $Me(AlO_2)_2$
Sulfur, phosphorus, arsenic, and their compounds	1000	No reaction
Carbon	2000	Reacts, forming Al_4C_3
SiO [1; 450, Vol. 1]		
Aluminum sulfide	1100	Reacts, forming sulfide
Water vapor	500	Reacts: $SiO + H_2O = SiO_2 + H_2$
Air	20	Reacts slowly
Air	500	Reacts, oxidizing to SiO_2
Alkali-metal hydroxides (solutions)	–	React: $SiO + 2MeOH = Me_2SiO_3 + H_2$
Oxides	–	Reacts: $SiO + MgO \cdot CaO + CaO = Ca_2SiO_4 + Mg$

1	2	3
Sulfur trioxide	800	Reacts
Carbon dioxide	400	Reacts, oxidizing to SiO_2
Chlorine	800	Reacts: $2SiO + 2Cl_2 = SiCl_4 + SiO_2$
SiO_2 [1; 448, Vol. 1; 450, Vol. 1]		
Hydrogen	1100	Reacts, reducing to silicon monoxide
Water	–	No reaction
Alkali-metal hydroxides and carbonates (melts)	–	React, forming silicates
Alkali-metal hydroxides and carbonates (solutions)	–	React upon boiling, to form soluble alkali metal silicates
Nitric, hydrochloric, or sulfuric acid	–	No reaction
Hydrofluoric acid	20	Reacts, forming silicon tetrafluoride or fluosilicic acid
Silicon or ferrosilicon	1130–1370	Reacts, reducing to silicon monoxide
Silicon in nitrogen medium	1450	Reacts, forming oxynitride Si_2N_2O
Carbon	2000	Reacts, forming carbide
P_2O_3 (P_4O_6) [1; 448, Vol. 1; 450, Vol. 1]		
Water	20	Reacts, forming phosphorous acid
Hydrogen chloride (gaseous)	–	Reacts: $P_4O_6 + 6HCl = 2H_3PO_3 + 2PCl_3$
Air	With heating	Reacts, forming phosphorus pentoxide
Chlorine, bromine, or sulfur	150	Reacts
P_2O_5 (P_4O_{10}) [1; 448, Vol. 1; 450, Vol. 1]		
Ammonium fluoride	135	Reacts, forming salts of hexafluorophosphoric, monofluorophosphoric, and difluorophosphoric acids

1 – Continued

1	2	3
Water	–	Reacts: $P_2O_5 + H_2O = 2HPO_3$, $P_2O_5 + 2H_2O = H_4P_2O_7$, and $P_2O_5 + 3H_2O = 2H_3PO_4$
Nitric acid	–	Reacts: $P_2O_5 + 2HNO_3 = N_2O_5 + 2HPO_3$
Perchloric acid	–	Reacts: $P_2O_5 + 2HClO_4 = Cl_2O_7 + 2HPO_3$
Carbon	–	Reacts at high temperature: $P_2O_5 + 5C = 5CO + 2P$
Phosphorus pentachloride	–	Reacts, forming phosphorus oxychloride $POCl_3$
SO_2 [1; 448, Vol. 1; 450, Vol. 1]		
Water	–	Reacts
Atmospheric oxygen	20	No reaction
Water	–	Reacts, forming sulfuric acid
Ferric oxide	–	Reacts: $3SO_2 + Fe_2O_3 = Fe_2(SO_3)_3$ and $Fe_2(SO_3)_3 = FeSO_3 + FeS_2O_6$
Potassium thiosulfate	–	Reacts, forming potassium trithionate
Oxygen	400	Reacts, oxidizing to SO_3 in presence of catalysts
Nitric acid	–	Reacts, forming nitrosyl bisulfate $(NO)(HSO_4)$
Manganese dioxide	–	Reacts: $2SO_2 + MnO_2 = MnS_2O_6$ and $SO_2 + MnO_2 = MnSO_4$
Sodium bicarbonate	–	Reacts, forming sodium pyrosulfite $Na_2S_2O_5$
Sulfur	–	Reacts in aqueous solutions: $SO_2 + S + H_2O = H_2S_2O_3$
Hydrogen sulfide	20	Reacts: $3SO_2 + H_2S = H_2S_4O_6$ and $SO_2 + 2H_2S = 2H_2O + 3S$
Hydrogen sulfide + sulfur	–	React: $3SO_2 + H_2S + S = H_2S_5O_6$
Hydrogen sulfide + caustic solution	–	React: $4SO_2 + 2H_2S + 6NaOH = 3NaS_2O_3 + 5H_2O$
Carbon monoxide	500	Reacts in presence of catalyst (bauxite), with liberation of sulfur

1	2	3
Chlorine	–	Reacts, forming sulfuryl chloride SO_2Cl_2
Fluorine	650	Reacts, forming sulfur hexafluoride
Alkali-metal fluorides	–	React with liquid SO_2, forming fluorosulfonates $MeSO_2F$
SO_3 [1; 448, Vol. 1; 450, Vol. 1]		
Water	–	Reacts, forming sulfuric acid
Hydrogen chloride (dry)	–	Reacts, forming chlorosulfonic acid HSO_3Cl
Potassium bifluoride + sulfuric acid	–	Reacts, forming fluosulfonic acid
Sulfur dichloride	–	Reacts, forming thionyl chloride $SOCl_2$
Perchloric acid	–	Reacts, forming explosive compound $(ClO_3) \cdot (HS_2O_7)$
Chloric acid	–	Reacts, forming explosive compound $(ClO_2)_2S_3O_{10}$
Hydrofluoric acid	35	Reacts, forming fluosulfonic acid
Carbon tetrachloride	–	Reacts, forming pyrosulfuryl chloride $S_2O_5Cl_2$
Cl_2O [1; 448, Vol. 1; 450, Vol. 1]		
Water	–	Reacts: $Cl_2O + H_2O = 2HOCl$
Carbon tetrachloride	–	Reacts, dissolving readily
Organic substances	–	Explodes upon contact
ClO_2 [1; 448, Vol. 1; 450; Vol. 1]		
Water	4	Readily soluble
Alkaline-earth metal hydroxides + hydrogen peroxide	–	Reacts: $2ClO_2 + Ba(OH)_2 \cdot 8H_2O + H_2O_2 = Ba(ClO_2)_2 + 10H_2O + O_2$ $10H_2O + O_2$
Ozone	–	Reacts: $2ClO_2 + 2O_3 = Cl_2O_6 + 2O_2$
Lead monoxide + sodium hydroxide	–	React: $2ClO_2 + PbO + 2NaOH = PbO_2 + 2NaClO_2 + H_2O$
Fluorine	50	Reacts, forming fluorine chloride oxide $FClO_2$

1 – Continued

1	2	3
	Cl_2O_7 [1; 448, Vol. 1; 450, Vol. 1]	
Water	20	Reacts, forming perchloric acid
Iodine	–	Reacts explosively upon contact
Sulfur or phosphorus	20	No reaction
Carbon tetrachloride	–	Miscible in all proportions
	K_2O [446, Vol. 1; 450, Vol. 2]	
Ammonia (liquid)	–	Reacts: $K_2O + NH_3 = KOH + KNH_2$
Water	–	Dissolves in water, forming hydroxide; absorbs water vigorously
Hydrogen	250	Reacts: $K_2O + H_2 = KOH + KH$
Iodine	–	Reacts upon heating, to form KI
Sulfur	–	Reacts upon heating, to form K_2S
Carbon monoxide	–	Is absorbed vigorously and reacts, forming K_2CO_3
Fluorine or chlorine	–	Reacts upon heating, to form KF and KCl
	K_2O_2 [446; Vol. 1; 450, Vol. 2]	
Potassium amide	–	Reacts, forming explosive mixture
Water	–	Reacts: $K_2O_2 + H_2O = 2KOH + 1/2O_2$
	KO_2 [446, Vol. 1; 450, Vol. 2]	
Nitric oxide	–	Reacts: $2KO_2 + 3NO = KNO_3 + KNO_2 + NO_2$
Water	–	Reacts: $2KO_2 + 2H_2O = 2KOH + H_2O_2 + O_2$
Sulfuric acid (dilute)	–	Reacts: $2KO_2 + H_2SO_4 = K_2SO_4 + H_2O_2 + O_2$
Carbon	–	Reacts: $2KO_2 + C = K_2CO_3 + 1/2O_2$
Carbon monoxide	–	Reacts: $2KO_2 + CO = K_2CO_3 + O_2$

1	2	3
Carbon dioxide	–	Reacts: $2KO_2 + CO_2 = K_2CO_3 + 3/2O_2$
Carbon dioxide + sodium peroxide	–	React: $2KO_2 + 2CO_2 + Na_2O_2 = Na_2CO_3 + K_2CO_3 + 2O_2$
Organic compounds	–	React (are oxidized)
	CaO [1; 446, Vol. 1]	
Bromine	–	Reacts, forming bromide
Water	–	Reacts vigorously, forming calcium hydroxide
Oxyhydrogen gas	–	In oxyhydrogen flame, CaO emits a strong light (limelight)
Sulfur	–	Reacts, forming sulfide
Carbon	–	Reacts, forming carbide
Carbon dioxide	–	Reacts, forming calcium carbonate
Phosphorus	–	Reacts upon heating
Glycerol	100	Reacts: $CaO + C_3H_5O_3H_3 = CaC_3H_5O_3H + H_2O$
Metals (Na, K, Be, Mg, Al), silicon, ferrosilicon	–	React at high temperature, reducing CaO to metallic calcium
Oxides MgO, Al_2O_3, B_2O_3, Bi_2O_3, SiO_2, SnO_2, TiO_2, ZrO_2, V_2O_5, MoO_3, WO_3, MnO, Fe_2O_3, Cr_2O_3	–	React upon heating, to form corresponding salts of the type $CaSiO_3$, $CaZrO_3$, $Ca(Al_2O_4)$, $Ca(Cr_2O_4)$
Methanol	80	Reacts: $CaO + 2CH_3OH = Ca(OCH_3)_2 + H_2O$
	CaO_2 [1; 446, Vol. 1]	
Water	–	Difficultly soluble
	CaO_4 [1; p. 337; 446, Vol. 1]	
Water	–	Difficultly soluble
Acids (dilute)	–	React, with evolution of oxygen
	Sc_2O_3 [1; 446, Vol. 1]	
Boron	1900	Reacts, forming borides

1 – Continued

1	2	3
Bromine	1200	Reacts in presence of carbon, forming bromide
Acids (dilute)	–	Dissolves slowly
Sulfuric acid (concentrated)	–	Reacts upon heating: $Sc_2O_3 + 3H_2SO_4 = Sc_2(SO_4)_3 + 3H_2O$
Carbon disulfide	–	Reacts, forming sulfide
Carbon	1800	Reacts, forming carbide
Carbon + nitrogen	1300	React, forming nitride
Chlorine	1000–1300	Reacts in presence of carbon, forming chloride
TiO [446, Vol. 2; 449]		
Water	–	No reaction
Air	800	Reacts, forming TiO_2
Sulfuric or hydrochloric acid (dilute)	–	Reacts: $2TiO + 3H_2SO_4 = Ti_2(SO_4)_3 + 2H_2O + H_2$; $2TiO + 6HCl = 2TiCl_3 + 2H_2O + H_2$
Ti_2O_3 [446, Vol. 2; 449]		
Air	1000	Reacts, oxidizing to TiO_2
Nitric acid	–	Reacts upon boiling, to form TiO_2
Sulfuric acid	–	Reacts, forming trivalent titanium sulfates
Dilute acids and caustic solutions	–	No reaction
TiO_2 [1, pp. 337–338; 446, Vol. 2; 449]		
Nitrogen + carbon	1200	React, forming nitride
Water	–	No reaction
Air	–	No reaction
Alkali-metal hydroxides and carbonates (melts)	–	Reactions: $TiO_2 + 2NaOH = Na_2TiO_3 + H_2O$; $TiO_2 + 4NaOH = Na_4TiO_4 + 2H_2O$; $TiO + Na_2CO_3 = Na_2TiO_3 + CO_2$; $TiO_2 + 2Na_2CO_3 = Na_4TiO_4 + 2CO_2$
Alkali-metal bisulfates (melts)	–	React, forming titanium sulfates or oxysulfates
Potassium pyrosulfate	–	Reacts in melt, forming titanyl sulfate
Calcium or aluminum	750	Reacts, reducing to metallic titanium

1	2	3
Calcium hydride	–	Reacts, forming titanium hydride
Sulfuric acid (concentrated)	–	Reacts upon heating: $TiO_2 + 2H_2SO_4 = Ti(SO_4)_2 + 2H_2O$
Magnesium	–	Reacts, reducing to lower oxides
Sodium	900–950	Reacts, reducing to lower oxides
Metal oxides (ZnO, NiO)	–	React: $TiO_2 + MeO = MeTiO_3$
Alloy of 30% Na and 70% Ca	900–950	Reacts, reducing to metallic titanium
Carbon	1100–1150	Reacts in vacuo, forming carbide
Carbon monoxide	800	Reacts, reducing to Ti_2O_3, or at higher temperatures to TiO and TiC
Phosgene	–	Reacts, forming chloride
Carbon tetrachloride (vapor)	–	Reacts: $TiO_2 + CCl_4 = TiCl_4 + CO_2$
Chlorine	800–1000	Reacts, forming chloride
Chlorine	700–800	Reacts in presence of carbon, forming chloride
	VO [1; 446, Vol. 2]	
Water	–	No reaction
Acids (dilute)	–	React (dissolve), forming corresponding bivalent vanadium salts
	V_2O_3 [1; 446, Vol. 2; 450, Vol. 1]	
Bromine (vapor)	–	Reacts upon heating, to form vanadyl tribromide
Hydrogen	2500	Reacts under pressure of 500 kPa, reducing to vanadium
Air	–	Reacts slowly, oxidizing to VO_2 at room temperature, and to V_2O_5 upon heating
Calcium	900–950	Reacts, reducing to metallic vanadium
Acids	–	React, forming corresponding salts

1 – Continued

1	2	3
Hydrofluoric acids	–	Reacts, forming fluoride
Selenium + hydrogen	500–600	React, forming vanadium selenide
Thionyl chloride	200	Reacts, forming chloride
Hydrogen sulfide	700	Reacts, forming sulfide
Sulfur chloride	300	Reacts, forming chloride
Carbon	1200	Reacts, forming carbide
Chlorine	500–600	Reacts, forming vanadium oxytrichloride
VO_2 [446, Vol. 1; 447, Vol. 3; 450, Vol. 1]		
Hydrogen	530–600	Reacts, reducing to V_2O_3
Air	–	Reacts upon heating; oxidizes to V_2O_5
Alkali metal hydroxides	–	React, forming vanadates
Nitric acid	–	Reacts upon heating (oxidizes)
Hydrochloric acid	–	Reacts, forming vanadyl chloride $VOCl_2$
Orthophosphoric acid	–	Reacts, forming blue-colored compounds $VO_2 \cdot 2H_3PO_4$ and $2VO_2 \cdot 6H_3PO_4 \cdot H_2O$
Alkaline-earth metal oxides	–	React in melt, forming vanadates $MeVO_3$, $MeVO_4$, and Me_3VO_5
V_2O_5 [1; 446, Vol. 2; 447, Vol. 3; 450, Vol. 1]		
Water	–	Reacts weakly
Hydrogen	600	Reacts, reducing to V_2O_3
Hydrogen	1700	Reacts, reducing to VO
Hydrogen peroxide	–	Reacts in concentrated caustic medium, forming blue-violet ions VO_8^{3-}, in near-neutral solutions yellow ions VO_6^{3-}, in acidic solutions the red peroxide cation VO_3^{3+}, and with very high acidity the blue cation VO^{2+}
Hydrogen chloride (dry)	–	Reacts in presence of phosphorus pentoxide, forming vanadium oxytrichloride

1	2	3
Alkali-metal hydroxides (solutions)	–	React at room temperature, forming ortho- and pyro-vanadates (Me_3VO_4 and $Me_4V_2O_7$) and upon heating, trimetavanadates ($Me_3V_3O_9$)
Alkali-metal carbonates	–	React, in melt, forming orthovanadates
Sulfuric acid (concentrated)	280	Reacts, reducing to salts of quadrivalent vanadium, or forming oxysulfate
Hydrochloric acid (concentrated)	–	Reacts upon heating: $V_2O_5 + 6HCl = 2VOCl_2 + Cl_2 + 3H_2O$
Oxalic acid	–	Reacts, forming vanadyl oxalate
Silicon	–	Reacts upon heating to form silicide
Sulfur chloride	300	Reacts, forming chloride
Sulfur dioxide	–	Reacts upon heating: $V_2O_5 + SO_2 = 2VO_2 + SO_3$
Thionyl chloride	73	Reacts, forming vanadium oxytrichloride
Carbon + bromine vapor	–	Reacts upon heating, forming vanadium tribromide
Chlorine	650	Reacts: $V_2O_5 + 3Cl_2 = 2VOCl_3 + 3/2O_2$
Zinc	–	Reacts in acidic medium: $V_2O_5 + Zn + 6HCl = 2VOCl_2 + ZnCl_2 + 3H_2O$
	CrO [1; 446, Vol. 2]	
Hydrogen	–	Reacts upon heating; reduces to metallic chromium
Air	100	Reacts, oxidizing to Cr_2O_3
Nitric or sulfuric acid	–	No reaction
Hydrochloric acid (dilute)	–	Reacts, forming chromium dichloride and trichloride
Carbon dioxide	1000	Reacts, forming Cr_2O_3
	Cr_2O_3 [1; 446, Vol. 2; 447, Vol. 3; 450, Vol. 1]	
Aluminum, silicon, calcium, magnesium, sodium, or potassium	1600	Reacts at high temperature, reducing to metal
Hydrogen	1600	Reacts, reducing to metallic chromium

1 – Continued

1	2	3
Alkali-metal hydroxides	–	React in melt: Cr_2O_3 ++ $2MeOH = 2MeCrO_2 + H_2O$
Potassium hydroxide + potassium chorate		Reacts in melt: $Cr_2O_3 + 4KOH + KClO_3 = 2K_2CrO_4 + KCl + 2H_2O$
Potassium pyrosulfate or potassium bisulfate (melt)	–	Reacts in melt, forming sulfate
Potassium permanganate	–	Reacts in alkaline medium: $Cr_2O_3 + 2KMnO_4 + 2KOH = 2K_2CrO_4 + 2MnO_2 + H_2O$
Lithium or sodium carbonate	–	Reacts upon calcining to form chromates
Sodium bromate	–	Reacts: $5Cr_2O_3 + 6NaBrO_3 + 2H_2O = 3Na_2Cr_2O_7 + 2H_2Cr_2O_7 + 3Br_2$
Sodium carbonate + nitrate (melt)	–	Reacts in melt: $Cr_2O_3 + 3NaNO_3 + 2Na_2CO_3 = 2Na_2CrO_4 + 3NaNO_2 + 2CO_2$
Bivalent metal oxides	–	React in melt, forming binary spinel-type compounds
Hydrogen sulfide or carbon disulfide	–	Reacts, forming sulfide
Carbon	1000	Reacts, forming carbides
Carbon tetrachloride	–	Reacts upon heating, forming chromous chloride
Chlorine	800	Reacts in the presence of carbon or carbon monoxide, forming chromous chloride
Benzene, hexane, or cyclohexane	–	Reacts, reducing to metallic chromium
	CrO_2 [446, Vol. 2]	
Water	100	Reacts, converting to Cr_2O_3 and CrO_3
Alkali metal hydroxides or carbonates	–	React in melt, forming chromates
Hydrochloric acid	–	Reacts, evolution of chlorine chlorine

1	2	3
		CrO_3 [1; 446, Vol. 2; 447, Vol. 3; 450, Vol. 1]
Water	–	Dissolves, forming chromic acid
Hydrogen peroxide	–	Reacts at low temperatures in methyl ether solution, forming compound $CrO_3 \cdot (CH_3)_2O$
Hydrogen chloride (gaseous)	–	Reacts, forming chromyl chloride
Univalent metal iodates	–	React, forming compounds $MeCrIO_6$
Hydrobromic, hydroiodic, hypophosphorous, hypophosphoric, or organic acids	–	Reacts (are oxidized)
Alcohols	–	React, forming esters of chromic acid
Carbon monoxide, carbon, arsenic, sulfur, selenium, or iodine	–	Reacts (are oxidized)
Rubidium carbonate	–	Reacts, forming rubidium chromate or dichromate
		MnO [1; 446, Vol. 2]
Ammonium sulfide	100	Reacts, forming MnS
Water	–	No reaction
Hydrogen	–	Reacts at high temperatures, reducing to metallic manganese
Air	200–300	Reacts, oxidizing to MnO_2
Acids	–	React, forming the corresponding salts
Silicon tetrachloride	–	Reacts upon heating to form $MnCl_2$, SiO_2, and Cl_2
Sodium hydroxide	800	Reacts in presence of oxygen, forming sodium hypomanganate Na_3MnO_4

1 – Continued

1	2	3
	Mn_2O_3 [1; 446, Vol. 2]	
Aluminum	–	Reacts upon heating; reduces to metal
Water	–	No reaction
Hydrogen	300	Reacts, reducing to MnO
Air	950–1100	Reacts (oxidizes)
Nitric or sulfuric acid (dilute)	20	Reacts: $Mn_2O_3 + 2HNO_3 = MnO_2 + Mn(NO_3)_2 + H_2O$ and $Mn_2O_3 + H_2SO_4 = MnO_2 + MnSO_4 + H_2O$
Sulfuric acid (concentrated)	–	Reacts, forming $Mn_2(SO_4)_3$
Hydrofluoric acid	–	Reacts, forming fluoride
	MnO_2 [1; 446, Vol. 2; 447, Vol. 3; 448, Vol. 1]	
Ammonia (gaseous)	–	Reacts: $6MnO_2 + 2NH_3 = 3Mn_2O_3 + N_2 + 3H_2O$
Ammonium nitrate	–	Reacts: $MnO_2 + 4NH_4NO_3 = Mn(NO_3)_2 + 3N_2 + 8H_2O$
Air	350–900	Reacts, forming Mn_2O_7
Water	–	No reaction
Hydrogen	170	Reacts, reducing to MnO
Alkali-metal hydroxides	–	React, forming unstable salts Me_2MnO_3; in presence of oxygen, forms Me_2MnO_4
Ferrous sulfate (anhydrous)	–	Reacts upon calcining to form sulfate: $2MnO_2 + 2FeSO_4 = 2MnSO_4 + Fe_2O_3 + 1/2O_2$
Potassium nitrate + potassium carbonate (melt)	–	Reacts: $MnO_2 + KNO_3 + K_2CO_3 = K_2MnO_4 + KNO_2 + CO_2$
Potassium bromide + acetic acid	–	Reacts: $MnO_2 + 2KBr + 4CH_3COOH = Mn(CH_3COO)_2 + 2KCH_3COO + Br_2 + 2H_2O$
Calcium oxide (melt)	–	Reacts, forming manganates: $2CaO \cdot MnO_2$, $CaO \cdot MnO_2$, $CaO \cdot 2MnO_2$, $CaO \cdot 3MnO_2$, $CaO \cdot 5MnO_2$

1	2	3
Sulfuric acid (concentrated)	110	Reacts: $MnO_2 + H_2SO_4 = MnSO_4 + H_2O + 1/2O_2$
Sulfurous acid	–	Reacts, forming manganese dithionate
Hydrochloric acid (concentrated)	90	Reacts: $MnO_2 + 4HCl = MnCl_2 + 2H_2O + Cl_2$
Hydrofluoric acid	450–500	Reacts: $2MnO_2 + 4HF = 2MnF_2 + 2H_2O + O_2$
Mixture of sulfuric acid and hydrogen peroxide	–	Reacts: $MnO_2 + H_2O_2 + H_2SO_4 = MnSO_4 + 2H_2O + O_2$
Sodium chloride + sulfuric acid	–	Reacts: $MnO_3 + 2NaCl + 3H_2SO_4 = MnSO_4 + 2NaHSO_4 + 2H_2O + Cl_2$
Magnesium chloride	–	Reacts: $MnO_2 + 2MgCl_2 = MnCl_2 + 2MgO + Cl_2$
Hydrogen sulfide	–	Reacts in presence of oxygen: $2MnO_2 + 3H_2S + 3O_2 = MnSO_4 + MnS_2O_3 + 3H_2O$
Sulfur dioxide	10	Reacts, forming dithionate
Chlorine in presence of potassium hydroxide	–	Reacts: $2MnO_2 + 3Cl_2 + 8KOH = 2KMnO_4 + 6KCl + 4H_2O$
Acetyl chloride	0	Reacts: $MnO_2 + 4CH_3COCl = MnCl_4 + 2(CH_3CO)_2O$
	Mn_2O_7 [1; 446, Vol. 2]	
Water	–	Reacts, forming permanganic acid
Combustible materials	–	React; are oxidized and burned
Acetic acid	–	Interacts, dissolving without decomposition
	FeO [1; 446, Vol. 2]	
Water	–	Reacts upon heating; oxidizes to Fe_2O_3
Hydrogen	–	Reacts upon heating; reduces to metallic iron
Air	200–250	Reacts; oxidizes to Fe_2O_3

1 – Continued

1	2	3
Acids	–	React, forming the corresponding salts
Silicon oxide	–	Reacts in melt, forming silicates
Carbon or carbon monoxide	–	Reacts, reducing to metallic iron
	Fe_2O_3 [1; 446, Vol. 2; 450, Vol. 2]	
Water	–	No reaction
Hydrogen (dry)	400	Reacts, reducing to metallic iron
Hydrogen (moist)	400	Reacts, forming Fe_3O_4
Hydrogen (moist)	800	Reacts, forming FeO
Iron fluoride	950	Reacts, forming oxofluoride
Sulfuric acid (concentrated)	–	Reacts, forming sulfate
Hydrochloric acid	–	Reacts, forming chloride
Chlorine	400	Reacts in presence of carbon, forming chloride
	CoO [1; 446, Vol. 2]	
Aluminum oxide	–	Reacts, forming binary spinel-type salts
Aluminum, boron, hydrogen, iron, silicon, carbon, or carbon monoxide	–	Reacts upon heating; reduces to metallic cobalt
Air	900	Reacts; oxidizes to Co_3O_4
Alkali-metal hydroxides (melts)	–	React, forming cobaltites Me_2CoO_2
Alkali-metal hydroxides (concentrated solutions)	–	React upon heating to form hydroxocobaltites $Me_2[Co(OH)_4]$
Potassium carbonate + silicon oxide	–	Reacts, forming potassium and cobalt orthosilicates
Oxygen	> 100	Reacts, oxidizing to Co_3O_4
Oxygen + sulfur dioxide	600	Reacts, forming $CoSO_4$
Nitric, sulfuric, or hydrochloric acid	–	Reacts, forming the corresponding salts
Metal oxides	800–1500	React, forming compounds or solid solutions

1	2	3
Sulfur or hydrogen sulfide	–	Reacts upon heating to form CoS
Chlorine or carbon tetrachloride	–	Reacts upon heating to form $CoCl_2$
	Co_3O_4 [1; 446, Vol. 2]	
Aluminum, hydrogen, potassium, sodium, carbon, or carbon monoxide	–	Reacts upon heating, reduces to metallic cobalt
Nitric or sulfuric acid	–	Reacts, with evolution of oxygen: $Co_3O_4 + 3H_2SO_4 = 3CoSO_4 + 3H_2O + 1/2O_2$ and $Co_3O_4 + 6HNO_3 = 3Co(NO_3)_2 + 3H_2O + 1/2O_2$
Hydrochloric acid	–	Reacts, with evolution of chlorine: $Co_3O_4 + 8HCl = 3CoCl_2 + 4H_2O + Cl_2$
	NiO [1; 446, Vol. 2]	
Barium oxide	–	Reacts in oxygen medium, forming binary salts
Water	–	No reaction
Silicon, boron, hydrogen, or carbon	–	Reacts upon heating; reduces to metallic nickel
Acids (dilute)	–	React, forming the corresponding salts
Metals (Al, Mg, Zn, Be, Cu, Pb, Fe, Co)	–	React upon heating; reduce to metal
Acidic oxides	–	React upon heating
	Cu_2O [1; 446, Vol. 2; 450, Vol. 2]	
Ammonia (solution)	–	Reacts, forming complex $[Cu(NH_3)]OH$
Water	–	No reaction
Hydrogen	155	Reacts, reducing to metal
Hydrogen bromide (gaseous)	–	Reacts: $Cu_2O + 2HBr = 2CuBr + H_2O$
Hydrogen halides	–	Reacts, forming compounds $H[CuI_2]$ and CuI
Arsenic triiodide	–	Reacts: $3Cu_2O + 2AsI_3 = 6CuI + As_2O_3$

1 – Continued

1	2	3
Oxygen	1000	Reacts, oxidizing to CuO
Sulfuric acid (dilute)	–	Reacts: $Cu_2O + H_2SO_4 = CuSO_4 + H_2O + Cu$
Hydrochloric acid (concentrated)	–	Reacts: $Cu_2O + 8HCl = 2H_3[CuCl_4] + H_2O$
Hydrocyanic acid	–	Reacts: $Cu_2O + 2HCN = 2CuCN + H_2O$
Cuprous chloride (solution)	–	Reacts: $3Cu_2O + 4CuCl_2 + 3H_2O = 6CuCl + CuCl_2 \cdot 3Cu(OH)_2$
Alkali metals and Mg, Zn, Al	–	React upon heating; reduce to metallic copper or alloy of copper with other metal
Mercuric chloride	–	Reacts: $3Cu_2O + 3HgCl_2 + 3H_2O = 2CuCl_2 + CuCl_2 \cdot 3Cu(OH)_2 + 3Hg$
Selenium dioxide	–	Reacts: $4Cu_2O + 7SeO_2 = Cu_2Se + 6CuSeO_3$
Carbon monoxide	200	Reacts, reducing to metal
Diethylsulfate	–	Reacts: $Cu_2O + (C_2H_5)_2SO_4 = Cu_2SO_4 + (C_2H_5)_2O$
CuO [1; 446, Vol. 2; 447, Vol. 3; 450, Vol. 2]		
Aluminum carbide	–	Reacts: $12CuO + Al_4C_3 = 12Cu + 2Al_2O_3 + 3CO_2$
Ammonia	–	Reacts: $3CuO + 2NH_3 = 3Cu + N_2 + 3H_2O$
Ammonium iodide	–	Reacts, dissolving
Alkali-metal arsenites	–	React upon heating: $2CuO + Me_3AsO_3 = Cu_2O + Me_3AsO_4$
Water	–	No reaction
Hydrogen	250	Reacts, reducing to metal
Ferrous chloride	–	Reacts upon heating: $3CuO + 2FeCl_2 = 2CuCl + CuCl_2 + Fe_2O_3$
Acids	–	Reacts, forming the corresponding salts
Calcium carbide	–	Reacts: $5CuO + CaC_2 = 5Cu + CaO + 2CO_2$

1	2	3
Metals K, Na, Be, Mg, Ca, Al	–	React, reducing to metal
Sodium arsenite	–	Reacts: $2CuO + Na_3AsO_3 = Cu_2O + Na_3AsO_4$
Nitrosyl perchlorate	200	Reacts, forming cupric perchlorate
Carbon, carbon monoxide, or hydrocarbons	–	Reacts, reducing to metal
Arsenous oxide + acetic acid	–	Reacts, forming Schweinfurth green $3Cu(AsO_2)_2 \cdot Cu(CH_3COO)_2$
Stannous chloride	–	Reacts upon heating: $2CuO + SnCl_2 = 2CuCl + SnO_2$
Phosphorus trichloride	–	Reacts upon heating: $17CuO + 5PCl_3 = 2Cu_3(PO_4)_2 + CuCl_2 + 10CuCl + POCl_3$
	Cu_2O_3 [1]	
Alkali-metal hydroxides	–	React, forming tetrahydroxocuprates $Me[Cu(OH)_4]$
Hydrochloric acid	–	Reacts, with evolution of chlorine
	ZnO [1; 446, Vol. 2]	
Barium oxide	1100	Reacts, forming binary oxide or oxozincate $BaZnO_2$
Water	–	No reaction
Alkali-metal hydroxides (solutions)	–	React: $ZnO + MeOH + H_2O$ $Me[Zn(OH)_3]$
Calcium carbide	–	Reacts: $ZnO + CaC_2 = CaO + Zn + 2C$
Hydrochloric acid	–	Reacts, forming chloride
Phosphoric acid	–	Reacts, forming zinc orthophosphate
Metal oxides	800–900	React, forming zincates
Nitrosyl perchlorate	250	Reacts, forming zinc perchlorate
Sulfur dioxide	–	Reacts with aqueous suspension of ZnO, forming sulfite
Carbon or carbon monoxide	1100	Reacts, reducing to metal

1 – Continued

1	2	3
Ferrosilicon	–	Reacts: $2ZnO + FeSi = 2Zn + Fe + SiO_2$
Zinc	150	Reacts, forming zinc hydroxychloride
	Ga_2O [1; 450, Vol. 2]	
Air (dry)	20	No reaction
Air (dry)	700	Reacts, forming Ga_2O_3 and Ga
Sulfuric acid	–	Reacts; reduced to H_2S
	GaO [1]	
Water	20	No reaction
Acids	–	React (dissolve)
	Ga_2O_3 [1; Vol. 1; 450, Vol. 2]	
Ammonia	900	Reacts, forming nitride
Water	20	No reaction
Hydrogen	700–830	Reacts, reducing to GaO
Hydrogen	830–900	Reacts, reducing to metal
Gallium	–	Reacts, forming lower oxide
Gallium halides	–	Reacts, forming oxyhalides
Alkali-metal hydroxides (solutions)	–	React, forming salts
Acids	–	React (dissolve)
Hydrogen sulfide	1200	Reacts, forming sulfide
Thionyl chloride	200	Reacts, forming gallium trichloride
Zinc oxide or magnesium oxide	–	Reacts in melt, forming compounds $Ga_2[MeO_4]$
	GeO [1; 446, Vol. 1; 450, Vol. 1]	
Water	–	No reaction; strong jet converts to colloidal state
Hydrogen peroxide (ammoniacal solution)	–	Reacts, oxidizing to GeO_2
Air	20	No reaction
Air	500	Reacts, oxidizing to GeO_2
Alkali-metal hydroxides	–	No reaction

1	2	3
Nitric acid (fuming)	–	Reacts, oxidizing slightly
Hydrochloric acid	–	No reaction
Stannic chloride	–	Reacts; reduced to bivalent tin
	GeO_2 [1; 446, Vol. 1; 450, Vol. 1]	
Aluminum, magnesium, or carbon	–	Reacts upon heating; reduces to germanium
Ammonia	700	Reacts, forming nitride
Barium chloride (concentrated solution) + hydrofluoric acid	–	Reacts: $GeO_2 + BaCl_2 + 6HF = Ba[GeF_6] + 2HCl + 2H_2O$
Water	25	Dissolves hexagonal modification but not tetragonal
Hydrogen	450	Reacts (reduces)
Alkali-metal hydroxides or carbonates	–	React: $GeO_2 + 4MeOH = Me_4GeO_4 + 2H_2O$ and $GeO_2 + 2Me_2CO_3 = Me_4GeO_4 + 2CO_2$
Hydrobromic or hydroiodic acid	280	Reacts, forming $GeBr_4$ and GeI_4
Nitric acid or sulfuric acid	–	Reacts slowly
Hydrochloric acid or hydrofluoric acid	100	Reacts, dissolving hexagonal modification but not tetragonal
Hydrofluoric acid + potassium chloride (solution)	–	Reacts, forming potassium germanium fluoride K_2GeF_6
Sodium hydroxide (solution)	100	Reacts, dissolving hexagonal modification to form alkali metal germanates; slowly dissolves tetragonal modification
Hydrogen sulfide	800	Reacts, forming germanium sulfide
Lead oxide	500	Reacts, forming series of compounds in which $GeO_2 : PbO$ ratio varies from 1 : 6 to 1 : 0.33

1 – Continued

1	2	3
Thallium carbonate (solution)	–	Reacts: $5GeO_2 + Tl_2CO_3 = Tl_2Ge_5O_{11} + CO_2$
As_2O_3 [1; 448, Vol. 1; 450, Vol. 1]		
Water	–	Reacts (dissolves)
Hydrogen	–	Reacts in acidic solution at moment of evolution, reducing to AsH_3
Hydrogen fluoride (anhydrous)	140	Reacts, forming arsenic tetrafluoride
Nitric acid	70	Reacts: $As_2O_3 + 2HNO_3 + 2H_2O = 2H_3AsO_4 + N_2O_3$
Oxidizers	–	React, oxidizing to arsenic acid
Tin chloride	–	Reacts, precipitating arsenic from hydrochloric acid solutions
Carbon or potassium cyanide	–	Reacts, reducing to metal
As_2O_5 [1]		
Water	20	Reacts vigorously forming arsenic acid
SeO_2 [1; 448, Vol. 1; 450, Vol. 1; 453]		
Water	20	Reacts, forming selenous acid
Hydrogen peroxide	–	Reacts, forming selenic acid
Hydrogen chloride	–	Reacts with dry SeO_2, forming compound $SeO_2 \cdot 2HCl$
Sulfuric acid (concentrated)	–	Reacts, forming selenous acid
Selenium tetrachloride	–	Reacts upon heating to form oxychloride $SeOCl_2$
Selenium + hydrogen chloride (dry)	–	Reacts, forming selenium chloride
Selenium + hydrogen bromide	–	Reacts, forming selenium bromide
Selenium + chlorine	–	Reacts, forming oxychloride
SeO_3 [1; 450, Vol. 1; 453]		
Water	20	Reacts, forming selenic acid

1	2	3
Hydrogen chloride + sulfur dioxide (liquid)	–	Reacts, forming chloroselenic acid $HSeO_3Cl$
Selenic acid	–	Reacts, forming selenium pyro-acids $H_2Se_2O_7$, $H_4Se_3O_{11}$, and $H_2Se_3O_{10}$
Hydrofluoric acid	–	Reacts, forming fluoroselenic acid
Sulfur	–	Reacts, reducing to selenium
Rb_2O [446, Vol. 1; 448, Vol. 1]		
Hydrogen	–	Reacts upon passage through heated Rb_2O, forming a mixture of RbH and RbOH
Oxygen	–	Reacts upon heating to form RbO_2
Rb_2O_2 [446, Vol. 1; 448, Vol. 1]		
Water	–	Reacts: $Rb_2O_2 + H_2O = 2RbOH + {}^1/_2O_2$
Oxygen	–	Reacts upon heating to form RbO_2
Rubidium	–	Reacts in silver crucible, forming Rb_2O
RbO_2 [446, Vol. 1; 448, Vol. 1]		
Water	–	Reacts: $2RbO_2 + 2H_2O = 2RbOH + H_2O_2 + O_2$
Rubidium	–	Reacts in silver crucible, forming Rb_2O
SrO [446, Vol. 1]		
Ammonium chloride	180–300	Reacts: $SrO + 2NH_4Cl = SrCl_2 + 2NH_3 + H_2O$
Water	–	Reacts, dissolving with absorption of heat
Calcium cyanide	700	Reacts: $SrO + KCN = Sr + KCNO$
Dilute acids	–	React: $SrO + 2HCl = SrCl_2 + H_2O$ and $SrO + 2HNO_3 = Sr(NO_3)_2 + H_2O$
Metals (K, Mg, Al, W)	–	React, reducing to metallic Sr

1 – Continued

1	2	3
Methanol	–	Reacts, forming $Sr(OH)OCH_3$ and $Sr(OCH_3)_2$
	SrO_2 [1; p. 446, Vol. 2]	
Hydrogen peroxide	20	Reacts, forming $SrO_2 \cdot 2H_2O_2$
	Y_2O_3 [1; p. 446, Vol. 2]	
Ammonium chloride or iodide	–	Reacts upon heating: $Y_2O_3 + 6NH_4I = 2YI_3 + 6NH_3 + 3H_2O$
Hydrogen	200	Reacts, reducing to metallic yttrium
Hydrogen sulfide	1450	Reacts, forming sulfide
Carbon disulfide (vapor)	–	Reacts, forming sulfide
Carbon	1800	Reacts, forming carbide
Chlorine	–	Reacts in presence of carbon, forming chloride
	ZrO_2 [1; 446, Vol. 2]	
Ammonium bifluoride	–	Reacts, forming fluoride
Alkali-metal hydroxides or carbonates (solutions)	–	React, forming zirconates Me_2ZrO_3 and Me_4ZrO_4
Calcium or magnesium alloys Ca–Na and Mg–Na	–	React upon heating, reducing to metallic zirconium
Nitric or hydrochloric acid	–	No reaction
Sulfuric acid (concentrated)	–	Reacts upon heating, forming sulfate
Sulfuric acid (dilute)	–	No reaction
Hydrofluoric acid	500	Reacts: $ZrO_2 + 6HF = H_2[ZrF_6] + 2H_2O$
Oxides of germanium, silicon, or titanium		React in melt, forming silicates, titanates, and germanates
Rare-earth metal oxides	–	React in melt, forming zirconates of the type Me_2ZrO_7
Sulfur trioxide	400	Reacts, forming $Zr(SO_4)_2$

1	2	3
Alkaline-earth metals	–	React, forming zirconates $MeZrO_3$
Alkali-metal oxides and chlorides		React, forming Me_2ZrO_3
Carbon disulfide (vapor)	–	Reacts, forming sulfide
Carbon	1900–2000	Reacts, forming carbide
Carbon in nitrogen medium	1100–1200	Reacts, forming nitride
	NbO [446, Vol. 2]	
Air or oxygen	–	Reacts upon heating; oxidizes to Nb_2O_5
Acids HCl, HF, or $HF + H_2SO_4$	–	Reacts
Chlorine	–	Reacts upon heating: $NbO + {}^3/_2Cl_2 = NbOCl_3$
	NbO_2 [446, Vol. 2]	
Air	–	Reacts upon heating; oxidizes to Nb_2O_5
Water	–	No reaction
Hydrogen	1300	Reacts (reduces)
Alkali-metal hydroxides (solutions)	–	No reaction
Acids	–	No reaction
Sodium nitrite	250–360	Reacts: $3NbO_2 + 3NaNO_2 = Na_3Nb_3O_9 + 3NO$
Sodium sulfite	450–600	Reacts: $12NbO_2 + 3Na_2SO_3 = 2Na_3Nb_3O_9 + 3Nb_2O_5 + 3S$
Sodium sulfate	800–1100	Reacts: $18NbO_2 + 3Na_2SO = 2Na_3Nb_3O_9 + 6Nb_2O_5 + 3S$
Chlorine	200–400	Reacts: $10NbO_2 + 5Cl_2 = 4Nb_2O_5 + 2NbCl_5$
	Nb_2O_5 [1; 446, Vol. 2; 447, Vol. 3, 450, Vol. 1]	
Ammonia	500–800	Reacts, forming nitride
Aluminum, calcium, or magnesium	–	Reacts, reducing to monoxide

1 – Continued

1	2	3
Aluminum bromide	200	Reacts, forming niobium pentabromide
Aluminum chloride	400	Reacts, forming chloride
Bromine (liquid) + carbon	–	Reacts, forming niobium pentabromide
Hydrogen	800–1000	Reacts, reducing to dioxide
Hydrogen	1200	With extended contact (50 h), reduces to monoxide
Hydrogen peroxide	–	Reacts with aqueous solution of Nb_2O_5/KOH melt, forming peroxy salt K_3NbO_3
Water	–	No reaction
Hydrogen	–	Reacts, reducing to NbO_2 and NbO
Hydrogen chloride (gaseous)	–	Reacts, forming volatile chloride
Hydrogen fluoride (gaseous)	–	Reacts, forming volatile fluoride
Potassium hydroxide (solution)	120–200	Reacts, forming potassium hexaniobate ($K_8Nb_6O_{19}$) or metaniobate ($KNbO_3$)
Potassium bifluoride + hydrofluoric acid	–	Reacts, forming heptafluoroniobate K_2NbF_7
Calcium hydride; aluminum, calcium, or magnesium silicide	–	Reacts, reducing to metal
Alkali-metal carbonates (melts)	–	React, forming orthoniobates Me_3NbO_4
Hydrochloric acid		Reacts, depending on acid concentration and oxide calcination temperature
Sulfuric acid	–	Reacts, depending on acid concentration and oxide calcination temperature
Silicon	–	Reacts, reducing to metallic niobium

1	2	3
Magnesium	600	Reacts in vacuo, reducing to dioxide, monoxide, or metallic niobium, with magnesium metaniobate as a by-product
Niobium	1200–1250	Reacts, reducing to dioxide or monoxide
Sulfur monochloride	220–300	Reacts, forming niobium pentachloride
Thionyl chloride	220	Reacts, forming niobium oxychloride
Carbon	900	Reacts, reducing to dioxide
Carbon	1200	Reacts, reducing to monoxide
Carbon	2100	Reacts, reducing to metallic niobium
Carbon	1700–1800	Reacts in vacuo, forming niobium carbide
Carbon monoxide	1100	Reacts, reducing to monoxide
Carbon tetrachloride	200–225	Reacts, forming volatile chloride $NbCl_5$ and oxychloride $NbOCl_3$
Hydrogen sulfide or carbon disulfide	–	React, forming metaniobates Nb_2OS_3
Alkali-metal sulfates, chromates, molybdates, tungstates	–	Reacts, forming metaniobates $MeNbO_3$
Phosphorus pentachloride	180–235	Reacts, forming chloride
Chlorine	800–850	Reacts, forming $NbOCl_3$
Chlorine + carbon	1000–1050	Reacts, forming niobium chloride (volatile), and oxychloride
MoO_2 [1; 446, Vol. 2; 447, Vol. 3; 450, Vol. 1]		
Water	> 100	Reacts, oxidizing to MoO_3
Hydrogen	950–1100	Reacts, reducing to metallic molybdenum
Air	–	Reacts, oxidizing to MoO_3

1 – Continued

1	2	3
Alkaline-earth metal hydroxides	–	React upon heating without access to air, forming colored molybdates $MeMoO_3$
Oxidizing acids	–	React, oxidizing to MoO_3
Sulfur	–	Reacts, forming sulfide
Carbon dioxide	–	Reacts, oxidizing to MoO_3
Chlorine	300	Reacts, forming molybdenum oxychloride
MoO_3 [1; 446, Vol. 2; 447, Vol. 3; 450, Vol. 1]		
Ammonia in solution	–	React, forming molybdates
Ammonia (gaseous)	–	Reacts upon heating: $3MoO_3 + 2NH_3 = 3MoO_2 + 3H_2O + N_2$
Water	–	Dissolves very slightly
Hydrogen	450–470	Reacts, reducing to MoO_2
Hydrogen	800–900	Reacts, reducing to metallic molybdenum
Hydrogen chloride (dry)	150–200	Reacts, forming yellow needles of $H_2[MoO_3Cl_2]$
Alkali-metal hydroxides and carbonates (solutions)	–	React, forming molybdates $Me_2[MoO_4]$
Alkali-metal hydroxides, carbonates, and borates (melts)	–	Reacts, forming molybdates and isopolymolybdates
Hydrochloric or nitric acid	–	No reaction
Sulfuric acid (concentrated)	–	Reacts, forming molybdenum dioxosulfate
Hydrofluoric acid	–	Reacts, forming MoO_2F_2
Lithium fluoride	–	Reacts upon heating to form MoO_2F_2
Molybdenum	580	Reacts, reducing to Mo_4O_{11}
Oxides of iron, calcium, or copper	500–600	React, forming molybdates $MeMoO_4$
Carbon monoxide	400	Reacts: $MoO_3 + CO = MoO_2 + CO_2$
Hydrogen sulfide	–	Reacts upon heating to form MoS_2

1	2	3
Alkali-metal fluorides (other than LiF)	–	React, forming oxofluoromolybdates of the type $Me_3[MoO_3F_3]$
Chlorine, alkali-metal chlorides, carbon, phosphorus, iron	–	Reacts upon heating to form oxychloride MoO_2Cl_2
	TcO_2 [446, Vol. 2; 450, Vol. 1]	
Air or oxygen	–	Reacts, oxidizing to Tc_2O_7
Chlorine	300	Reacts
	Tc_2O_7 [446, Vol. 2; 450, Vol. 1]	
Water	–	Reacts, forming pertechnetic acid $HTcO_4$
Hydrogen	–	Reacts upon heating; reduces to metal
Technetium	–	Reacts upon heating; reduces to TcO_3
	RuO_2 [1; 446, Vol. 2]	
Hydrogen or carbon monoxide	–	Reacts upon heating, reduces to metallic ruthenium
Air	930	Reacts, causing dissociation to ruthenium and oxygen
Acids	–	No reaction
	RuO_4	
Water	20	Dissolves
Alkali-metal hydroxides	–	React: $RuO_4 + 2MeOH = Me_2RuO_4 + H_2O + 1/2O_2$
Hydrochloric acid (concentrated)	–	Reacts, with evolution of chlorine and formation of of oxychloride, which with excess is reduced to $RuCl_4$
Organic substances	–	React vigorously, reducing to RuO_2
	RhO [446, Vol. 2]	
Water	–	No reaction
Acids	–	No reaction

1 – Continued

1	2	3
		Rh_2O_3 [446, Vol. 2]
Water	–	No reaction
Hydrogen	–	Reacts upon heating; reduces to metal
Hydrochloric + nitric acid	–	No reaction
		PdO [1; p. 354, 446, Vol. 2]
Hydrogen	20	Reacts, reducing to metallic palladium
Air	Up to 700	No reaction
Oxygen	Up to 800	No reaction
Acids	–	No reaction
Hydrobromic acid (concentrated)	–	Reacts
Mixture of nitric and hydrochloric acids		No reaction
Carbon monoxide	–	Reacts; oxidized to CO_2
		Ag_2O [1; 446, Vol. 2; 450, Vol. 2]
Ammonia (solution)	–	Reacts, forming complex compound $[Ag(NH_3)_2]OH$
Ammonium carbonate	–	Reacts: $Ag_2O + 2(NH_4)_2CO_3 = [Ag(NH_3)_2]_2CO_3 + 2H_2O + CO_2$
Water	18	Dissolves $1.1 \cdot 10^{-4}$ mole/liter
Hydrogen	100	Reacts, reducing to metallic silver
Hydrogen peroxide	20	Reacts: $Ag_2O + H_2O_2 = 2Ag + H_2O + O_2$
Halogens (solutions)	–	Reacts, forming salts $AgIO_3$
Potassium hydroxide + potassium permanganate		Reacts, forming silver peroxide
Nitric acid	–	Reacts, forming $AgNO_3$
Hydrofluoric acid	–	Reacts, forming fluoride AgF; with excess acid, binary compounds $AgF \cdot HF$ and $AgF \cdot 3HF$ separate out

1	2	3
Chloric acid	–	Reacts: $Ag_2O + 2HClO_3 = 2AgClO_3 + H_2O$
Sodium oxide	400	Reacts, forming sodium orthoargentite Na_3AgO_2
Alkali metal cyanides	–	React in solutions: $Ag_2O + 4MeCN + H_2O = 2Me[Ag(CN)_2] + 2MeOH$
Alkali metal sulfides, sulfur	–	React, forming Ag_2S
Chromic oxide	–	Reacts: $5Ag_2O + Cr_2O_3 = 2Ag_2CrO_4 + 6Ag$
Chromic hydroxide (alkaline solution)	–	Reacts: $3Ag_2O + 2Cr(OH)_3 + 4MeOH = 2Me_2CrO_4 + 6Ag + 5H_2O$
AgO [1; 446, Vol. 2; 450, Vol. 2]		
Ammonia, alkali metal nitrides	–	React (oxidized)
Water	–	No reaction
Nitric, sulfuric, or perchloric acid (concentrated)	–	React (dissolve)
Zinc + potassium hydroxide (solution)	–	Reacts: $AgO + Zn + 2KOH + H_2O = Ag + K_2[Zn(OH)_4]$
CdO [1; 446, Vol. 2]		
Oxides of aluminum, iron, or silicon	–	React upon heating to form aluminates, ferrites, or silicates
Hydrogen	300	Reacts, reducing to metallic cadmium
Cadmium chloride (solution)	210	Reacts, forming cadmium hydroxychloride Cd(OH)Cl
Acids	–	React, forming corresponding salts
Silicon dioxide	60–800	Reacts, forming silicate $CdO \cdot SiO_2$
Silicon dioxide	900	Reacts, forming monosilicate $2CdO \cdot SiO_2$
Nitrosyl perchlorate	60–110	Reacts, forming cadmium perchlorate $Cd(ClO_4)_2$

1 — Continued

1	2	3
Carbon	600	Reacts, reducing to metallic cadmium
Sulfur (melt)	300	Reacts, forming sulfide
Hydrogen sulfide	—	Reacts upon heating to form CdS
Chlorine	—	Reacts upon heating to form chloride
Zinc sulfate (solution)	—	Reacts: $CdO + ZnSO_4 + H_2O = CdSO_4 + Zn(OH)_2$
In_2O_3 [1; 446, Vol. 1; 450, Vol. 2]		
Ammonia (gaseous)	200–300	Reacts, reducing to metallic indium
Ammonia (solution)	—	No reaction
Hydrogen	200–300	Reacts, reducing to metallic indium
Alkali-metal hydroxides	—	No reaction
Acids	—	React (dissolve)
Magnesium, sodium, or carbon	—	Reacts upon heating; reduces to metallic indium
Carbon disulfide	700	Reacts, forming indium sulfide
Thionyl chloride	300	Reacts, forming indium trichloride
Fluorine	500	Reacts, forming indium trifluoride
Bivalent metal oxides	—	React in melt, forming spinel-type compounds
SnO [1; 446, Vol. 1; 450, Vol. 1]		
Nitrogen dioxide	—	Reacts: $SnO + NO_2 = SnO_2 + NO$
Boron, hydrogen, or silicon	—	Reacts upon heating; reduces to metallic tin
Water (vapor)	—	Reacts: $SnO + H_2O = SnO_2 + H_2$
Air	550	Reacts, oxidizing to SnO_2
Alkali-metal hydroxides (concentrated solutions)	—	React: $2SnO + 2MeOH + 2H_2O = Me_2[Sn(OH)_6] + Sn$
Acids	—	React, forming corresponding salts

1	2	3
Oxygen	550	Reacts, forming SnO_2
Sulfur	–	Reacts: $2SnO + 5S = 2SnS_2 + SO_2$
Sulfur dioxide	–	Reacts: $2SnO + SO_2 = 2SnO_2 + S$
Carbon dioxide	–	Reacts: $SnO + CO_2 = SnO_2 + CO$
Chlorine	–	Reacts: $2SnO + 2Cl_2 = SnO_2 + SnCl_4$
Ethanol (vapor)	350	Reacts: $6SnO + C_2H_5OH = 6Sn + 2CO_2 + 3H_2O$
SnO_2 [1; 446, Vol. 1; 448, Vol. 1; 450, Vol. 1]		
Aluminum, zinc, or magnesium	300	Reacts, reducing to metallic tin
Hydrogen	750	Reacts, reducing to metallic tin
Alkali-metal hydroxides or oxides	–	React in melt: $SnO_2 + Me_2O = MeSnO_3$ and $SnO_2 + 2MeOH = Me_2SnO_3 + H_2O$
Potassium cyanide	–	Reacts: $SnO_2 + 2KCN = Sn + 2KCNO$
Sulfuric acid (concentrated)	–	Reacts upon heating: $SnO_2 + 2H_2SO_4 = Sn(SO_4)_2 + 2H_2O$
Sodium carbonate + sulfur (melt)		Reacts, forming thiostannates Na_2SnS_3
Carbon		Reacts, reducing to metallic tin
Carbon monoxide	450	Reacts, reducing to SnO
Carbon monoxide	750	Reacts, reducing to metallic tin
Carbon in stream of chlorine	–	Reacts upon calcining to form tin tetrachloride
Ethanol (vapor)	650	Reacts: $SnO_2 + 4C_2H_5OH = Sn + 2CH_2CHO + 2C_2H_4 + 4H_2O$
Sb_2O_3 [1; 446, Vol. 1; 448, Vol. 1; 450, Vol. 1]		
Water	–	Reacts slowly
Hydrogen, magnesium, sodium, or carbon	700	Reacts, reducing to metallic antimony
Air	–	Reacts upon calcining, oxidizing to Sb_2O_4
Alkali-metal hydroxides (solutions)	–	React, forming antimonites

1 – Continued

1	2	3
Hydrofluoric acid	–	Reacts, forming antimony trifluoride
Oxygen	900	Reacts, forming SbO_2
Nitric or sulfuric acid (dilute)	–	No reaction
Hydrochloric acid (concentrated)	–	Reacts upon heating
Tartaric acid	–	Reacts (dissolves)
	SbO_2 [446, Vol. 1; 448, Vol. 1]	
Aluminum, hydrogen, magnesium, carbon, or carbon monoxide	700	Reacts (reduces)
Water	–	No reaction
Hydrochloric acid + potassium iodide	–	Reacts upon heating: $2SbO_2 + 8HCl + 2KI = 2SbCl_3 + I_2 + 2KCl + 4H_2O$
Alkali-metal hydroxides or carbonates (melts)	–	React, forming pyroantimony of the type $Me_3Sb_2O_7$
Antimony	–	Reacts upon heating to form Sb_2O_3
	Sb_2O_5 [446, Vol. 1]	
Water	–	Dissolves slowly, reacting: $Sb_2O_5 + 7H_2O = 2H[Sb(OH)_6]$
Hydrogen, carbon	–	Reacts at high temperatures, reducing to metallic antimony
Salts of tartaric acid	–	React: $Sb_2O_5 + 2KHC_4O_6H_4 = 2K[SbO_2H_2]C_4O_6H_2 + H_2O$
Potassium hydroxide (solution)	–	Reacts: $Sb_2O_5 + 2KOH + 5H_2O = 2K[Sb(OH)_6]$
Alkali-metal carbonates; oxides of uni-, bi-, and trivalent metals	–	React upon extended fusion, forming mixed oxides $Sb_2O_5 \cdot Me_2O$, $Sb_2O_5 \cdot MeO$, $Sb_2O_5 \cdot MeO_2$, $Sb_2O_5 \cdot Me_2O_3$
Hydrochloric acid (concentrated)	–	Reacts: $Sb_2O_5 + 12HCl = 2H[SbCl_6] + 5H_2O$
Sulfur	–	Reacts upon heating: $2Sb_2O_5 + 11S = 2Sb_2S_3 + 5SO_2$

1	2	3
Chlorine	–	Reacts upon heating: $Sb_2O_5 + 3Cl_2 = 2SbCl_3 + 5/2O_2$
TeO_2 [1; 448, Vol. 1; 450, 1; 453]		
Water	–	Very little solubility (1 in 150,000)
Alkali-metal hydroxides (solutions)	–	React, forming tellurites Me_2TeO_3
Acids (concentrated)	–	React, forming quadrivalent tellurium salts
TeO_3 [1; 448, Vol. 1; 450, Vol. 1; 450]		
Water	–	No reaction
Alkali-metal hydroxides (solutions)	–	React, forming the corresponding tellurates
I_2O_5 [448, Vol. 1; 450, Vol. 1]		
Water	–	Reacts, forming iodic acid
Carbon monoxide	–	Reacts: $I_2O_5 + 5CO = 5CO_2 + I_2$
Cs_2O [446, Vol. 1; 448, Vol. 1]		
Ammonia (liquid)	–	Reacts, forming cesium hydroxide and cesium amide
Water	–	Reacts vigorously, forming hydroxide
Hydrogen	180	Reacts upon slight heating: $Cs_2O + H_2 = CsOH + CsH$
Oxygen	150	Reacts, forming superoxide
Fluorine or chlorine	150–200	Reacts, forming CsF and CsCl
Cs_2O_2 [446, Vol. 1; 448, Vol. 1]		
Water	–	Reacts: $Cs_2O_2 + 2H_2O = 2CsOH + H_2O_2$
CsO_2 [446, Vol. 1; 448, Vol. 1]		
Water	–	Reacts: $2CsO_2 + 2H_2O = 2CsOH + H_2O_2 + O_2$
BaO [1; 357; 446, Vol. 1]		
Ammonium chloride	–	Reacts upon heating: $BaO + 2NH_4Cl = BaCl_2 + 2NH_3 + H_2O$

1 – Continued

1	2	3
Water	–	Reacts, forming $Ba(OH)_2$
Air	500	Oxidizes up to formation of peroxide BaO_2
Oxygen	330	Reacts, forming peroxide BaO_2
Metals (Mg, Zn, Al), silicon	1100–1250	React, reducing to metallic barium
Oxides	–	React: $BaO + SiO_2 = BaSiO_3$; $2BaO + PbO_2 = Ba_2PbO_4$; $BaO + Al_2O_3 = Ba(AlO_2)_2$; and $BaO + Fe_2O_3 = Ba(FeO_2)_2$
Sulfur	–	Reacts: $2BaO + 3S = 2BaS + SO_2$
Sulfur dioxide	230	Reacts: $BaO + SO_2 = BaSO_3$
Carbon disulfide	–	Reacts: $3BaO + CS_2 = 2BaS + BaCO_3$
Carbon	–	Reacts, forming carbide BaC_2
Chlorine	–	Reacts upon heating: $4BaO + 3Cl_2 = BaO_2 + 2BaCl_2 + Ba$ $Ba(ClO)_2$
	BaO_2 [1; 446, Vol. 1]	
Water	–	Difficultly soluble, with formation of hydrate $BaO \cdot 8H_2O$
Air	700	Reacts, splitting out oxygen: $BaO_2 = BaO + 1/2O_2$
Hydrogen peroxide	–	Reacts, forming $BaO_2 \cdot H_2O_2$
Bivalent iron salts	–	React, oxidized to trivalent iron salts
Acids, dilute	–	React: $BaO_2 + 2HCl = BaCl_2 + H_2O_2$ and $BaO_2 + H_2SO_4 = BaSO_4 + H_2O_2$
Mercurous chloride	–	Reduced to metallic mercury
Alcohol, ether	–	No reaction
	La_2O_3 [1; 446, Vol. 2]	
Nitrogen in presence of carbon	1200	Reacts, forming nitride
Aluminum iodide	450	Reacts: $La_2O_3 + 2AlI_3 = 2LaI_3 + Al_2O_3$

1	2	3
Ammonium salts	250–350	React, with evolution of ammonium
Water	–	Dissolves readily, forming hydroxide
Nitric or hydrochloric acid	–	Reacts, forming the corresponding salts
Sulfuric acid	–	Reacts: $La_2O_3 + 3H_2SO_4 = La_2(SO_4)_3 + 3H_2O$
Lanthanum fluoride	900	Reacts in vacuo: $La_2O_3 + LaF_3 = 3LaOF$
Hydrogen selenide	1000	Reacts, forming selenide
Sulfur monochloride	–	Reacts: $2La_2O_3 + 6S_2Cl_2 = 4LaCl_3 + 3SO_2 + 9S$
Carbon disulfide	–	Reacts: $La_2O_3 + 3CS_2 = La_2S_3 + 3CO + 3S$
Carbon	1800	Reacts, forming carbide
Carbon and nitrogen	1200	Reacts, forming nitride
Carbon monoxide	–	Reacts, forming $La_2(CO_3)_3$
Chlorine	–	Reacts in presence of carbon, forming chloride
	Ce_2O_3 [1; 447, Vol. 3]	
Air	–	Reacts (oxidizes)
Acids	–	React
	CeO_2 [1; 447, Vol. 3]	
Calcium chloride (melt)	900–1000	Reacts, with formation of $CeCl_3$
Hydrochloric acid + dioxane	–	Reacts, with formation of dioxane derivatives of hexachloroceric acid $H_2CeCl_6 \cdot 4C_4H_8O_2$
Acids	–	Reacts in presence of reducing agents
Sodium, barium, or strontium oxide	450–600	Reacts, forming cerate Na_2CeO_3, $BaCeO_3$, or $SrCeO_3$
Sodium hydroxide	900	Reacts, forming cerate
Carbon	1800	Reacts, forming carbide
Hydrofluoric acid	–	Reacts, forming cerium trifluoride

1 – Continued

1	2	3
	HfO_2 [1; 446, Vol. 2]	
Water	–	No reaction
Carbonates and alkalis (melts)	–	React, forming alkali metal hafnates Me_2HfO_3 and Me_4HfO_4
Hydrofluoric acid	–	Reacts: $HfO_2 + 6HF = H_2[HfF_6] + 2H_2O$
Acids	–	No reaction
Germanium, silicon, or titanium oxide (melts)	–	Reacts, forming hafnium germanate, silicate, or titanate
Alkaline-earth metal oxides (melts)	–	React, forming hafnates $MeHfO_3$
	TaO_2 [446, Vol. 2]	
Water	–	No reaction
Air	–	Reacts, oxidizing to Ta_2O_5
Alkali-metal hydroxides	–	React upon fusion, forming tantalates
Potassium nitrate	–	Reacts upon heating, oxidizes to Ta_2O_5
Acids	–	React slowly
	Ta_2O_5 [1; 4 .6, Vol. 2; 44 , Vol. 3, 450, Vol. 1]	
Ammonia	–	Reacts, forming nitride
Aluminum, calcium, or magnesium silicide	–	Reacts, reducing to metallic tantalum
Aluminum tribromide	200	Reacts: $3Ta_2O_5 + 10AlBr_3 = 6TaBr_5 + 5Al_2O_3$
Aluminum chloride	400	Reacts, forming chloride
Water	–	No reaction
Hydrogen	1200	No reaction
Hydrogen peroxide	–	Reacts, forming pertantalic acid $HTaO_4$; reaction of melts of Ta_2O_5 and KOH with aqueous solutions forms the peroxide salt K_3TaO_8
Bromine + carbon	–	Reacts, forming bromide

1	2	3
Hydrogen chloride or bromide	–	No reaction
Hydrogen fluoride	–	Reacts, forming volatile fluoride
Alkali-metal hydroxides in carbonates; metal oxides (MgO, CaO, SrO, BaO, MnO, and FeO)	–	React upon fusion, forming tantalates
Alkali-metal hydroxides (solutions)	150–200	React, forming metatantalates $MeTaO_3$
Calcium hydride	–	Reacts, forming tantalum hydride
Mineral acids	–	No reaction
Hydrofluoric acid	–	Reacts, forming potassium heptafluorotantalate
Sulfuric acid + hydrogen peroxide	–	Reacts, forming peroxotantalic acid
Sulfur + hydrogen sulfide	–	No reaction
Carbon disulfide + hydrogen sulfide	1000	Reacts, forming disulfide
Thionyl chloride	230–240	Reacts: $Ta_2O_5 + 5SOCl_2 = 2TaCl_5 + 5SO_2$
Carbon	1100	Reacts in hydrogen medium, reducing to TaO_2 and TaO
Carbon	1300	Reacts in vacuo, forming carbide
Carbon + nitrogen	–	Reacts at high temperatures, forming nitride
Carbon monoxide	1100	Reacts, reducing to tantalum monoxide
Carbon tetrachloride	200–250	Reacts, forming chloride
Alkali-metal fluorides (solutions)	–	Reacts in aqueous solutions: $Ta_2O_5 + 12KF + 3H_2O = 2K_3[TaOF_6] + 6KOH$
Phosphorus pentachloride	180–245	Reacts: $Ta_2O_5 + 5PCl_5 = 2TaCl_5 + 5POCl_3$

1 – Continued

1	2	3
Chlorine	1250	Reacts, forming volatile chloride
WO_2 [1; 446, Vol. 2; 450, Vol. 1]		
Nitrous oxide	500	Reacts: $4WO_2 + 3N_2O = W_4O_{11} + 3N_2$
Nitrogen dioxide	300	Reacts: $WO_2 + NO_2 = WO_3 + NO$
Hydrogen	800–860	Reacts, reducing to metal
Hydrogen fluoride	500	Reacts, forming oxyfluoride WOF_2
Air, oxygen	500	Reacts, oxidizing to WO_3
Alkali-metal hydroxides (solutions)	–	React: $WO_2 + 2MeOH = Me_2WO_4 + H_2$
Hydrochloric acid	–	Reacts: $WO_2 + 2HCl = WO_2Cl_2 + H_2$
Hydrofluoric acid	600	Reacts: $WO_2 + 2HF = WOF_2 + H_2O$
Chlorine	–	Reacts, forming WO_2Cl_2, or WCl_6 at high temperature
WO_3 [1; 446, Vol. 2; 447, Vol. 3; 450, Vol. 1]		
Alkali-metal hydroxides (solutions)	–	React, forming tungstate Me_2WO_4
Water	–	No reaction
Hydrogen	300–500	Reacts, reducing to W_4O_{11}
Hydrogen	575–600	Reacts, reducing to WO_2
Hydrogen	800–860	Reacts, reducing to metallic tungsten
Hydrogen chloride (gaseous)	–	Reacts upon heating: $WO_3 + 2HCl = WO_2Cl_2 + H_2O$
Tungsten	800	Reacts, reducing to $W_{18}O_{49}$
Metal oxides	–	React upon heating to form tungstates and polytungstates
Sulfur	–	Reacts in layer of vapor: $2WO_3 + 7S = 2WS_2 + 3SO_2$
Thionyl chloride	200	Reacts, forming tunsten oxytetrachloride $WOCl_4$
Carbon monoxide	800	Reacts, reducing to W_4O_{11}

1	2	3
Carbon tetrachloride	200–280	Reacts, forming chloride
Phosphorus	500	Reacts: $10WO_3 + 8P_4 = 10WP_2 + 3P_4O_{10}$
Chlorine	–	Reacts: forming oxychlorides WO_2Cl_2 and $WOCl_4$
ReO_2 [1; 446, Vol. 1]		
Bromine, chlorine, or iodine	–	Reacts upon heating to form oxyhalide
Bromine or chlorine water	–	Reacts, forming perrhenic acid $HReO_4$
Water	–	No reaction
Hydrogen	800	Reacts, reducing to metallic rhenium
Air	–	Reacts upon heating; oxidizes to Re_2O_7
Hydrogen peroxide	–	Reacts, forming perrhenic acid
Hydrogen chloride (gaseous)	–	Reacts upon heating to form oxyhalide
Oxygen	–	Reacts, oxidizing to Re_2O_7
Nitric acid	–	Reacts, forming perrhenic acid
Hydrochloric acid (concentrated)	–	Reacts, forming $H_2[ReCl_6]$
ReO_3 [1; 446, Vol. 2; 450, Vol. 1]		
Water	–	No reaction
Hydrogen, tin dichloride	–	Reacts upon heating (reduces)
Air	110	Reacts, oxidizing to Re_2O_7
Alkali-metal hydroxides (concentrated solutions)	–	React upon heating: $3ReO_3 + 2MeOH = 2MeReO_4 + ReO_2 + H_2O$
Alkali-metal hydroxides (melts)	–	React in presence of oxidizing agents, forming salts of mesorhenic acid Me_3ReO_5
Nitric acid	–	Reacts, forming perrhenic acid $HReO_4$

1 – Continued

1	2	3
Sulfuric acid or hydrochloric acid	–	No reaction
	Re_2O_7 [1; 446, Vol. 2]	
Water	–	Reacts, forming perrhenic acid
Hydrogen	300	Reacts, reducing to ReO_2
Hydrogen	500	Reacts, reducing to metallic rhenium
Rhenium	650	Reacts in vacuo, forming ReO_2
Rhenium pentachloride	–	Reacts, forming rhenium oxychloride ReO_3Cl; with excess $ReCl_5$, forms oxytetrachloride $ReOCl_4$
Hydrogen sulfide	80	Reacts, forming Re_2S_7
Ethyl or methyl alcohol, ether, chloroform, acetone, acetic acid	–	React (dissolve)
Carbon monoxide	250	Reacts: $Re_2O_7 + CO = 2ReO_3 + CO_2$ or under pressure of 20 MPa: $Re_2O_7 + 17CO = Re_2(CO)_{10} + 7CO_2$
	OsO_2 [1; 446, Vol. 2]	
Water	–	No reaction
Hydrogen	–	Reacts, reducing to metal
Air	–	Reacts, oxidizing to OsO_4
Acids	–	No reaction
	OsO_4 [1; 446, Vol. 2]	
Water	–	Dissolves without decomposition
Hydrogen	800	Reacts, decomposing OsO_4, forming a mirror
Potassium hydroxide	–	Reacts, forming $K_2[OsO_4(OH)_2]$
Hydrochloric acid (dilute)	–	No reaction
Hydrochloric acid (concentrated)	–	Reacts, with evolution of chlorine and formation of $OsCl_4$

1	2	3
Mixture of potassium chloride and hydrochloric acid	–	Reacts, forming chloroxo-osmate $K_2[OsO_2Cl_4]$
Mixture of potassium nitrate and nitric oxide	–	Reacts, forming nitrooxo-osmate $K_2[OsO_2(NO_2)_4]$
	IrO_2 [1; 446, Vol. 2]	
Water	–	No reaction
Hydrogen	–	Reacts upon heating; reduces to metal
Nitric or sulfuric acid	–	No reaction
Hydrochloric acid	–	Reacts, forming complex acid H_2IrCl_6
	PtO [1; 446, Vol. 2]	
Hydrogen	–	Reacts upon heating; reduces to platinum sponge
Nitric + hydrochloric acid	20	Reacts
	Au_2O [446, Vol. 2; 450, Vol. 2]	
Ammonia (concentration solution)	–	Reacts, forming $Au_3N \cdot NH_3$
	Au_2O_3 [446, Vol. 2; 450, Vol. 2]	
Hydrogen iodide	–	Reacts: $Au_2O_3 + 6HI + 2AuI + 2I_2 + 3H_2O$
Potassium hydroxide (solution)	–	Reacts, forming $K[Au(OH)_4] \cdot H_2O$
	HgO [1; 446, Vol. 2]	
Ammonia (solution)	–	Reacts, forming Millon's base $[(HOHg)_2NH_2]OH$
Bromine (vapor or solution)	–	Reacts, forming oxybromide $HgBr_2 \cdot 3HgO$
Hydrogen	50	Reacts, reducing to metal
Hydrogen cyanide (solution)	–	Reacts, forming mercuric cyanide $Hg(CN)_2$

1 – Continued

1	2	3
Hydrogen peroxide		Reacts, forming peroxide HgO_2
Ferrio ferrocyanide	–	Reacts: $9HgO + Fe_4[Fe(CN)_6]_3 + 9H_2O = 9Hg(CN)_2 + 4Fe(OH)_3 + 3Fe(OH)_2$
Iodine	100	Reacts, forming HgI_2 and $Hg(IO_3)_2$
Nitric or hydrochloric acid	–	Reacts, forming the corresponding salt
Sulfur monochloride	20	Reacts: $2HgO + 2S_2Cl_2 = 2HgCl_2 + SO_2 + 3S$
Chlorine	20	Reacts: $2HgO + 2Cl_2 = Hg_2OCl_2 + Cl_2O$, or upon heating: $HgO + Cl_2 = HgCl_2 + 1/2O_2$
Thionyl chloride	160	Reacts: $HgO + 5SOCl_2 = HgCl_2 + 3SO_2Cl_2 + S_2Cl_2$
Fluorine	–	Reacts upon heating to form HgF_2
Phosphorus or hypophosphorous acid	–	Reacts (oxidized)
	Tl_2O [460, Vol. 1; 450, Vol. 2]	
Bromine, chlorine, or iodine	–	Reacts, with formation of halide and evolution of oxygen
Water	–	Reacts (dissolves)
Hydrogen or carbon monoxide	–	Reacts upon heating, reducing to metallic thallium
Air or oxygen	–	Reacts upon heating to form Tl_2O_3
Nitric acid	–	Reacts: $Tl_2O + 2HNO_3 = 2TlNO_3 + H_2O$
Acetic acid	–	Reacts: $Tl_2O + 2CH_3COOH = 2TlCH_3COO + H_2O$
Silicon tetrachloride	360–370	Reacts, forming TlCl and SiO

1	2	3
Sulfur	–	Reacts in melt, forming thallium sulfide
Ethanol	–	Reacts: $Tl_2O + 2C_2H_5OH = 2TlOC_2H_5 + H_2O$
Tl_2O_3 [446, Vol. 1; 450, Vol. 2]		
Water	–	Reacts slowly (dissolves)
Alkali-metal hydroxides	–	React slowly (dissolve)
Acids	–	React (dissolve)
Carbon monoxide	–	Reacts upon heating; reduces to metal
Fluorine	–	Reacts, forming TlF_3
PbO [1; 446, Vol. 1; 448, Vol. 1; 450, Vol. 1]		
Ammonium chloride	–	Reacts upon heating: $4PbO + 2NH_4Cl = PbCl_2 \cdot 3PbO + 2NH_3 + H_2O$
Water	20	Dissolves 1.2 mg of yellow oxide per 100 g of water, and 0.6 mg of red oxide per 100 g of water
Hydrogen, boron, potassium, calcium carbide, sodium, carbon, or carbon monoxide	–	Reacts upon heating, reducing to metallic lead
Alkali-metal hydroxides (concentrated solutions)	–	React, forming $Me[Pb(OH)_3]$ and $Me[Pb(OH)_4]$ $Me_2[Pb(OH)_4]$
Calcium cyanide	–	Reacts in melt: $PbO + KCN = Pb + K + CO + 1/2N_2$
Acids	–	React, forming bivalent lead salts
Oxygen	400–500	Reacts, oxidizing to Pb_3O_4
Carbon or carbon monoxide	400–500	Reacts, reducing to metallic lead
Sulfur	–	Reacts upon heating to form sulfide

1 — Continued

1	2	3
Carbon disulfide	–	Reacts upon heating to form sulfide
Lead selenide	1100	Reacts: $2PbO + PbSe = 3Pb + SeO_2$; $3PbO + PbSe = 3Pb + PbSeO_3$ and $4PbO + PbSe = 4Pb + PbSeO_4$
Pb_3O_4 [1; 446, Vol. 1; 448, Vol. 1; 450, Vol. 1]		
Water	–	Dissolves slightly
Hydrogen peroxide in presence of nitric acid or sulfuric acid	–	Reacts: $Pb_3O_4 + H_2O_2 + 6HNO_3 = 3Pb(NO_3)_2 + 4H_2O + O_2$ and $Pb_3O_4 + H_2O_2 + 3H_2SO_4 = 3PbSO_4 + 4H_2O + O_2$
Hydrogen	250–350	Reacts, reducing to lead
Air	550	Reacts, with formation of PbO and evolution of oxygen
Potassium nitrate (molten)		Reacts
Nitric or sulfuric acid (dilute)		Reacts upon heating to form bivalent lead salts and lead dioxide
Sulfuric acid (concentrated)	–	Reacts upon heating: $Pb_3O_4 + 3H_2SO_4 = 3PbSO_4 + 1/2O_2 + 3H_2O$
PbO_2 [446, Vol. 1; 448, Vol. 1; 450, Vol. 1]		
Water	–	Reacts slowly
Hydrobromic acid	–	Reacts: $PbO_2 + 4HBr = PbBr_2 + Br_2 + 2H_2O$
Sodium hydroxide	–	Reacts upon heating: $PbO_2 + 2NaOH + 2H_2O = Na_2[Pb(OH)_6]$
Nitric acid + manganese sulfate		Reacts: $5PbO_2 + 6HNO_3 + 2MnSO_4 = 2PbSO_4 + 3Pb(NO_3)_2 + 2HMnO_4 + 2H_2O$
Hydroiodic acid	–	Reacts: $PbO_2 + 4HI = PbI_2 + I_2 + 2H_2O$
Sulfuric acid (concentrated)	–	Reacts: $PbO_2 + H_2SO_4 = PbSO_4 + 1/2O_2 + H_2O$
Hydrochloric acid (concentrated)	20	Reacts: $PbO_2 + 4HCl = PbCl_4 + 2H_2O$

1	2	3
Sulfuric acid + manganese sulfate	–	Reacts: $5PbO_2 + 3H_2SO_4 + 2MnSO_4 = 5PbSO_4 + 2HMnO_4 + 2H_2O$
Basic oxides	–	React in melt, forming orthoplumbates Ca_2PbO_4, Ba_2PbO_4, etc.
Sulfur dioxide	–	Reacts, forming $PbSO_4$
Fluorine	20	Reacts, forming PbF_2 and $PbF_2 \cdot PbO$
Chlorine	–	Reacts upon heating to form $PbCl_2$ and $PbCl_2 \cdot PbO$
	BiO [446, Vol. 1; 448, Vol. 1]	
Hydrogen	300	Reacts, reducing to metallic bismuth
Moist air	–	Reacts, oxidizing
Hydrochloric acid	–	Reacts: $3BiO + 6HCl = 2BiCl_3 + Bi + 3H_2O$
Carbon dioxide	–	Reacts: $5BiO + CO_2 = 2Bi_2O_3 + Bi + CO$
	Bi_2O_3 [446, Vol. 1; 448, Vol. 1]	
Aluminum, boron, hydrogen, potassium cyanide, calcium carbide, carbon, carbon monoxide, or alkali metals	–	React upon heating; reduce to metallic bismuth
Halogens	–	React, forming trihalides and oxyhalides
Alkali-metal hydroxides (solutions)	–	No reaction
Alkali-metal hydroxides + bromine	–	React, forming bismuthates $MeBiO_3$
Acids	–	React, forming corresponding bismuth salts
Sodium peroxide	–	Reacts upon heating: $Bi_2O_3 + 3Na_2O_2 = 2Na_3BiO_4 + 1/2O_2$
Lead oxide (melt)		Reacts, forming compounds $2PbO \cdot Bi_2O_3$; $2PbO \cdot 3Bi_2O_3$; $PbO \cdot 4Bi_2O_3$

1 – Continued

1	2	3
PoO [446, Vol. 1]		
Alkali-metal hydroxides (solutions)		React, forming salts
Acids		React, forming salts
PoO_2 [446, Vol. 1; 450, Vol. 1]		
Hydrogen	200	Reacts, reducing to metallic polonium
Hydrogen chloride (gaseous)	200	Reacts: $PoO_2 + 4HCl = PoCl_4 + 2H_2O$
Alkali-metal hydroxides	–	React, forming polonites Me_2PoO_3
Carbon tetrachloride	200	Reacts: $PoO_2 + CCl_4 = PoCl_4 + CO_2$
ThO_2 [1; 450, Vol. 2]		
Nitrogen + carbon	200	Reacts, forming nitride
Boron	1100	Reacts, forming borides
Hydrogen fluoride	–	Reacts, forming fluoride
Alkali-metal hydroxides (solutions or melts)	–	No reaction
Potassium acid sulfate (melt)	–	Reacts, forming sulfate
Calcium	1100–1300	Reacts, reducing to metallic thorium
Sulfuric acid (concentrated)	–	Reacts, forming sulfate
Acids (dilute)	–	No reaction
Sulfur	1200–1300	Reacts, forming sulfides
Mixture of nitric acid with fluoride		Reacts
Carbon	1800	Reacts, forming carbides
Carbon tetrachloride, phosgene, or phosphorus pentachloride	–	Reacts, forming chloride
Pa_2O_5 [450, Vol. 2]		
Hydrogen	1550	Reacts, reducing to PaO_2

1	2	3
Sulfuric acid	–	Reacts upon heating
Hydrofluoric acid	–	Reacts, forming complex compound $K_2[PaF_7]$
Sodium carbonate (melt)	–	No reaction
Fluoride	–	Reacts upon heating to form PaF_5
Phosgene	–	Reacts at high temperatures: $Pa_2O_5 + 5COCl_2 = 2PaCl_5 + 5CO_2$
	UO_2 [1; 450, Vol. 2]	
Water	20	No reaction
Water	800	Reacts, forming U_3O_8
Air	–	Reacts, forming U_3O_8
Alkali-metal hydroxides (solutions)	–	No reaction
Oxygen	700	Reacts, oxidizing to UO_3
Nitric acid	25	Reacts, forming uranyl nitrate $UO_2(NO_3)_2 \cdot 6H_2O$
Sulfuric or hydrochloric acid	–	Reacts slowly, forming salts $U(SO_4)_2$ and UCl_4
Hydrofluoric acid	–	Reacts, forming hydrate, which is converted to anhydrous salt upon heating
Sulfur chloride	450	Reacts, forming chloride
Carbon tetrachloride	450	Reacts, forming chloride
Chlorine	–	Reacts, forming uranyl chloride UO_2Cl_2
	U_3O_8 [1; 450, Vol. 2]	
Nitrogen dioxide (liquid)		Reacts, forming $UO_2(NO_3)_2 \cdot 2NO_2$
Hydrogen	1500	Reacts, forming UO_2
Hydrogen fluoride (gaseous)	550	Reacts, forming uranyl fluoride UO_2F_2 and fluoride UF_4
Air	900	Reacts, with splitting out of oxygen and partial conversion of U_3O_8 to UO_2, with which U_3O_8 forms solid solutions
Nitric acid	25	Reacts, forming uranyl nitrate $UO_2(NO_3)_2 \cdot 6H_2O$

1 – Continued

1	2	3
Sulfuric or hydrochloric acid	25	No reaction
Sulfuric acid + alcohol	–	Reacts in sunlight, forming sulfate $U(SO_4)_2 \cdot 4H_2O$
Sulfur chloride	450	Reacts, forming chloride UCl_4
Carbon monoxide	750	Reacts, forming UO_2
Carbon tetrachloride	400	Reacts, forming mixture of UCl_4 and UCl_5
Chlorine	800	Reacts in presence of carbon, forming UCl_4
Fluorine	650	Reacts, forming hexafluoride
UO_3 [1 450, Vol. 2]		
Nitrogen dioxide (liquid)	–	Reacts, forming $UO_2(NO_3)_2 \cdot 2NO_2$
Ammonia + hydrogen fluoride	–	Reacts, forming fluoride UF_4
Water	–	Reacts, forming $H_2UO_4 \cdot H_2O$
Hydrogen fluoride	350	Reacts, forming uranyl fluoride UO_2F_2
Dichlorodifluoromethane	400	Reacts: $UO_3 + 2CF_2Cl_2 = UF_4 + Cl_2 + COCl_2 + CO_2$
Potassium fluoride	850	Reacts, forming KUO_3F
Nitric acid	25	Reacts, forming uranyl nitrate
Sulfuric acid	25	Reacts, forming uranyl sulfate
Acetic acid	–	Reacts, forming uranyl acetate
Sulfur chloride	450	Reacts, forming UCl_4
Carbon monoxide	350	Reacts, reducing to UO_2
Carbon tetrachloride	400	Reacts, forming mixture of UCl_2 and UCl_5
Carbon tetrabromide	–	Reacts, forming UBr_3 and $UOBr_3$
NpO_2 [1; 450, Vol. 2]		
Nitrogen dioxide	400	Reacts, oxidizing to Np_3O_8
Aluminum bromide	–	Reacts: $3NpO_2 + 4AlBr_3 = 3NpBr_4 + 2Al_2O_3$
Hydrogen + hydrogen fluoride	500	Reacts, forming fluoride NpF

1 — Continued

1	2	3
Acids (concentrated)	–	React in presence of oxidizing agents (potassium bromate)
Hydrogen sulfide + carbon disulfide	1000	Reacts, forming oxosulfide NpOS
	Np_3O_8	
Perchloric acid	–	Reacts, forming ions of quinquevalent and sexivalent neptunium
Carbon tetrachloride	–	Reacts, forming $NpCl_4$
	PuO_2 [1; 450, Vol. 2]	
Hydrogen fluoride (gaseous)	600	Reacts in presence of hydrogen: $PuO_2 + 3HF + 1/2H_2 = PuF_3 + 2H_2O$, and in presence of oxygen: $PuO_2 + 4HF = PuF_4 + 2H_2O$
	AmO_2 [1; 450, Vol. 2]	
Hydrogen	–	Reacts upon heating, reducing to Am_2O_3
Aluminum bromide	550	Reacts: $3AmO_2 + 4AlBr_3 = 3AmBr_3 + 2Al_2O_3 + 3/2Br_2$
Aluminum + iodine	500	Reacts, forming triiodide
Carbon tetrachloride	800	Reacts: $AmO_2 + 2CCl_4 = AmCl_3 + 2COCl_2 + 1/2Cl_2$

2. CATALYTIC PROPERTIES [1; 380–393, 455–465]

Catalyst composition	Conditions of experiment	Reactions catalyzed
1	2	3
	H_2O, H_2O_2 [1]	
H_2O	–	Hydrolysis of chlorobenzene by caustic soda
H_2O	–	Hydrolysis of dihalogen and polyhalogen derivatives of fatty series
$H_2O + Ca(OH)_2$	–	Hydrolysis of methyl chloride
$H_2O_2 + V_2O_5$	25–30°C	Oxidation of unsaturated aldehydes, alcohols, and acids (epoxylation, oxylation)
$H_2O_2 + OsO_4$	25–30°C	Oxidation of unsaturated aldehydes, alcohols, and acids (epoxylation, oxylation)
$H_2O_2 + CrO_3$	25–30°C	Oxidation of unsaturated aldehydes, alcohols, and acids (epoxylation, oxylation)
	Li_2O [1]	
$Li_2O + NiO$, promoted with In_2O_3		Decomposition of N_2O_4
	BeO [1]	
BeO + MgO on pumice, promoted with CeO_2	400–800°C	Dehydrogenation of secondary alcohols
BeO	400–600°C 30–60 MPa	Isosynthesis
BeO	–	Dehydration of methyltetrahydrofuran
BeO	400–450°C	Isomerization of polyethylene hydrocarbons
$BeO + UO_2$	500–600°C	Oxidation of methane to formaldehyde

1	2	3
	B_2O_3 [1]	
B_2O_3	–	Cracking of C_7–C_{16} paraffinic hydrocarbons
B_2O_3 + MgO	180–200°C	Oxidation of ammonia to nitrous oxide
B_2O_3 + MgO + acid phosphates		Hydration of olefins
	NO [1]	
NO	–	Thermal chlorination of chloroform
NO	–	Manufacture of sulfuric acid
	NO_2 [1]	
NO_2	–	Oxidation of benzene to phenol
	Na_2O_2 [1]	
Na_2O_2	–	Oxidation of ammonia
	MgO [1]	
MgO	250–400°C	Reduction of unsaturated aldehydes and ketones
MgO	400–450°C	Hydration of acetylene
MgO + CaO	–	Decomposition of nitrous oxide
MgO + SnO_2	–	Decomposition of hydrogen peroxide
MgO + Co	200°C	Fischer–Tropsch reaction
MgO + NiO	800°C	Cracking of hydrocarbons
MgO + ZrO_2	400°C 500 kPa	Conversion of hydrocarbons
MgO + B_2O_3	180–200°C	Oxidation of ammonia to nitrous oxide
MgO + ZnO on pumice, promoted with CeO_2	–	Dehydrogenation of secondary alcohols
MgO + Fe_2O_3 + CuO + K_2O	550–630°C	Dehydrogenation of ethylbenzene to styrene
MgO + SiO_2 + C(sp)	–	Dehydration and dehydrogenation of alcohols

2 – Continued

1	2	3
$MgO + SiO_2 + ZrO_2$(sp)	–	Isomerization of paraffinic hydrocarbons
$MgO + SnO_2 + Al_2O_3$	–	Production of butadiene from ethyl alcohol (Lebedev method)
$MgO + SiO_2$	–	Cracking of hydrocarbons
Al_2O_3 [1; 455–458, 460, 462, 463, 465]		
Al_2O_3	–	Formation of aromatic hydrocarbons
Al_2O_3	260°C	Dehydration of alcohols and glycerol
Al_2O_3	350–360°C	Polymerization of propylene, isobutylene, and cycloolefins
Al_2O_3	325°C	Hydrolysis of ethers
Al_2O_3	530–540°C	Ketonization of primary alcohols
Al_2O_3 on carborundum	360–380°C	Oxidation of furfural
Al_2O_3	300°C	Shift in double bond of homologs with alkyl chains
Al_2O_3	195°C	Decomposition of HBr
Al_2O_3	–	Dehydration of ethyl alcohol, with condensation
Al_2O_3	–	Catalytic cracking of hydrocarbons
$Ag + Al_2O_3$(sp)	565°C	Oxidation of ethylene
Al_2O_3	275°C	Esterification of ethyl alcohol to ethyl acetate
$Al_2O_3 + SiO_2$	600°C	Dehydration of carbonyl and carboxyl compounds
$Al_2O_3 + SiO_2$	120–370°C 2.6–42 MPa	Hydration of olefins
$Al_2O_3 + Mo_2O_3$	460°C	Isomerization of n-pentane
$Al_2O_3 + V_2O_5$	400–500°C	Isomerization of cyclohexene to methylcyclopentenes
$Al_2O_3 + Cr_2O_3$	560–580°C	Dehydrogenation of paraffins and olefins

1	2	3
$Al_2O_3 + K_2O$	400–550°C	Ammonia synthesis
$Al_2O_3 + Cr_2O_3 + Na_2O$	560°C	Dehydrogenation of n-butane, propane, and ethylbenzene
$Al_2O_3 + Cr_2O_3 + CeO_2$	575–700°C	Dehydrogenation of n-butane
$Al_2O_3 + Cr_2O_3 + Na_2O$	550°C	Dehydrogenation of n-hexane, n-heptane, and cyclohexane
$Al_2O_3 + Cr_2O_3 + CeO_2$	475–530°C	Naphtha reforming
$Fe + K_2O + Al_2O_3$	–	Ammonia synthesis
$Al_2O_3 + SiO_2$, promoted with CeO_2	300°C	Decarboxylation of fatty acids and esters
$Al_2O_3 + SiO_2 + TiO_2 + ZrO_2$	–	Cracking of individual hydrocarbons
$Al_2O_3 + SiO_2$	250–400°C	Synthesis of hydrogen cyanide from formamide
Ni on Al_2O_3(sp)	250–400°C	Reduction of carbonyl compounds
Ni on Al_2O_3(sp)	–	Replacement of halogen or sulfo group
Ni on Al_2O_3(sp)	–	Oxidation of sodium hypophosphite to phosphite
Ni on Al_2O_3(sp)	–	Conversion of formaldehyde
Ni on Al_2O_3(sp)	–	Dehydrogenation of fatty acids
Ni on Al_2O_3(sp)		Preparation of amines by reductive alkylation
Ni on Al_2O_3(sp)	–	Dehalogenation
Ni on Al_2O_3(sp)	–	Dehydrogenation of benzene
Pd on Al_2O_3(sp)	–	Hydrogenation of fats and oils
Pt on Al_2O_3(sp)	80°C, p_H = 15 MPa	Hydrogenation of terephthalonitrile and isophthalonitrile
(Pt + Pd) on Al_2O_3(sp)	–	Oxidation of carbon monoxide
$Al_2O_3 + V_2O_5 + K_2O + SiO_2$	400–600°C	Oxidation of sulfur dioxide

2 – Continued

1	2	3
Ni on Al_2O_3	800–1100°C	Methane conversion
(Pt + Pd) on Al_2O_3(sp)	–	Oxidation of carbon monoxide
Ni, Pt, Pd, Rh, Ru, Ir on Al_2O_3(sp)	300–350°C	Treatment of exhaust gas to remove NO_x
Pd on Al_2O_3(sp)	–	Treatment to remove NO_x
$Al_2O_3 + B_2O_3$	–	Cracking of C_7–C_{16} paraffinic hydrocarbons
$Al_2O_3 + SnO_2 + MgO$	–	Preparation of butadiene from ethanol (Lebedev reaction)
$Al_2O_3 + SiO_2$	–	Polymerization of hydrocarbons
$V_2O_5 + K_2O + Al_2O_5 + SiO_2$	485°C	Oxidation of sulfur dioxide
MnO_2 on Al_2O_3(sp)	350–450°C	Oxidation of methane
$Fe_2O_3 + Al_2O_3$	–	Decomposition of hydrogen peroxide
$Al_2O_3 + CuO$	250–350°C	Oxidation of hydrocarbons
$ZnO + Al_2O_3$	650°C	Dehydrogenation of ethylbenzene to styrene
$SrO + Al_2O_3$	–	Decomposition of nitrous oxide
$MoO_2 + Al_2O_3$	–	Reforming of hydrocarbons
$CeO_2 + Al_2O_3$	200–300°C	Oxidation of carbon monoxide
$Sm_2O_3 + Al_2O_3$	400–600°C	Dehydrogenation of C_2–C_{12} alcohols
SiO_2 [1, 455, 456, 458, 461, 462, 465]		
SiO_2	–	Decomposition of ammonia
$SiO_2 + Al_2O_3$	425–550°C	Cracking of hydrocarbons
$SiO_2 + MgO$	–	Cracking of hydrocarbons
$SiO_2 + Al_2O_3$	–	Polymerization of hydrocarbons
$SiO_2 + ZrO_2$	–	Polymerization of hydrocarbons
$SiO_2 + CaO$	–	Isomerization of hydrocarbons

1	2	3
$MgO + SiO_2 + Cu$	–	Dehydration and dehydrogenation of alcohols
Ni on (SiO_2 + Al_2O_3)(sp)	250–350°C 2.5 MPa	Isomerization of hydrocarbons
Pt on (SiO_2 + Al_2O_3)(sp)	–	Isomerization of aromatic hydrocarbons
Pt on SiO_2(sp)	900°C 13 MPa	Hydrogenation of terephthalonitrile
SiO_2	560°C	Dehydrocyclization of n-n-heptane
$SiO_2 + Al_2O_3$	550–750°C	Partial oxidation of methane
$Al_2O_3 + SiO_2$	600°C	Dehydration of carbonyl and carboxyl compounds
$Al_2O_3 + SiO_2$	120–370°C	Hydration of olefins
$Al_2O_3 + SiO_2 + TiO_2 + ZrO_2$	–	Cracking of individual hydrocarbons
$MnO_2 + SiO_2$	410°C	Vapor-phase oxidation of methylpyridines to aldehydes
	P_2O_5 [1; 457, 464]	
P_2O_5	25–45°C	Isomerization of 1-phenyl-1-pentene
$P_2O_5 + H_2O$	25–45°C	Isomerization of hydrocarbons (1-butene)
$V_2O_5 + MoO_3 + P_2O_5 + TiO_2$	410°C	Vapor-phase oxidation of methylpyridines to aldehydes
$V_2O_5 + MoO_3 + P_2O_5 + TiO_2$	410°C	Dehydration of ethanol
$MoO_3 + V_2O_5 + P_2O_5 + NaO +$ NiO on ZrO_2	350–450°C	Oxidation of benzene to maleic anhydride
	SO_2 [1]	
SO_2	290°C	Polymerization of oils
SO_2	–	Polymerization of isobutylene, alcohol, and methyl methacrylate
SO_2	–	Polymerization of linseed oil

2 – Continued

1	2	3
SO_2	–	Copolymerization of linseed oil with alcohol
SO_2	140–160°C	Polymerization of vinyl-acetic acid
SO_2	–	Decomposition of hydroper-oxides
SO_2	–	Cis-trans conversion of esters of oleic acid and other fatty acids
	K_2O [1, 465]	
$Fe + K_2O + Al_2O_3$(sp)	–	Ammonia synthesis
$BeO + Fe_2O_3 + K_2O$, pro-moted with CeO_2	400–800°C	Dehydrogenation of secon-dary alcohols
$MgO + Fe_2O_3 + CuO + K_2O$	550–630°C	Dehydrogenation of ethyl-benzene to styrene
$V_2O_5 + K_2O + Al_2O_3 + SiO_2$	485°C	Oxidation of sulfur dioxide
$Co_2O_3 + Al_2O_3 + K_2O$	–	Methane synthesis
$ZrO_2 + Fe_2O_3 + K_2O$	650°C	Dehydrogenation of butene to butadiene
	CaO [1, 465]	
CaO	400–450°C	Hydration of acetylene
CaO	–	Cyanide synthesis
CaO	–	Decomposition of ammonia
CaO	–	Oxidation of acetic acid to acetone
MgO + CaO	–	Decomposition of nitrous oxide
SiO_2 + CaO	–	Isomerization of hydro-carbons
$V_2O_5 + K_2O + CaO + SiO_2$	485°C	Oxidation of sulfur dioxide
	Sc_2O_3 [455]	
Sc_2O_3	400°C	Dehydration and dehydro-genation of ethanol

1	2	3
Sc_2O_3	175–275°C	Para-ortho conversion of hydrogen
Sc_2O_3	350–450°C	Conversion of isopropyl alcohol
	TiO_2 [1; 464]	
TiO_2	–	Dehydrocyclization of paraffins
TiO_2	–	Dehydration of ethanol and o-aminophenylethanol
TiO_2	–	Dehydrogenation of alcohols
TiO_2	–	Isomerization of olefins
TiO_2	–	Hydrolysis of ethers
TiO_2	300°C	Hydrolysis of esters
TiO_2	150–400°C	Oxidation of carbon monoxide
TiO_2	–	Decomposition of formic acid
TiO_2	–	Preparation of aliphatic amines from alcohols and ammonia
TiO_2 on activated carbon	–	Esterification of acetic acid with ethanol
TiO_2	510°C	Dehydrocyclization of n-heptane
$Al_2O_3 + SiO_2 + TiO_2 + ZrO_2$	–	Cracking of individual hydrocarbons
$V_2O_5 + MoO_3 + P_2O_5 + TiO_2$	410°C	Vapor-phase oxidation of methylpyridines to aldehydes
	VO_2 [1]	
VO_2	–	Oxidation of benzene to phenol
	V_2O_4 [1]	
$V_2O_4 + V_2O_5$	500°C	Oxidation of sulfur dioxide
	V_2O_5 [1, 458]	
V_2O_5 on asbestos or pumice	300–500°C	Oxidation of benzene, toluene, phenanthrene, and cyclohexane

1	2	3
V_2O_5 on alundum, or fused V_2O_5 + SnO_2	410–470°C	Oxidation of naphthalene to naphthoquinone, phthalic anhydride, CO_2, and H_2O
V_2O_5 + SnO_2 on pumice	–	Oxidation of naphthalene to phthalic anhydride
V_2O_5; or V_2O_5 on asbestos	400–500°C	Oxidation of olefinic hydrocarbons
V_2O_5 on asbestos or pumice	400°C	Oxidation of methane to CO_2 and H_2O
V_2O_5	400°C	Oxidation of butane to maleic and acetic acids and formaldehyde
V_2O_5 on pumice	300–400°C	Oxidation of pyridine and furfural
V_2O_5	300°C	Oxidation of methanol to formaldehyde
V_2O_5	400°C	Hydrogenation of cresol, olefins, diolefins, and acetylene
V_2O_5	–	Dehydrogenation of nitrogen-containing compounds
V_2O_5	400–500°C	Hydration of acetylene
V_2O_5	380–400°C p_{H_2} = 4 MPa	Conversion of alcohols to paraffinic hydrocarbons
V_2O_5 on Al_2O_3	475–510°C	Cyclization of heptane
V_2O_5	200–380°C	Oxidation of furfural, furoic acid, tetrahydrofuran, and furan
V_2O_5 + MoO_3; V_2O_5 + MoO_3 + P_2O_5 + TiO_2	410°C	Vapor-phase oxidation of methylpyridines to aldehydes
V_2O_5 + K_2O + Al_2O_3 (CaO) + SiO_2	485°C	Oxidation of sulfur dioxide
H_2O_2 + V_2O_5	25–30°C	Oxidation of unsaturated aldehydes, alcohols, and acids
Al_2O_3 + V_2O_5 + K_2O + SiO_2	400–500°C	Isomerization of cyclohexene to methylcyclopentenes

1	2	3
$Al_2O_3 + V_2O_5$	400–600°C	Oxidation of sulfur dioxide
	Cr_2O_3 [1; 458, 459. 462]	
$Co + Ni + Cr_2O_3$	–	Dehydration and dehydrogenation of alcohols
Cr_2O_3	–	Dehydrogenation of alkanes and cyclanes
Cr_2O_3 on Al_2O_3	490°C	Dehydrogenation and dehydrocyclization of methylcyclohexane
$Cr_2O_3 + ZnO + MnO_2$	400°C	Decarbonylation of furfural
Cr_2O_3	200–350°C	Hydrogenation of olefins
Cr_2O_3	400–450°C	Hydration of acetylene
Ni on Cr_2O_3(sp)	140°C	Hydrogenation of benzene
Ni on Cr_2O_3(sp)	–	Hydrogenation of carbon monoxide
Cr_2O_3	500–550°C	Aromatization of paraffinic hydrocarbons
Cr_2O_3 on Al_2O_3	490°C	Isomerization of paraffinic hydrocarbons
$Cr_2O_3 + SiO_2 + Al_2O_3$	–	Polymerization of ethylene to polyethylene
$Cr_2O_3 + MnO_2$	–	Synthesis of alcohols from CO and H_2
Cr_2O_3	–	Production of hydrogen from conversion gas
Cr_2O_3	275–300°C	Conversion of alcohols to ketones
Cr_2O_3	275°C	Esterification of ethanol to ethyl acetate
Cr_2O_3 on Al_2O_3	350–450°C	Oxidation of methane
Cr_2O_3	–	Oxidation of aldehyde to acetaldehyde, formaldehyde, CO_2, and H_2O
Cr_2O_3	–	Oxidation of sulfur dioxide
Cr_2O_3	475°C	Cyclization of heptane
Cr_2O_3	250°C	Shift of double bond in benzene homologs with alkyl chain

2 – Continued

1	2	3
Ni on Cr_2O_3(sp)	100–130°C	Hydrogenation of fats and oils
Ni on Cr_2O_3(sp)	135°C, p_{H_2} = 1.5 MPa	Hydrogenation of cottonseed oil
(Ni + Cu) on Cr_2O_3(sp)	300–700°C	Sorption of hydrogen
Ni on Cr_2O_3(sp)	20–60°C	Hydrogenation of organic compounds (cyclohexane, furfural) in 96% ethanol
Cr_2O_3 on Al_2O_3	500–650°C	Dehydrogenation of butane
Cr_2O_3 + ZnO + MnO	300–500°C	Vapor-phase decarbonylation of furfural and mixtures of methylfurfural with furfural
H_2O_2 + Cr_2O_3	25–30°C	Oxidation of unsaturated aldehydes, alcohols, and acids
Al_2O_3 + Cr_2O_3	560–580°C	Dehydrogenation of paraffins and aldehydes
MnO_2 + Cr_2O_3	–	Oxidation of acetylene to acetone
Fe_2O_3 + Cr_2O_3	450–500°C	Carbon monoxide conversion
Fe_2O_3 + Cr_2O_3 + K_2O	560–630°C	Dehydrogenation of ethylbenzene to styrene
(CuO + Cr_2O_3) on Al_2O_3	165°C	Treatment of exhaust gas to remove NO_x
ZnO + Cr_2O_3		Methanol synthesis
ZnO + Cr_2O_3		Carbon monoxide conversion
	MnO [1]	
MnO	–	Dehydrogenation of ethanol
MnO	400–460°C	Preparation of ketones from carboxylic acids and their mixtures
MnO	–	Decomposition of formic acid
MnO	–	Dehydration of organic acids
Fe + MnO	440°C	Hydration of acetylene
MnO + B_2O_3 + acid phosphates		Hydration of olefins

1	2	3
$MnO + SiO_2$	410°C	Vapor-phase oxidation of methylpyridines to aldehydes
	Mn_2O_3 [1; 464]	
Mn_2O_3 on Al_2O_3	350–450°C	Oxidation of methane
Mn_2O_3	–	Decomposition of nitrous oxide
$Mn_2O_3 + MnO_2$	–	Oxidation of carbon monoxide
	MnO_2 [1; 459]	
MnO_2	150–200°C	Oxidation of acetylene
MnO_2 on pumice	400–450°C	Oxidation of benzene to benzaldehyde
$MnO_2 + Cr_2O_3$	–	Oxidation of acetylene to acetone
MnO_2	–	Decomposition of hydrogen peroxide
MnO_2	–	Production of hydrogen from conversion gas
MnO_2	–	Oxidation of ammonia
MnO_2	360–380°C	Oxidation of propylene
MnO_2	–	Decomposition of potassium hypochlorite
MnO_2	275–300°C	Conversion of alcohols to ketones
$Cr_2O_3 + MnO_2$	–	Synthesis of alcohols from carbon monoxide and hydrogen
$U_2O_3 + ThO_2 + MnO_2 + Ni + Co$ on kieselguhr	–	Synthesis of gasoline from CO and H_2O
	FeO [1]	
FeO	–	Synthesis of methane and formaldehyde
FeO	–	Hydrogenation of naphthalene
FeO	–	Hydrogenation of heavy oils and light oils [from coal tar]

2 — Continued

1	2	3
$FeO + Fe_2O_3$	450–500°C	Ammonia synthesis
	Fe_2O_3 [1; 457–459; 462; 465]	
$Fe_2O_3 + MgO + K_2O$	–	Ammonia synthesis
$Fe_2O_3 + FeO$	500–550°C	Cracking of petroleum products
$Fe_2O_3 + Cr_2O_3$	450–500°C	Carbon monoxide conversion
$Fe_2O_3 + Cr_2O_3 + K_2O$	560–630°C	Dehydrogenation of ethylbenzene to styrene
Fe_2O_3	680°C	Oxidation of sulfur dioxide
Fe_2O_3	210–340°C	Oxidation of carbon monoxide and propylene
$Fe + Fe_2O_3$	500°C	Oxidation of carbon monoxide
$Fe_2O_3 + Bi_2O_3$	550–600°C	Oxidation of ammonia
$Fe_2O_3 + ZrO_2$	240–275°C, 4.9–7.0 MPa	Oxidation of p-toluic acid to terephthalic acid
$Fe_2O_3 + Al_2O_3$	–	Decomposition of hydrogen peroxide
Fe_2O_3	250°C	Decomposition of hydrogen bromide
Fe_2O_3 on Al_2O_3	–	Chlorination of propylene
$Fe_2O_3 + ZnO$	420–470°C	Hydration of acetylene
$Fe_2O_3 + Al_2O_3$	150–500°C	Oxidation of carbon monoxide
$Fe_2O_3 + Na_2O + nSiO_2$	680–720°C	Oxidation of sulfur dioxide
$ZrO_2 + Fe_2O_3 + K_2O$	650°C	Dehydrogenation of butene to butadiene
	Fe_3O_4 [1; 457]	
Fe_3O_4	700–800°C	Production of hydrogen from conversion gas
	CoO [1]	
CoO	–	Decomposition of nitrous oxide
CoO	150–260°C	Hydrogenation of anthracene
CoO	–	Hydrogenation of naphthalene, wood tar, heavy oils, and light oils

1	2	3
CoO	–	Synthesis of methane and formaldehyde
CoO	–	Decomposition of methane
	Co_2O_3 [1]	
Co_2O_3	–	Oxidation of carbon monoxide
Co_2O_3	–	Decomposition of hydrogen peroxide
Co_2O_3	–	Oxidation of propylene
Co_2O_3	275–300°C	Conversion of alcohols to ketones
Co_2O_3	–	Hydrogenation of aniline
$Co_2O_3 + CuO + MnO_2$	–	Synthesis of olefin-rich hydrocarbons from carbon monoxide and hydrogen
$Co_2O_3 + Al_2O_3 + K_2O$	–	Methane synthesis
	Co_3O_4 [1; 457–459]	
Co_3O_4	350–450°C	Oxidation of methane
$Co_3O_4 + Fe_2O_3$	–	Oxidation of ammonia
$Co_3O_4 + CuO + MnO_2 + NiO$	–	Total oxidation of hydrogen, carbon monoxide, and methane
	NiO [1]	
NiO on pumice	400–450°C	Oxidation of benzene to maleic acid
NiO	150–200°C	Oxidation of acetylene
NiO on asbestos	100–350°C	Oxidation of methane
NiO	180–250°C	Oxidation of carbon monoxide and propylene
NiO	–	Decomposition of methane
NiO	–	Decomposition of hydrogen peroxide
NiO	–	Decomposition of nitrous oxide
NiO	–	Hydrogenation of unsaturated compounds
NiO	240°C	Hydrogenation of anisole
NiO	–	Hydrogenation of aromatic acids and amines

2 – Continued

1	2	3
NiO	–	Hydrogenation of carbon monoxide
NiO	–	Hydrogenation of naphthalene, wood tar, heavy oils, and light oils
NiO	–	Cracking of petroleum products
NiO on aluminosilicate	–	Polymerization of light olefinic hydrocarbons
NiO + Li_2O, promoted with In_2O_3	–	Decomposition of N_2O_4
MgO + NiO	400°C	Cracking of hydrocarbons
MoO_3 + NiO + Al_2O_3; WO_3 + NiO + Al_2O_3	550–590°C	Hydrocracking of n-paraffinic hydrocarbons
	Cu_2O [1, 464]	
Cu_2O	150–250°C	Oxidation of carbon monoxide
Cu_2O	–	Oxidation of ethylene
Cu_2O on pumice or silicon carbide	350–450°C	Oxidation of propylene to acrolein, CO_2, and H_2O
Cu_2O	200–300°C	Decomposition of nitrous oxide
	CuO [1, 459, 464, 465]	
CuO	150–400°C	Oxidation of carbon monoxide
CuO on asbestos	250–350°C	Oxidation of methane to CO_2, H_2O, and formaldehyde
CuO	150–200°C	Oxidation of acetylene
CuO	150°C	Oxidation of benzene by hydrogen peroxide to benzoquinone and maleic acid
CuO on pumice	–	Oxidation of phenanthrene to phthalic acid
CuO	–	Decomposition of nitrous oxide
CuO	–	Methanol synthesis
CuO	200–210°C, 1.4–1.8 MPa	Hydrogenation of anthracene

1	2	3
CuO	–	Dehydration of alcohols
CuO	–	Dehydrogenation of alkanes and cyclanes
CuO on activated carbon	–	Conversion of arsenic trioxide to arsenic pentoxide
$CuO + Cr_2O_3$	300–400°C, 2 MPa	Reduction of esters of carboxylic acids
$CuO + WO_3$ on activated carbon	–	Hydration of ethylene hydrocarbons
$CuO + Al_2O_3$ (1 : 1)	250–350°C	Oxidation of hydrocarbons
$(CuO + Cr_2O_3)$ on Al_2O_3	165°C	Treatment of exhaust gas to remove NO_x
$MgO + Fe_2O_3 + CuO + K_2O$	550–630°C	Dehydrogenation of ethylbenzene to styrene
$ThO_2 + CuO$	200–450°C	Esterification of ethanol
	ZnO [1; 455; 461; 462]	
Cu + ZnO	–	Methanol synthesis
$ZnO + Cr_2O_3$	–	Methanol synthesis
ZnO	–	Methanol synthesis
$ZnO + Cr_2O_3 + K_2O$	490°C; 2.4 MPa	Synthesis of methanol and higher alcohols from CO and H_2
$ZnO + MnO_2 + Cr_2O_3 + K_2O$	–	Synthesis of methanol and higher alcohols from CO and H_2
$ZnO + Cr_2O_3$	–	Methane synthesis
$ZnO + Cr_2O_3$	–	Carbon monoxide conversion
$ZnO + SnO_2$	–	Oxidation of methane to formaldehyde, CO_2, and H_2O
ZnO	–	Oxidation of acetylene to acetone
ZnO	–	Oxidation of alcohols to aldehydes and ketones
ZnO	–	Decomposition of acetic acid to acetone
$ZnO + Cr_2O_3$	–	Hydrogenation of saturated compounds
$ZnO + Ga_2O_3$	–	Dehydrogenation of butenes

2 – Continued

1	2	3
$ZnO + Al_2O_3$	650°C	Dehydrogenation of ethylbenzene to styrene
Ni on ZnO(sp)	350°C, 30 MPa	Hydrogenation of benzene
ZnO	–	Dehydrogenation of alcohols
ZnO	–	Dehydration of acetic acid
ZnO	400–450°C	Hydration of acetylene
ZnO	275–300°C	Conversion of alcohols to ketones
ZnO	0–150°C	Deuterium/hydrogen exchange
ZnO	–	Water gas conversion
Ni on ZnO(sp)	250–450°C	Hydrogenation of allyl alcohol
ZnO	513°C	Dehydrocyclization of n-heptane
$ZnO + ZnCrO_4 + H_2O$	300°C, 2.5 MPa	Methanol synthesis
$Cr_2O_3 + ZnO + MnO$	300–500°C	Vapor-phase decarbonylation of furfural or mixtures of methylfurfural with furfural
	Ga_2O_3 [1]	
$Ga_2O_3 + ZnO$	–	Dehydrogenation of butenes
Ga_2O_3	–	Decomposition of nitric oxide
Ga_2O_3	–	Synthesis of hydrogen peroxide
Ga_2O_3	–	Oxidation of ammonia
	GeO [1]	
GeO	–	Interaction of alcohols with ammonia and aniline
	GeO_2 [1]	
GeO_2	265°C	Polymerization of ethyleneterephthalate
GeO_2	–	Interaction of carbon monoxide with ammonia
GeO_2	250–400°C	Oxidation of carbon monoxide

1	2	3
	As_2O_5 [1]	
As_2O_5	–	Oxidation of sulfur dioxide by oxygen to sulfur trioxide
	SeO_2 [1]	
SeO_2	–	Hydrolysis of nitrogen-containing compounds of the naphthalene series
	Pb_2O [1]	
Fe + Pb_2O(pr)	–	Fischer–Tropsch synthesis
Fe + Pb_2O(pr)	–	Ammonia synthesis
Fe_2O_3 + Pb_2O(pr)	–	Dehydrogenation of butylene to butadiene
Fe_2O_3 + Pb_2O(pr)	–	Dehydrogenation of ethylbenzene to styrene
	SrO [1]	
SrO	400–450°C	Hydration of acetylene
SrO + Al_2O_3		Decomposition of nitrous oxide
	Y_2O_3 [1; 455]	
Y_2O_3 + Dy_2O_3 + Gd_2O_3 + Er_2O_3 + Eu, Tb, Ho, Tm, Yb	400°C	Decomposition of ethyl alcohol to C_2H_4 and CH_3CHO
Y_2O_3	400–430°C	Dehydrogenation of methyl alcohol to CO and H_2
Y_2O_3	–	Dehydrogenation and dehydration of n-propyl, isopropyl, and n-butyl alcohols
Y_2O_3	350–360°C	Dehydrogenation and dehydration of ethyl and methyl alcohols
$Y_2O_3 \cdot H_2O$	–	Dehydrogenation of tetralin to naphthalene
Y_2O_3	110–270°C	Para-ortho conversion of hydrogen

1	2	3
Y_2O_3, Y_2O_3 on Al_2O_3	500–560°C	Dehydrogenation and dehydrocyclization of hydrocarbons
Y_2O_3	350–450°C	Conversion of isopropyl alcohol
Y_2O_3	385–430°C	Ketonization of n-butyl alcohol
	ZrO_2 [1]	
ZrO_2	600–1900°C	Decomposition of nitric oxide
ZrO_2	475°C	Cyclization of heptane
ZrO_2	400–450°C	Production of ketones from carboxylic acids and their mixtures
ZrO_2	500–1200°C	Hydrocracking of mineral oils
ZrO_2	300–390°C	Dehydration of alcohols
ZrO_2	400–450°C	Conversion of alcohol to butadiene
ZrO_2	200–500°C	Condensation of $C_6H_5COCH_3$ to styrene
ZrO_2	–	Hydrolysis of C_6H_5Cl and C_6H_5Br
$ZrO_2 + Fe_2O_3 + K_2O$	650°C	Dehydrogenation of butene to butadiene
ZrO_2 on SiO_2	300°C	Synthesis of butadiene from ethanol and acetaldehyde
$ZrO_2 + SiO_2$	204–260°C	Polymerization of monoolefins with alcohol
ZrO_2 on SiO_2	–	Oxidation of naphthalene to naphthoquinone
$ZrO_2 + Fe_2O_3 + Co_2O_3$	200–425°C	Treatment of naphtha to remove N_2, S_2, O_2, and As
$Ni + ZrO_2 + ThO_2$	225–275°C	Hydrogenation of diisobutylene
$Co + ZrO_2$ on pumice	150–500°C	Hydrogenation of polymer naphthas
$Pt + ZrO_2$	570°C	Hydrogenation of cyclohexane

1	2	3
$Pt + ZrO_2$	–	Isomerization of naphthalenes
$Pt + ZrO_2$	500–1200°C	Oxidation of ethylene
	NbO_2 [1]	
NbO_2	–	Oxidation of benzene to phenol
	MoO_2 [1; 457]	
MoO_2	–	Oxidation of toluene to benzoic acid
MoO_2	–	Destructive hydrogenation of hydrocarbons
MoO_2	120–200°C	Hydrogenation of anthracene
$MoO_2 + Al_2O_3$	–	Reforming of hydrocarbons
	MoO_3 [1; 457; 458]	
MoO_3	–	Hydrogenation of saturated compounds
MoO_3	–	Isomerization of hydrocarbons
MoO_3, promoted with CeO_2	–	Oxidation of methanol
$Co + MoO_3$	–	Hydrodesulfurization
MoO_3 on Al_2O_3	490–530°C	Reforming of hydrocarbons
MoO_3 on Al_2O_3	460–535°C	Decomposition of nitrous oxide
MoO_3	300–400°C, 20 MPa	Oxidation of benzene
MoO_3	370°C	Oxidation of methyloctatetraene and furfural
MoO_3 on Al_2O_3	490°C	Dehydrocyclization of n-heptane
MoO_3	440–480°C 0.8–1.0 MPa	Desulfurization of oxygen-containing substances
$MoO_3 + NiO + MgO$	4500°C 20 MPa	Vapor-phase hydrogenation of coal
$MoO_3 + NiO + Al_2O_3$	550–590°C	Hydrocracking of n-paraffinic hydrocarbons
$MoO_3 + V_2O_5 + P_2O_5 + Na_2O + NiO$ on ZrO_2	350–450°C	Oxidation of benzene to maleic anhydride

2 — Continued

1	2	3
$MoO_3 + Co_2O_3$ (Bi_2O_3)	400—500°C	Oxidative ammonolysis of propylene
	RuO [1]	
RuO	—	Hydrogenation of unsaturated compounds
	RuO_2 [1]	
RuO_2	20—100°C	Hydrogenation of organic compounds
	PdO [1]	
PdO on SiO_2	300—450°C	Oxidation of methane
PdO on aluminosilicate	—	Polymerization of light olefinic hydrocarbons
	Ag_2O [1]	
$Ag + Ag_2O$	—	Oxidation of ethylene
Ag_2O	80—140°C	Oxidation of carbon monoxide
Ag_2O on SiO_2	275—380°C	Oxidation of naphthalene
Ag_2O	400—450°C	Oxidation of naphthenes and terpenes
Ag_2O on carborundum	360—380°C	Oxidation of furfural
$Ag_2O + V_2O_5 + MoO_3 + Al_2O_3$ on SiO_2	450—530°C	Oxidation of benzene
Ag_2O on Al_2O_3	—	Hydration of acetylene
	CdO [1; 462]	
CdO	275—300°C	Conversion of alcohols to ketones
$CdO + In_2O_3$	—	Decomposition of nitrous oxide
$CdO + CaO + P_2O_5$	—	Synthesis of acetaldehyde from acetylene
	In_2O_3 [1]	
In_2O_3	25°C	Oxidation of ammonia in carbon tetrachloride solution

1	2	3
In_2O_3	–	Decomposition of nitrous oxide
In_2O_3	400°C	Dehydrogenation and dehydration of ethanol
	SnO, SnO_2 [1; 457]	
SnO	400–450°C	Hydration of acetylene
SnO_2	–	Decomposition of ammonia
$SnO_2 + Sb_2O_3$	480°C	Partial oxidation of propylene
$SnO_2 + Sb_2O_3$	–	Oxidative ammonolysis of propylene
$MgO + SnO_2$	–	Decomposition of hydrogen peroxide
$V_2O_5 + SnO_2$ on pumice	–	Oxidation of naphthalene to phthalic anhydride
$SnO_2 + ZnO$	–	Oxidation of methane to formaldehyde, CO_2, and H_2O
	Sb_2O_3 [1]	
$Sb_2O_3 + SnO_2$	480°C	Partial oxidation of propylene
Sb_2O_3	–	Decomposition of hydrogen bromide
	TeO_2 [1]	
TeO_2	310–340°C	Oxidation of propylene
	BaO, Ba_2O_2 [1; 462]	
BaO	400–450°C	Hydration of acetylene
$BaO + SiO_2 + K_2O + V_2O_5 + Al_2O_3 + Fe_2O_3$	–	Oxidation of sulfur dioxide
$BaO + Ba_2O_2$	–	Oxidation of ammonia
$Ba_2O_2 + Ag(pr)$	400–450°C	Oxidation of ethylene
	La_2O_3 [1; 455]	
La_2O_3	360–350°C	Dehydrogenation and dehydration of ethyl and isopropyl alcohols
La_2O_3	460–500°C	Decomposition of nitrous oxide

2 – Continued

1	2	3
La_2O_3	–	Dehydration of methyltetrahydrofuran
Co + La_2O_3(pr) on kieselguhr	–	Hydrogenation of carbon monoxide to hydrocarbons
La_2O_3	–	Oxidation of carbon monoxide
La_2O_3, La_2O_3 on Al_2O_3	543–584°C	Dehydrogenation of cyclohexane
La_2O_3	20 to –78°C	Hydrogenation of ethylene
La_2O_3 + Al_2O_3 + SiO_2	575°C	Dehydrogenation of n-butane
La_2O_3(pr)	–	Ketonization of n-butyl alcohol
La_2O_3 on quartz or asbestos	275–325°C	Esterification of acids with alcohols
	CeO_2 [1; 455; 459]	
CeO_2	–	Oxidation of carbon monoxide
CeO_2 + Al_2O_3	200–300°C	Oxidation of carbon monoxide
CeO_2 + Pd + ThO_2	480–800°C	Methane conversion
CeO_2 + Ni	–	Methane conversion
CeO_2 + Cu(pr)	200°C	Oxidation of methanol to aldehyde
CeO_2	–	Oxidation of benzene to phenol
CeO_2	450°C	Fischer–Tropsch synthesis to produce isobutane
CeO_2 + MgO	300–600°C	Dehydrogenation of ethylbenzene to styrene
CeO_2	600–675°C	Dehydrogenation of ethylnaphthalene
CeO_2	350–360°C	Dehydrogenation and dehydration of ethanol
CeO_2 on SiO_2	140–400°C	Alkylation of thiophene with olefins
CeO_2	–	Esterification of acetic acid with ethanol
CeO_2 + WO_3	760°C	Conversion of benzene to biphenyl

1	2	3
$CeO_2 + Al_2O_3$	500–700°C	Synthesis of hydrogen cyanide from carbon monoxide and ammonia
CeO_2	–	Isomerization of propylene oxide to allyl alcohol
CeO_2	–	Polymerization of glycol terephthalates
CeO_2	−196°C	Ortho-para conversion of hydrogen
CeO_2	–	Hydrogen/deuterium exchange reaction
CeO_2	–	Ketonization of acetic acid
$Ni + CeO_2$(pr) on pumice	–	Hydrogenation of carbon dioxide
$Ni + CeO_2$(pr) on asbestos		Hydrogenation of heterocyclic unsaturated compounds
CeO_2	25–45°C	Decomposition of hydrogen peroxide
CeO_2	510°C	Cyclization of heptane
CeO_2	350–450°C	Oxidation of hydrogen
$CeO_2 + Al_2O_3$	450°C	Treatment of exhaust gas to remove NO_X
CeO_2, $CeO_2 + ZrO$, $CeO_2 + Al_2O_3$, CeO_2 on Al_2O_3	450–540°C	Conversion of n-heptane and n-butane
$CeO_2 + Al_2O_3 + MoO_3$	500–555°C	Dehydrogenation and dehydrocyclization of hydrocarbons
$Ce_2O_3 + CeO_2$	333–471°C	Conversion of ethyl and isopropyl alcohol
$Ce_2O_3 + CeO_2$	385–430°C	Ketonization of n-butyl alcohol
$MoO_3 + CeO_2$(pr)	–	Oxidation of methanol
$PbO_2 + CeO_2$ on asbestos	350–500°C	Oxidation of methane to CO_2 and H_2O
$ThO_2 + CeO_2$ on fireclay	500°C	Oxidation of hydrogen

1	2	3
	Pr_2O_3, Pr_6O_{11} [1, 455]	
Pr_2O_3	350–360°C	Dehydration and dehydrogenation of alcohols
Pr_2O_3	350–450°C	Oxidation of hydrogen
Pr_2O_3 on Al_2O_3	290–610°C	Dehydrogenation of cyclohexane
Pr_2O_3, Pr_6O_{11} on Al_2O_3	450–550°C	Conversion of n-butane
Pr_6O_{11}	359–397°C	Dehydration and dehydrogenation of isopropyl alcohol
Pr_6O_{11}	385–430°C	Ketonization of n-butyl alcohol
Pr_6O_{11} on quartz or asbestos	250–325°C	Esterification of acids with alcohols
	Nd_2O_3 [1, 455]	
Nd_2O_3 promoted with Al_2O_3	–	Dehydrogenation of paraffinic and cyclic hydrocarbons
Nd_2O_3 + Al_2O_3	500°C	Dehydrogenation of propane and n-haptane, with parallel cyclization
Nd_2O_3 + Al_2O_3	400–600°C	Dehydrogenation of alcohols
Nd_2O_3	350–360°C	Dehydration of ethanol
Nd_2O_3, Nd_2O_3 on SiO_2	196–259°C	Ortho-para conversion of hydrogen
Nd_2O_3	350–450°C	Oxidation of hydrogen
Nd_2O_3	–	Oxidation of carbon monoxide
Nd_2O_3	–	Deuterium/hydrogen exchange in cyclohexane
Nd_2O_3, NdO_3 on Al_2O_3	–	Conversion of cyclohexane and n-heptane
Nd_2O_3	–	Hydrogenation of ethylene
Nd_2O_3 + Al_2O_3 + Cr_2O_3(pr)	500°C	Conversion of n-octane and n-butane
Nd_2O_3	385–430°C	Ketonization of n-butyl alcohol
Nd_2O_3 on quartz or asbestos	250–325°C	Esterification of acids with alcohols

1	2	3
	Sm_2O_3 [1; 455]	
Sm_2O_3	600°C	Oxidation of methanol to CO_2 and H_2O
Sm_2O_3	–	Dehydrogenation of cyclic hydrocarbons
Sm_2O_3	370–425°C	Dehydration of ethyl and isopropyl alcohols
Sm_2O_3, promoted with Al_2O_3	500°C	Dehydrogenation of n-heptane, with parallel cyclization
$Sm_2O_3 + Al_2O_3$	400–600°C	Dehydrogenation of C_2–C_{12} alcohols
Sm_2O_3	350–450°C	Oxidation of hydrogen
Sm_2O_3	–	Oxidation of carbon monoxide
Sm_2O_3	400–590°C	Conversion of cyclohexane and n-butane
Sm_2O_3	20 to –55°C	Low-temperature hydrogenation of ethylene
Sm_2O_3	385–430°C	Ketonization of n-butyl alcohol
Sm_2O_3	200–335°C	Conversion of aliphatic acids
Sm_2O_3 on quartz or asbestos	250–325°C	Esterification of acids with alcohols
	Eu_2O_3 [455]	
Eu_2O_3	–	Oxidation of carbon monoxide
Eu_2O_3	20 to –50°C	Low-temperature hydrogenation of ethylene
Eu_2O_3, Eu_2O_3 on Al_2O_3	545–560°C	Conversion of n-heptane
	Gd_2O_3 [1; 455]	
Gd_2O_3	350–360°C	Dehydrogenation and dehydration of alcohols
Gd_2O_3	–196 to –250°C	Ortho-para conversion of hydrogen
Gd_2O_3	–	Oxidation of carbon monoxide
Gd_2O_3	–	Deuterium/hydrogen exchange in cyclohexane

2 – Continued

1	2	3
Gd_2O_3	530–590°C	Dehydrogenation of cyclohexane
$Gd_2O_3 + Al_2O_3 + SiO_2$(pr)	575°C	Dehydrogenation of n-butane
Gd_2O_3	400–550°C	Conversion of n-butane
Gd_2O_3	385–430°C	Ketonization of n-butyl alcohol
	TbO_x [455]	
TbO_x	500–550°C	Conversion of n-butane
	Dy_2O_3 [1; 455]	
Dy_2O_3	350–360°C	Dehydrogenation and dehydration of alcohols
$Dy_2O_3 + Al_2O_3$	240°C 1.4 MPa,	Polymerization of ethylene
Dy_2O_3	350–460°C	Oxidation of hydrogen
Dy_2O_3	–	Oxidation of carbon monoxide
Dy_2O_3	400–550°C	Conversion of n-butane
Dy_2O_3	385–430°C	Ketonization of n-butyl alcohol
Dy_2O_3	250–325°C	Esterification of acids by alcohols
	Ho_2O_3 [455]	
Ho_2O_3	350–450°C	Oxidation of hydrogen
Ho_2O_3	–	Oxidation of carbon monoxide
Ho_2O_3	–	Dehydrogenation and dehydrocyclization of hydrocarbons
Ho_2O_3	450–550°C	Conversion of n-butane
Ho_2O_3	385–430°C	Ketonization of n-butyl alcohol
	Er_2O_3 [1; 455]	
Er_2O_3	545–590°C	Dehydrogenation of cyclohexane
Er_2O_3	–	Oxidation of carbon monoxide
Er_2O_3, Er_2O_3 on Al_2O_3	474–518°C	Dehydrogenation and dehydrocyclization of hydrocarbons

1	2	3
Er_2O_3	400–530°C	Conversion of cyclohexane, n-butane, and 1-alkyl-cyclopentenes
Er_2O_3	385–430°C	Ketonization of n-butyl alcohol
	Tm_2O_3 [1; 455]	
Tm_2O_3	–	Dehydrogenation of cyclohexane
Tm_2O_3	490–520°C	Dehydrogenation of tetralin
Tm_2O_3	400–550°C	Conversion of n-butane
Tm_2O_3	316–354°C	Conversion of isopropyl alcohol
	Yb_2O_3 [1; 455]	
Yb_2O_3	–	Dehydrogenation of cyclohexane
$Yb_2O_3 + Al_2O_3$	240°C, 40 MPa	Polymerization of ethylene
Yb_2O_3	350–450°C	Oxidation of hydrogen
Yb_2O_3	–	Oxidation of carbon monoxide
Yb_2O_3	490–530°C	Dehydrogenation of tetralin
Yb_2O_3	450–550°C	Conversion of n-butane
Yb_2O_3	385–430°C	Ketonization of n-butyl alcohol
	Lu_2O_3 [455]	
Lu_2O_3	350–450°C	Oxidation of hydrogen
Lu_2O_3	–	Oxidation of carbon monoxide
Lu_2O_3	519–566°C	Dehydrogenation of tetralin
	HfO_2 [1]	
HfO_2 on SiO_2	300–400°C	Synthesis of butadiene from ethanol and acetaldehyde
HfO_2 on SiO_2	–	Alkylation of thiophene with carboxylic acid chloride and anhydrides
$HfO_2 + Al_2O_3$	–	Cracking of hydrocarbons
	TaO_2 [1]	
TaO_2	–	Oxidation of benzene to phenol

2 — Continued

1	2	3
TaO_2	–	Oxidation of toluene to benzoic acid
	WO_2, W_2O_5 [1]	
WO_2	–	Hydrogenation of heavy oils and light oils [from coal tar]
W_2O_5	–	Dehydration of ethyl and allyl alcohols
W_2O_5	360–380°C, (7–13) MPa	Hydration of olefins
	WO_3 [1]	
WO_3	300–400°C, 200 MPa	Oxidation of benzene and toluene
$WO_3 + MoO_3$ on pumice	–	Oxidation of toluene to benzaldehyde, benzoic acid, CO_2, and H_2O
WO_3 on SiO_2	350°C, 1.4 MPa	Hydration of ethylene
WO_3	300–400°C, 0.7–1.0 MPa	Isomerization of n-hexane
WO_3	–	Destructive hydrogenation of fuels
WO_3	–	Preparation of aliphatic amines from alcohols and ammonia
WO_3 on charcoal	–	Dissolution of cellulose
$WO_3 + NiO + Al_2O_3$	550–590°C	Hydrocracking of n-paraffins
	OsO_4 [1]	
OsO_4	–	Oxidation of alcohol with air
	PtO, PtO_2 [1]	
Pt + PtO	–	Oxidation of ammonia
PtO_2	–	Hydrogenation of dimethylcyclohexane and acids
PtO_2	200°C	Hydrogenation of aniline, pyridine, and quinolines

1	2	3
PtO_2	–	Hydrogenation of oleic acid to stearic acid
PtO_2	35°C	Deuterium exchange with methanol
PtO_2	–	Reduction of carbonyl-containing compounds in the presence of ammonia
Pd + PtO_2	–	Production of diamines and polyamines by catalytic reduction of nitrile group
	Tl_2O_3, TlO_2 [1]	
Tl_2O_3 on asbestos	350°C	Oxidation of ethylene to azobenzene and phenazine
Tl_2O_3	–	Decomposition of ozone
Tl_2O_3	–	Oxidation of sulfur dioxide
TlO_2	400–450°C	Hydration of acetylene
	PbO, PbO_2 [1]	
PbO	–	Production of hydrogen from conversion gas
PbO_2	–	Oxidation of ammonia
PbO_2 + CeO_2 on asbestos	350–500°C	Oxidation of methane to CO_2 and H_2O
	Bi_2O_3 [1]	
Bi_2O_3 + MoO_3	450°C	Oxidative ammonolysis of propylene
	ThO_2 [1]	
ThO_2	–	Dehydration of alcohols
ThO_2 + Co on kieselguhr	–	Gasoline synthesis
ThO_2 + Cu(pr)	300–380°C	Dehydrogenation and dehydration of ethanol
ThO_2	–	Dehydrogenation of cyclohexane
ThO_2	400–500°C, 2.5–20 MPa	Hydration of ethylene and acetylene
ThO_2	–	Hydrogenation of acetone

2 – Continued

1	2	3
Ni + ThO_2(sp)	–	Hydrogenation of castor oil
ThO_2 on SiO_2	–	Oxidation of naphthalene to naphthoquinone
ThO_2 + CeO_2 on fireclay	500°C	Oxidation of hydrogen
ThO_2 + Co	–	Oxo synthesis
ThO_2	400–450°C	Hydrolysis of esters
ThO_2	400–440°C	Isomerization of olefins
ThO_2	–	Interaction of alcohols with ammonia and aniline to form amines
ThO_2	–	Cyanide synthesis
ThO_2 + CuO	200–450°C	Esterification of ethanol
ThO_2 on activated carbon	–	Esterification of acetic acid with ethanol
Pd + ThO_2(sp)	480–800°C	Oxidation of methane with water vapor
ThO_2	–	Polymerization of acetaldehyde
ThO_2	300°C	Condensation of acetylene with ammonia
ThO_2 on SiO_2	–	Alkylation of aromatic hydrocarbons with olefins
ThO_2	300–800°C	Isotope exchange O^{16}/O^{18}
ThO_2	–78 to 300°C	Deuterium/hydrogen exchange
ThO_2	250–260°C	Reduction of carboxylic acids to aldehydes
ThO_2 on SiO_2	555°C	Cyclization of heptane
ThO_2 on kaolin	–	Splitting of aliphatic monobromo derivatives
ThO_2	400–450°C	Isosynthesis
Co + ThO_2(pr)	–	Synthesis of hydrocarbons from carbon monoxide and hydrogen
	U_2O_3, UO_2 [1]	
U_2O_3 + ThO_2 + MnO_2 + Ni on kieselguhr		Gasoline synthesis

1	2	3
$UO_2 + BeO$	500–600°C	Oxidation of methane to formaldehyde
UO_2	450°C	Oxidation of o-xylene
UO_2	–	Oxidation of benzene to phenol
UO_2	–	Oxidation of toluene to benzoic acid

REFRACTORY PROPERTIES

1. SOLID-PHASE INTERACTION OF OXIDES WITH OTHER SUBSTANCES [4; 222; 305; 426; 451; 468–564; 624–628]

Substance	Temperature, °K	Type of interaction
1	2	3
		BeO
Be*[1]	< 1623	No reaction
B	1373	Powders react, forming beryllium boride
C*[1]	1588–2573	At temperatures above 1773°K, BeO is reduced
C*[2]	1573–1673	No reaction when held for 1500 h
C*[1]–*[3]	1723–2843	Forms beryllium carbide and carbon monoxide; above 2373°K, carbide vaporizes and then (at 2473°K and above) dissociates into Be (vapor) and carbon
Al	< 873	No reaction
Si*[2]	1673	No reaction
Ca	< 1123	No reaction
Ti*[1]	1673	Mixed powders react, with reduction of BeO to Be
Ti*[1]	1673–1873	Compact specimens do not react
Cr	1273–1473	Mixed powders react, with reduction of BeO
Mn	1273–1473	Mixed powders react weakly, with reduction of BeO
Fe*[2]	1273–1473	Mixed powders react weakly, with reduction of BeO
Ni*[2]	1273–1723	No reaction
Zr*[1]	1703	Mixed powders react, with reduction of BeO to high-purity metal

1	2	3
Zr*2	2073	In compact specimens, zirconium penetrates along grain boundaries of oxide, with some corrosion of the oxide and formation of ZrO_2 at the interface
Nb*1	1363–1813	No reaction
Nb*1	1973–2173	Forms niobium beryllide
Nb*2	1673–2073	Weak reaction, with adhesion of specimens; at 2073°K, reaction proceeds along oxide grain boundaries, with formation of Nb_2O_5
Mo*1	1973–2173	Forms molybdenum beryllide
Mo*2	2073	No reaction
Ta*1	1363–1813	No reaction
Ta	1873	Initial reaction (threshold temperature)
Ta*1	1973–2473	Forms new phase
W*1	2073–2373	Forms volatile tungsten oxides; at 2073–2173°K, reaction is observed within 0.5–1 h, and at 2273°K within 2 min
Pb	573	No reaction
Th*1	1273–2073	Reacts, beginning at 1373°K, with formation of thorium oxide
TiC	1773–2473	Forms beryllium carbide
MgO*1	1873–2073	Forms liquid phase
MgO*4	2173	Vigorous reaction; mixture becomes molten
Al_2O_3*4	2173	Vigorous reaction
Cr_2O_3	1673	Mixed powders react weakly, with formation of $BeO \cdot Cr_2O_3$
ZrO_2	1873	Initial reaction (threshold temperature)
ZrO_2*1	2073	Reacts weakly; adhesion and melting are observed

1 – Continued

1	2	3
ZrO_2*[1]	2173–2273	Specimens adhere after holding 2 min; at 2273°K, a liquid phase is formed
La_2O_3	2573	Forms $2BeO \cdot La_2O_3$
CeO_2	1973	Initial reaction (threshold temperature)
Gd_2O_3	1763	Forms eutectic
Sm_2O_3	1693	Forms eutectic
ThO_2*[1]	2373–2474	Reacts, with complete conversion to melt at 2473°K
ThO_2*[4]	2173	Vigorous reaction
UO_2	2073	No reaction
	MgO	
Be*[2]	1873–2073	Metal is oxidized; oxide is severely corroded at 2073°K
B*[1]	–	Forms magnesium borides
C*[1]	1723–2073	MgO is reduced
C*[2]	1273	No reaction
C*[4]	2073	MgO is reduced
Si*[2]	1673, 1873	Forms Mg_2SiO_4
Ti	1123	Titanium penetrates along grain boundaries of oxide; MgO is slightly attacked at the interface
Ti*[2]	1673–2073	Weak reaction at 1673°K; severe corrosion of oxide and formation of Mg_2TiO_4 at 1873–2073°K
Ni*[2]	1673–2073	No reaction
Zr*[1]	2173–2273	Forms new phase
Zr*[2]	2073	Severe corrosion of oxide
Nb*[1]	1363–1813	No reaction
Nb*[1]	2173–2273	Forms new phase
Nb*[2]	1673–2073	Slight reaction beginning at 2073°K, with formation of Nb_2O_3 along grain boundaries of oxide
Mo*[1]	2273	Silvery film formed on MgO
Mo*[2]	2073	No reaction

1	2	3
Ta*[1]	1363–2373	Reacts above 1873°K, with formation of new phase
W*[1]	2273	MgO is reduced; weak adhesion of specimens
TiB_2*[1]	1473–2273	MgO is reduced, and volatile B_2O_2 is formed
ZrB_2*[1]	1373–2273	Forms new phase
SiC	1873	Vigorous reaction
TiC*[1]	2073–2573	MgO is reduced; forms compounds of the type Me(C, O) and also magnesium carbide, which decomposes to magnesium and carbon
TiC*[2]	1873–2173	Forms TiO or Mg_2TiO_4
ZrC*[1]	2273–2573	MgO is reduced; forms magnesium carbide and solid solution Zr(C, O)
NbC*[1]	2073–2573	Forms metallic niobium and solid solution Nb(C, O)
Mo_2C*[1]	2073–2573	Slight reaction
HfC*[1]	2273	No reaction
TaC*[1]	2473–2573	No reaction
WC*[1]	2073–2573	Slight reaction
TiN*[1]	1573–2273	MgO is reduced; forms magnesium nitride and oxynitride
ZrN*[1]	1673–2273	Slight reaction
Al_2O_3*[4]	2173	Forms spinels on surface of Al_2O_3 crucible; without adhering
FeO*[4]	2173	Vigorous reaction and melting
ZrO_2*[1]	2273	ZrO_2 and MgO adhere, and a liquid phase is formed
ZrO_2*[4]	2273	Vigorous reaction
CeO_2	1973	Initial reaction (threshold temperature)
HfO_2	1673–1873	Reaction begins at 1773°K, with formation of solid solution
ThO_2*[1]	2573	Reaction in pores, with severe breakdown of ThO_2, no adhesion

1 — Continued

1	2	3
ThO_2 *2	2173	Vigorous reaction
UO_2	673–2073	No reaction
PuO_2 *2–*3	1773	No reaction
	Al_2O_3	
Be *2	Up to 1623	No reaction in solid phase
C *1	1673–2073	Al_2O_3 is reduced
C *2	1273–1473	No reaction
C *4	2223	Al_2O_3 is reduced
C	1973–2273	At 1973°K, forms aluminum oxycarbide (Al_4O_4C); and at 2073–2273°K, forms aluminum carbide
Ti	1123	Titanium penetrates along oxide grain boundaries
Ti *1	973–1973	Forms solid solutions
Ti *2	1873	Slight reaction
Ni *2	1673	No reaction
Nb *1	1363–1913	No reaction
Nb *2	1873–2073	Weak reaction, with formation of Nb_2O_5
Mo *1	1473–2223	Reaction starts at 2173°K, with formation of Mo oxides and pores in metal
Mo *2	2073	No reaction
Ta *1	1363–2173	Reacts above 1873°K without touching; oxide is decomposed
W *1	973–2223	Reaction starts at 2173°K, with formation of tungsten oxides and pores in metal
UC_2	2173	Weak reaction
BN *2	2073	Forms aluminum nitride
ZrO_2 *4	2173	Vigorous reaction
MoO_3 *2–*4	873–1073	Forms MoO_{3-x} ($x = 0.11-1.0$)
CeO_2	1973	Initial reaction (threshold temperature)
ThO_2 *4	2173	Vigorous reaction
UO_2	2073	No reaction

1	2	3
	SiO_2	
C*1, *3, *4	1523–1973	Reacts, with reduction of SiO_2 and formation of SiC; in vacuo above 1523°K, in air above 1753°K, and in a reducing medium above 1873°K
Al	723	Forms crystalline silicon at interface
SiC*2	1873–1973	Mixed powders react, forming SiO and CO
MoO_3*2–*4	873–1073	No reaction
UO_2	1923	Forms eutectic
	CaO	
C*1	1573	CaO is reduced
C*4	2403	CaO is reduced, and CaC_2 is formed
UC_2	2173	Weak reaction
ZrO_2	–	Forms $CaZrO_3$
	Se_2O_3	
MnO*4	973–1773	Forms $SeMnO_3$ and solid solutions
UO_2*4	1653	Forms solid solution
UO_2*1	2023	No reaction
	TiO_2	
C*3	1573	Reacts, with reduction of TiO_2 through intermediate oxides to titanium carbide
Si*2	1673	Reacts weakly, with surface decolorization of oxide
Ti*2	1123–2073	Reacts weakly, with surface decolorization of oxide and interdiffusion of TiO_2 and Ti, and at 2073°K with severe corrosion of oxide
Nb*2	1873–2073	Oxide is corroded
Mo*2	2073	No reaction
TiC*1	1273–2073	Reaction starts at 1473°K

1 – Continued

1	2	3
WC*1	1573–1673	TiO_2 is reduced (60% reduced in 32 min at 1573°K, or 85% reduced in 30 min at 1673°K), with formation of W–Ti
BN*2	1673–2073	Forms titanium nitride
TiN	1273–1473	Reacts
NiO	293–1273	Reaction starts at 1023–1073°K, with formation of $NiTiO_3$
UO_2	1373	Reacts
	V_2O_3	
VC*1	1273–1873	Forms vanadium monoxide and carbides; free vanadium appears at 1673°K
WC*1	1573	V_2O_3 is reduced (60% reduced in 10 min)
WC*1	1673	Reacts, forming W–V alloy (100% in 30 min)
	V_3O_4	
BN*2	2073–2273	Forms VB_2
	Cr_2O_3	
Cr_3C_2*1	1273–1973	Reaction starts at 1473°C, with formation of Cr_7C_3 and Cr at 1473–1873°K; specimen melts at 1973°K
WC*1	1573–1673	Cr_2O_3 is reduced (100% in 30 min), and W–Cr alloy is formed
BN*2	1873	Forms CrB
	Mn_2O_3	
La_2O_3*4	973–1573	Forms $LaMn^{3+}_{x}Mn^{4+}_{1-x}O_{3+y}$
La_2O_3*1	973–1473	Forms $LaMnO_3$
	MnO_2	
C*2	773–873	Reaction starts at 823°K, with reduction of MnO_2; reaction rate increases sharply at 863°K

1	2	3
		FeO*5
C*1	1193–1473	FeO is reduced; at 1423°K, metal is carburized and melted
ZrN	1573	Forms ZrO_2 and Fe, with evolution of N_2
		Fe_2O_3*5
C*1	773–973	Forms lower oxides of iron and CO_2
MoO_3*2–*4	873–1073	No reaction
		CoO
C*1	1193–1473	CoO is reduced to metal
		NiO
C*1	1193–1473	NiO is reduced to metal
ZrN	1573	Forms ZrO_2 and Ni, with evolution of N_2
		Ni_2O_3
BN*2	1723	Ni_2O_3 is reduced to Ni
		SrO
SiC*1, *2, *4	> 1273	Forms SrO, SiO_2, and CO_2 (in air), or $2SrO \cdot SiO_2$, SrC_2, and Sr (in vacuo or in argon)
		Y_2O_3
Nb*1	1363–1813	Reacts
Mo*1	1923	Weak adhesion of specimens
W*1	1373	Forms $3Y_2O_3 \cdot WO_3$
Re*1	1363–1813	No reaction
UO_2*2	1873	Forms solid solution
		ZrO_2
Be*2	Up to 1623	No reaction in solid phase
C*2	1323–1423	No reaction
C*1	1573–2073	ZrO_2 is reduced
C*4	1673	ZrO_2 is reduced
Al	1873	No reaction
Si*2	1673	No reaction in solid phase
Ti*2	1673–1873	Weak reaction at 1873°K

1 – Continued

1	2	3
Ti	1223	Titanium penetrates along grain boundaries of oxide; edges of crystals are attacked at interface
Ni*2	–	No reaction in solid phase
Zr*2	2073	No reaction
Nb*1	2273	Weak reaction
Nb*2	1673	Weak reaction
Nb*2	2073	Reacts along grain boundaries
Mo*1	1723–2573	Reaction starts at 2173°K, with formation of molybdenum oxides and pores in metal; specimens adhere slightly at 2473°K, and ZrO_2 decomposes at 2573°K
Mo*2	2073	No reaction
Ta	1873	Initial reaction (threshold temperature)
Ta*1	2073–2373	Forms new phase
W*1	1723–2573	Yellow film appears at 1873°K; reaction begins at 2173°K, forming tungsten oxides and pores in metal
W*2	2798	Reaction with ZrO_2 stabilized with 9% Y_2O_3 gives a slight change in color
Re,*1 (60 Mo + 40 Re)*1	2623	Reaction with ZrO_2 stabilized with Y_2O_3 gives a slight change in color
TiB_2*1	1373–2073	ZrO_2 is reduced, and a volatile oxide of boron is formed
ZrB_2*1	1573–2073	Forms a porous phase at interface
SiC*3	1573, 1873	Reacts at 1873°K, with formation of zirconium carbide
TiC*1	1673–2673	Forms oxycarbides, with evolution of carbon monoxide (content of bound carbon decreases)

1	2	3
Cr_3C_2*[1]	1573–2273	Forms Cr_7C_3 and ZrC
ZrC*[1]	1473–2373	Reaction starts at 2073°K, with formation of zirconium oxycarbides
ZrC*[1]	1673–2673	Forms new phase, with a decrease in content of bound carbon
NbC*[1]	1573–2673	Forms complex oxides and carbides
NbC*[1]	2373	Forms intermediate layer $Zr_{0.7}Nb_{0.3}$, $Zr_{0.68-0.40}Nb_{0.32-0.60}$, $NbZr_{0.05}$, Nb_2C
Mo_2C*[1]	2273–2573	Forms frontal layer of friable phase
HfC*[1]	2273–2573	No reaction
TaC*[1]	2273–2573	Reacts at 2573°K, with formation of new phase
WC*[1]	1673	ZrO_2 is reduced (39% in 30 min); W–Zr alloy is formed
WC*[1]	2573	Weak reaction
UC*[2]	2723	No reaction
UC_2	2173	Weak reaction
BN*[2]	1873–2173	Forms ZrB, ZrN + ZrB_2
TiN*[1]	1673–2073	Reacts, with reduction of ZrO_2, formation of zirconium nitride, and evolution of nitrogen dioxide
ZrN*[1]	1573–2073	Forms nonstoichiometric zirconium nitride
CeO_2	1973	Initial reaction (threshold temperature)
Nd_2O_3*[4]	1673–1873	Forms solid solution Nd_2ZrO_7–ZrO_2
Tl_2O_3*[4]	673–1273	No reaction
ThO_2*[1]	2573	Weak reaction, with slight adhesion of specimens; no liquid phase
ThO_2*[4]	2173	Very weak reaction
UO_2	2073	Forms solid solution

1 – Continued

1	2	3
$ZrSi_2$	2223	Forms solid solution
$MoSi_2$*[3]	1573–1873	No reaction
	Nb_2O	
C	1433	Forms Nb_2C
C	1533	Forms NbC
	Nb_2O_3	
BN*[2]	2023	Forms Nb_3B_4
	Nb_2O_5	
NbC*[1]	1273–1773	Reaction begins at 1573°K, with formation of new phase
WC*[1]	1673	Forms W–Nb alloy (100% in 30 min)
	MoO_3	
C*[1]	773–973	Forms lower oxides of molybdenum
	BaO	
UO_2	2073	Forms solid solutions
	La_2O_3	
W*[1]	1373	Forms $3La_2O_3 \cdot WO_3$
BN*[2]	2023	Forms new phase LaB_2
	CeO_2	
W*[1]	1973	Vigorous reaction
UO_2	1273	Forms continuous series of solid solutions
	Nd_2O_3	
UO_2	2473	Forms solid solutions
	HfO_2	
W*[1], Re*[1] (60 Mo + 40 Re)*[1]	2623	Slight color change
TiC	1773–2473	Forms oxycarbides
HfC*[1]	1273–2273	No reaction
HfN*[2]	–	No reaction
Ti_2O_3*[4]	673–1273	No reaction

1	2	3
	Ta_2O_3	
BN*[2]	1873–2273	Forms TaB, and at 2273°K forms TaB_2
	Ta_2O_5	
WC*[1]	1673	Forms W–Ta alloy (100% in 30 min)
UO_2	1473	Forms U_2TaO_7 and TaO_8
	WO_3	
BN*[2]	1673	Forms tungsten
	ThO_2	
Be*[2]	Up to 1623	No reaction in solid phase
B*[1]	1873–2273	Forms thorium borides
C*[1]	1673, 2273	ThO_2 is reduced
C*[4]	1873	ThO_2 is reduced
Si*[2]	Up to 1693	Silicon penetrates along grain boundaries; oxide is corroded
Ti*[2]	2073	Titanium penetrates along grain boundaries; no new phase detected
Ni*[2]	1723	No reaction in solid phase
Zr*[2]	1673, 1873	Weak reaction; oxide is corroded
Zr*[2]	2073	Zirconium penetrates along grain boundaries; ZrO_2 is formed
Nb*[2]	1873–2073	Weak reaction; at 2073°K, niobium penetrates along grain boundaries, and oxide is corroded
Nb*[1]	2373	No reaction
Mo*[2]	2073	No reaction
Mo*[1]	2173–2573	Reacts, with formation of film and weak adhesion of specimens; reaction weak at 2573°K
Ta*[1]	2173	Initial reaction (threshold temperature)

1 – Continued

1	2	3
W*[1]	2473–2573	Reacts slightly, with reduction of ThO_2; strong adhesion of specimens at 2573°K
Re,*[1] (60 Mo + 40 Re)*[1]	–	No reaction when held 2 h
WB*[1]	2373	No reaction
ThB_4*[1]	2373	No reaction
ThB_6*[1]	2373	Forms ThB_4
TiC	1773–2273	Forms ThC
UC*[2]	2723	No reaction
UO_2	2073	Forms solid solution
	UO_2	
Be	873	Slight reaction
C	1473–2473	Reacts, with formation of uranium monocarbide up to 2273°K, and dicarbide above 2473°K
Al	773	Forms UAl_3 and UAl_4
Si	1603	Reacts
Fe	1073	Reacts when pressed
Ni	1673	Weak reaction
Ni*[1]	1773	No reaction
Zr	673–723	Weak reaction
Nb	1273	Forms solid solution
Nb*[1], *[2]	2773	No reaction
Mo*[1], *[2]	2893	No reaction
Ta*[1], *[2]	2203–3033	No reaction
Ta*[1]	2423	Forms dark gray phase of tantalum oxides that volatilizes upon prolonged heating
W*[1], *[2]	2203–3033	No reaction
Au	Up to 1336	No reaction in solid phase
WB_2*[1]	3033	No reaction
Mo_2C*[1]	2693	No reaction
HfC*[1], *[2]	2203–3033	No reaction
UC	1723	Forms oxycarbides and free uranium
TaN*[1]	3033	No reaction
UN	1473–1873	No reaction

1	2	3
PuO_2	1773	Forms solid solutions
$CaO-ZrO_2$*3	2073	Forms solid solutions
UP	2663	Forms eutectic
	PuO_2	
C*1	1273–1473	Reacts, with reduction of PuO_2 to Pu_2O_3, and at 1473°K with formation of plutonium carbide
Nb*1	1673, 1773	No reaction when held for periods up to 100 h
Mo*1	1673, 1773	No reaction at 1673°K when held for 110 h; at 1773°K, forms a new phase
Ta*1	1673, 1773	No reaction when held for 110 h; at 1773°K, a new phase (40–60 μm) is formed when held for 1000 h
Ta	2003–2403	Forms Pu_2O_3 and TaO
	Cm_2O_3	
V	1523	No reaction
Zr	1523	Reacts
Nb	1523	No reaction
Mo	1923, 2133	No reaction
Ta	1523, 1923, 2123	No reaction
W	1923, 2123	No reaction
(Nb + 1% Zr)	1523	No reaction
(Ta + 10% W)	1523, 1923, 2123	Reaction starts at 1923°K
(W + 26% Re)	1923, 2123	No reaction

*1 Vacuum. *2 Neutral medium (argon, helium, or nonoxidizing gas mixtures). *3 Reducing or protective medium (hydrogen, nitrogen, or carbon monoxide). *4 Air. *5 For more detailed information, see [639].

Liquid metal	Temperature, °C	Atmosphere	Contact angle θ, deg
1	2	3	4
		BeO [565, 566, 260]	
Al	1000	Vacuum	Metal soaks through ceramic
Si	1450	Hydrogen	88
Si	1450	Helium	76
Ni	1500	Vacuum	128
Ni	1500	Hydrogen	152
Ni	1500	Helium	146
Ni+0,1%O_2	1500	"	100
Fe	1550	Hydrogen	147
Fe	1550	Helium	127
Pb	600—900	Hydrogen	138—132
Ni	1500	Vacuum	132
Ni	1500	Hydrogen	152
Ni	1500	Helium	132
Ni	1500	Argon	125
Pb	600—900	Hydrogen	136—127
Cu	1100—1300	Argon	140—122
Ag	1075	"	137
Ag	1075	Air	90
Ag	1100	"	115—120
Sn	800—1100	Vacuum	158—139
V	1800	"	35
Cr	1900	"	100
Pt	1790	"	125
		MgO [565, 567, 260]	
Si	1450	Hydrogen	101
Si	1450	Helium	95
Fe	1550	"	130
Fe	1550	Vacuum	123
Ni	1500	Helium	152
Cu	1150	"	160
Sn	1100	"	139
Ag	1235	Vacuum	136
		(MgNi)O [568]	
Sn	700—1000	Vacuum	130—10
Ag	1000—1300	Argon	144—53
Cu	1000—1300	"	134—24

1	2	3	4
	Al_2O_3 [565, 566, 567, 569, 1]		
Al	940	Argon	170
Al	1255	"	48
	Al_2O_3 [565, 567, 569]		
Ag	1000	Vacuum	144
Ag	1100	"	148
Ag+5%C	980	Air	28
Ag+10%C	980	"	14
Pb	400	Hydrogen	144
Pb	900	"	133
Si	1450	Helium	80
Si	1450	Hydrogen	82
Co	1500	"	125
Cu	1200	Vacuum	138
Cu	1200	Hydrogen	136
Cu	1200	Argon	141
Cu+1,5% O_2	1300	Helium	38
Bi	T_m	Vacuum	90
Ge	T_m	"	>90
In	T_m	"	>90
	Al_2O_3 [565, 566, 567, 1]		
Ba	T_m	Vacuum	90, metal soaks through ceramic
Sn	800	"	90
Sn	800	Hydrogen	150
	Al_2O_3 [565, 566, 1]		
Sn+10%In	1100	Vacuum	174
Sn+10%In	800	"	138
Ni	1500	Helium	150
Ni	1500	Hydrogen	133
Ni	1500	Vacuum	128
Ni	T_m	Hydrogen	115
Ni+5%V	T_m	"	95
Ni+10%Mn	T_m	"	95
	Al_2O_3 [565, 566, 1, 407]		
Ni+20%Cr	T_m	Hydrogen	85
Ni+8,7%Cr	1500	Vacuum	108

2 — Continued

1	2	3	4
Ni+0,8%Ti	1500	Vacuum	90
Ni+3%In	1500	Helium	132
Fe	1550	Vacuum	141
Fe	1550	Nitrogen	139
Fe+8%Si	1550	Helium	114
Fe+0,7%O_2	1570	»	109
Fe+0,47%C	1570	»	106
Fe+2%S	1570	»	101
Cr	1900	»	65
		Al_2O_3 [570]	
Cr	1900	Reducing	~1—10
Cr	1950	Inert	~45
		Al_2O_3 [565, 566, 1, 407]	
Cr	1950	Air	~1—10
Cr	1950	Carbon-containing	~45
Mn	1300	Vacuum	~0—10
Mn	1300	Reducing	~90
Mn	1300	Inert	~45
Mn	1300	Air	~45
Mn	1300	Carbon-containing	~45
Mn	1300	Vacuum	~45
Fe	1580	Reducing	~90
Fe	1580	Inert	~90
Fe	1580	Air	~45
Fe	1580	Carbon-containing	~45
Fe	1550	Vacuum	~90
Au	1100	Helium	138
Cu	1100	Reducing	>90
Cu	1100	Inert	>90
Cu	1100	Air	0
Cu	1100	Carbon-containing	0—45
Cu	1100	Vacuum	>90
Co	1550	Reducing	>90
Co	1550	Inert	>90
Co	1550	Air	0—20
Co	1550	Carbon-containing	>90
Co	1550	Vacuum	>90
Co	1450	Reducing	~90
Ni	1450	Inert	~90
Ni	1450	Air	~90
Ni	1450	Carbon-containing	~90
Ni	1450	Vacuum	~45
Fe+Mn	1500	Reducing	~1—45
Fe+Mn	1500	Inert	~1—45
Fe+Mn	1500	Air	~45

1	2	3	4
Fe+Mn	1550	Carbon-containing	~45
Fe+Mn	1550	Vacuum	~45
Fe+Cr	1650	Reducing	~40
Fe+Cr	1650	Inert	~40
Fe+Cr	1650	Air	~30
Fe+Cr	1650	Carbon-containing	~45
Fe+Cr	1650	Vacuum	~0—40
Mn—Ti	1600	"	~45
Cu—Ti	1600	"	~45
Co—Ti	1600	"	~45
Ni—Ti	1550	"	~90
Sn	250	Reducing	>90
Pb	350	Reducing	>90
Pb	450	"	90
		Al_2O_3—Cr_2O_3 (solid solution) [571, 569, 1] 0%—100%	
Fe	1550	Argon	91—88,5
		90%—10%	
Fe	1550	Argon	108—105
		95%—5%	
Fe	1550	"	126—124
		SiO_2 [569, 1]	
Hg	Room temp.	Air	140
Cu	1100	Vacuum	148
Cu	1100	Helium	134
Fe	1550	Nitrogen	115
Ni	1550	"	125
Ag	1000	Helium	139
Sn	900	"	127
Pb	800	"	164
Au	1100	Helium	140
		SiO_2 (opaque quartz)	
Ferro-manganese	1550	Nitrogen	0

2 — Continued

1	2	3	4
		CaO [572, 573]	
Fe	1550	Vacuum	132
Ni	1500	"	135
		TiO_2 [565, 566]	
Si	1450	Helium	107
Ni	1500	Vacuum	104
Ni	1500	Hydrogen	143
Ni	1500	Helium	120
Ni	1500	Argon	112
Ni	1550	Vacuum	104
Ni	1500	Hydrogen	143
Ni	1500	Helium	121
Co	1500	Helium	109
Fe	1550	Vacuum	72
Fe	1550	Hydrogen	84
Fe	1550	Helium	0
Fe	1550	Argon	0
Ag	1200	Air	73
		V_2O_5 [574]	
Ag	1000	Inert	>90, metal soaks through
		Cr_2O_3 [575]	
Fe	1550	Argon	88
Ni	1450	Vacuum	65
Ni+2%C	1450	"	103
Ni+3,8%C	1350	"	98
Ni+7,5%C	1450	"	90
Ni+1%Cr	1450	"	85
Ni+5%Cr	1450	"	96
Ni+10%Cr	1450	"	98
Ni+15%Cr	1450	"	103
Ni+25%Cr	1450	"	105
		Fe_3O_4 [566]	
Sn	1000	Argon	52
Cu	1200	"	57
Sn	1000	Vacuum	0

1	2	3	4
		CoO [566, 573]	
Ni	1500	Argon	58
Co	1500	"	70
Sn+0,4%Ti	900	"	87
		NiO [566]	
Cu	1200	Argon	68
Ag	1125	"	66
Sn	920	"	27
Ni	1500	"	70
		NiO [566, 573]	
Fe	1550	Argon	88
		Cu_2O [1]	
Ag	980	Nitrogen + 2% O_2	0
		ZnO [567]	
Ag	1160	Argon	106
Ag	1200	Air	90
		ZrO_2 [565]	
Si	1450	Hydrogen	71
Si	1450	Helium	71
Ni	1500	Vacuum	118
Ni	1500	Hydrogen	130
Ni	1500	Helium	120
Fe	1550	Vacuum	92
Fe	1550	Hydrogen	111
Fe	1550	Helium	102
		CdO [567]	
Ag	970	Argon	112
Ag	970	Hydrogen	90
		TiO_2 [565, 567, 260]	
Ni	1500	Helium	134
Fe	1550	Hydrogen	111

2 – Continued

1	2	3	4
Fe	1550	Helium	100
Ag	1000	Air	90
Ag	1300	"	90
Cr	1900	Hydrogen	92
	UO_2 [567, 260]		
Na	200—500	Hydrogen	152—40
Al	1100	Argon	130
Al	1100, after holding 22 h	"	46
Si	1420	"	90
Sn	500—1100	"	140—110
Pb	400—700	"	137—124
Bi	500—1100	"	118—95
	Glass [1]		
Hg	Room temp.	Air	149
Cu	1150	Helium	60
Cu	1150	Hydrogen	60
Cu	1150	Air	0
Cu	1150	Nitrogen	0
Ag	1100	Helium	70
Ag	1100	Hydrogen	73
Ag	1100	Air	0
Ag	1100	Nitrogen	0
Au	1100	Helium	60
Au	1100	Hydrogen	45
Au	1100	Air	55
Au	1100	Nitrogen	53
Ni	1500	Helium	55
Ni	1500	Hydrogen	60
Ni	1500	Air	0
Ni	1500	Nitrogen	0
Pd	1600	Helium	55
Pd	1600	Hydrogen	40
Pd	1600	Air	25
Pd	1600	Nitrogen	20
Pt	1700	Helium	60
Pt	1700	Hydrogen	43
Pt	1700	Air	0
Pt	1700	Nitrogen	0
	Quartz glass [577]		
Fe	1600	Air	52
Fe	1600	Nitrogen	111

1	2	3	4
	Chrome-magnesite [577]		
Fe	1600	Air	63
Fe	1600	Nitrogen	102
	Magnesite [577]		
Fe	1600	Air	90
Hg	25	Nitrogen	113
	Chromite [577]		
Fe	1600	Air	72
Fe	1600	Nitrogen	81
	Glass [1]		
Hg	Room temp.	Air	137
Na	300	Vacuum	90
	Black marble [579]		
Hg	Room temp.	Air	149
	Diabase [579]		
Hg	Room temp.	Air	142
	German granite [579]		
Hg	Room temp.	Air	134
	Calcite [579]		
Hg	Room temp.	Air	141
	Pyrite [579]		
Hg	Room temp.	Air	142
	Mica [1]		
Hg	Room temp.	Air	126

2 — Continued

1	2	3	4
Ferrite (42,8% Fe_2O_3; 42,07% MnO; 14,3% MgO [566])			
Cu	1200	Vacuum	95
Ferrite (40% Fe_2O_3; 40% MgO; 20% [MnO; (CoO + 5% MgO)]			
Cu	1100	Vacuum	67
$MgO \cdot Cr_2O_3 \cdot 3MgO \cdot 2Cr_2O_3$ [566]			
Sn	1000	Argon	155
Cu	1100	"	164
Cu	1500	"	160
Ni	1530	"	133
Co	1530	"	113
Fe	1560	"	90
Magnesite [1]			
Fe	1560	Nitrogen	126
Ferro-manganese	1560	Nitrogen	79
Silico-manganese	1560	—	105
Scorified magnesite			
Fe	1560	Nitrogen	100
Chrome-magnesite [1]			
Steel (1,6%C, 0,2%Mn)	T_m	CO atmosphere	166
Slag*[1]			
Matte (Ni—21,8%; Fe—49,9; S—25,3%)	1050	Nitrogen	35
Slag*[2]			
Matte (Ni-21,8%; Fe-49,9%; S-25,3%)	1050	Nitrogen	26

*[1]Slag composition: 16% FeO; 20% CaO; 39.7% SiO_2; 9% Al_2O_3; 10.2% MgO [1]. *[2]Slag composition: 16% FeO; 20% CaO; 39.7% SiO_2; 9% Al_2O_3; 10.25% MgO.

Reagent	Temperature, °C	Nature of interaction
1	2	3
		BeO [1]
Ni	1600	No reaction (hel)
Ni	1800	No reaction (hel)
Be	1800	Beryllium does not soak into BeO and apparently does not react (hel)
Si	1600	Slight interaction; molten silicon soaks through BeO along grain boundaries (hel)
Nb	1800	A thin transition layer appears; slight disintegration of crucible occurs (hel)
Ti	1800	Corrosion of BeO; absorption of Ti; dark coloration of crucible (hel)
Zr	1800	Absorption of Zr and disintegration of crucible (hel)
Mo	1800	No reaction (hel)
C	2300	Reaction begins; carbon is covered with a metallic film (vac)
N_2	2000	Yellow film on crucible (vac)
N_2	2000	Silvery film on crucible (vac)
ThO_2	2100	Melts, with complete liquefaction at 2200°C (vac)
ZrO_2	1800	Slight adhesion and glazing of walls (vac)
MgO	1800	Liquid phase appears; strong adhesion of melt to BeO crucible (vac)
Al_2O_3	1900	Strong reaction (rdx)
MgO	1900	Strong reaction and fusion (rdx)
ZrO_2	1900	Strong reaction (rdx)
Spinel	1900	Strong reaction (rdx)
ThO_2	1900	Strong reaction (rdx)
		MgO [1]
Ni	1800	Rather strong interaction and fusion of transition layer (hel)
Mo	1800	No reaction (hel)
Be	1600	Reacts and permeates MgO crucible (hel)

3 – Continued

1	2	3
Si	1410	Start of reaction on crucible surface (hel)
Si	1600	Active reaction, forming Mg_2SiO_4 (hel)
Zr	1800	Strong interaction (hel)
Nb	1600	No reaction (hel)
Nb	1800	Weak reaction (hel)
C	1800	Start of reaction; no reduction (vac)
N	2000	Reduction of MgO; slight adhesion (vac)
Mo	1600	Silvery film on MgO (vac)
BeO	1800	New phase appears; MgO and BeO adhere (vac)
		Al_2O_3 [1]
Mo	1800	No reaction (hel)
Ni	1410	No reaction (hel)
Ni	1600	No reaction (hel)
Ni	1800	Weak reaction (hel)
Nb	1600	Weak reaction (hel)
Nb	1800	Weak reaction, with formation of Nb_2O_3 (hel)
Ti	1600	Weak reaction, with formation of TiO_2 (hel)
Be	1900	Strong reaction (hel)
Be	1600	No reaction (hel)
Be	1800	Weak reaction, with formation of BeO (hel)
Si	1410	Weak reaction, with formation of SiO_2 and $3Al_2O_3 \cdot 2SiO_2$ (hel)
Si	1600	Strong reaction, with formation of SiO_2 and $3Al_2O_3 \cdot 2SiO_2$ (hel)
Zr	1600	Weak reaction, with formation of ZrO_2 (hel)
Zr	1900	Strong reaction, with formation of ZrO_2 (hel)
BeO	1900	Formation of spinel on crucible surface (rdx)
Spinel	1900	Very weak reaction (rdx)
ZrO_2	1900	Strong reaction (rdx)
ThO_2	1900	Strong reaction (rdx)

1	2	3
		Al_2O_3 [305]
W	2080	Chemical reaction, with formation of Al and WO_3 (vac)
Mo	2080	Reacts, with formation of Al and MoO_3 (vac)
Ta	2080	Active chemical reaction, with decomposition of Al_2O_3 (vac)
		Al_2O_3 [1]
Nb	2080	Active chemical reaction (vac)
		SiO_2 [305]
W	2700	Active chemical reaction, with formation of WO_3 (vac)
		CaO [1]
C	2000	Formation of metallic calcium and carbide (carb, CO)
		TiO_2 [1]
Mo	1600	No reaction (hel)
Ni	1450	No reaction (hel)
Ni	1600	No reaction (hel)
Ti	1600	Weak reaction (hel)
Be	1600	Strong reaction (hel)
Si	1410	Weak reaction (hel)
Si	1610	Strong reaction (hel)
Zr	1600	Strong reaction (hel)
W	1800	Chemical reaction, with formation of WO_3 and lower oxides of titanium (vac)
Mo	1880	Chemical reaction, with formation of MoO_3 (vac)
Ta	1880	Strong chemical reaction, with formation of tantalum oxides (vac)
Nb	1880	Active chemical reaction (vac)
		V_2O_5 [305]
W	700	Active chemical reaction, with formation of tungsten oxides (vac)
Mo	700	Active chemical reaction
Ta	700	Active chemical reaction
Nb	700	Active interaction (vac)

3 – Continued

1	2	3
		CuO [1]
Solutions of trivalent iron salts	–	Forms corresponding salts of bivalent copper and $Fe(OH)_3$ (vac)
		Cu_2O [1]
Solutions of trivalent iron salts	–	Forms ferrite, stable above 1373°K (vac)
		ZrO_2 [1]
Mo	1600	No reaction (hel)
Mo	1800	No reaction (hel)
Ni	1410	No reaction (hel)
Ni	1600	No reaction (hel)
Ni	1800	No reaction (hel)
Nb	1600	Weak reaction (hel)
Nb	1800	Strong reaction (hel)
Ti	1400	No reaction (hel)
Ti	1600	Weak reaction, with formation of TiO_2 (hel)
Be	1400	No reaction (hel)
Be	1600	Weak reaction, with formation of BeO (hel)
Si	1410	No reaction (hel)
Si	1600	Strong reaction, with formation of SiO_2 (hel)
Zr	1600	No reaction (hel)
Zr	1800	No reaction (hel)
W	1600	Yellow film on crucible; little change up to 2000°C, with appreciable decomposition at higher temperatures (vac)
Mo	2200	Slight adhesion; decomposition at 2300°C (vac)
ThO_2	2200	Slight adhesion; no liquid phase at 2300°C (vac)
BeO	1900	Slight adhesion as a result of fusion; complete melting at 2000°C (vac)
Graphite	1600	Reduction of ZrO_2 and formation of carbide (vac)

1	2	3
MgO	2000	Liquid phase, welding at 2100°C (vac)
Al_2O_3	1900	Strong reaction (rdx)
MgO	1800	Very weak reaction (rdx)
	1900	Weak reaction (rdx)
	1900	Very weak reaction (rdx)
Spinel	1900	Very weak reaction (rdx)
Carbon, hydrogen, sulfur vapor	2000	Almost no attack (rdx)
Iron silicates	Above 2000	Interacts (rdx)
Sodium and potassium sulfates, acid salts, fluorides		Act very slowly (rdx)
N_2	~2000	Forms nitrides
C		Forms carbides (rdx)
		Nb_2O_5 [305]
W	1460	Active chemical reaction, with formation of WO_3 (vac)
Mo	1460	Active reaction, with formation of MoO_3 (vac)
Ta	1460	Active interaction (vac)
Nb	1460	Chemical reaction (vac)
		MoO_3 [305]
W	825	Active chemical reaction, with formation of tungsten oxides WO_3 and $W_{18}O_{49}$ (vac)
Mo	825	Active chemical reaction (vac)
Ta	825	Active interaction (vac)
		CeO [1]
ThO_2		Gives mixture melting above 2600°C (vac)
Be		Gives eutectics
MgO		Gives eutectics
CaO		Gives eutectics

3 – Continued

1	2	3
Al_2O_3 and other oxides		Gives eutectics
Cr_2O_3		Forms $Cr_2O_3 \cdot CeO_2$
		Ta_2O_5 [305]
W	1910	Active reaction, with formation of TaO and WO_3 (vac)
Mo	1910	Active reaction, with formation of MoO_3 (vac)
Ta	1910	Active chemical reaction (vac)
Nb	1910	Active interaction (vac)
		WO_3 [305]
W	1490	Active interaction (vac)
Mo	1490	Active reaction, with formation of oxides MoO_3 (vac)
Ta	1490	Active interaction (vac)
Nb	1490	Active chemical reaction (vac)
		ThO_2 [1]
Pt	1800	ThO_2 crucibles do not undergo any breakdown, and will carry a platinum melt in an oxidizing medium (vac)
Mo	1800	No reaction (hel)
Ni	1600	No reaction (hel)
Nb	1800	Weak reaction (hel)
Ti	1670	No reaction (hel)
Ti	1800	No reaction (hel)
Be	1600	Weak reaction
Si	1600	Active reaction (hel)
Zr	1600	No reaction (hel)
Zr	1800	No reaction (hel)
Graphite	2000	Reduction of ThO_2, with formation of carbides (vac)
BeO	2100	Fusion; complete melting at 2200°C (vac)
MgO	2200	Breakdown of ThO_2; no adhesion to crucible (vac)
W	2200	Very slight reduction of ThO_2; weak reaction at 2300°C, severe welding (vac)

1	2	3
Mo	1900	Slight reaction below 2300°C, adhesion to crucible (vac)
Al_2O_3	1600	No reaction (rdx)
MgO	1800	No reaction (rdx)
BeO	1800	Strong reaction (rdx)
ZrO_2	1800	No reaction (rdx)
Spinel	1800	No reaction (rdx)
		UO_2 [1]
C	1500	Reaction begins at 1350°C, continuing effectively under conditions of monovariant equilibrium (rdx)

Note: Abbreviations in parentheses indicate the atmosphere in which the interaction between reagent and oxide takes place: vac = vacuum; hel = helium; rdx = typical gas-furnace reducing/oxidizing atmosphere; carb = carbon.

CHAPTER X

PHASE DIAGRAMS OF BINARY ELEMENT/OXYGEN SYSTEMS

Phase diagrams of binary systems are presented in this chapter, along with descriptions of some of these systems.

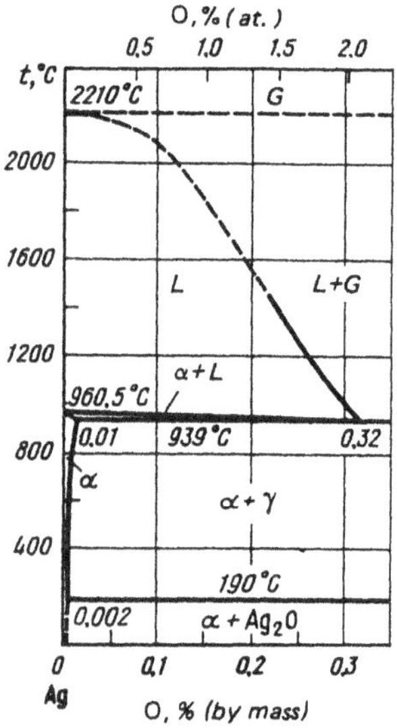

Fig. 40. Phase diagram of Ag–O system [16].

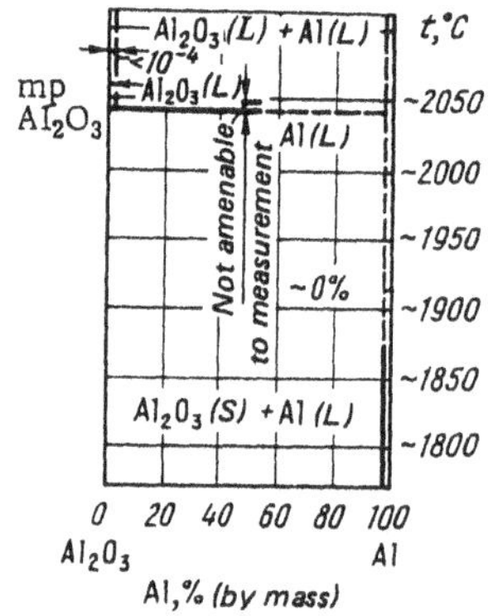

Fig. 41. Phase diagram of Al–O system [580].

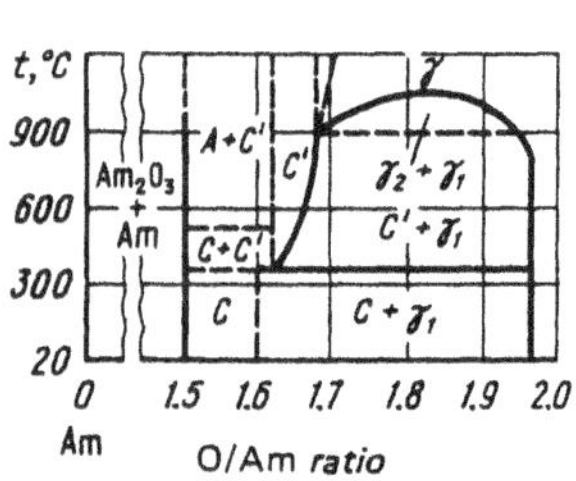

Fig. 42. Phase diagram of Am–O system [581]. A) Hexagonal Am_2O_3; C) low-temperature Am_2O_3; C') high-temperature Am_2O_3; γ, γ_1, γ_2) AmO_{2-x}. Diagram covers states in Am/AmO_2 region. With ratio $O/Am < 1.5$, a two-phase region $Am + Am_2O_3$ is observed. The Am_2O_3 exists in two modifications: a low-temperature form C (body-centered cubic) and a high-temperature form A (hexagonal). At 350°C, as a result of interaction of the oxygen-rich phase C with $AmO_{1.98}$, the compound $AmO_{1.62}$ is formed (phase C'). At higher temperatures, the region of this phase extends to a ratio $O/Am = 1.68$. At temperatures above 1020°C and in the region of compositions $1.7 \leq O/Am \leq 2.0$, there is a face-centered cubic phase γ (AmO_{2-x}). At lower temperatures there is a two-phase region $\gamma_1 + \gamma_2$ consisting of two phases AmO_{2-x} of different composition (also face-centered cubic). At still lower temperatures, the two-phase region consists of the phases C' and γ_1; and at room temperature, the γ_1 phase with the composition $AmO_{1.98}$ is in equilibrium with C (Am_2O_3).

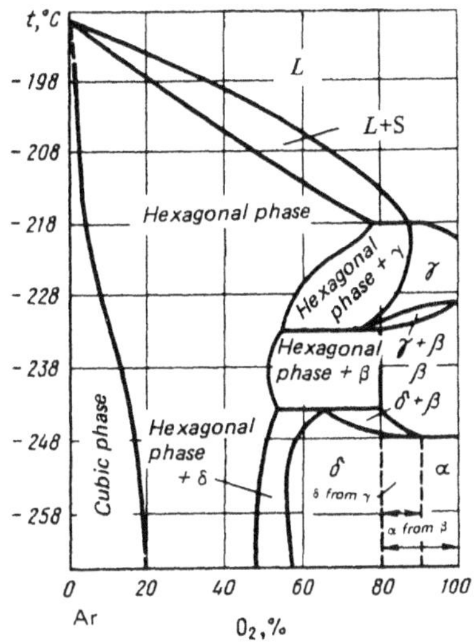

Fig. 43. Phase diagram of Ar–O system [1]. Below the solidus line are single-phase and two-phase regions of solid solutions. In the argon-rich part of the system, directly under the solidus line, there is a broad region of homogeneity of a phase with a densely packed hexagonal structure. This phase is metastable in pure argon, but even small quantities of oxygen stabilize the phase. When the temperature is lowered, it changes over to a face-centered cubic phase containing from 1% to 20% O_2. With contents of O_2 from 20% to 50%, the hexagonal phase is preserved down to −273°C.

On the oxygen-rich side, the diagram is complicated because of conversions taking place in oxygen. The γ-O_2 dissolves up to 12.5% Ar, after which there is formed a broad two-phase region (from 55 to 87.5% O_2) consisting of a solution of Ar in γ-O_2 plus a solution of oxygen in hexagonal Ar.

β-O_2 dissolves up to 21% Ar; at higher contents, the Ar is also in equilibrium with the hexagonal phase, forming a two-phase region. The α-oxygen may contain up to 20% Ar.

Between the regions of homogeneity of the α and hexagonal phases, we find a structure designated by the letter δ. The position of the boundary of this phase on the side of higher oxygen contents cannot be determined unambiguously, since the region between 80% and 90% O_2 may pertain either to phase δ or α, depending on the conditions of specimen preparation. If a specimen with a content of 80% O_2 (for example) that has the structure γ is rapidly cooled below the $\beta - \alpha$ conversion temperature (23.5°K), the conversion product is δ; if this specimen is cooled from a region in which it consists of phase β, the pure α-structure is obtained. Therefore, the diagram shows as dashed lines the two positions of the boundary between the region α and the presumed two-phase region $\alpha + \delta$. Phase δ may exist between 55% and 90% O_2, but when cooled from the β region it can exist only between 55% and 80% O_2.

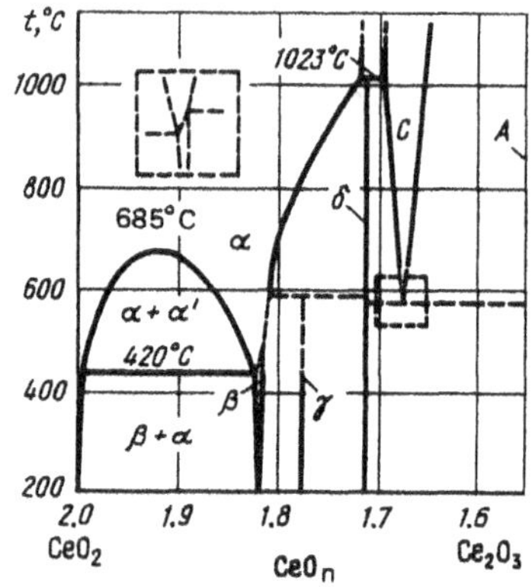

Fig. 44. Phase diagram of Ce–O system (from Ce_2O_3 to CeO_2) [1]. In the Ce–O system, the following phases are known: A, Ce_2O_3 (hexagonal); C, Ce_2O_3 (cubic), homogeneous from $CeO_{1.651}$ to $CeO_{1.688}$ at 1023°C (more precisely, 1028°C); δ phase ($CeO_{1.74}$); γ phase ($CeO_{1.778}$); β phase ($CeO_{1.818}$), and α phase (CeO_2). Shown in the figure is an example diagram in the CeO_2–Ce_2O_3 region. Phase δ is stable up to 1023°C; at higher temperatures it decomposes, with the formation of C and α. Between 636 and 1023°C, the highly ordered phase δ exists in equilibrium with an unordered phase $CeO_{1.72+x}$ and a phase C($CeO_{1.70-y}$). Below 600°C, two types of equilibrium may exist (shown in the diagram by dashed lines): 1) Phase C decomposes through a eutectoid reaction to δ and $CeO_{1.5}$; 2) the defective phase C is ordered, with the formation of a stoichiometric phase with the composition $CeO_{1.67}$.

Thus, below 600°C, two two-phase regions may coexist: $CeO_{1.74}$ + $CeO_{1.67}$ and $CeO_{1.66}$ + $CeO_{1.5}$. The ordered phase β ($CeO_{1.818}$) changes over at 420°C to the cubic α'-phase. The two-phase region α' + α becomes smaller with increasing temperature, and disappears at 685°C. Above this temperature, phases α' and α are identical. The ordered phase γ ($CeO_{1.778}$) decomposes at 600°C, with the formation of phases α and δ.

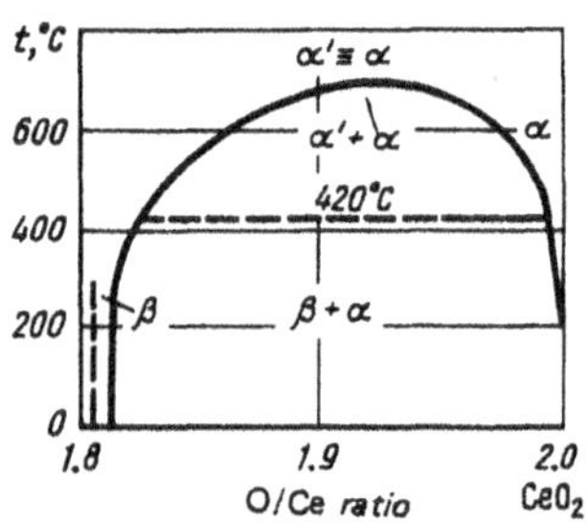

Fig. 45. Partial phase diagram of Ce–O system [1] in composition region $CeO_{1.8}$–CeO_2 and temperature region 0 – 800°C.

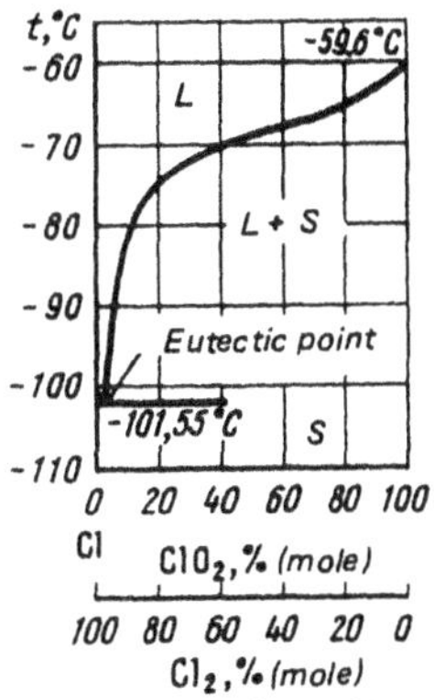

Fig. 46. Partial phase diagram of Cl–O system [582].

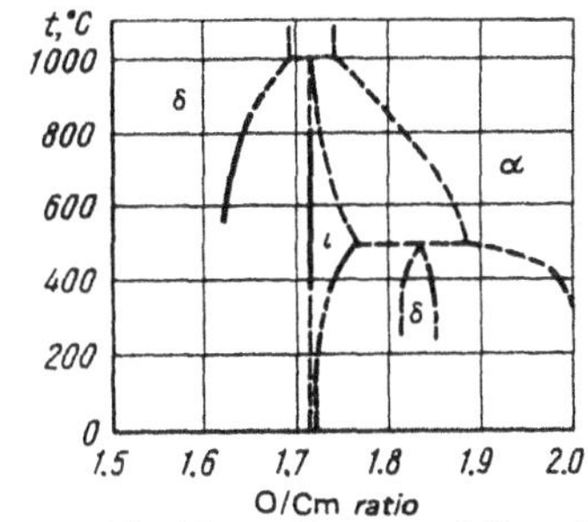

Fig. 47. Phase diagram of Cm–O system [583] is hypothetical, constructed by analogy with diagrams of the lanthanoids Ce–O and Pr–O, since the experimentally obtained temperature/composition isobars are very similar to the corresponding curves for the rare-earth oxides. The dashed lines are very indeterminate; the peritectoid decomposition temperatures of 1000°C for phase ι and 500°C for phase δ were chosen arbitrarily.

Phase ι corresponds approximately to $CmO_{1.72}$, phase δ to $CmO_{1.82}$. (The terminology used in designating the phases is that accepted for the lanthanoid oxides.) Phase δ has a rather broad region of homogeneity. The phase with the composition somewhat below O/Cm = 2.0 is apparently similar to the nonstoichiometer α phase of the lanthanoid oxides. The nonstoichiometric phase found at high temperatures is regarded as being similar to the σ phase in lanthanoid oxides.

The great width of the region of phase ι may be due to the presence of the difficutly removable Am as an impurity in the test specimen.

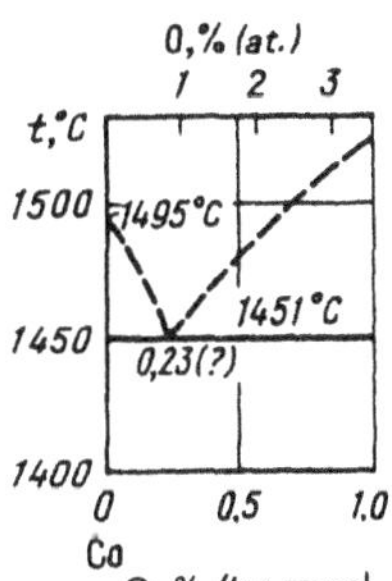

Fig. 48. Phase diagram of Co–CoO system [16].

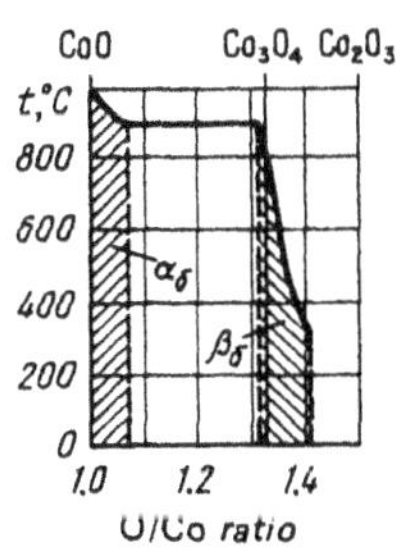

Fig. 49. Phase diagram of Co–O system [256].

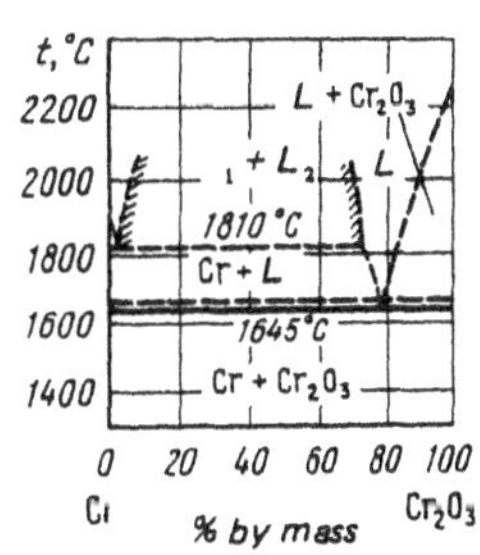

Fig. 50. Phase diagram of Cr–O system [584].

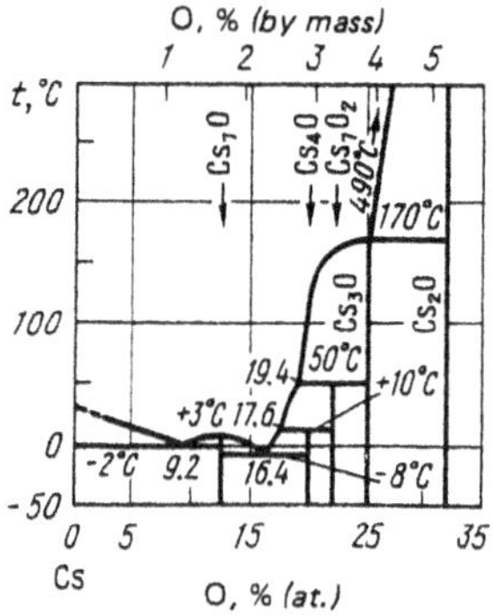

Fig. 51. Phase diagram of Cs–O system [586, Vol. 1].

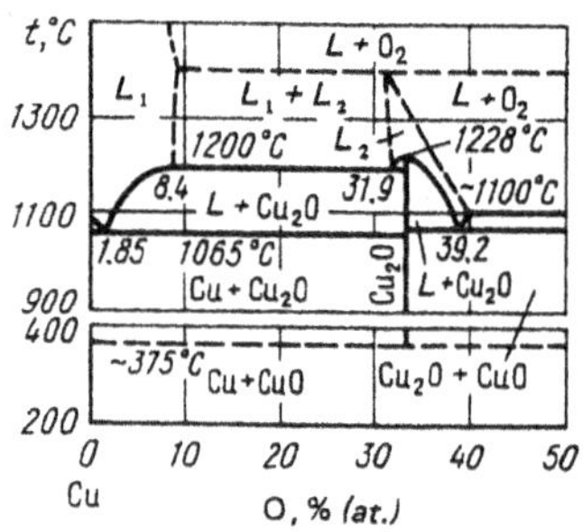

Fig. 52. Phase diagram of Cu–O system [587].

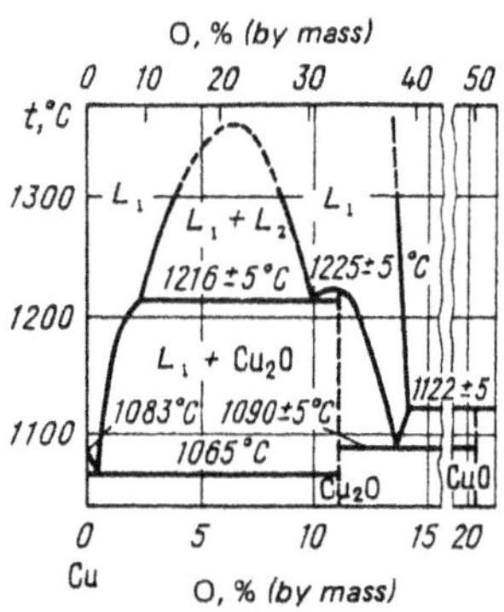

Fig. 53. Phase diagram of Cu–O system [588].

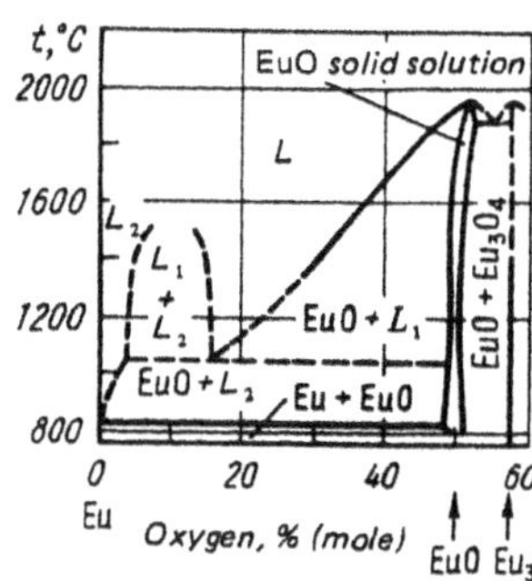

Fig. 54. Phase diagram of Eu–O system [589].

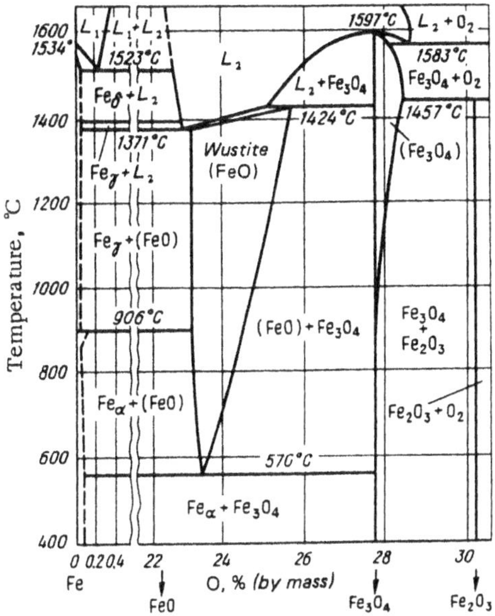

Fig. 55. Phase diagram of Fe–O system [639]. The oxide compounds of iron represented on the diagram are wustite (solid solutions of oxygen in ferrous oxide), magnetite (Fe_3O_4), and hematite (Fe_2O_3). The vacant lattice sites in wustite, upon additional oxidation, may be filled with ions of oxygen and trivalent iron; hence, wustite proves to be a solution of variable concentration, and is represented on the diagram not by a vertical line, but rather by a region (FeO). The minimal possible content of oxygen in wustite is almost invariant with temperature; the maximal content increases markedly with increasing temperature. Below the temperature of 570°C, at which the eutectoid conversion $Fe_\alpha + Fe_3O_4 \rightarrow 4FeO$ takes place, ferrous oxide cannot exist, even in solution.

At elevated temperatures, additional dissolution of oxygen in Fe_2O_3 is pos sible, leading to the formation of the region (Fe_3O_4).

The solubility of oxygen in all modifications of iron is low, so that the presence of oxygen does not bring about any change in the temperature for the transition of iron from one modification to the other. On the diagram, these solid solutions are merely nominal, are not to scale, and are bounded by dashed lines. The solubility of oxygen in Fe_δ is about 0.005% by mass, and the solubility in Fe_γ is somewhat lower. The conversion $Fe_\gamma \rightarrow Fe_\alpha$ is accompanied by an increase in oxygen solubility. At 900°C, the maximal content of oxygen in Fe_α approaches 0.03%, decreasing with decreasing temperature.

In the molten state, a broad region of immiscible liquids is observed, i.e., a melt of an oxygen solution in iron L_1 and a melt of oxides L_2. The melting point of wustite and the temperature for melting of wustite and its conversion to L_2 increases from 1371 to 1424°C as the oxygen content is increased. Magnetite (Fe_3O_4) melts congruently at 1597°C, being converted to L_2.

The homogeneous region of existence of L_2 is limited on the one side by compositions of liquid wustite with the lowest oxygen content, and on the other side by the line of greatest possible oxygen content in L_2 in the presence of gaseous oxygen under a pressure of 100 kPa.

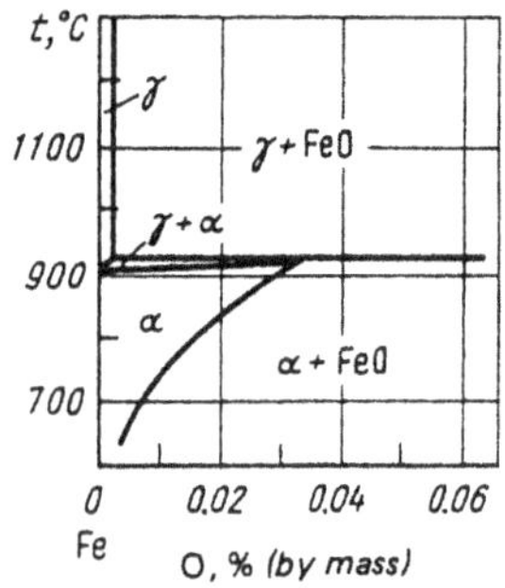

Fig. 56. Partial phase diagram of Fe–O system [1], characterizing transitions to solid state in temperature interval 600–1300°C, with oxygen content up to 0.06% by mass.

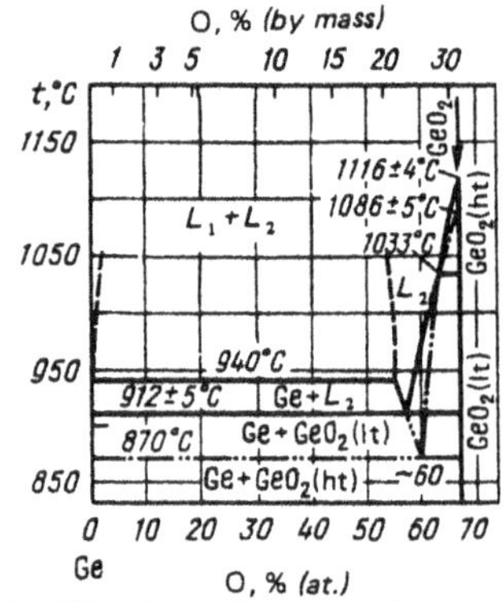

Fig. 57. Phase diagram of Ge–O system [586, Vol. 2]: — Stable equilibrium –·· – Metastable equilibrium with GeO_2(ht) –·– Metastable equilibrium with GeO_2 (lt).

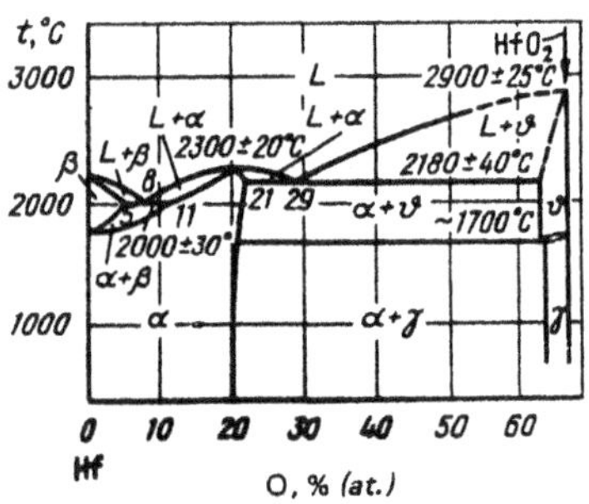

Fig. 58. Phase diagram of Hf–O system [1].

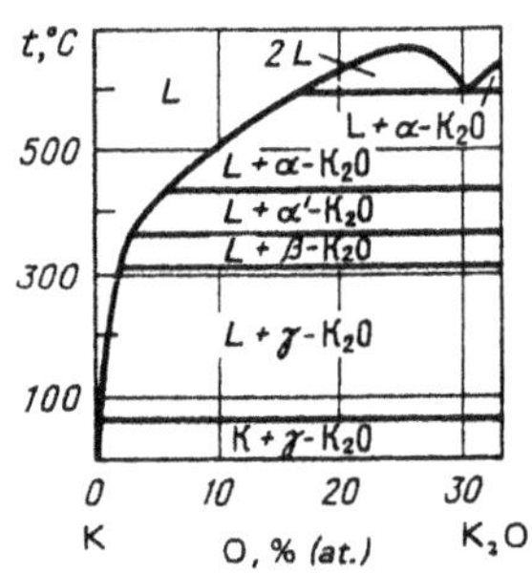

Fig. 59. Phase diagram of K–O system [590].

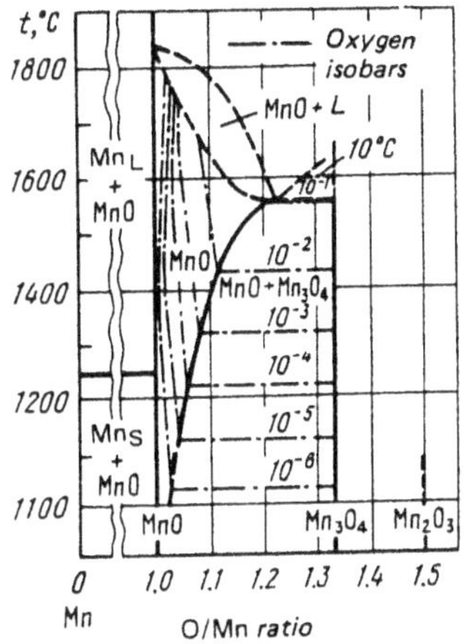

Fig. 60. Phase diagram of Mn–O system [591].

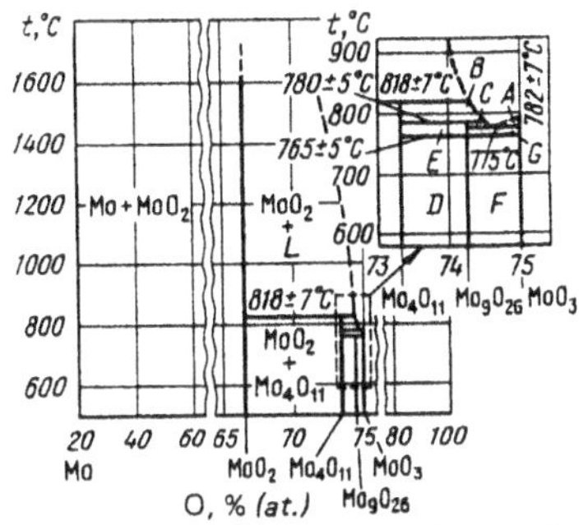

Fig. 61. Phase diagram of Mo–O system [593]: A) $MoO_3 + L$; B) $Mo_4O_{11} + L$; C) $Mo_9O_{26} + L$; D) $Mo_4O_{11} + Mo_9O_{26}$ (ζ); F) Mo_9O_{26}(ζ) + MoO_3; G) Mo_9O_{26}(β') + MoO_3.

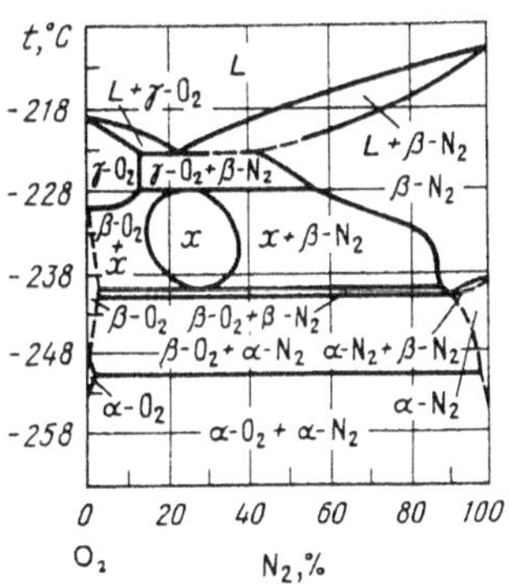

Fig. 62. Phase diagram of N–O system [594].

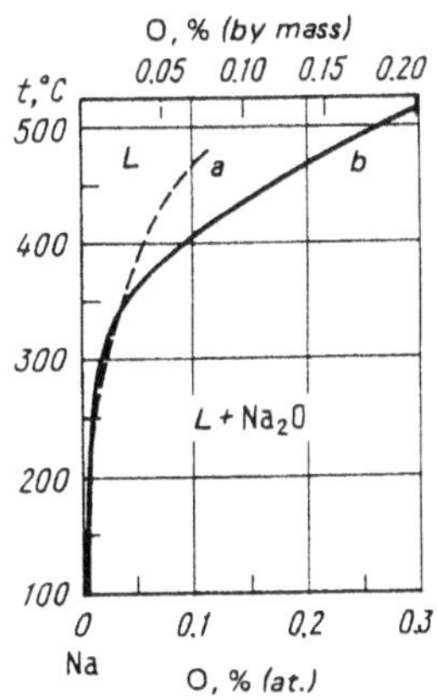

Fig. 63. Solubility of oxygen in liquid sodium, according to data of various investigators [586, Vol. 2].

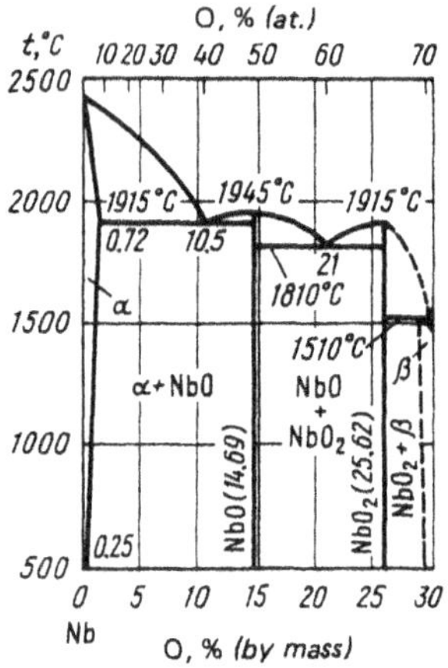

Fig. 64. Phase diagram of Nb–O system [595].

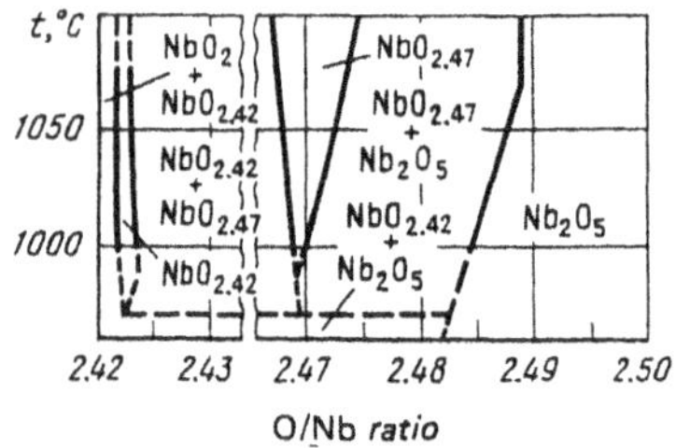

Fig. 65. Phase diagram of Nb–O system [596].

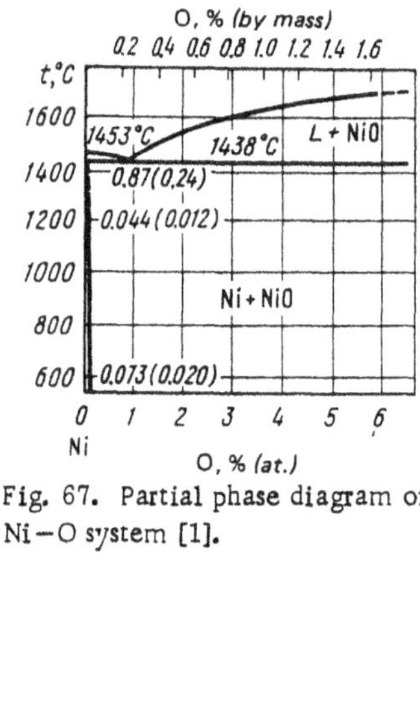

Fig. 67. Partial phase diagram of Ni–O system [1].

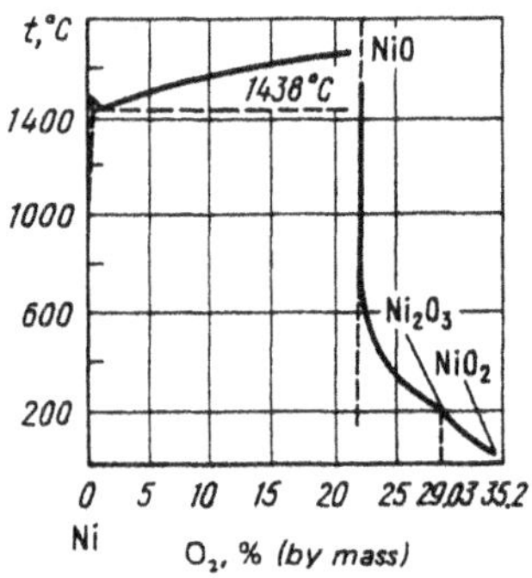

Fig. 66. Phase diagram of Ni–O system [597].

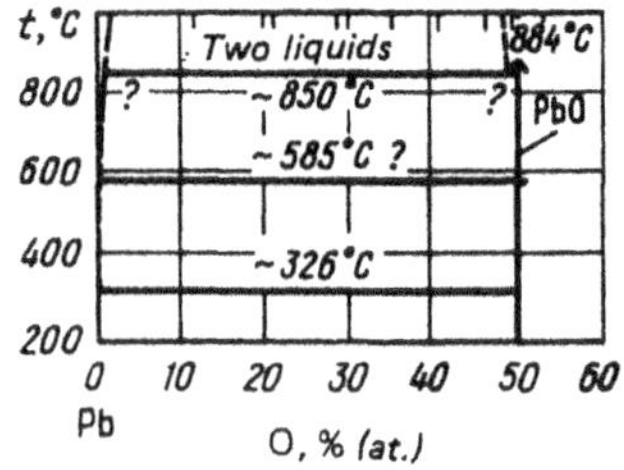

Fig. 68. Phase diagram of Pb–O system (in Pb–PbO region) [1].

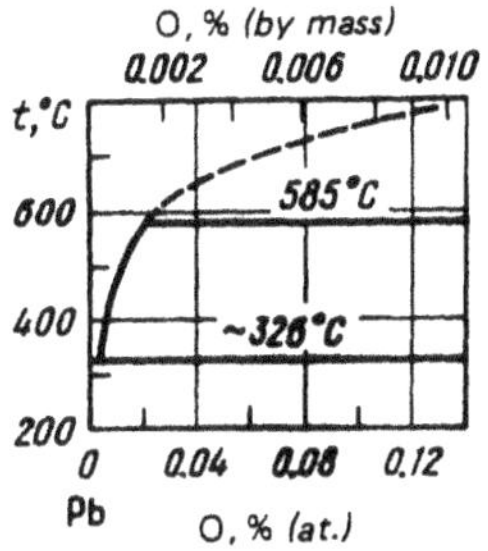

Fig. 69. Partial phase diagram of Pb–O system [1].

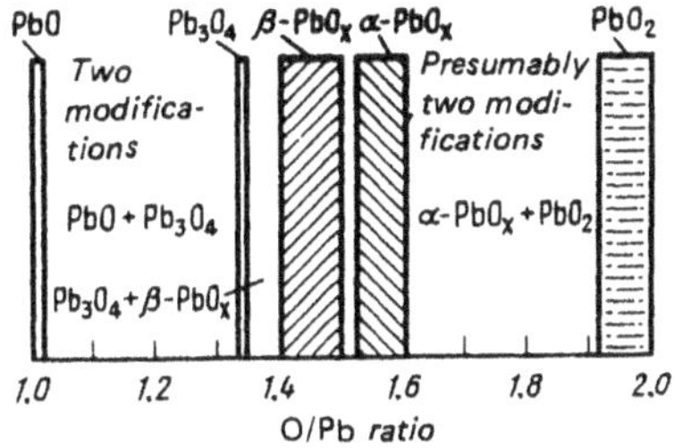

Fig. 70. Width of phase regions in Pb–O system (from PbO to PbO_2) [598].

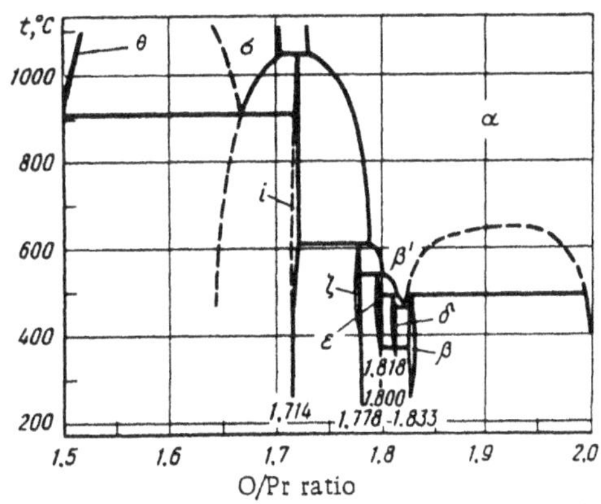

Fig. 71. Phase diagram of Pr–O system [599]. Between sesquioxide Pr_2O_3 and dioxide PrO_2 there are five ordered phases that are stable at low temperatures, with a narrow region of homogeneity, assigned to the homologous series Pr_nO_{2n-2}. These oxides are as follows: $Pr_7O_{12}(PrO_{1.714})$ – phase ι ; Pr_9O_{16}-$(PrO_{1.778})$ – phase ; $Pr_{10}O_{18}(PrO_{1.800})$ – phase ε; $Pr_{11}O_{20}(PrO_{1.818})$ – phase δ; $Pr_{12}O_{22}(PrO_{1.833})$ – phase β. There are also two nonstoichiometric phases at high temperatures and pressures: σ with composition below $PrO_{1.68}$ and α with composition above $PrO_{1.72}$. Phase β' is unordered phase β. Phase θ is the high-temperature form A of the sesquioxide.

For the majority of the ordered phases, a peritectoid conversion is observed, with transition to an unordered state.

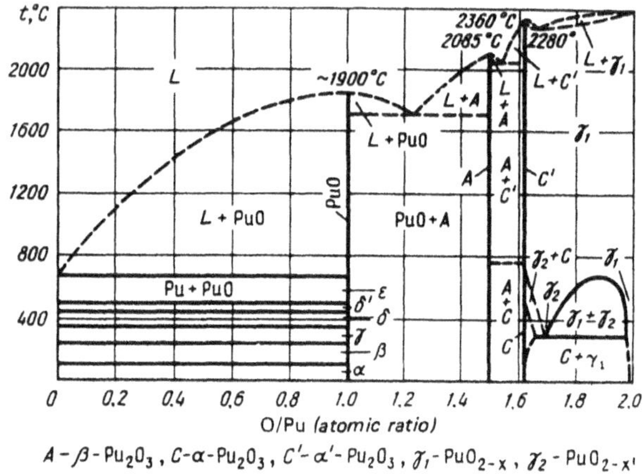

Fig. 72. Phase diagrams of Pu–O system [1] in Pu–PuO_2 region. For metallic plutonium, six polymorphic forms are known: $\alpha \xrightleftharpoons{120^\circ C} \beta \xrightleftharpoons{210^\circ C} \gamma \xrightleftharpoons{315^\circ C} \delta \xrightleftharpoons{460^\circ C} \delta' \xrightleftharpoons{475^\circ C} \varepsilon$ [52]. Plutonium forms the oxides PuO, Pu_2O_3, and PuO_2. The composition of the sesquioxide varies from $PuO_{1.51}$ to $PuO_{1.61}$. The Pu_2O_3 exists in two modifications: α-Pu_2O_3, the body-centered cubic C-form, changes at elevated temperatures into the high-temperature α'-Pu_2O_3 (C'-form), and β-Pu_2O_3 changes into the hexagonal A-form.

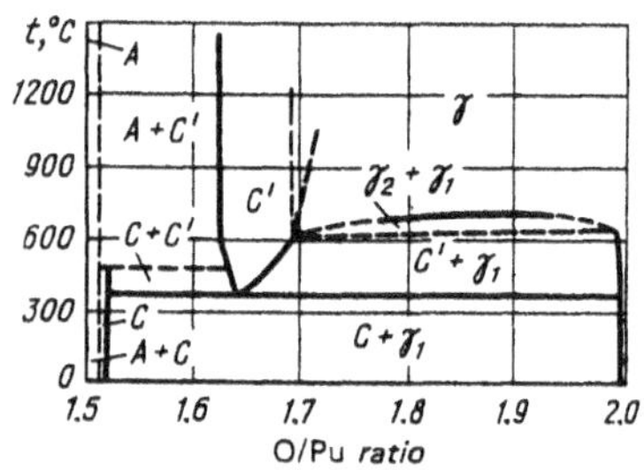

Fig. 73. Phase diagram of Pu–O system in Pu_2O_3 – PuO_2 region [581]. A is hexagonal Pu_2O_3; C is low-temperature Pu_2O_3; C' is high-temperature Pu_2O_3; γ, γ_1, and γ_2 are PuO_{2-x}. In the region of ratios O/Pu > 1.70 at high temperatures, there is a homogeneous region of a phase with variable composition PuO_{2-x} (γ).

Plutonium dioxide characteristically forms oxides with a deficiency of oxygen, in contrast to uranium dioxide, which forms oxides with an excess of oxygen UO_{2+x} (see Fig. 84). In the indicated region of compositions at low temperatures, there is a two-phase region PuO_{2-x} + $PuO_{2-x'}$ (or $\gamma_1 + \gamma_2$). Above 300°C, α-Pu_2O_3 (C) absorbs oxygen and forms a new cubic phase α'-Pu_2O_3 or $PuO_{1.61}$ (C'). The eutectoid decomposition of $PuO_{1.71}$ that is indicated in Fig. 72 has not been confirmed by more recent studies. The existence of phase A (β-Pu_2O_3) at room temperature has not been proven conclusively; it is possible that the only phase existing at room temperature is type C (α-Pu_2O_3).

In the Pu_2O_3 – PuO_2 region, the diagram is similar to that of the Am–O system (see Fig. 42).

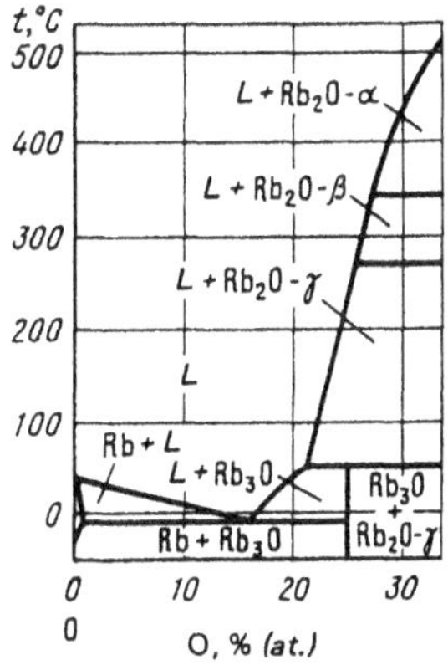

Fig. 74. Phase diagram of Rb–O system [600].

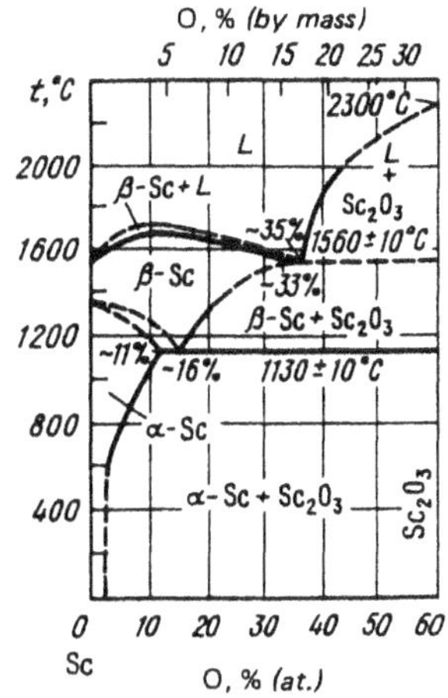

Fig. 75. Phase diagram of Sc–O system [252].

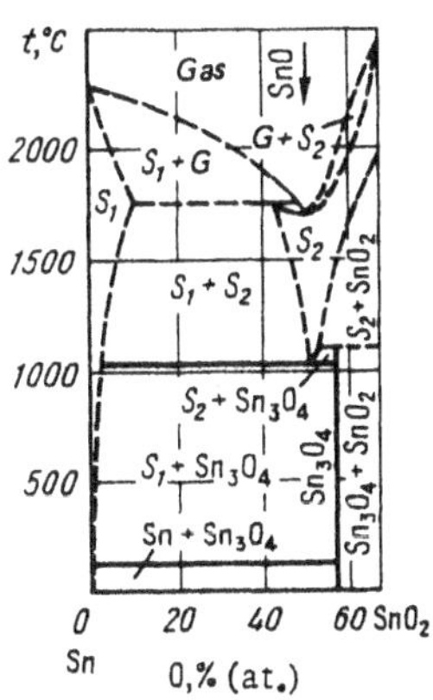

Fig. 76. Phase diagram of Sn–O system [1], schematic in Sn–SnO_2 region. The dashed lines show the limits of the hypothetical regions. The stable compounds are SnO_2 (up to melting point of 2000°C) and Sn_3O_4 (even though some investigators state that this oxide does not exist). SnO exists in the gaseous and molten states. Above 1040°C, tin reacts with SnO_2, forming SnO. Upon cooling to 1040°C, SnO solidifies and simultaneously dissociates into Sn and Sn_3O_4. Thus, there is no stable crystalline form of SnO. Below 400°C, SnO may be obtained, but in a metastable state; upon heating above 400°C it decomposes in accordance with the equation $4SnO \rightarrow Sn_3O_4 + Sn$.

The boiling point of tin is 2270°C, that of tin dioxide ~ 2500°C; the boiling point of SnO is estimated at approximately 1700°C.

In the liquid state, there is a region of immiscibility of two liquids, the tin-rich S_1 and oxygen-rich S_2, this region extending between oxygen contents of 3% and 50% (at.) at 1040°C.

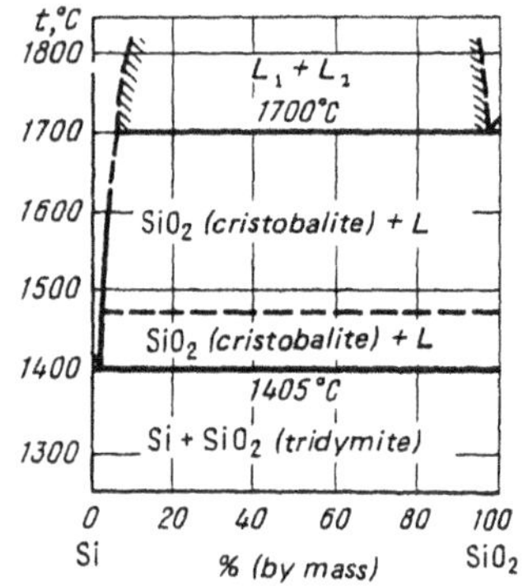

Fig. 77. Phase diagram of Si–O system [584].

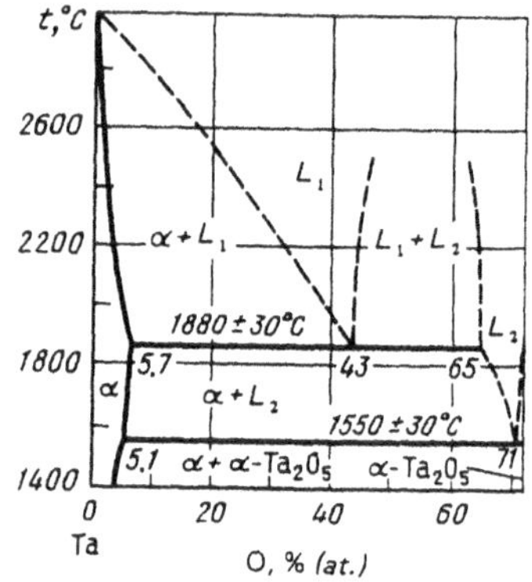

Fig. 78. Phase diagram of Ta–O system [253].

Fig. 79. Phase diagram of Tb–O system [601]. The existence of three phases has been established. Phase δ ($TbO_{1.818}$) is stable at low temperatures. In the interval between 469 and 490°C, it changes over to the phase ι ($TbO_{1.714}$); phase ι has a narrow region of homogeneity. In the interval between 787 and 833°C, the transition ι → δ is established. Phase δ is stable at high temperatures, above 700°C.

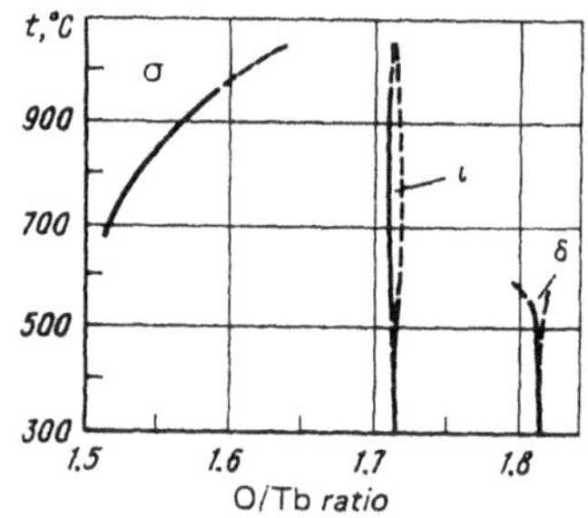

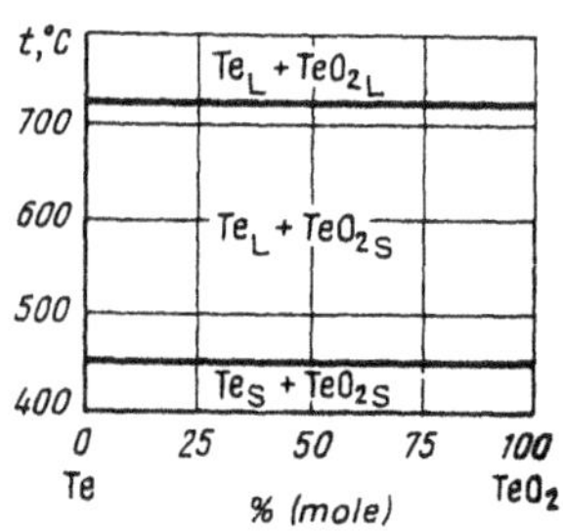

Fig. 80. Phase diagram of Te–O system [1].

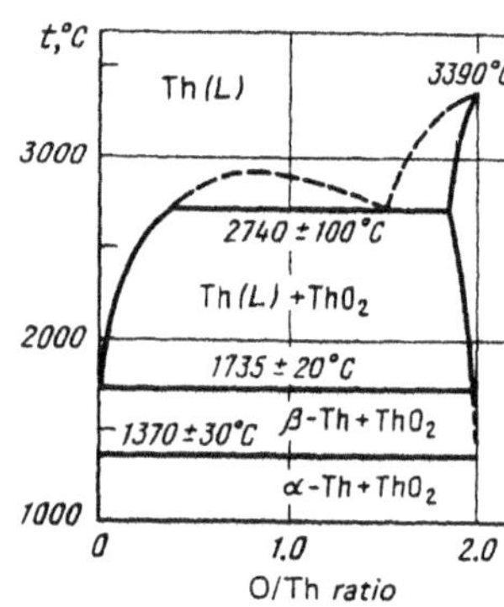

Fig. 81. Phase diagram of Th–O system [254].

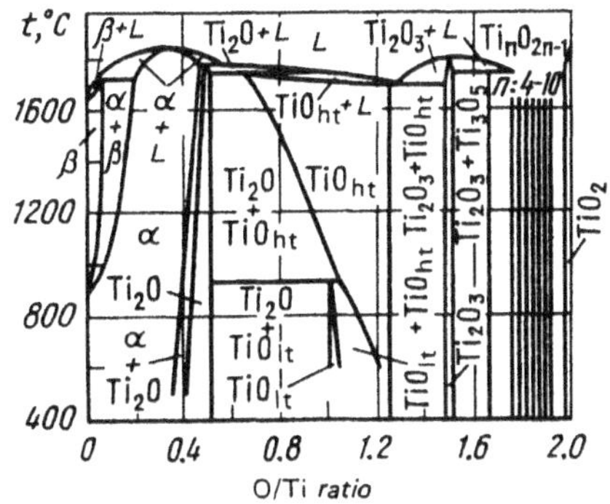

Fig. 82. Phase diagram of Ti–O system [602]. The TiO_{ht} is high-temperature TiO, and TiO_{lt} is low-temperature TiO. The region Ti– Ti_2O contains metallic phases α (hexagonal) and β (cubic) and a phase Ti_2O of variable composition. The melting point of β-Ti is 1660 ± 10°C, and the temperature of the transition $\alpha \rightleftharpoons \beta$ is 882 ± 3°C. The phase Ti_2O (Ti_2O_{1-y}) has a region of homogeneity that becomes broader at lower temperatures. In the region between Ti_2O and Ti_2O_3, we find the high and low-temperature modifications of TiO (TiO_{ht} and TiO_{lt}). The temperature of the transition $TiO_{ht} \rightleftharpoons TiO_{lt}$ is between 950 and 925°C. The region Ti_2O_3–TiO_2 contains Ti_2O_3, Ti_3O_5, seven discrete phases of the homologous series Ti_nO_{2n-1} (n from 4 to 10), and TiO_2. The Ti_2O_3 melts at 1839 ± 10°C. The Ti_3O_5 melts incongruently; its peritectic temperature is 1774 ± 10°C. The melting point of TiO_2 is 1870°C.

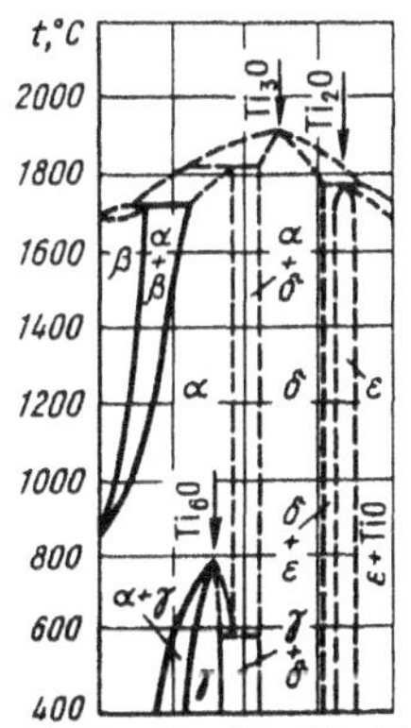

Fig. 83. Phase diagram of Ti–O system, with suboxides taken into account [603]. In addition to the suboxide Ti_2O, which is included in the phase diagram of Fig. 82, existence has been established for the suboxides Ti_6O and Ti_3O. The Ti_6O is formed from the α-solid solution, and the Ti_3O is formed upon crystallization. In the part of the phase diagram that is shown schematically for the Ti–O system, all three of the known suboxides have been introduced; these are the γ, δ, and ε phases of variable composition, based, respectively, on Ti_6O, Ti_3O, and Ti_2O.

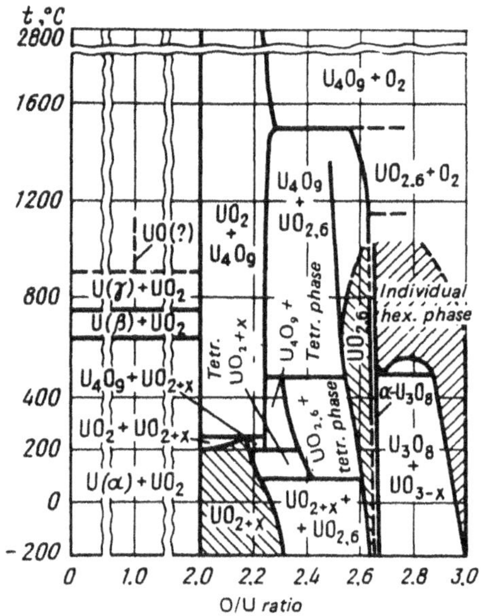

Fig. 84. Phase diagram of U–O system [1]. A large number of oxide phases of uranium have been discovered in the U–O system; the existence of 16 phases has been established reliably, and at least 12 more have been reported. Hence the phase diagram of the U–O system is very complex, and has not yet been fully studied. As a result of the numerous studies that have been made, many different versions of the phase diagrams have been constructed, primarily within the composition limits U–UO_3.

Here we present three diagrams (Figs. 84–86), which give a fairly full picture of the available information on the oxide phases of uranium within the indicated region. The diagram shown here has been constructed from the results of a number of different studies; it covers the region of compositions defined by O/U ratios from 0 to 3, and a wide region of temperatures, from 200 to 2800°C.

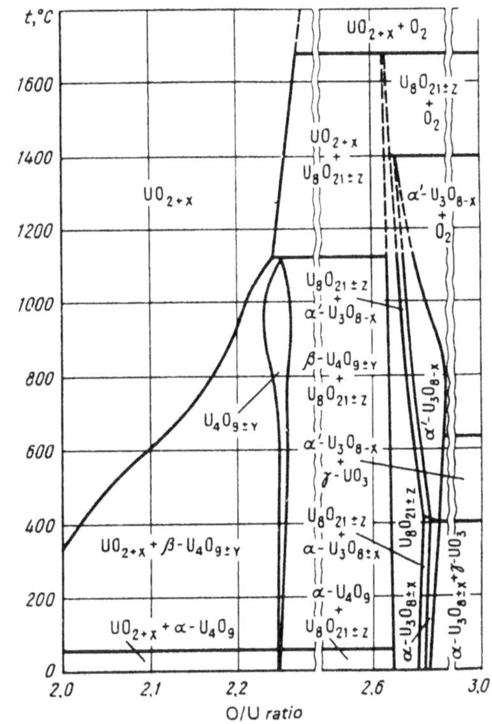

Fig. 85. Phase diagram of U–O system [251], with O/U from 2.0 to 3.0, temperature region 0–1800°C; the boundaries of the phase regions at atmospheric pressure have been refined. Existence has been established for a second-order phase transition for U_4O_9 at 65°C, caused by a change in symmetry of the cell from rhombohedral to cubic. It has been found that only one stable modification of uranium trioxide γ-UO_3 exists at atmospheric pressure.

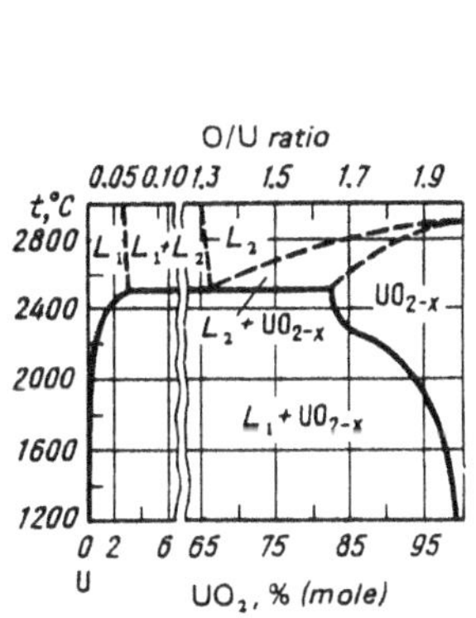

Fig. 86. Phase diagram of U–UO_2 system in high-temperature region [604]; approximate. In the liquid state, a region of layer separation has been detected. The monotectic region corresponds to a composition of 65 ± 5% (mole) UO_2 and a temperature of 2500 ± 30°C.

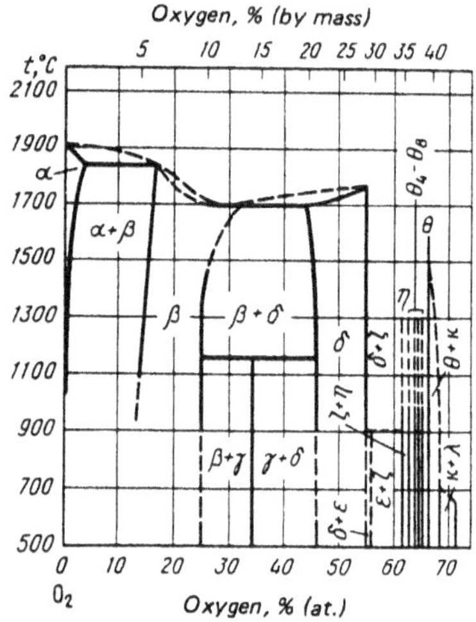

Fig. 87. Phase diagram of V–O system [605]. The system has not yet been studied conclusively. A large number of oxides of vanadium are known; some of them have a more or less wide region of homogeneity. The phase diagram shown here, which has been constructed from the results of a number of studies, includes the following phases:

Phase	Region of Concentrations
α (solid solutions in α-V)	$VO_{0.00}$–$VO_{0.01}$
β	$VO_{0.18}$–$VO_{0.33}$
γ (V_2O)	$VO_{0.53}$
δ ($VO_{1\pm x}$)	$VO_{0.85}$–$VO_{1.24}$
ε ($VO_{1.27}$)*	–
ζ (V_2O_3)	–
η (V_3O_5)	–
θ_n ($4 \leq n \leq 8$)	V_nO_{2n-1}†
θ (VO_2)	–
κ (V_6O_{13})	–
λ	$VO_{2.47}$–$VO_{2.5}$

As can be seen from the diagram, phase β is formed through a peritectic reaction between VO and vanadium-rich liquid at 1690°C; phase γ, which lies between VO and β, is formed at a temperature between 1100 and 1200°C Later studies have established the peritectoid conversion $\alpha + \beta \rightleftharpoons V_9O$ at a temperature of about 510°C.

* Formed at temperatures below 900°C. The existence of this phase has not been demonstrated conclusively.

† Five compounds with this general formula.

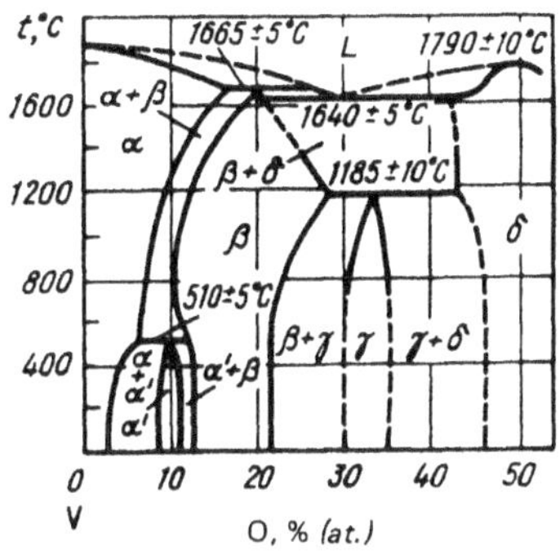

Fig. 88. Phase diagram of V–VO system [606]. Regions of homogeneity have been shown for the phase V_9O (α') and the phase V_2O (γ), the temperature for peritectoid formation of which has been determined to be 1185°C. Existence has been established for a eutectic reaction between phases β and δ at 1640°C and 29% (at.) oxygen. The temperature of the peritectoid reaction of formation of phase β has been defined more precisely (1665°C). The solubility of oxygen in vanadium has also been defined more precisely; it was found to be considerably greater than indicated in previous studies. The temperature of congruent melting of phase δ (VO) has been determined to be 1790°C.

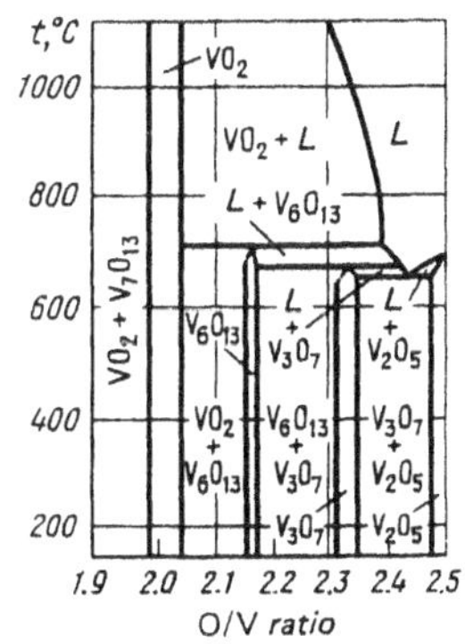

Fig. 89. Phase diagram of V–O system [607] in region of higher oxides of vanadium, from approximately 65.5% to 71.5% O (at.). Here the regions of homogeneity have been defined more precisely for the phases VO_2(θ), V_6O_{13}(κ), and V_2O_5(λ), and also for the phase V_3O_7, which did not appear in the previous diagram (Fig. 87). It has been shown which reactions form these phases.

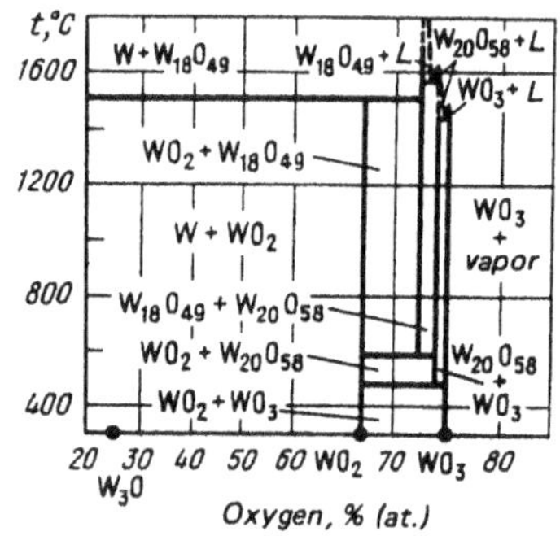

Fig. 90. Phase diagram of W–O system [608]. Diagram is schematic. Shown on this diagram are the compounds WO_2, $W_{18}O_{49}$, $W_{20}O_{58}$, and WO_3, for which stability at high temperatures has been confirmed; the regions of homogeneity of these phases have not been determined.

The question of the existence of the oxide W_3O has not been resolved; this oxide is apparently not a stable phase.

The upper temperature of stability of WO_2 is 1530 ± 5°C, above which it dissociates into W and $W_{18}O_{49}$. No maximal temperature of stability has been established for the phase $W_{18}O_{49}$; however, it is known that this phase is stable at 1700°C. The phase $W_{20}O_{58}$ is stable to at least 1550°C, at which temperature a certain amount of liquid phase has already been formed. The melting point of the WO_3 phase is 1472°C; but possibly this represents melting of a phase of nonstoichiometric composition with a deficiency of oxygen.

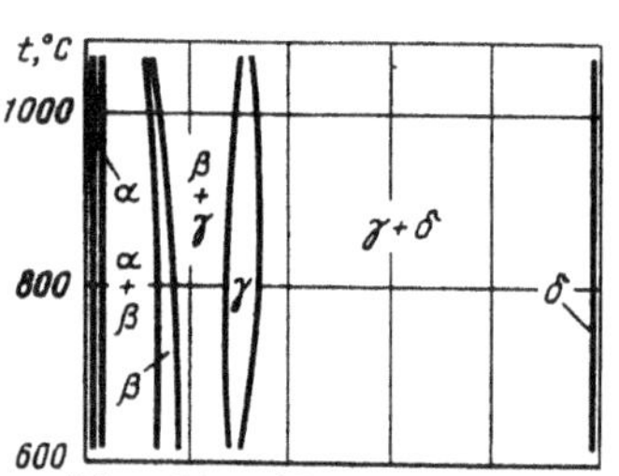

Fig. 91. Regions of phases $\alpha(WO_3)$, $\beta(W_{20}O_{58})$, $\gamma(W_{18}O_{49})$, and $\delta(WO_2)$ in W–O system [1]. The view has been expressed that there exists a homologous series of compounds with the general formula $W_n O_{3n-2}$. Members of this series are oxides with the compositions $W_{20}O_{58}(WO_{2.90})$, $W_{40}O_{118}$-$(WO_{2.95})$, $W_{50}O_{148}(WO_{2.96})$, and $WO_{2.97}$, $WO_{2.98}$, obtained by various investigators. All these discrete compounds are included in a narrow region of phase β, adjacent to WO_3.

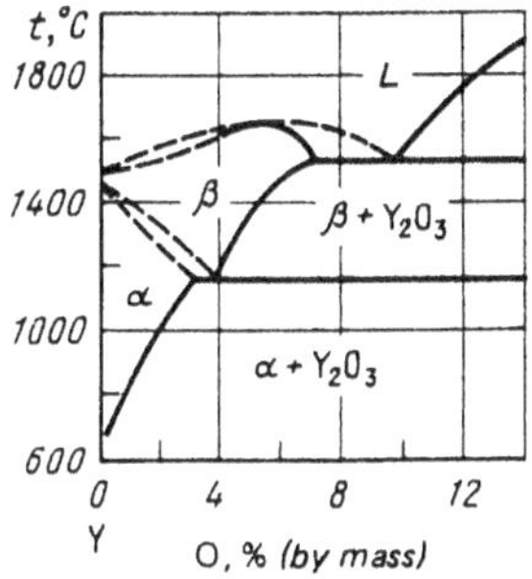

Fig. 92. Phase diagram of Y–O system [255].

Fig. 93. Phase diagram of Zr–O system [1]. The diagram includes one chemical compound, zirconium dioxide, for which the existence of two modifications has been established, i.e., a stable monoclinic form and a metastable tetragonal form. The transition temperature is about 1000°C. Dissolution of oxygen in zirconium raises the temperature of the transition α ⇌ β. The limiting solubility of oxygen in β-Zr is 10.4% (at.), and in α-Zr 29.2% (at.). The peritectic temperature of formation of the β-phase is 1940°C. The temperature of the eutectic reaction between the α-solid solution and ZrO_2 is 900°C; the eutectic composition is 41% O (at.).

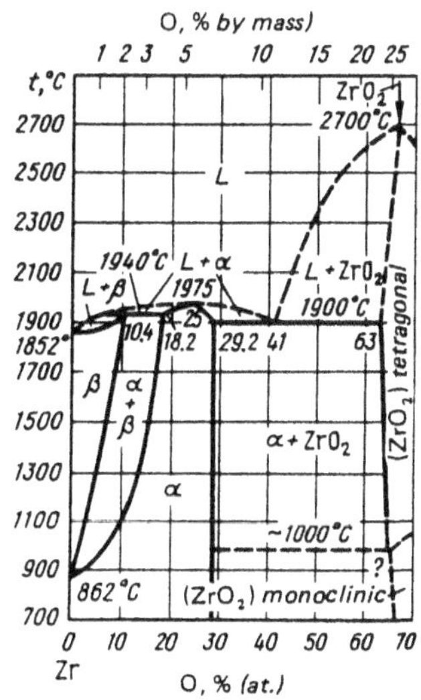

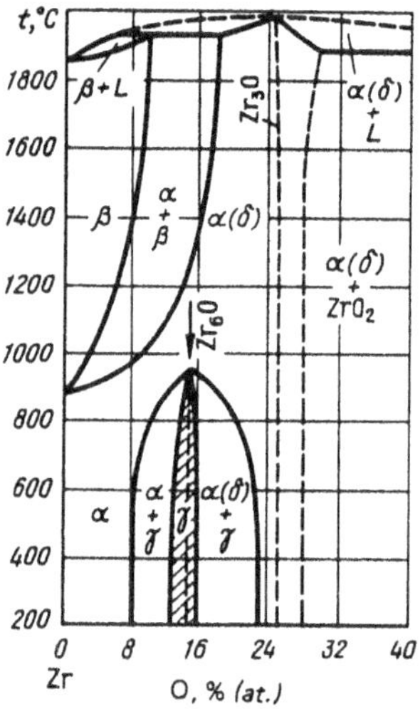

Fig. 94. Phase diagram of Zr–O system, with suboxides taken into account [2]. In the more recent studies, existence has been demonstrated for the suboxides Zr_6O and Zr_3O, and this has introduced changes in the diagram in the region of α-solid solutions. The suboxide Zr_6O is formed from the α-solid solution and is stable up to 940°C. The region of homogeneity of phase γ (based on Zr_6O) is represented by the shaded area on the diagram.

On the two sides of this region of homogeneity are the two-phase regions α + γ and α(δ) + γ; here, α(δ) is oxygen-rich phase α. The Zr_3O is formed by crystallization from the melt; upon heating, it is stable up to the melting point, 1975°C.

Fig. 95. Phase diagram of Zr–O system in high-temperature region, from study [609] of this system in the region of 50–66.7% O (at.) at high temperatures. The existence of the cubic high-temperature modification of ZrO_2 was confirmed. The transition temperature for ZrO_2 (cub.) ZrO_2 (tetr.) was found to be approximately 1490°C in the two-phase region and 2285°C in pure ZrO_2.

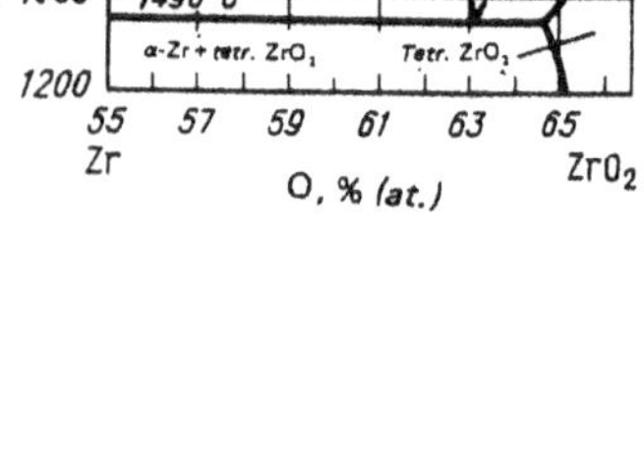

REFERENCES

1. Samsonov, G. V., ed., Physical Chemical Properties of Oxides, 1st edition [in Russian], Izd. Metallurgiya, Moscow (1969) [English translation: The Oxide Handbook, IFI/Plenum, New York (1973)].
2. Kornilov, I. I., and Glazova, V. V., Interaction of Refractory Metals of The Transition Groups with Oxygen [in Russian], Izd. Nauka, Moscow (1967).
3. Mandelkern, L., ed., Non-Stoichiometric Compounds [Russian translation], Izd. Khimiya, Moscow (1971).
4. Kotel'nikov, R. B., Bashlykov, S. N., Galiakbarov, Z. G., and Kashtanov, A. I., Superrefractory Elements and Compounds (Handbook) [in Russian], Izd. Metallurgiya, Moscow (1969).
5. Index to the X-ray Powder Data File, Am. Soc. Testing Mater., Philadelphia (1962).
6. Kubaschewski, O., and Evans, E. L. Metallurgische Thermochemie, Berlin (1956) [English editions: Metallurgical Thermochemistry, Butterworth-Springer, London (1951); 2nd ed., Wiley, New York (1956)].
7. Hausner, H. H., and Friedeman, H. C., High-Temperature Compounds (Data Book), General Astrometals, Yonkers, N. Y. (1962).
8. Tresvyatskii, S. G., and Cherepanov, A. M., High-Temperature Materials and Articles Made of Oxides [in Russian], Izd. Metallurgiya, Moscow (1957).
9. Glushko, V. P., ed., Thermal Constants of Substances (Handbook) [in Russian], Izd. Akad. Nauk SSSR, VINITI, IVT, Nos. 1-7 (1965-74).
10. Gerasimov, Ya. I., Chemical Thermodynamics in Nonferrous Metallurgy [in Russian], Vol. 3, Metallurgizdat, Moscow (1963).
11. Wyckoff, R. W. G., Crystal Structures, 2nd ed., Interscience, New York (1963).
12. Bokii, G. B., Introduction to Crystal Chemistry [in Russian], Izd. Mosk. Gos. Univ., Moscow (1954).
13. Samsonov, G. V., and Konstantinov, V. I., Tantalum and Niobium [in Russian], Metallurgizdat, Moscow (1959).
14. Vol, A. E., Structure and Properties of Binary Metal Systems [in Russian], Vol. 2, Fizmatgiz, Moscow (1962) [English translation: Handbook of Binary Metal Systems, Structure and Properties, U.S. Dept. Commerce Clearing House Fed. Sci. Tech. Inform., Springfield, Va. (1966)].
15. Butler, G. M., and Hausner, H. H., eds., Data Book of Uranium Dioxide, Gladding, McBean & Co., Los Angeles (1960).
16. Hanson, M., and Anderko, K., Structures of Binary Alloys [Russian translation], Metallurgizdat, Moscow (1962) [English edition: Constitution of Binary Alloys, 2nd ed., McGraw-Hill, New York (1962)].
17. Kaye, G. W. C., and Laby, T. H., Tables of Physical and Chemical Constants [Russian translation, 2nd ed.], Fizmatgiz, Moscow (1962) [English edition: Tables of Physical and Chemical Constants and Some Mathematical Functions, 12th ed., Longmans Green, London/New York (1959)].

18. Campbell, I. E., High-Temperature Technology [Russian translation], IL, Moscow (1959).
19. Krzhizhanovskii, R. E., and Shtern, Z. Yu., Thermophysical Properties of Nonmetallic Materials [in Russian], Énergiya, Leningrad (1973).
20. Serebrennikov, V. V., and Alekseenko, A. A., Course in Chemistry of Rare-Earth Elements [in Russian], Izd. Tomsk. Gos. Univ., Tomsk (1963).
21. Goldsmith, A., Handbook of Thermophysical Properties of Materials, Vol. 3, Macmillan, New York (1961).
22. Zefirov, A. P., ed., Thermodynamic Properties of Inorganic Substances (Handbook) [in Russian], Atomizdat, Moscow (1965).
23. Krishna Rao, K. V., Nagender Naidu, S. V., and Iyengar, L., J. Appl. Crystallogr., 6(2):136 (1973).
24. Lundy, T. S., Padgett, R. A., and Banus, M. D., Metall. Trans., 4(4):1179 (1973).
25. Harding, B. C., Phys. Status Solidi B, 56(2):645 (1973).
26. Narayan, J., and Washburn, J., Acta Metall., 21(5):533 (1973).
27. Ozhegov, P. I., Myasoedov, B. F., and Zakharov, E. A., Dokl. Akad. Nauk SSSR, 212(5):1122 (1973).
28. Nekrasov, L. I., Zh. Fiz. Khim., 46(3):749 (1972).
29. Chizhikov, D. M., Tsvetkov, Yu. V., Kazenas, E. K., et al., Zh. Fiz. Khim., 46(3):806 (1972).
30. Pak, T. A., Zh. Fiz. Khim., 46(8):2121 (1972).
31. Navrotsky, A., and Kleppa, O. J., J. Am. Ceram. Soc., 50(11):626 (1967).
32. Bobyrenko, Yu. Ya., Zholnin, A. B., and Konovalov, V. K., Zh. Fiz. Khim., 46(5):1305 (1972).
33. Semenov, G. A., Frantseva, K. E., and Zhelkova, E. K., Vestn. Leningr. Gos. Univ., No. 3, Issue 16, p. 82 (1970).
34. Yashvili, T. S., and Gvelesiani, G. G., Zh. Fiz. Khim., 45(4):983 (1971).
35. Kolbin, N. I., Samoilov, V. M., and Shuvalov, B. N. Zh. Fiz. Khim., 45(8):2114 (1971).
36. Vol'nov, I. I., Zh. Fiz. Khim., 45(11):2905 (1971).
37. Petukhov, V. A., Chekhovskii, V. Ya., and Bagdasarov, Kh. S., Teplofiz. Vys. Temp., 11(5):1083 (1973).
38. Chechel'nitskii, A. Z., Teplofiz. Vys. Temp., 10(2):285 (1972).
39. Lyon, W. L., and Baily, W. E., J. Nucl. Mater., 22(3):332 (1967).
40. Mortlock, A. J., and Price, D. M., J. Chem. Phys., 58(2):634 (1973).
41. Yagodovskaya, T. V., and Nekrasov, L. I., Zh. Fiz. Khim., 44(7):1718 (1970).
42. Makatun, V. N., and Pechkovskii, V. V., Zh. Fiz. Khim., 44(10):2667 (1970).
43. Sabirzyanov, T. G., Zh. Fiz. Khim., 44(5):1313 (1970).
44. Aivazov, M. I., Muranevich, A. Kh., Domashnev, I. A., et al., Teplofiz. Vys. Temp., 8(3):546 (1970).
45. Aivazov, M. I., Muranevich, A. Kh., Domashnev, I. A., et al., Teplofiz. Vys. Temp., 9(1):90 (1971).
46. Hildenbrand, D. L., and Murad, E., J. Chem. Phys., 53:3403 (1970).
47. Gusarev, A. V., and Gorokhov, L. N., Teplofiz. Vys. Temp., 9(3):505 (1971).
48. Tsagareishvili, D. Sh., and Gvelesiani, G. G., Teplofiz. Vys. Temp., 9(3):643 (1971).

49. Shpil'rain, É. É., Kagan, D. N., and Barkhatov, L. S., Teplofiz. Vys. Temp., 9(5):926 (1971).
50. Aivazov, M. I., Muranevich, A. Kh., Domashnev, I. A., et al., Teplofiz. Vys. Temp., 9(6):1177 (1971).
51. Chekhovskii, V. Ya., Stavroskii, G. I., and Ivanov, A. B., Teplofiz. Vys. Temp., 9(6):1182 (1971).
52. Toropov, N. A., Barsagovskii, V. P., Bondar', I. A., and Udalov, Yu. P., Phase Diagrams of Silicate Systems (Handbook) [in Russian], No. 2, Izd. Nauka, Leningrad (1970).
53. Smith, D. L., J. Less-Common Met., 31(3):345 (1973).
54. Dudarev, V. Ya., Tsentsiper, A. B., and Dobrolyuvoba, M. S., Kristallografiya, 18(4):759 (1973).
55. Sjöstrand, M. E., and Keesom, P. H., Phys. Rev. Bull., 7(8):3558 (1973).
56. Chandrashekhar, G. V., Barros, H. L. C., and Honig, J. M., Mater. Res. Bull., 8(4):369 (1973).
57. Hoch, M., High Temp. High Pressures, 4(5):493 (1972).
58. Eckert, L. J., and Bradt, R. C., Mater. Res. Bull., 8(4):375 (1973).
59. Wuensch, B. J., Steele, W. C., and Wasilos, T., J. Chem. Phys., 58(12):5258 (1973).
60. Shirasaki, S., and Hama, M., Chem. Phys. Lett., 20(4):361 (1973).
61. Kroingol'd, F. I., and Kulinkin, B. S., Opt. Spektrosk., 33(4):706 (1972).
62. Men', A. N., Vorob'ev, Yu. P., and Chufarov, G. I., Physicochemical Properties of Nonstoichiometric Oxides [in Russian], Izd. Khimiya, Leningrad (1973).
63. Coillot-Demay, M.-F., Langouet, L., and Decamps, E.-A., C. R. Acad. Sci., 276(18):B757 (1973).
64. Inagaki, Hirotake, Bull. Inst. Chem. Res. Kyoto Univ., 51(1):27 (1973).
65. Popov, D. P., and Styrov, S. M., Nauchn. Tr. Krasnoyarsk. Politekh. Inst., No. 15, p. 63 (1973).
66. Piacente, V., Bardi, G., Malaspina, L., et al., J. Chem. Phys., 59(1):31 (1973).
67. Winchell, A. N., and Winchell, G., The Microscopical Characters of Artificial Inorganic Solid Substances. The Optical Properties of Artificial Minerals [Russian translation], Izd. Mir., Moscow (1967) [English edition: 3rd ed., Academic Press, New York (1964)].
68. Alyamovskii, S. I., Zainulin, Yu. G., Shveikin, G. P., et al., Izv. Akad. Nauk SSSR Neorg. Mater., 9(10):1837 (1973).
69. Shpil'rain, É. E., Yakimovich, K. A., and Tsitsarkin, A. F., Teplofiz. Vys. Temp., 11(5):1001 (1973).
70. Deb, S. K., Solid State Commun., 11(5):713 (1972).
71. Mokerov, V. G., Fiz. Tverd. Tela, 15(8):2393 (1973).
72. Drapak, I. T., Izv. Vyssh. Uchebn. Zaved. Fiz., No. 7, p. 126 (1969).
73. Vargaftik, N. B., Volyak, L. D., and Volgov, B. N., in: Thermophysical Properties of Liquids [in Russian], Izd. Nauka, Moscow (1970), p. 182.
74. Chekhovskii, V. Ya., Inzh.-Fiz. Zh., 5(8):62 (1962).
75. Wicks, C. E., and Blok, F. E., Thermodynamic Properties of 65 Elements and Their Oxides, Halides, Carbides, and Nitrides [Russian translation], Metallurgiya, Moscow (1965) [English edition: U.S. Bur. Mines Bull. 605, U.S. Govt. Print. Office, Washington, D. C. (1963)].

76. Voronkova, E. M., Grechushnikov, B. N., Distler, G. I., and Petrov, I. P., Optical Materials for Infrared Technology [in Russian], Izd. Nauka, Moscow (1965).
77. Tumanov, A. T., and Portnyi, K. I., eds., Refractory Materials in Machinery Construction (Handbook) [in Russian], Izd. Mashinostroenie, Moscow (1967).
78. Mishchenko, K. P., and Ravdel', eds., Concise Handbook of Physicochemical Constants [in Russian], 7th ed., Izd. Khimiya, Leningrad (1974).
79. Glushko, V. P., Gurvich, L. V., et al., eds., Thermodynamic Properties of Individual Substances (Handbook) [in Russian], 2nd ed., Vols. 1 and 2, Izd. Akad. Nauk SSSR, Moscow (1962).
80. Goroshchenko, Ya. G., Chemistry of Niobium and Tantalum [in Russian], Izd. Naukova Dumka, Kiev (1966).
81. Concise Chemical Encyclopedia [in Russian], Vols. 1-5, Izd. Sovetskaya Éntsiklopediya, Moscow (1961-67).
82. Valiev, K. A., Kopaev, Yu. V., Mokerov, V. G., et al., in: 7th Ural Conference on Spectroscopy [in Russian], Sverdlovsk (USSR), (1971), No. 3, p. 147.
83. Kuz'mina, I. P., Lobachev, A. N., Predtechenskii, B. S., et al., Kristallografiya, 18(3):635 (1973).
84. Nekrasov, B. V., Course in General Chemistry [in Russian], 14th ed., Izd. Goskhimizdat, Moscow (1962).
85. Shaldin, Yu. V., Smirnov, S. P., Belyaev, L. M., et al., Kristallografiya, 18(3):570 (1973).
86. Hughes, A. E., and Webb, A. P., Solid State Commun., 13(2):167 (1973).
87. Tazenkov, B. A., and Khalturin, A. S., in: 25th Hertzen Reader – Physical and Semiconductor Electronics [in Russian], Leningrad (1972), p. 95.
88. Rose, B. H. and Hensley, E. B., Phys. Rev. Lett., 29(13):861 (1972).
89. Evans, B. D., and Kemp, J. C., Phys. Rev. Bull., 2(10):4179 (1970).
90. Gavrish, D. I., ed., Refractory Production (Handbook) [in Russian], Vol. 1, Izd. Metallurgiya, Moscow (1965).
91. Vasil'eva, I. A., and Shaulova, É. Yu., Zh. Fiz. Khim., 43(12):3047 (1969).
92. Vasil'eva, I. A., and Granovskaya, Zh. V., Zh. Fiz. Khim., 48(6):1536 (1974).
93. Sheindlin, A. E., ed., Radiation Properties of Solid Materials (Handbook) [in Russian], Izd. Énergiya, Moscow (1974).
94. Volzhenskii, D. S., and Savitskii, V. G., Izv. Vyssh. Uchebn. Zaved. Fiz., No. 11, p. 151 (1974).
95. Slyusar', N. P., Krivorotenko, A. D., Fomichev, E. N., et al., Zh. Fiz. Khim., 47(10):2706 (1973).
96. Matveev, M. A., Matveev, G. M., and Kharitonov, F. Ya., Izv. Akad. Nauk SSSR Neorg. Mater., 2(2):395 (1966).
97. Rivkin, S. L., Thermodynamic Properties of Gases [in Russian], 3rd ed., Izd. Énergiya, Moscow (1973).
98. Fomchenkov, L. P., Maier, A. A., and Gracheva, N. A., Izv. Akad. Nauk SSSR Neorg. Mater., 10(11):2020 (1974).
99. Solov'eva, A. E., Gavrish, A. M., and Zoz, E. I., Zh. Neorg. Khim., 19(6):1446 (1974).
100. Bragg, W. L., and Claringbull, G. F., Crystal Structure of Minerals [Russian translation], Izd. Mir, Moscow (1967) [English edition: Cornell Univ. Press (1965)].

101. Krasnov, K. S., Timoshinin, V. S., Danilova, T. G., et al., Molecular Constants of Inorganic Compounds [in Russian], Izd. Khimiya, Leningrad (1968).
102. Chikalla, T. D., McNeilly, C. E., and Roberts, F. P., J. Am. Ceram. Soc., 55(8):428 (1972).
103. Chikalla, T. D., McNeilly, C. E., Bates, J. L., et al., Colloq. Int. CNRS, No. 205, p. 351 (1972).
104. Sata, T., and Takahashi, T., Colloq. Int. CNRS, No. 205, p. 331 (1972).
105. Ghezzo, M., and Brown, D. M., J. Electrochem. Soc., 120(1):146 (1973).
106. Chen, W. K., Peterson, N. L., and Robinson, L. C., J. Phys. Chem. Solids, 34(4):705 (1973).
107. Harding, B. C., Philos. Mag., 27(2):481 (1973).
108. Mimkes, J., and Wuttig, M., J. Am. Ceram. Soc., 54(1):65 (1971).
109. Lay, K. W., J. Am. Ceram. Soc., 54(1):18 (1971).
110. Gruenwald, T. B., and Gordon, G., J. Inorg. Nucl. Chem., 33(4):1151 (1971).
111. Perkins, R. A., and Rapp, R. A., Metall. Trans., 4(1):193 (1973).
112. Murarka, S. P., and Swalin, R. A., J. Phys. Chem. Solids, 32(9):2015 (1971).
113. Karyakin, A. V., Anikina, L. I., Pavlenko, L. I., and Laktionova, N. V., Spectral Analysis of Rare-Earth Oxides [in Russian], Izd. Nauka, Moscow (1974).
114. Shpil'rain, É. É., Yakimovich, K. A., and Tsitsarkin, A. F., High Temp. High Pressures, 4(1):67 (1972).
115. Contamin, P., Bacmann, J. J., and Marin, J. F., J. Nucl. Mater., 42(1):54 (1972).
116. Kakimoto, M., Murabayashi, M., and Takahashi, Y., J. Nucl. Mater., 46(3):309 (1973).
117. Vinokurov, I. V., Izv. Akad. Nauk SSSR Neorg. Mater., 6(1):31 (1970).
118. Igant'eva, N. I., and Bamburov, V. G., Izv. Akad. Nauk SSSR Neorg. Mater., 6(1):154 (1970).
119. Reimann, D. K., and Lundy, T. S., J. Nucl. Mater., 28(2):218 (1968).
120. Dwivedi, G. L., and Subbarao, E. S., J. Am. Ceram. Soc., 56(8):443 (1973).
121. Berard, M. F., Wirkus, C. D., and Wilder, D. R., J. Am. Ceram. Soc., 51(11):643 (1968).
122. Wircus, C. D., Berard, M. F., and Wilder, D. R., J. Am. Ceram. Soc., 52(8):456 (1969).
123. Volpe, M. L., and Reddy, J., J. Chem. Phys., 53(3):1117 (1970).
124. Martens, H., and Ziegenbalg, S., Z. Anorg. Allgem. Chem., 401(2):145 (1973).
125. Mordovin, O. A., Timofeeva, N. I., and Drozdova, L. N., Izv. Akad. Nauk SSSR Neorg. Mater., 3(1):187 (1967).
126. Whitney, W. P., and Stubican, V. S., J. Phys. Chem. Solids, 32(2):305 (1971).
127. Kirchner, H. P., J. Am. Ceram. Soc., 52(7):379 (1969).
128. Sokolova, I. D., and Voskresenskaya, N. K., Izv. Akad. Nauk SSSR Neorg. Mater., 6(7):1358 (1970).
129. Sheasby, J. S., and Cox, B., J. Less-Common Met., 15(2):129 (1968).
130. Harding, B. C., Price, D. M., and Mortlock, A. J., Philos. Mag., 23(182):399 (1971).
131. Kundra, K. D., and Ali, S. Z., J. Appl. Crystallogr., 3(6):543 (1970).

132. Singh, S., Davenport, J. C., and Mills, N. D., J. Am. Ceram. Soc., 53(3):169 (1970).
133. Krishna Rao, K. V., Nagender Naidu, S. V., and Iyengar, L., J. Am. Ceram. Soc., 53(3):124 (1970).
134. Filatov, S. K., and Frank-Kamenetskii, V. A., Kristallografiya, 14(5):804 (1969).
135. Marsh, S. P., High Temp. High Pressures, 5(5):503 (1973).
136. Chizhikov, D. M., Pavlov, Yu. A., Tsvetkov, Yu. V., et al., Izv. Vyssh. Uchebn. Zaved. Chern. Metall., No. 7, p. 5 (1970).
137. Samsonov, G. V., and Gil'man, I. Ya., Poroshk. Metall., No. 11, p. 73 (1974).
138. Braun, S. M., Poroshk. Metall., No. 6, p. 82 (1970).
139. Borokova, L. B., Lukin, E. S., Poluboyarinov, D. N., et al., Ogneupory, No. 11, p. 39 (1970).
140. Popil'skii, R. Ya., and Smirnov, V. A., Ogneupory, No. 12, p. 31 (1970).
141. Barkhatov, L. S., Kagan, D. N., Tsitsarkin, A. F., et al., Teplofiz. Vys. Temp., 11(6):1188 (1973).
142. Men', A. A., and Chechel'nitskii, A. Z., Teplofiz. Vys. Temp., 11(6):1309 (1973).
143. Malyutin, S. A., Samplavskaya, K. K., and Karapetyants, M. Kh., Zh. Neorg. Khim., 16(6):1475 (1971).
144. Paukov, I. E., Rakhmenkulov, F. S., Dobrolyubova, M. S., et al., Izv. Akad. Nauk SSSR Ser. Khim., No. 9, p. 2135 (1970).
145. Zhuze, V. P., Novruzov, O. N., and Popova, A. A., Izv. Akad. Nauk SSSR Neorg. Mater., 4(9):1493 (1968).
146. Shakhtin, D. M., Levintovich, É. V., Pivovar, T. L., et al., Izv. Akad. Nauk SSSR Neorg. Mater., 4(9):1603 (1968).
147. Kovalev, N. N., Petrov, A. V., and Sorokin, O. V., Fiz. Tverd. Tela, 13(1):291 (1971).
148. Koloskova, L. A., Kruzhkin, I. N., Manzhelii, V. G., et al., Fiz. Tverd. Tela, 15(6):1913 (1973).
149. Asamoto, R. R., Anselin, F. L., and Conti, A. E., J. Nucl. Mater., 29(1):67 (1969).
150. Vishnevskii, I. I., and Skripak, V. N., Ogneupory, No. 11, p. 16 (1970).
151. Wolf, M. W., and Martin, J. J., Phys. Status Solidi A, 17:215 (1973).
152. Millstein, J., J. Phys. Chem. Solids, 31(4):886 (1970).
153. Aring, K., and Sievers, A. J., J. Appl. Phys., 38(3):1496 (1967).
154. Glushkova, V. B., and Bogdanov, A. G., Izv. Akad. Nauk SSSR Khim., No. 7, p. 1131 (1965).
155. Moore, J. P., and McElroy, D. L., J. Am. Ceram. Soc., 54(1):40 (1971).
156. Gordienko, S. P., Fenochka, B. V., and Fesenko, V. V., Rare-Earth Metals and Their Refractory Compounds (Handbook) [in Russian], Izd. Naukova Dumka, Kiev (1971).
157. Diffusion Data, Vol. 5, Nos. 1-4 (1971).
158. Goldsmith, L. A., and Douglas, J. A. M., J. Nucl. Mater., 47(1):31 (1973).
159. Kolyadin, V. I., Il'in, É. P., Kharlamov, A. G., et al., At. Énerg., 36(1):59 (1974).
160. Müller, E., Atomwirtsch.-Atomtech., 17:37 (1972).

161. Pavlov, Yu. A., Kryukov, S. N., Sheboldaev, S. B., et al., Izv. Akad. Nauk SSSR Neorg. Mater., 11(4):661 (1975).
162. Forman, R., J. Appl. Phys., 44(1):66 (1973).
163. Tsagareishvili, D. Sh., and Gvelesiani, G. G., Teplofiz. Vys. Temp., 11(2):300 (1973).
164. Slyusar', N. P., Krivorotenko, A. D., Fomichev, E. N., et al., Teplofiz. Vys. Temp., 11(1):213 (1973).
165. Zumsteg, A., Ziegler, M., Bosch, M., et al., Helv. Phys. Acta, 46(1):15 (1973).
166. Kiparisov, S. S., and Fistul', A. D., Izv. Akad. Nauk SSSR Neorg. Mater., 11(2):378 (1975).
167. Kiparisov, S. S., Kuznetsov, S. A., Kozlov, V. G., et al., Izv. Akad. Nauk SSSR Neorg. Mater., 7(6):1073 (1971).
168. Diffusion Data, Vol. 6, Nos. 1-4 (1972).
169. Klassen-Neklyudova, M. V., and Bagdasarov, Kh. S. (principal authors), Ruby and Sapphire [in Russian], Izd. Nauka, Moscow (1974).
170. Gauthier, M., Duclot, M., Hammou, A., et al., J. Solid State Chem., 9(1):15 (1974).
171. Baybarz, R. D., J. Inorg. Nucl. Chem., 35(12):4149 (1973).
172. Diffusion Data, Vol. 7, Nos. 1-3 (1973).
173. Jeapes, A. P., Leadbetter, A. J., Waterfield, C. G., et al., Philos. Mag., 29(4):803 (1974).
174. Tret'yakov, Yu. D., Chemistry of Nonstoichiometric Oxides [in Russian], Izd. Mosk. Gos. Univ., Moscow (1974).
175. Deaton, R. L., and Wiedenheft, C. J., J. Inorg. Nucl. Chem., 35(2):649 (1973).
176. Aivazov, M. I., and Muranevich, A. Kh., Izv. Akad. Nauk SSSR Neorg. Mater., 9(12):2156 (1973).
177. Williams, E. L., J. Am. Ceram. Soc., 48(4):190 (1965).
178. Popel', S. I., Solov, V. I., and Esin, O. A., Zh. Fiz. Khim., 43(12):3175 (1969).
179. Mosley, W. C., J. Inorg. Nucl. Chem., 34(2):539 (1972).
180. Gulbransen, E. A., Andrew, K. F., and Brassart, F. A., J. Electrochem. Soc., 110(3):242 (1963).
181. Wakihara, Masataka; and Katsura, Takashi, Metall. Trans., 1(2):363 (1970).
182. Golubenko, A. N., and Rezukhina, T. N., Izv. Akad. Nauk SSSR Neorg. Mater., 3(1):101 (1967).
183. Semenov, G. A., Zh. Neorg. Khim., 10(10):2390 (1965).
184. Chizhikov, D. M., Tsvetkov, Yu. V., and Kasenas, E. K., Izv. Akad. Nauk SSSR Met., No. 3, p. 55 (1971).
185. Appen, A. A., Heat-Resistant Inorganic Coatings [in Russian], Izd. Khimiya, Leningrad (1967).
186. Mitin, B. S., and Nagibin, Yu. A., Izv. Akad. Nauk SSSR Neorg. Mater., 7(5):814 (1971).
187. Jungermann, E., and Plieth, K., Z. Phys. Chem. (Frankfurt), 53(1-6):215 (1967).
188. Kulikov, I. S., Izv. Akad. Nauk SSSR Met., No. 3, p. 48 (1971).
189. Haschke, J. M., and Eick, H. A., J. Phys. Chem., 73(2):374 (1969).

190. Wick, O. J., ed., The Plutonium Handbook [Russian translation], Vol. 2, Izd. Atomizdat, Moscow (1973) [English edition: Gordon & Breach, New York (1967)].
191. Justice, V. N., Westrum, E. F., Chang, J. E., et al., J. Phys. Chem., 73(2):333 (1969).
192. Berezovskii, G. A., Ikorskii, V. P., and Pakov, I. E., Zh. Fiz. Khim., 49(1):261 (1975).
193. Kleykamp, H., and Paneth, L. J., J. Inorg. Nucl. Chem., 35(2):477 (1973).
194. Tsagareishvili, D. Sh., and Gvelesiani, G. G., Izv. Akad. Nauk SSSR Neorg. Mater., 9(11):1936 (1973).
195. Kulikov, I. S., Thermal Dissociation of Compounds [in Russian], Izd. Metallurgiya, Moscow (1966).
196. Poulton, D. J., and Smeltzer, W. W., J. Electrochem. Soc., 117(3):378 (1970).
197. Balducci, G., De Maria, G., Guido, M., et al., J. Chem. Phys., 56(7):3422 (1972).
198. Arkhipov, V. A., Kutina, É. A., Dobretsov, V. N., et al., Radiokhimiya, 16(1):123 (1974).
199. Dityat'ev, A. A., Tret'yakov, Yu. D., Bulgakova, T. I., et al., Izv. Akad. Nauk SSSR Neorg. Mater., 9(11):1999 (1973).
200. Paladino, A. E., and Kingery, W. D., J. Chem. Phys., 37(5):957 (1962).
201. Haul, R. and Just, D., J. Appl. Phys., 33(1):487 (1962).
202. Bénard, G., Oxidation of Metals [Russian translation], Vol. 2, Izd. Metallurgiya, Moscow (1969) [French edition: Oxydation des métaux, Gauthier-Villars, Paris (1964)].
203. Ignatov, D. V., Model', M. S., Sokiryanskii, A. F., et al., in: Chemistry of Metal Alloys [in Russian], Izd. Nauka, Moscow (1973), p. 208.
204. Mori, Kazumi, and Suzuki, Kanae, Trans. Iron Steel Inst. Jpn., 12(6):464 (1972).
205. Orszagh, J., and Bouillon, E., Mem. Sci. Rev. Metall., 70(4):319 (1973).
206. Chatterji, D., and Vest, R. W., J. Am. Ceram. Soc., 55(11):575 (1972).
207. Baur, J. P., Bartlett, R. W., Ong, J. N., et al., J. Electrochem. Soc., 110(3):185 (1963).
208. Elyutin, V. P., Mitin, V. S., and Samoteikin, V. V., Izv. Akad. Nauk SSSR Met., No. 3, p. 227 (1971).
209. Vargaftik, N. B., Handbook of Thermophysical Properties of Gases and Liquids [in Russian], 2nd ed., Izd. Nauka, Moscow (1972).
210. Kunugi, M., Soga, N., Sawa, H., et al., J. Am. Ceram. Soc., 55(11):580 (1972).
211. Hite, H. E., and Kearngey, R. J., J. Appl. Phys., 38(13):5424 (1967).
212. Haglund, J. A., and Hunter, O., J. Am. Ceram. Soc., 56(6):327 (1973).
213. Nekrasov, B. V., ed., Chemist's Concise Handbook [in Russian], Goskhimizdat, Moscow (1954).
214. Povarennykh, A. S., Hardness of Minerals [in Russian], Izd. Akad. Nauk UkrSSR (1963).
215. Chung, D. H., and Buessem, W. R., J. Appl. Phys., 38(6):1967 (1967).
216. Skidan, B. S., Ogneupory, No. 12, p. 36 (1970).
217. Hallberg, J., and Hanson, R. C., Phys. Status Solidi, 42(1):305 (1970).
218. Powell, R., Hunter, O., and Manning, W. R., J. Am. Ceram. Soc., 54(10):487 (1971).

219. Cherepanov, A. M., and Tresvyatskii, G. S., Refractory Materials and Articles Made of Oxides [in Russian], 2nd ed., Izd. Metallurgiya, Moscow (1964).
220. Kornilov, I. I., Glazova, V. V. and Ruda, G. I., Izv. Akad. Nauk SSSR Neorg. Mater., 4(12):2106 (1968).
221. Gust, W. H., and Royce, E. B., J. Appl. Phys., 42(2):276 (1971).
222. Johnson, J. R., et al., Am. Ceram. Soc. Bull., 36:112 (1957).
223. Ryshkewitch, E., Oxide Ceramics, Academic Press, New York (1960).
224. Kingery, W. D., J. Am. Ceram. Soc., 42(1):6 (1959).
225. Mitrdorf, U., and Helmreich, D., J. Acoust. Soc. Am., Vol. 49, Pt. 2, No. 3, p. 723 (1971).
226. Krushchov, M. M., and Berkovich, E. S., Microhardness (Proceedings of Conference on Microhardness) [in Russian], Izd. Akad. Nauk SSSR, Moscow (1951).
227. Baranova, T. F., Popil'skii, R. Ya., and Klimova, G. F., Ogneupory, No. 9, p. 60 (1972).
228. Romashin, A. G., and Prudikova, N. I., Ogneupory, No. 4, p. 43 (1971).
229. Popil'skii, R. Ya., Adushkin, L. E., Pivenskii, Yu. E., et al., Ogneupory, No. 4, 45 (1971).
230. Tresvyatskii, S. G., Nazarenko, N. D., Dubok, V. A., and Nechitailo, V. F., Ukr. Fiz. Zh., 10(6):681 (1965).
231. Methods of Testing for Microhardness [in Russian], in collection edited by M. M. Khrushchov, Izd. Nauka, Moscow (1965).
232. Mott, B. W., Micro-Indentation Hardness Testing [Russian translation], Metallurgizdat, Moscow (1960) [English edition: Butterworths, New York (1956)].
233. Ablova, M. S., Golubkov, A. V., Zhuze, V. P., et al., Fiz. Tverd. Tela, 13(12):3711 (1971).
234. Vishnevskii, I. I., Aksel'rod, E. I., and Tal'yanskaya, N. D., Izv. Akad. Nauk SSSR Neorg. Mater., 8(9):1641 (1972).
235. Eliezer, D., and Brandon, D. G., Thin Solid Films, 12:319 (1972).
236. Samsonov, G. V., Koval'chenko, M. S., and Fen', E. K., Phys. Sintering, 4(1):21 (1972).
237. Yasuda, Eiich; Ootsuka, Motoya; Kimura, Shiushishi; et al., Bull. Tokyo Inst. Technol., No. 108, p. 113 (1972).
238. Chung, D. H., and Buessem, W. R., J. Appl. Phys., 39(6):2777 (1968).
239. Marklund, K., and Mahmoud, S. A., J. Phys. Soc., 3(2):75 (1971).
240. Manning, W. R., Marlowe, M. O., and Wilder, D. R., J. Am. Ceram. Soc., 49(4):227 (1966).
241. Stavrolakis, J. A., and Norton, F. H., J. Am. Ceram. Soc., 33(9):263 (1950).
242. Butyshkin, V. P., Romanovich, I. V., and Timofeeva, N. I., Izv. Akad. Nauk SSSR Neorg. Mater., 8(9):1638 (1971).
243. Son, P. R., and Bartels, R. A., J. Phys. Chem. Solids, 33(4):819 (1972).
244. Manghnani, M. H., J. Phys. Chem. Solids, 33(11):2149 (1972).
245. Oliver, D. W., J. Appl. Phys., 40(2):893 (1969).
246. Johnston, D. L., Thrasher, P. H., and Kearney, R. J., J. Appl. Phys., 41(1):427 (1970).
247. Pace, N. G., Saunders, G. A. Sumengen, Z., and Thorp, J. S., J. Mater. Sci., 4(12):1106 (1969).
248. Uchida, Naoya; and Ohmachi, Yoshiro; J. Appl. Phys., 40(12):4692 (1969).
249. Arlt, G., and Schweppe, F., Solid State Commun., 6:783 (1968).

250. Anderson, O. L., and Andreatch, P., J. Am. Ceram. Soc., 49(8):404 (1966).
251. Kovba, L. M., Dokl. Akad. Nauk SSSR, 194(1):98 (1970).
252. Kuprashvili, I. S., Naumkin, O. P., and Savitskii, E. M., Izv. Akad. Nauk SSSR Neorg. Mater., 5(12):2123 (1969).
253. Jehn, H., and Olzi, E., J. Less-Common Met., 27(3):297 (1972).
254. Benz, R., J. Nucl. Mater., 29(1):43 (1969).
255. Terekhova, V. F., and Savitskii, E. M., Yttrium [in Russian], Izd. Nauka, Moscow, (1967).
256. Rode, E. Ya., Zh. Neorg. Khim., 1(6):1430 (1956).
257. Shartsis, L., and Canga, R., J. Res. Nat. Bur. Stand., 43:221 (1949).
258. Shartsis, L., Spinner, S., and Smock, A. W., J. Res. Nat. Bur. Stand., 40(1):60 (1948).
259. Kozakevitch, P., Rev. Metall., 46:505 (1949).
260. Kostikov, V. I., Levin, V. Ya., Maurakh, M. A., and Mitin, B. S., in: High-Temperature Materials [in Russian], Izd. Metallurgiya, Moscow (1968), p.143.
261. Encyclopedic Dictionary of Physics [in Russian], Vol. 3, Izd. Sovetskaya Éntsiklopediya (1963).
262. Technical Encyclopedia – Handbook of Physical, Chemical and Engineering Constants [in Russian], Vol. 5, OGIZ RSFSR, Sovetskaya Éntsiklopediya (1930).
263. Marmer, É. N., Gurvich, O. S., and Mal'tsev, L. F., in: High-Temperature Materials [in Russian], Izd. Metallurgiya, Moscow (1967), p. 215.
264. Sparks, J. T., and Komoto, T., J. Appl. Phys., 37:1040 (1966).
265. Shakhtin, D. M., Lavintovich, É. V., Pivovar, T. L., and Eliseeva, G. G., Ogneupory, No. 7, p. 37 (1965).
266. Smith, H. G., Nicklow, R. M., Raubenheimer, L. J., and Wilkison, M. K., J. Appl. Phys., 37:1047 (1966).
267. Okinaka, Hideyuki; Nagasawa, Kasuge Koji; Bando, Yoshahika; Kachi, Sukeji; and Takada, Toshio, J. Phys. Soc. Jpn., 28:798 (1970).
268. Zubkov, V. G., Matveenko, I. I., Tallerchik, B. A., and Gel'd, P. V., Fiz. Tverd. Tela, 13(11):3426 (1971).
269. Ford, C. J., Segel, S. L., Seymour, E. F., and Hyland, G. J., Phys. Condens. Mater., 14(2):111 (1972).
270. Iliev, M., Angelov, S., Stoyanov, S., et al., Phys. Status Solidi B, 58(1):k27 (1973).
271. Horota, Eichi; and Kowanata, Tadshi, Dig. Intern. Conf. Kyoto, New York, No. 4, p. 21 (1973).
272. Kzendzov, Ya. M., and Makarov, V. V., Fiz. Tverd. Tela, 12(11):3166 (1970).
273. Rechtin, M. D., and Averbach, B. L., Phys. Rev. Solid State, 6(11):4294 (1972).
274. Kündig, W., Kobelt, M., Appel., H., et al., J. Phys. Chem. Solids, 30:819 (1969).
275. Tret'yakov, Yu. D., Thermodynamics of Ferrites [in Russian], Izd. Khimiya, Leningrad (1967).
276. Ekstrom, T., Acta Chem. Scand., 26(8):3381 (1972).
277. Adamyan, V. E., Vinokurov, I. V., et al., Fiz. Tverd. Tela, 13(8):2309 (1971).
278. Aivazov, M. I., and Turov, S. V., Izv. Akad. Nauk SSSR Neorg. Mater., 10(5):861 (1974).

279. Querel-Ambrunar, S., and Bertant, E. F., Solid State Commun., 11(5):605 (1972).
280. MacChesney, J. B., Williams, H. J., Sherwood, R. C., et al., J. Appl. Phys., 37:1435 (1966).
281. J. Solid State Chem., 5(2):262 (1972).
282. Toropov, N. A., ed., Chemistry of High-Temperature Materials [in Russian], Izd. Nauka, Leningrad (1967).
283. Aivazov, M. I., Domaev, I. A., and Sarkisyan, A. G., Izv. Akad. Nauk SSSR Néorg. Mater., 6(5):846 (1970).
284. Zhuze, V. P. Andreev, A. A., and Shelykh, A. I., Fiz. Tverd. Tela, 10:3674 (1968).
285. Kwan, Clarence C. Y., Criffiths, C. H., and Eastwood, H. K., J. Appl. Phys. Lett., 20(2):93 (1972).
286. Vinogradov, A. A., and Shelykh, A. I., Fiz. Tverd. Tela, 13(11):3310 (1971).
287. Banus, M. D., Reed, T. B., and Strauss, A. J., J. Phys. Rev. Bull. Solid State, 5(8):2775 (1972).
288. Musikhin, V. I., Bastukhov, É. A., Vatolin, N. A., and Leninskikh, B. M., Izv. Akad. Nauk SSSR Neorg. Mat., 7(1):99 (1971).
289. Zuev, K. P., and Dolgintsev, V. D., Izv. Vyssh. Uchebn. Zaved. Fiz., No. 2, p. 110, (1971).
290. Gulyaeva, G. P., and Korzh, P. D., Izv. Akad. Nauk SSSR Neorg. Mater., 5:966 (1969).
291. Djerassi, H., and Sorriaux, A. P., J. Appl. Phys., 43(3):1275 (1972).
292. Jarzebski, Z. M., Acta Phys. Pol. A, 42(4):371 (1972).
293. Hayashi, Shintaro, Rev. Elec. Commun. Lab., 20(78):968 (1972).
294. Keezer, R. C., Bowman, D. L., and Becker, J. H., J. Appl. Phys., 39:2062 (1968).
295. Stuart, M., Brit. J. Appl. Phys., 18:1637 (1967).
296. Bull. Fac. Educ. Kanazawa Univ. Nat. Sci., No. 21, p. 57 (1972).
297. Chandhyry, A. K., and Rao, K. V., Phys. Status Solidi, 32:731 (1969).
298. F. L. Weichman, Can. J. Phys., 51(6):680 (1973).
299. Robinson, M. L. A., and Roetschi, H. A. C., J. Phys. Chem. Solids. 29:1503 (1968).
300. Van Daal, H. J., J. Appl. Phys., 39:4467 (1968).
301. O'clock, G. D., J. Appl. Phys. Lett., 19(10):403 (1971).
302. Tateno, J., Solid State Commun., 8:1043 (1970).
303. Livi, L. and Lubert, L. Phys. Condens. Matter, 7:368 (1968).
304. Osburn, C. M., and Vest, R. W., J. Am. Ceram. Soc., 54(9):428 (1971).
305. Elyutin, V. P., ed., High-Temperature Materials [in Russian], Izd. Metallurgiya, Moscow (1968).
306. Chernenko, I. M., in: Radiophysics and Semiconductor Electrical Equipment [in Russian], Izd. Dnepropetrovsk. Univ., Dnepropetrovsk (USSR) (1967).
307. Eror, N. G., and Wagner, J. B., J. Electrochem. Soc., 118(10):1665 (1971).
308. Kozheurov, V. A., and Mikhailov, G. T., Zh. Fiz. Khim., 41(11):2892 (1967).
309. Samara, G. A., Phys. Rev. Lett., 21:795 (1968).
310. Bruck, A., and Tahnhauser, D. S., J. Appl. Phys., 38:2520 (1967).
311. Young, A. P., and Schwartz, C. M., J. Phys. Chem. Solids, 30:249 (1969).

312. Pankov, Yu. D., Postov, Yu. V., et al., Izv. Vyssh. Uchebn. Zaved. Fiz., No. 3, p. 126 (1963).
313. Shkol'nik, A. L., Izv. Akad. Nauk SSSR Ser. Fiz., 31(12):2050 (1967).
314. Rogers, D. B., Shannon, R. D., and Gillson, J. L., J. Solid State Chem., 3(2):314 (1971).
315. Kreingol'd, F. I., and Kulinkin, B. S., Fiz. Tekh. Poluprovodn., 4(12):2353 (1970).
316. Jarzebski, Z. M., Bull. Acad. Polon. Sci. Ser. Sci. Chem., 17(4):221 (1969).
317. Schwab, G.-M., and Bohla, F., Z. Naturforsch. Teil. A, 23:1550 (1968).
318. Rao, C. N., Rao, R., Subba, G. V., and Romdas, S., J. Chem. Phys., 73:672 (1969).
319. Ladd, L., and Paul, W., Solid State Commun., 7:425 (1969).
320. Hoeneisen, B., Mead, C. A., and Nicolet, M. A., Solid-State Electron., 14(10):1057 (1971).
321. Arsen'eva, A. M., and Kurchatov, B. V., Zh. Éksp. Teor. Fiz., 4:576 (1934).
322. Ryden, W. D., Lawson, A. W., and Sartein, C. C., Phys. Lett. A, 26:209 (1968).
323. Korzo, V. F., Izv. Vyssh. Uchebn. Zaved. Fiz., No. 9, p. 86 (1967).
324. Groth, R., Phys. Status Solidi, 14(1):69 (1966).
325. Friedrich, E., and Sittig, L., Z. Anorg. Chem., 145:127 (1925).
326. Poluboyarinov, D. N., Shapira, E. Ya., Bakunov, V. S., and Akopov, F. A., Izv. Akad. Nauk SSSR Neorg. Mater., 2(2):336 (1966).
327. Ahn, K. J., and Schafer, M. W., J. Appl. Phys., 38:1197 (1967).
328. Ozkan, O. T., and Monlson, A. J., J. Phys. D (Appl. Phys.), 3:983 (1970).
329. De Wit, H. J., and Crevcocur, C., Phys. Lett. A, 25(5):393 (1967).
330. Okinaka, Hideyuki; Nagasawa, Koichi; and Kosuge, Koji, J. Phys. Soc. Jpn., 29(1):245 (1970).
331. Zuev, K. P., and Dolgintsev, V. D., Izv. Akad. Nauk SSSR Neorg. Mater., 4(8):1498 (1968).
332. Nagels, P. D., Solid State Commun., 5(2):193 (1967).
333. Huntley, D. J., Can. J. Phys., 44:2952 (1966).
334. Dorfman, Ya. G., Magnetic Properties and the Structure of Matter [in Russian], Khimizdat, Moscow (1955).
335. Selwood, P. W., Magnetochemistry [Russian translation], IL, Moscow (1949) [English edition: Interscience, New York (1943)].
336. Goronovskii, I. T., Nazarenko, Yu. I., and Nekryach, E. F., Concise Handbook of Chemistry [in Russian], Izd. Akad. Nauk UkrSSR, Kiev (1962).
337. Samokhvalov, A. A., Morozov, Yu. N., Volkenshtein, N. V., et al., Fiz. Tverd. Tela, 12(10):2865 (1970).
338. Yadav, H. S., Murty, D. S., Verma, S. N., et al., J. Appl. Phys., 44(5):2197 (1973).
339. Gruner, H., Optik, 39(4):443 (1974).
340. Judge, D. L., and Lee, L. C., J. Chem. Phys., 58(1):104 (1973).
341. Lee, L. C., and Judge, D. L., Can. J. Phys., 51(4):378 (1973).
342. Lee, L. C., and Judge, D. L., J. Chem. Phys., 57(10):4443 (1972).
343. Bernas, A., and Truong, Thu-Ba, C. R. Acad. Sci. Ser. B., 277(14):391 (1973).
344. Mironova, N. A., in: Radiation Defects in Semiconductors [in Russian], Izd. Beloruss. Gos. Univ., Minsk (1972), p. 192.
345. Bradford, A. P., Hass, G., and McFarland, M., Appl. Opt., 11(10):2242 (1972).

346. Zakis, Yu. R., and Trukhin, A. N., Uch. Zap. Latv. Univ. (Riga), 182:49 (1973).
347. Lagarde, P., and Narenberg, M. A., Phys. Status Solidi B, 64(2):567 (1974).
348. Schirmer, O. F., Koidl, P., and Reik, H. G., Phys. Status Solidi B, 62(2):385 (1974).
349. Mikho, V. V., Fedchuk, A. P., and Shaikina, O. F., Pis'ma Zh. Éksp. Teor. Fiz., 17(5):241 (1973).
350. Chou, H., and Fan, H., Phys. Rev. B, 10(3):901 (1974).
351. Prikhod'ko, L. V., and Bagdasarov, Kh. S., Opt. Spektrosk., 34(6):1210 (1973).
352. Rusin, J. M., and Mueller, J. I., Am. Ceram. Soc. Bull., 52(9):708 (1973).
353. Motovilov, O. A., and Rozhdestvenskii, V. M., Opt.-Mekh. Promst., No. 8, p. 42 (1958).
354. Kharif, Ya. L., Galaktionov, S. S., Dergacheva, N. M., et al., Fiz. Tverd. Tela, 17(4):987 (1975).
355. Shalimova, K. V., and Nikitenko, V. A., Zh. Prikl. Spektrosk., 22(4):667 (1975).
356. Blasse, G., and Bril, A., J. Phys. Chem. Solids 31(4):704 (1970).
357. Shalimova, K. V., Margulova, T. Kh., Malova, T. I., et al., At. Énerg., 36(4):312 (1974).
358. Vos, K., and Krusemeyer, H. J., Solid State Commun., 15:945 (1974).
359. Kaminskii, B. T., Plygunov, A. S., and Prokof'eva, G. N., Ukr. Khim. Zh., 39(9):946 (1973).
360. Klingshirn, C., Ostertag, E., and Levy, R., Solid State Commun., 15(5):883 (1974).
361. Jensen, G. H., Phys. Status Solidi B, 64(1):K51 (1974).
362. Kearney, R. J., Cottini, M., Grilli, E., et al., Phys. Status Solidi B, 64(1):49 (1974).
363. Bogomolov, V. N., Bryskin, V. V., Sitnikova, A. A., et al., Fiz. Tverd. Tela, 15(8):2347 (1973).
364. Mokerov, V. G., and Galiev, G. B., Fiz. Tverd. Tela, 16(1):266 (1974).
365. Knyazev, A. S., Zakharov, V. P., Mityureva, I. A., et al., Fiz. Tverd. Tela, 15(8):2371 (1973).
366. Izvozchikov, V. A., Shamba, E. M., and Kishmariya, S. R., Zh. Prikl. Spektrosk., 20(5):902 (1974).
367. Novak, L., Solid State Commun., 9(23):2129 (1971).
368. Ghosh, A. X., Lauer, R. B., and Addise, R. R., Phys. Rev. Bull., 8(10):4842 (1973).
369. Tubbs, M. R., Phys. Status Solidi A, 21(1):253 (1974).
370. Kreingol'd, F. I., Izv. Akad. Nauk SSSR Neorg. Mater., 5(9):1639 (1969).
371. Val'tere, A. Ya., Bondar', I. A., and Ovsyankin, V. V., Izv. Akad. Nauk SSSR Neorg. Mater., 9(3):413 (1973).
372. Chen, Y., Unruh, W. P., Abraham, M. M., et al., J. Am. Ceram. Soc., 56(8):438 (1973).
373. Finkenrath, H., Fricke, W., and Uhle, N., Phys. Status Solidi B, 60(1):341 (1973).
374. Andrianov, G. O., Aronov, A. G., and Smirnova, T. V., Phys. Status Solidi B, 60(1):79 (1973).

375. Nikolaev, V. I., Samokhvalov, A. A., and Sever, G. N., Fiz. Tekh. Poluprovodn., 9(2):379 (1975).

376. Klucker, R., Nelkowski, H., Park, Y. S., et al., Phys. Status Solidi B, 45(1):265 (1971).

377. Crawford, J. R., Dragsdorf, R., and Dean, P. J., J. Appl. Phys., 44(1):385 (1973).

378. Yagodovskaya, T. V., and Nekrasov, L. I., Zh. Fiz. Khim., 45(9):2329 (1971).

379. Whited, R. C., Christopher, J. F., and Walker, W. C., Solid State Commun., 13(11):1903 (1973).

380. Hughes, A. E., and Pells, G. P., J. Phys. C, 5(17):2543 (1972).

381. Ratinen, H., Phys. Status Solidi A, 15(2):K109 (1973).

382. Heitmann, W., Appl. Opt., 12(2):394 (1973).

383. Buchenau, U., and Müller, I., Solid State Commun., 11(9):1291 (1972).

384. Satybaev, N. M., and Malov, M. M., Tr. Mosk. Énerg. Inst. (Gosénergoizdat, Moscow–Leningrad), No. 12, p. 102 (1972).

385. Pandey, K. N., Kanari, P. S., and Singh, V. B., Labdev J. Sci. Technol. A, 9(3-4):220 (1971).

386. Hvam, J. M., Solid State Commun., 12(2):95 (1973).

387. Riemann, E., and Young, L., J. Appl. Phys., 44(3):1044 (1973).

388. Miloslavskii, V. K., Opt. Spektrosk., 7(2):244 (1959).

389. Wolffing, B., and Hurych, Z., Phys. Status Solidi A, 16(2):K161 (1973).

390. Tomotika, T., Ftany, Y., and Tomioka, K., J. Phys. Soc. Jpn., 33(2):409 (1972).

391. Linares, R. C., J. Phys. Chem. Solids, 28(7):1285 (1967).

392. Satyanarayana, M., and Ramakrisharao, V., Indian J. Pure Appl. Phys., 10(11):844 (1972).

393. Afanas'eva, L. A., Mikho, V. V., Shchegol'kov, S. V., et al., Izv. Vyssh. Uchebn. Zaved. Fiz., No. 11, p. 102 (1972).

394. Thomas, J. H., Appl. Phys. Lett., 22(8):406 (1973).

395. Deb, S. K., Philos. Mag., 27(4):801 (1973).

396. Feinbeib, J., Scouler, W. J., and Ferretti, A., Phys. Rev. Bull., 165(3):765 (1968).

397. Hanafi, Z. M., and Ismail, F. M., Z. Phys. Chem. (Frankfurt), 79(5-6):263 (1972).

398. Dibierzewska-Mozrzymas, E., Marcinow, T., and Pietrasik, J., Opt. Appl., 2(1):23 (1972).

399. Izvozchikov, V. A., Phys. Status Solidi A, 14(1):161 (1972).

400. Fidrya, A. K., in: Solid State Physics [in Russian], Izd. Naukova Dumka, Kiev (1972), p. 34 [possibly same as Fiz. Tverd. Tela (Khar'kov)].

401. Harvey, P. J., Childs, B. G., and Moerman, G., J. Am. Ceram. Soc., 56(3):134 (1973).

402. Mokhnyak, Ya., Zh. Prikl. Spektrosk., 18(5):917 (1973).

403. Hulin, D., Benoit a la Guillaume, C., and Hanus, J., Magnetism and Magnetic Materials, 17th AJP Annual Conf., New York (1972), p. 850.

404. Dulenov, E. V., Batsanov, S. S., and Kustova, G. N., Zh. Strukt. Khim., 13(5):935 (1972).

405. Federov, N. D., Concise Handbook for Engineering Physicists – Nuclear Physics and Atomic Physics [in Russian], Gosatomizdat, Moscow (1961).

406. Gordeev, I. V., Kardashev, D. A., and Malyshev, A. V., Handbook of Nuclear Physics Constants for Reactor Design [in Russian], Gosatomizdat, Moscow (1960).
407. Handbook of Nuclear Physics [Russian translation, L. A. Artsimovich, ed.], Fizmatgiz, Moscow (1963).
408. Wilks, R. S., J. Nucl. Mater., 26(2):137 (1968).
409. Beryllium Oxide; Proceedings of 1st International Conference on Beryllium Oxide, Sydney, 1963 [Russian translation, R. A. Belyaev and V. I. Kushakovskii, eds.], Atomizdat, Moscow (1968).
410. Keilholtz, G. W., Lee, J. E., and Moore, R. E., Nucl. Sci. Eng., 26(3):329 (1966).
411. React. Mater., 10(2):87 (1967).
412. Hickman, B. S., and Walker, D. G., Philos. Mag., 11(114):1101 (1965).
413. Wilks, R. S., Desport, G. A., and Bradley, R., Proc. Brit. Ceram. Soc., No. 7, p. 403 (1967).
414. Keilholtz, F. W., and Moore, R. E., Nucl. Appl., 3(11):686 (1967).
415. Collins, C. G., Proceedings of Symposium on Radiation Damage and Reactor Materials, IAEA, Vienna (1963), p. 725.
416. Crawford, J. H., and Wittells, M., in: Nuclear Fuel and Reactor Materials [in Russian], Vol. 6, Gosatomizdat, Moscow (1959), p. 435.
417. MacChesney, J. B., and Johnson, G. E., J. Appl. Phys., 35(9):2784 (1964).
418. Wullaert, R. A., Burian, R. J., Melehan, J. B., et al., in: Effect of Radiation on Materials and Elements of Electronic Circuits [Russian translation], Atomizdat, Moscow (1967), p. 142.
419. Lastman, B., Radiation Phenomena in Uranium Dioxide [in Russian], Atomizdat, Moscow (1964).
420. Rau, R. C., and Chase, G. A., Phys. Status Solidi, 19(2):645 (1967).
421. Hickman, B. S., and Walker, D. G., J. Nucl. Mater., 10(3):243 (1963).
422. Hickman, B. S., J. Nucl. Mater., 17(3):270 (1965).
423. Elston, J., and Labbe, C., J. Nucl. Mater., 4(2):143 (1961).
424. Cooper, M. K., Palmer, A. R., and Stolarski, G. Z. A., J. Nucl. Mater., 9(3):320 (1963).
425. Budylin, B. V., and Vorob'ev, A. A., Effect of Radiation on Ionic Structures [in Russian], Gosatomizdat, Moscow (1962).
426. Belyaev, R. A., Beryllium Oxide [in Russian], Gosatomizdat, Moscow (1962).
427. Brekhovskikh, S. M., Viktorova, Yu. N., Grinshtein, Yu. L., and Landa, L. M., Principles of Radiation Material Science of Glass and Ceramics [in Russian], Stroiizdat, Moscow (1971).
428. React. Mater., 9(2):94 (1966).
429. React. Mater., 9(4):221 (1966-67).
430. Rau, R. C., and Lacefield, K., Phys. Status Solidi, 21(1):K69 (1967).
431. Slavuta, V. I., and Shinakov, A. A., in: Radiation Damage in Solids (Summaries of Papers from Republic Conference) [in Russian], Izd. IF Akad. Nauk UkrSSR, Kiev (1974), p. 129.
432. Groves, G. W., and Kelly, A., Philos. Mag., 8(93):1437 (1963).
433. Kostyukov, N. S., Kharitonov, F. Ya., and Antonova, N. P., Radiation and Corrosion Resistance of Electroceramics [in Russian], Atomizdat, Moscow (1973).

434. Dan, G. J., and Davis, M. V., Nucl. Sci. Eng., 21(1):30 (1965).
435. Gus'kov, Yu. K., and Sachkov, V. F., At. Énerg., 6(2):204 (1959).
436. Gulyamov, U. G., Kostyukov, N. S., and Sokolov, A. P., in: Questions of Atomic Science and Technology; Series "Radiation Material Science and Radiation Procedures and Equipment [in Russian], No. 1, Izd. Nauchno-Issled. Inst. At. Reakt., Dmitrograd (USSR) (1973), p. 35.
437. Rau, R. S., and Bartram, S. F., J. Am. Ceram. Soc., 48(5):233 (1965).
438. McVay, T. N., J. Am. Ceram. Soc., 45(1):48 (1962).
439. Martin, D. G., J. Phys. Chem. Solids, 10(1):64 (1959).
440. Veevers, K., J. Nucl. Mater., 40(3):289 (1971).
441. Sabine, T. M., Pryor, A. W., and Hickman, B. S., Philos. Mag., 8(85):43 (1963).
442. Rau, R. C., Philos. Mag., 16(142):663 (1967).
443. Bowen, D. H., and Clarke, F. J. P., Philos. Mag. 9(99):413 (1964).
444. Walker, D. G., and Hickman, B. S., Philos. Mag., 12(117):445 (1965).
445. Khudyakov, A. V., Marakueva, N. A., Klimenkov, V. I., and Balandin, G. S., At. Énerg., 24(5):483 (1968).
446. Ripan, R., and Chetyants, L., Inorganic Chemistry, Parts 1 and 2 [Russian translation from Romanian], Izd. Mir., Moscow (1972).
447. Cotton, F. A., and Wilkinson, J., Modern Inorganic Chemistry, Vols. 1-3 [Russian translation], Izd. Mir, Moscow (1969), [English edition: Advanced Inorganic Chemistry, Interscience, New York (1966)].
448. Remy, H., Course in Inorganic Chemistry, Vols. 1 and 2 [Russian translation], IL, Moscow (1963), [English edition: Treatise on Inorganic Chemistry, Am. Elsevier, New York (1956)].
449. Zelikman, A. N., and Meerson, G. A., Metallurgy of Rare Metals [in Russian], Izd. Metallurgiya, Moscow (1973).
450. Nekrasov, B. V., Principles of General Chemistry, Parts I and II [in Russian], Izd. Khimiya, Moscow (1973).
451. Samsonov, G. V., Markovskii, L. Ya., Zhigach, A. F., and Valyashko, M. G., Boron and Its Compounds and Alloys [in Russian], Izd. Akad. Nauk UkrSSR, Kiev (1960).
452. Nikitin, I. V., and Rosolovskii, V. Ya., Usp. Khim., 40(11):1913 (1971).
453. Bagnall, K. W., Chemistry of Selenium, Tellurium, and Polonium [Russian translation], Atomizdat, Moscow (1971) [English edition: Topics in Inorganic and General Chemistry, Vol. 7, Elsevier, New York (1966)].
454. Dratovski, M., and Pachesova, L., Usp. Khim., 37(4):537 (1968).
455. Minachev, Kh. M., Khodakov, Yu. S., et al., Rare Earths in Catalysis [in Russian], Izd. Nauka, Moscow (1972).
456. Davidyants, A. A., and Pervushkin, N. I., Production of Cracking Catalysts and High-Activity Silica Gels [in Russian] Izd. Khimiya, Moscow (1972).
457. Boreskov, G. K., Catalysis, Parts I and II [in Russian], Izd. Sib. Otd. Akad. Nauk SSSR, Novosibirsk (1971).
458. Mukhlenov, I. P., ed., Fluid-Bed Catalysis [in Russian], Izd. Khimiya, Moscow (1971).
459. Sokol'skii, D. V., and Popova, N. M., Catalytic Treatment of Exhaust Gas [in Russian], Izd. Nauka, Alma-Ata (USSR) (1970).

460. Sokol'skii, D. V., and Zhubanov, K. A., Hydrogenation of Vegetable Oils [in Russian], Izd. Nauka, Alma-Ata (USSR) (1972).
461. Giller, S. A., et al., eds., Heterogeneous Catalysis in Reactions of Preparation and Conversion of Hetrocyclic Compounds [in Russian], Izd. Zinatne, Riga (1971).
462. Mukhlenov, I. P. ed., Technology of Catalysis [in Russian], Izd. Khimiya, Leningrad (1974).
463. Skarchenko, V. K., Aluminosilicate Catalysts [in Russian], Izd. Akad. Nauk UkrSSR, Kiev (1963).
464. Thomas, J. M., and Thomas, W. J., Introduction to the Principles of Heterogeneous Catalysis [Russian translation], Izd. Mir, Moscow (1969) [English edition: Academic Press, New York (1967)].
465. Atroshchenko, V. I., ed., Kinetics of Heterogeneous Catalytic Processes under Pressure [in Russian], Izd. Khar'kov. Gos. Univ., Khar'kov (1974).
466. Roiter, V. A., ed., Catalytic Properties of Substances (Handbook) [in Russian], Izd. Naukova Dumka, Kiev (1968).
467. Manual on Catalysts for Ammonia and Hydrogen Production [Russian translation, V. P. Semenov, ed.], Izd. Khimiya, Leningrad (1973).
468. Markovskii, L. Ya., Orshanskii, D. L., and Pryanishnikov, V. P., Chemical Electrothermy [in Russian], Goskhimizdat, Leningrad – Moscow (1952).
469. Johnson, P., J. Am. Ceram. Soc., .33(5):168 (1950).
470. Kroll, W., and Schlechter, A., Trans. Electrochem. Soc., 93(5):247 (1948).
471. Mallet, M., Durbin, E., Udy, M., et al., J. Electrochem. Soc., 101:298 (1954).
472. Dayton, R., Oxley, J., and Townlei, C., J. Nucl. Mater., 11(1):36 (1963).
473. Norton, F., and Kingery, W., Study of Metal-Ceramic Interactions at Elevated Temperatures, USAEC, N-YO-3137, Mass. Inst. Technol., Cambridge, Mass. (1952).
474. Economos, G., Ind. Eng. Chem., 45(2):458 (1953).
475. Economos, G., and Kingery, W., J. Am. Ceram. Soc., 36(12):403 (1953).
476. Buddery, J., and Trachray, R., J. Inorg. Nucl. Chem., 3(3/4):190 (1956).
477. Kroll, W., Z. Anorg. Allgem. Chem., 240(4):331 (1939).
478. Samsonov, G. V., Ogneupory, No. 3, p. 122 (1956).
479. Fornwalt, D. E., Gourley, B. R., and Manzione, A. V., in: The Electron Microprobe, Wiley, New York (1966), p. 581.
480. Kieffer, R., and Benesovsky, F., Metallurgie, 58(347):119 (1958).
481. Johnson, P., J. Am. Ceram. Soc., 32(10):316 (1949).
482. Avgustinik, A. I., Kozlorskii, L. V. and Klimashin, G. M., Izv. Vyssh. Uchebn. Zaved. Khim. Khim. Tekhnol., 9(4):528 (1966).
483. Weiz, C., and Valkenberg, A., J. Res. Nat. Bur. Stand., 64(1):103 (1960).
484. Geller, R., and Yavorsky, P., J. Res. Nat. Bur. Stand., 35(1):87 (1945).
485. Zueva, L. S., Godina, N. A., and Keler, É. K., Ogneupory, No. 8, p. 368 (1960).
486. Tresvyatskii, S. G., and Kushakovskii, V. I., At. Énerg., 9(1):30 (1960).
487. Corwin, R., and Eyerly, G., J. Am. Ceram. Soc., 36(4):137 (1953).
488. Gopienko, V. G., Ogneupory, No. 6, p. 55 (1971).
489. Samsonov, G. V., Burykina, A. L., Strashinskaya, L. V., et al., Izv. Akad. Nauk SSSR Metall. Gorn. Delo, No. 4, p. 106 (1964).

490. Karyakin, L. L., and Degtyareva, É. V., Byull. Nauchno-Tekh. Inform. Ukr. Nauchno-Issled. Inst. Ogneuporov, No. 4, p. 91 (1958).
491. Osvald, M., Met. Corros. Ind., 75:318 (1952).
492. Kingery, W. D., in: High-Temperature Research [Russian translation], IL, Moscow (1962), p. 126.
493. Lambertson, W., and Mueller, M., J. Am. Ceram. Soc., 36(10):329 (1953).
494. Albaugh, F., React. Core Mater., 3(4):15 (1960).
495. Foster, L., Long, G., and Hunter, M., J. Am. Ceram. Soc., 39(1):1 (1956).
496. Elyutin, V. P., Pavlov, Yu. A., and Chelnokov, V. S., Izv. Akad. Nauk SSSR Neorg. Mater., 9(8):1365 (1973).
497. Zhmud', E. S., and Shmelev, A. E., Izv. Akad. Nauk SSSR Neorg. Mater., 9(10):1733 (1973).
498. Étinger, I. A., Mal'tseva, L. F., Savratskaya, L. A., et al., in: Refractory Carbides [in Russian], Izd. Naukova Dumka, Kiev (1970), p. 260.
499. Navias, L., Am. Ceram. Soc. Bull., 38(5):256 (1959).
500. Osipov, K. A., Galkin, B. D., and Urazaliev, U. S., Izv. Akad. Nauk SSSR Neorg. Mater., 9(10):1738 (1973).
501. Wilkinson, W., Uranium Metallurgy, Interscience, New York (1962).
502. O'Connor, T. E., U.S. Pat. 3,261,667 (1966).
503. Kotera, Y., and Ueno, A., Bull. Chem. Soc. Jpn., 46(11):3431 (1973).
504. Kharitonov, V. I., Manukhin, M. S., and Shamrai, F. I., Izv. Akad. Nauk SSSR Met., No. 6, p. 227 (1968).
505. Ryabchikov, I. V., Khrushchev, M. S., and Maksimov, Yu. S., Izv. Akad. Nauk SSSR Metall. Gorn. Delo, No. 6, p. 58 (1964).
506. Ryabchikov, I. V., Gorokh, A. V., Khrushchev, M. S., et al., Metally, No. 4, p. 38 (1966).
507. Ugai, Ya. A., Tverdokhlebova, L. Ya., Anokhin, V. Z., et al., Izv. Akad. Nauk SSSR Neorg. Mater., 10(6):1033 (1974).
508. Borisov, V. G., and Yudin, B. F., Ogneupory, No. 3, p. 37 (1968).
509. Williams, J., Livey, D. T., Barnes, E., et al., J. Nucl. Energ., 4(2):164 (1957).
510. Tsynkina, V. M., Byull. Nauchno-Tekh. Inform. Vses. Nauchno-Issled Proektn Rab. Ogneuporn. Promsti., No. 3, p. 98 (1958).
511. Nadler, M. P., and Fitzsimmons, E. S., J. Am. Ceram. Soc., 38(8):214 (1955).
512. Shveikin, G. P., Poroshk. Metall., No. 6, p. 67 (1962).
513. Meerson, G. A., Izv. Akad. Nauk SSSR Metall. Topl., No. 3, p. 33 (1962).
514. Norton, F. H., Refractories, McGraw-Hill, New York (1949).
515. Kosolanova, T. Ya., Fedorus, V. B., and Kuz'ma, Yu. B., Izv. Akad. Nauk SSSR Neorg. Mater., 2(8):1516 (1966).
516. Gilles, J., Rev. Hautes Temp. Refract., 2(3):237 (1965).
517. Gol'dshtein, L. M., Kalinichenko, I. I., Sheinkman, A. I., and Turlakov, V. N., Zh. Neorg. Khim., 18(12):3185 (1973).
518. Manojlovic, L., and Ristic, M., Institute of Nuclear Sciences Boris Kidic., NR-012-0373, Vinca, Belgrade (1962).
519. Rubinchik, Ya. S., Prokudina, S. A., and Pavlyuchenko, M. M., Izv. Akad. Nauk SSSR Neorg. Mater., 9(11):1951 (1973).
520. Ashin, A. K., and Rostovtsev, S. T., Izv. Vyssh. Uchebn. Zaved. Chern. Metall., No. 4, p. 11 (1964).
521. Kohl, H., and Marincek, V., Arch. Eisenhuettenwes., 38(7):493 (1967).

522. Khrustalev, B. N., and Leonovich, B. I., Nauchn. Tr. Chelyabinsk. Politekh. Inst., No. 133, p. 42 (1974).
523. Elyutin, V. P., Pavlov, Yu. A., and Polyakov, V. N., Izv. Vyssh. Uchebn. Zaved. Chern. Metall., No. 7, p. 6 (1966).
524. Suzuki, H., Iseki, T., and Nakamura, T., J. Nucl. Sci. Technol., 10(6):360 (1973).
525. White, J., Powder Metall., 8(15):64 (1965).
526. Levitskii, V. A., Chentsov, V. N., and Skolis, Yu. Ya., Zh. Fiz. Khim., 48(3):566 (1974).
527. Lang, S., Nat. Bur. Stand. (U.S.) Circ., No. 568 (1956).
528. Elyutin, V. P., Pavlov, Yu. A., Manukhin, A. V., and Melekhin, V. F., Izv. Vyssh. Uchebn. Zaved. Chern. Metall., No. 1, p. 5 (1970).
529. Liquid Metal Handbook, 2nd ed., New York (1954).
530. Campbell, J. B., Mater. Methods, 31(5):95 (1950).
531. Mal'kov, L. P., ed., Heat- and Corrosion-Resistant Metal/Ceramic Materials [in Russian], Oborongiz, Moscow (1959).
532. Kura, G., Jackson, J. H., Udy, M. C., et al., J. Met., 1(10):769 (1949).
533. Manning, C., and Stoops, R., J. Am. Ceram. Soc., 51(8):411 (1968).
534. Ere, Kekausu; and Yogyo, Kyokaishi, J. Ceram. Assoc. Jpn., 74(850):183 (1966).
535. Gropyanov, V. M., Yudin, B. F., and Avgustinik, A. I., Ogneupory, No. 10, p. 50 (1966).
536. Avgustnik, A. I., Gropyanov, V. M., Grozdetskaya, et al., Ogneupory, No. 12, p. 570 (1964).
537. Zverev, G. L., Metally, No. 4, p. 253 (1969).
538. Baskin, M. P., Tret'yakov, V. I., and Chaporova, I. N., Fiz. Met. Metalloved., 12(6):860 (1961).
539. Strakhov, V. L, Klyucharov, A. V., and Sergeev, G. G., Zh. Prikl. Khim., 46(9):2083 (1973).
540. Plyushchev, V. E., Kushakovskii, V. I., Zhidkov, B. A., et al., Izv. Akad. Nauk SSSR Neorg. Mater., 4(5):738 (1968).
541. Brewer, Z., and Krikorian, O., J. Electrochem. Soc., 103(1):38 (1956).
542. Sazhin, N. P., Kolchin, O. P., and Sumarokova, N. V., Izv. Akad. Nauk SSSR Metall. Topl., No. 6, p. 8 (1961).
543. Elyutin, V. P., Pavlov, Yu. A., and Sheboldaev, S. R., Izv. Akad. Nauk SSSR Met., No. 6, p. 21 (1967).
544. Magneli, A., and Kihlorg, D., Acta Chem. Scand., 5(4):578 (1951).
545. Pitman, D., and Das, D., J. Electrochem. Soc., 107(9):763 (1960).
546. Smithells, C., Metals Reference Book, 2nd ed., Vol. 2, Interscience, New York (1955).
547. Brewer, L., Sawyer, D. L., Tempeton, D. H., et al., J. Am. Ceram. Soc., 34(6):173 (1951).
548. Seddon, B., Miss. Reactor Group U.K. Atomic Energy Author. Rept., No. 842 (R) (1965).
549. React. Core Mater., 4(4):14 (1961).
550. Craven, B., and McCartney, E., J. Am. Ceram. Soc., 44(1):12 (1961).
551. Meerson, G. A., Kotel'nikov, R. B., and Bashlykov, S. N., At. Énerg., 9(5):387 (1960).

552. Nuclear Reactor Fuel Elements – Metallurgy and Fabrication, New York/ London (1962).
553. Scott, R., and Williams, L., Trans. Brit. Ceram. Soc., 57(3):199 (1958).
554. Nichols, R., Nucl. Eng., No. 29, p. 327 (1958).
555. Spengler, C., and Bratton, R., J. Nucl. Mater., 38(2):217 (1971).
556. Gangler, J., Nucl. Sci. Abstr., No. 8, p. 76 (1960).
557. Battelle Memorial Institute, Review, Metals Science of Reactor Materials [Russian translation], Vol. 1, Gosatomizdat, Moscow (1961), pp. 182, 205-206.
558. Amato, I., Colombo, R., and Petruccioli Balzari, A., J. Nucl. Mater., 19(2):190 (1966).
559. Kopel'man, B., Materials for Nuclear Reactors [in Russian], Atomizdat, Moscow (1962).
560. Stoops, R., and Hamme, J., J. Am. Ceram. Soc., 47(2):59 (1964).
561. Elston, J., and Guine, P., Bull. Inf. Sci. Tech. Commis. Energ. At. Fr., No. 100, p. 101 (1966).
562. Chikalla, T., J. Am. Ceram. Soc., 46(7):323 (1963).
563. Handwerk, I., White, G. D., and Hill, D. C., J. Am. Ceram. Soc., 46(1):29 (1963).
564. Baskin, Y., J. Am. Ceram. Soc., 48(12):652 (1965).
565. Humenik, M., and Kingery, W. D., J. Am. Ceram. Soc., 37(1):18 (1954).
566. Eremenko, V. N., and Naidich, Yu. V., Wetting of the Surfaces of Heat-Resistant Alloys with Liquid Metals [in Ukrainian], Izd. Naukova Dumka, Kiev (1958).
567. Livey, D. T., and Murray, P., Wärmefeste and Korrosionbestandige Sinterwerkstoffe, Reut, Tirol (1956).
568. Eremenko, V. N., and Naidich, Yu. V., Zh. Fiz. Khim., 33:1238 (1959).
569. Kingery, W. D., Am. Ceram. Soc. Bull., 35(4):108 (1956).
570. Tangermann, J., Neue Huette, 6:767 (1961).
571. Eremenko, V. N., Ukr. Khim. Zh., 23:573 (1957).
572. Levin, A. M., Tr. Dnepropetrovsk. Metall. Inst. (Izd. DMETI), No. 28, p. 89 (1958).
573. Tikkanen, M., and Jerkontorets, H., Ann. Phys., 147:37 (1963).
574. Lutton, F. A., and Cashmore, Brit. Pat. 729,342 (1955).
575. Kislyi, P. S., Panasyuk, A. D., and Belykh, A. B., Izv. Akad. Nauk SSSR Neorg. Mater., 11(9):1600 (1975).
576. Bradhurst, D. H., and Buchanan, A. S., J. Phys. Chem., 63:1486 (1959).
577. Paller, E., Neue Huette, 8:747 (1963).
578. Bondi, A., Chem. Rev., 52:417 (1953).
579. Loman, R., and Zwikker, N. P., Physica, 1:1181 (1934).
580. Yanagida, Hiroaki; and Kröger, F. A., J. Am. Ceram. Soc., 51(12):700 (1968).
581. Sari, C., and Zamorani, E., J. Nucl. Mater., 37(3):324 (1970).
582. Kepinski, J., and Kalucki, K., Can. J. Chem., 48(15):2320 (1970).
583. Chikalla, T. D., and Eyring, L., J. Inorg. Nucl. Chem., 31(1):85 (1969).
584. Johnson, R. E., and Muan, A., J. Am. Ceram. Soc., 51(8):430 (1968).
585. Ol'shanskii, Ya. I., and Shlepov, V. K., Dokl. Akad. Nauk SSSR, 91:561 (1953).

586. Elliott, R. P., Constitution of Binary Alloys [Russian translation, Vols. I and II, I. I. Novakov and I. L. Rogel'berg, eds.], Izd. Metallurgiya, Moscow (1970) [English edition: Elliott, R. P., Constitution of Binary Alloys, First Supplement, McGraw-Hill, New York (1975); may also refer to Hansen, M., Constitution of Binary Alloys (1958), and Shunk, M. A., Second Supplement (1969)].
587. Hofmann, W., and Klein, M., Z. Metallkd., 57(5):385 (1966).
588. Gerlach, J., Osterwald, J., and Stichel, W., Z. Metallkd., 59(7):576 (1968).
589. Shafer, M. W., Torrance, J. B., and Penney, T., J. Phys. Chem. Solids, 33(12):2251 (1972).
590. Natola, F., and Touzain, Ph., Can. J. Chem., 48(13):1955 (1970).
591. Hed, A. Z., and Tannhauser, D. S., J. Electrochem. Soc., 114(4):314 (1967).
592. Schmahl, N. G., and Hennings, Detlev, F. K., Arch. Eisenhuettenwes., 40(5):395 (1969).
593. Phillips, B., and Chang, L. L. I., Trans. Metall. Soc. AIME, 233(7):1433 (1965).
594. Barrett, C. S., Meyer, L., Greer, S. C., et al., J. Chem. Phys., 48(6):2670 (1968).
595. Zakharov, G. B., Popov, I. A., Zhorova, L. P. and Feddin, B. V., Niobium and Its Alloys [in Russian], Metallurgizdat, Moscow (1961).
596. Marucco, J.-F., C. R. Acad. Sci. C, 375(23):1391 (1972).
597. Bogatskii, D. P., Zh. Obshch. Khim., 21(1):3 (1951).
598. Stammler, M., Metall, 13(8):750 (1959).
599. Hyde, B. G., Bevan, D. J. M., and Eyring, L., Philos. Trans. Roy. Soc. London A, 259(1106):583 (1966).
600. Touzain, Ph., and Caillet, M., Rev. Chim. Minér., 8(2):277 (1971).
601. Burnham, D. A., Eyring, L., and Kordis, J., J. Phys. Chem., 72(13):4424 (1968).
602. Wahlbeck, Ph. G., and Gilles, P. W., J. Am. Ceram. Soc. 49(4):180 (1966).
603. Kornilov, I. I., in: A New Material of Construction – Titanium [in Russian], Izd. Nauka, Moscow (1972), p. 24.
604. Martin, A. E., and Edwards, R. K., J. Phys. Chem., 69(5):1788 (1965).
605. Stringer, J., J. Less-Common Met., 8(1):1 (1965).
606. Alexander, D. G., and Carlson, O. N., Metall. Trans., 2(10):2805 (1971).
607. McChesney, J. B., and Guggenheim, H. J., J. Phys. Chem. Solids, 30(2):225 (1969).
608. Phillips, B., and Change, L. L. I., Trans. Metall. Soc. AIME, 230(5):1203 (1964).
609. Ruh, R., and Garrett, H. J., J. Am. Ceram. Soc., 50(5):257 (1967).
610. Kalinenko, D. I., Kokhanenko, P. N., and Sonchik, V. K., Izv. Vyssh. Uchebn. Zaved. Fiz., No. 6, p. 116 (1974).
611. Freeouf, J. L., Phys. Rev. B, 7(8):3810 (1973).
612. Sorokin, O. M., Blank, V. A., and Lebedeva, G. A., Opt. Spectrosk., 35(3):501 (1973).
613. Zhuze, V. P., Lukirskii, D. P., and Startsev, G. P., Fiz. Tverd. Tela, 13(1):317 (1971).

614. Messick, L., Walker, W. C., and Glosser, R., Surface Sci., 37(1):267 (1973).
615. Popova, S. I., Tolstykh, T. S., and Ivlev, L. S., Opt. Spectrosk., 35(5):954 (1973).
616. Deiss, J. L., and Daunois, A., Surface Sci., 37(3):804 (1973).
617. Zolotarev, V. M., and Morozov, V. N., Opt. Spectrosk., 34(2):319 (1973).
618. Zhuze, V. P., ed., Physical Properties of Chalcogenides of Rare-Earth Elements [in Russian], Izd. Nauka, Leningrad (1973).
619. Liddell, H. M., J. Phys. D, 7(11):1588 (1974).
620. Popova, S. I., Tolstykh, T. S., and Vorov'ev, V. T., Opt. Spectrosk., 33(4):801 (1972).
621. McRoberts, R. D., Fonstad, C. G., and Hubert, D., Phys. Rev. B, 10(12):5213 (1974).
622. McIntyre, J. D. E., Surface Sci., 37(2):658 (1973).
623. Samsonov, G. V., Gil'man, I. Ya., and Andreeva, A. F., Izv. Akad. Nauk SSSR Neorg. Mater., 10(9):1645 (1974).
624. Grison, E., Plutonium 1960, Cleaver-Hume Press, London (1961).
625. Kuznetsova, V. G. (Kuznezowa, W. G.), Reshetnikov, R. G. (Reschetnikow), and Alekseev, O. A. (Alexeew), in: Plutonium Reactor Fuel, Vienna (1967), p. 357.
626. Distefano, I., and Link, H., Nucl. Technol., 19(1):34 (1973).
627. Veltz, H., J. Am. Ceram. Soc., 33(11):340 (1950).
628. Samsonov, G. V., Yasinskaya, G. A., and Shiller, É. A., Ogneupory, No. 7, p. 335 (1961).
629. Plotnikov, K. A., Stepanov, A. I., Krasil'nikova, A. A., et al., Izv. Akad. Nauk SSSR Neorg. Mater., 7(2):343 (1971).
630. Vol'nov, I. I., and Matveev, V. V., in: Rare Alkali Elements [in Russian], Izd. Nauka (Sib. Otd.), Novosibirsk (1967), p. 31.
631. Vogel, D. L., and Pieck, G. D., Acta Metall., 19(3):233 (1971).
632. Bondarenko, V. P., Fomichev, E. N., and Kandyuba, V. V., High Temp. High Pressures, 5(1):5 (1973).
633. Maglic, K., and Herak, R., Rev. Hautes Temp. Refrac., 7(3):247 (1970).
634. Vainshtein, V. M., and Fistul', V. I., in: Electronics and Its Application, Vol. 4 [in Russian], Izd. Mir, Moscow (1973), p. 108.
635. Darvoid, T. I., Dorofeeva, A. K., Popova, M. A., et al., Thallium and Its Applications in Modern Technology, Information from Institute "Tsvetmetinformatsya," Series 15, No. 6 (1972), p. 18.
636. Vainshtein, V. M., and Fistul', V. I., Fiz. Tekh. Poluprovodn., 4(8):1495 (1970).
637. Kikoin, I. K. (Ed.), Tables of Physical Constants (Handbook) [in Russian], Atomizdat, Moscow (1976).
638. Chemist's Handbook [in Russian], Vol. 1, 3rd ed., Izd. Khimiya, Leningrad (1971).
639. Filippov, S. S., Theory of Metallurgical Processes [in Russian], Izd. Metallurgiya, Moscow (1967).
640. Morozov, V. V., and Kunitskii, Yu. A., Electrode Materials [in Russian], Izd. Kiev. Politekh. Inst., Kiev (1977).

INDEX OF ELEMENTS AND THEIR OXIDES

GPSR Compliance

The European Union's (EU) General Product Safety Regulation (GPSR) is a set of rules that requires consumer products to be safe and our obligations to ensure this.

If you have any concerns about our products, you can contact us on ProductSafety@springernature.com

In case Publisher is established outside the EU, the EU authorized representative is:

Springer Nature Customer Service Center GmbH
Europaplatz 3
69115 Heidelberg, Germany

Batch number: 10712234

Printed by Printforce, the Netherlands